WETLAND PLANTS *of the* UPPER MIDWEST

A Full Color Field Guide to the
Aquatic and Wetland Plants of
Michigan, Minnesota and Wisconsin

STEVE CHADDE

WETLAND PLANTS OF THE UPPER MIDWEST

A Full Color Field Guide to the Aquatic and Wetland Plants of Michigan, Minnesota and Wisconsin

STEVE CHADDE

ISBN: 978-1951682682

Grateful acknowledgment is given to the Biota of North America Program (*www.bonap.org*) for permission to use their data to generate the distribution maps. Photographs were used under Creative Commons commercial use licenses.

The author can be reached via email: *steve@chadde.net*

VER. 1. (8/22/2022)

Contents

ANGIOSPERMS (Dicots), continued

Using this Book

MAJOR GROUPS
Presented first are the group of families comprising the **Ferns and Fern Allies** (true ferns, horsetails, clubmosses), followed by the **Gymnosperms** (conifers), and the **Angiosperms** (which make up the majority of the species described). The Angiosperms are subdivided into two groups: the **Dicots** and the **Monocots**. Dicots include many familiar trees, shrubs, and "wildflowers", such as those of the Aster Family; the Monocots include the grass, sedge, rush, and orchid families.

SPECIES DESCRIPTIONS
A standard format is used to describe each species: the lifeform (annual, perennial, herb, grass, etc.), stem and leaf characteristics, features of the flower and fruit, and a range of months when the species generally flowers are described for each species. The species' nativity in the Upper Midwest region is also noted (native to the region or introduced from elsewhere). Synonyms (other formerly accepted scientific names) are listed in italics. Habitats where the plant generally occurs are noted, as is the species' conservation status (endangered or threatened in each state). A regional map shows the verified presence (shown in green) or absence of each species within the counties of Minnesota, Wisconsin, and Michigan.

WETLAND INDICATOR STATUS (2018)
Wetland indicator ratings are listed for each species. Five categories are defined: **OBL, FACW, FAC, FACU** and **UPL** (defined below). Unlike earlier ratings, more than one rating may be applicable to a species, depending on the ecological region of the occurrence. The three states included here contain portions of three regions (see map, page 10):

- **NCNE** Northcentral and Northeast region
- **MW** Midwest
- **GP** Great Plains

· **OBL** (Obligate Wetland): Plants that almost always occur in wetlands (i.e. almost always in standing water or seasonally saturated soils).
· **FACW** (Facultative Wetland): Plants that usually occur in wetlands, but may occur in non-wetlands.
· **FAC** (Facultative): Plants that occur in wetlands and non-wetland habitats.
· **FACU** (Facultative Upland): Plants that usually occur in non-wetlands but may occur in wetlands.
· **UPL** (Obligate Upland): Plants that almost never occur in wetlands (or in standing water or saturated soils).

ADDITIONAL RESOURCES
Key to all Families included in the guide, page 410; **Key to Woody Plants** (trees, shrubs, vines), page 420; **Key to Aquatic Plants,** page 422. See page 428 for a **Glossary** of technical terms. **Appendices A** and **B** are comparisons of the region's species of *Potamogeton* and *Stuckenia* (pages 440-446). **Abbreviations** used throughout the text are listed on page 19. The **Index** includes one for common names (page 447), the other for scientific names (page 453).

Introduction

WETLAND PLANTS OF THE UPPER MIDWEST is a comprehensive manual on the identification, habitats, and distribution of the vascular plants found in the aquatic and wetland environments in the states of Michigan, Minnesota and Wisconsin. This full-color work replaces the author's *Wetland Plants of the Upper Midwest* (2019), and updates taxonomy to largely conform to that of the APG IV system (see *www.mobot.org*). Nomenclature of plant families, genera, and species generally follows that of the published volumes of *The Flora of North America* series (1993+), the *Synthesis of the North American Flora* (Kartesz 2014), and *ITIS, the Integrated Taxonomic Information System,* see *www.itis.gov*).

Over 900 plant species within 109 plant families are described. This work includes species and distribution information for aquatic and wetland plants found in Michigan, Minnesota, and Wisconsin. For convenience, the terms wetland or wetland plant are sometimes used to also include aquatic situations (the open water of lakes, ponds, rivers and streams) and aquatic plants.

The intended user of this book is the botanist, ecologist, natural resource manager, consultant, or other scientist engaged in wetland studies. However, this book was also written for the person simply interested in plants, or in the ecology, conservation, or aesthetic beauty provided by wetlands. To this end, the use of technical terminology has been minimized when possible, while still allowing the reader to accurately identify unknown plant specimens. A glossary of botanical terms is provided on page 428; abbreviations and symbols used are defined on page 19.

Plants were initially chosen for inclusion in this work according to their wetland indicator status, a classification system developed by the U.S. Fish and Wildlife Service (Reed 1988). As of 2022, a revised and updated National Wetland Plant List has been prepared and is incorporated into this work (U.S. Army Corps of Engineers 2020).

Species occurring in the Upper Midwest region with a wetland indicator status of Obligate or Facultative Wetland were initially included. In addition, a number of species more typical of drier habitats (Facultative species and Facultative Upland species) were included, based on published reports of their occurrence in wetlands, or on the author's own encounters with them in the region's wetlands.

The result is a floristic guide for the Upper Midwest region that is applicable to the aquatic plants of lakes, ponds, rivers, and streams, the plants of the adjoining wetlands, and to the plants found in the transition area, or ecotone, between wetlands and the surrounding uplands. Descriptions of the various wetland vegetation types in each of these settings begin on page 11.

For coverage of all vascular plants in the Upper Midwest region, see the author's *Michigan Flora: Upper Peninsula* (2016), *Minnesota Flora* (2013), and *Wisconsin Flora* (2013), *Field Manual of Michigan Flora* (Voss and Reznicek 2012) or the second edition of the *Manual of Vascular Plants of Northeastern United States and Adjacent Canada* (Gleason and Cronquist 1991), and the *Illustrated Companion to Gleason and Cronquist's Manual* (Holmgren and others 1998). Since the publication of the first edition of the author's *A Great Lakes Wetland Flora* in 1998, online species distribution data are available. Notable is the website of the Biota of North America Program (BONAP, *http://www.bonap.org*), which is the most comprehensive source of plant species distributions for all vascular plant species found in North America.

Natural heritage programs located in each state in the region provided information on uncommon species (listed as endangered or threatened), and sightings of these rare taxa should be reported to them. Non-vascular plants (mosses, liverworts) are not included in this work, although their importance as indicators of water chemistry and hydrologic conditions should be considered when conducting wetland studies. A standard reference for the region is *Mosses of Eastern North America* (Crum and Anderson 1981).

Wetlands Defined

In general, wetlands are lands where water saturation is the primary factor determining the nature of soil that will develop and the types of plant and animals that will occur there (Cowardin and others 1979). Wetlands occur as transitional areas between upland and aquatic ecosystems, and are covered by shallow water, or have a water table that is usually at or near the ground surface.

Wetlands will have at least one, and often all of the following three attributes regarding their vegetation, soils and hydrology:

· The land supports (at least periodically) plants that are predominantly "hydrophytes." Hydrophytes are plants growing in water, or on soil that is at least periodically deficient in oxygen as a result of high water content.

· The soils are largely undrained, "hydric" soils. Hydric soils are soils that are flooded or saturated long enough during the growing season so that anaerobic conditions develop in the upper portion of the soil.

· Water will either permanently or periodically cover the area during some or all of the growing season each year.

A wetland definition used by the U.S. Army Corps of Engineers in its regulatory program is: "Wetlands are those areas inundated or saturated by surface or groundwater at a frequency and duration sufficient to support, and that under normal circumstances do support, a prevalence of vegetation typically adapted for life in saturated soil conditions. Wetlands generally include swamps, marshes, bogs, and similar areas."

Aquatic plants

Plants of open water areas—aquatic plants—include those species found in lakes, ponds, rivers, streams, and springs. They occur as either free-floating plants on the water surface, such as the duckweeds (*Lemna*) and watermeal (*Wolffia*), or float below the water surface, as in coontail (*Ceratophyllum*) and common bladderwort (*Utricularia macrorhiza*). Aquatic species also occur as submergent plants anchored to the bottom substrate by roots or rhizomes. Submergent species may be entirely underwater as in water-milfoil (*Myriophyllum*), or have leaves floating on the water surface (water-lilies, *Nuphar*). Some aquatic species have differently shaped underwater and floating leaves, as in a number of pondweeds (*Potamogeton*).

Wetland Plants

Wetland species are intermediate between truly aquatic species and plants adapted to the moist or dry conditions of upland environments. Wetland species occur in habitats generally referred to as bogs, fens, marshes, swamps, thickets, and wet meadows, among others. Soils of these habitats may be organic (composed of decaying plant remains, or "peat"), of mineral origin, or a combination of both. Wetlands may be covered with shallow water for all or part of the year, or simply wet for a portion of the growing season.

On sites covered by shallow water, a group of wetland plant species termed emergent occur. These plants are rooted in soil but their stems and leaves are partly or entirely above the water surface. This large group includes many of the sedges (*Carex*), bulrushes (*Scirpus*), bur-reeds (*Sparganium*), rushes (*Juncus*), grasses (Poaceae) and the cat-tails (*Typha*).

A second large group of wetland species are those occurring on soils which are moist or even saturated throughout the year but are rarely covered by standing water. These sites include wetlands such as moist or wet meadows and low prairie. Many of the herbaceous dicots ("wildflowers") included in the Flora occur in these types of wetlands, especially on the drier margins. Species typically found in drier habitats are occasionally found in these wetlands as well, but are often restricted to hummocks slightly elevated above the surrounding lower, wetter areas.

Excluded from this Flora are species which occur only rarely in wetlands. For example, a number of upland plants are found on sites which may be wet for only a short time in the spring following snowmelt or runoff. Under the U.S. Fish and Wildlife Service classification system, these would generally be classified as either Facultative Upland (FACU) or Obligate Upland (UPL) species.

Plant Names

Each species of plant has a unique name made up of three parts. The first part is the genus to which a species belong; the second part is termed the specific epithet. The name is followed by the name of the person or persons who first named a species, and are often abbreviated. For example, "L." refers to Carolus Linnaeus, the 18th-century Swedish botanist considered the father of modern plant taxonomy. Together the two form the scientific, or botanical, name of the species.

Identifying Unknown Plants

The keys are the first step in identifying an unknown plant specimen. The keys are termed "dichotomous," meaning that the reader must choose between one of two leads which form a "couplet" at each step in the key. The lead is chosen which best describes the plant being identified and the user proceeds to the number following the lead in the key. The process is continued until the family, genus, and ultimately species are identified.

For a completely unknown plant (one in which even the family is unknown), start with the Family Keys (page 410). For families having only one genus or species in the Flora, these will be identified here as well. Once the family is identified, turn to the genus key. If the family is recognized for an unknown plant, begin identification with the genus key. Once the genus has been determined, use the species key to identify the specimen. In addition to the keys, use the description of the plant and its habitat and range information to verify your identification. With practice, the characters of many families and genera will be recognizable on sight, greatly reducing the time spent keying.

Equipment needs are minimal for successful identification—a 10–20 magnification hand lens and a ruler with mm scale (a ruler is also printed on the last page of the index). For some species, as in members of the Sedge Family (Cyperaceae), a more powerful dissecting microscope is useful for examining the smaller plant parts often needed for positive identification. When a microscope is available, plants may be collected, pressed and dried, and identified indoors at a later date.

Whenever possible, plants should be examined by leaving them rooted rather than uprooting or tearing off pieces to examine at eye level. This is especially important when working in wetlands, which are home to a disproportionate number of rare and uncommon plant species.

Conservation Status

The following terms are used to describe rarity within a state. Endangered and threatened status provide legal protection to a species.

· END Endangered: a species threatened with "extirpation" (extinction within a single state or area and not across its entire range).

· **THR** Threatened: a species whose survival within a state is not in immediate danger, but for which a threat exists; continued stress on the species may result in its becoming endangered.

Each state in the Upper Midwest region uses the endangered and threatened categories. Designations (and any federal listings) are noted in the description for each species.

Wetland Indicator Status

For each species described in the guide, the wetland indicator status is provided (see definitions page 6). The indicator status ratings are based on the 2018 National Wetland Plant List (see *https://wetland-plants.sec.usace.army.mil*) and are widely used in wetland delineation studies. Two species previously without indicator status ratings — *Aconitum columbianum* and *Platanthera praeclara* — have proposed updates and these are incorporated in their descriptions. Since 2012, the classification no longer uses positive or negative signs.

Regional delineations within the Upper Midwest region are based on those of the US Army Corps of Engineers (see map below). Three regions are present in the three states covered by this Flora:

· **NCNE**   Northcentral and Northeast region
· **MW**   Midwest
· **GP**   Great Plains

US Army Corps of Engineers regional delineations for the Upper Midwest region. Three regions are present in the three states covered by this Flora: **Great Plains** (a small portion of northwest Minnesota), **Midwest**, and **Northcentral and Northeast**.

Wetland Types of the Upper Midwest

WETLANDS OCCUR across the Upper Midwest region in a wide variety of physical settings and under a large range of climatic conditions. As a result, wetlands vary markedly in their size, appearance and composition. Wetlands of northern portions of the region differ from their southern counterparts as the vegetation changes from a "prairie-forest province" in the south, to a "northern forest province" in the north (Curtis 1971). Northern areas, for example, feature extensive areas covered by conifer bogs and swamps of black spruce (*Picea mariana*), tamarack (*Larix laricina*), and northern white cedar (*Thuja occidentalis*), and by various types of shrub-dominated and herbaceous wetlands, including alder thickets, sedge meadows, and open bogs.

Southward, floodplain forests along major rivers, and marshes and sedge meadows are common. In southern areas of the region, calcareous fens associated with upwelling groundwater are an unusual wetland type and home to a large number of uncommon plants. In north-central Minnesota, the Red Lake area of supports a vast peatland. Continuing west onto the eastern edge of the Great Plains, rolling grasslands are interspersed with low prairie and prairie pothole ponds.

The following key is a basic classification of the wetland types found in the Upper Midwest region. It is based in part on *Vegetation of Wisconsin* (Curtis 1971) and *Wetland Plants and Plant Communities of Minnesota and Wisconsin* (Eggers and Reed 1997).

Key to Wetland Types of the Upper Midwest Region

1   Open water areas of lakes, ponds, rivers and streams; vegetation is absent or aquatic plants with underwater or floating leaves are present . . . .
. . . . OPEN WATER COMMUNITIES

1   Wetland types of shallow water areas or on saturated or moist soil; trees, shrubs or herbaceous plants are dominant 2

2   Wetlands of shallow depressions, flats, or lakeshores; standing water may be present for several weeks each year, then drying; herbaceous plants dominant . .
. . . . SEASONALLY WET BASIN OR SHORE

2   Standing water present, or soils saturated for all or most of growing season 3

3   Woody plants (trees and shrubs) dominant (greater than about 50 percent canopy cover) 4

3   Herbaceous plants dominant, woody plants sparse; trees, if present, are stunted and mostly small in diameter (less than about 15 cm wide) and less than 5-6 m tall . . 11

4   Trees dominant; overstory canopy ± closed by mature trees 5

4   Shrubs dominant; mature trees absent or sparse and not forming a closed canopy 8

5   Hardwood trees dominant; soils are alluvial deposits, poorly drained mineral soils, or sometimes organic (WET HARDWOOD FOREST) 6

5   Conifers such as tamarack (*Larix laricina*), black spruce (*Picea mariana*), or northern white cedar (*Thuja occidentalis*) dominant; soils usually highly organic 7

6   Forests adjacent to rivers on alluvial soils, mostly central and southern portions of the region. Silver maple (*Acer saccharinum*), American elm (*Ulmus americana*), river birch, (*Betula nigra*) green ash (*Fraxinus pennsylvanica*), black willow (*Salix nigra*) or eastern cottonwood (*Populus deltoides*) are typical dominants . . . .
. . . . FLOODPLAIN FOREST

6 Forests of low-lying basins and kettles, with poorly drained mineral or organic soils, mostly central and northern areas. Black ash (*Fraxinus nigra*), red maple (*Acer rubrum*) or silver maple are typical dominants; northern white cedar may be common .... **HARDWOOD SWAMP**

7 Tamarack and/or black spruce dominant; sphagnum mosses form a nearly continuous carpet; soils are organic and acid .... **CONIFER BOG**

7 Northern white cedar or tamarack dominant; sphagnum mosses may be present but not a forming a continuous ground cover; soils with a high organic content, usually neutral to alkaline .... **CONIFER SWAMP**

8 Tall shrubs (mostly more than 1 m high) dominant; sphagnum mosses may be present but not forming a continuous mat (SHRUB WETLAND) **9**

8 Low shrubs (mostly less than 1 m high) dominant; sphagnum mosses present, or sparse or absent **10**

9 Speckled or tag alder (*Alnus incana*) dominant; occurring mostly in northern and central areas of region .... **ALDER THICKET**

9 Dogwoods (*Cornus*), willows (*Salix*), or other tall shrubs dominant; occurring throughout the region .... **SHRUB-CARR**

10 Evergreen shrubs of the Heath Family (Ericaceae) such as leatherleaf (*Chamaedaphne calyculata*), bog-rosemary (*Andromeda polifolia*), Labrador-tea (*Rhododendron groenlandicum*), or small cranberry (*Vaccinium oxycoccos*) dominant or common; sphagnum mosses forming a ± continuous carpet .... **OPEN BOG (POOR FEN)**

10 Deciduous shrubs, especially shrubby cinquefoil (*Dasiphora fruticosa*), often common; sites often sloping and maintained by a spring-fed supply of calcium-rich water; other calcium-indicating plants present; sphagnum mosses absent or sparse .... **CALCAREOUS FEN**

11 Dominant plants sedges (*Carex*) and other members of the Cyperaceae, cat-tails (*Typha*), and large emergent species such as bur-reed (*Sparganium*), bulrush (*Scirpus*), and water plantain (*Alisma*) **12**

11 Dominant plants are grasses or plants of calcium-rich sites **14**

12 Cat-tails, bulrushes (*Scirpus*), water plantain (*Alisma*), arrowheads (*Sagittaria*), pickerelweed (*Pontederia*), giant bur-reed, or emergent lake sedges (for example, *Carex lacustris, C. utriculata*) dominant; sites have standing water for all or most of growing season (sometimes drying to surface in summer or fall) ..... **MARSH**

12 Sedges (primarily *Carex*) dominant; water level often falling below ground surface during growing season **13**

13 Sedges dominant; sphagnum mosses may be present but not as a continuous carpet; soils saturated most of growing season, acid to alkaline **SEDGE MEADOW**

13 Sphagnum mosses forming a ± continuous mat; soils are acid peats; pitcher plants (*Sarracenia purpurea*), sundews (*Drosera*), sedges (such as *Carex oligosperma*), cottongrass (*Eriophorum*), and shrubs of the Heath Family (Ericaceae) typically present .... **OPEN BOG (POOR FEN)**

14 Wetlands often on sloping sites, with a spring-fed supply of calcium-rich water; species indicative of calcium-rich sites, such as fen star sedge (*Carex sterilis*), marsh-muhly (*Muhlenbergia glomerata*), and Ohio goldenrod (*Solidago ohioensis*) prominent .... **CALCAREOUS FEN**

14 Wetlands with water mainly from rainfall, springs, or surface drainage; calcium-indicating species not dominant; sites have standing water or are saturated during all or part of growing season **15**

15 Wetlands with saturated soils (rarely with standing water; plants such as bluejoint (*Calamagrostis canadensis*), redtop (*Agrostis stolonifera*), reed canarygrass (*Phalaris arundinacea*), and smooth goldenrod (*Solidago gigantea*) dominant . . . . **WET MEADOW**

15 Wetlands with standing water or saturated soil during the growing season; prairie grasses such as big bluestem (*Andropogon gerardii*), prairie cordgrass (*Spartina pectinata*), and prairie lowland species present . . . . **LOW PRAIRIE**

STEVE CHADDE

Typical northern peatland in Michigan's Upper Peninsula. This wetland can be classified as an **Open Bog** (**Poor Fen**); scattered, stunted trees of black spruce (*Picea mariana*) are present. Tawny cotton-grass (*Eriophorum virginicum*) is common; other members of the Sedge Family (Cyperaceae) such as running bog sedge (*Carex oligosperma*) and slender sedge (*Carex lasiocarpa*) are also present. Common shrubs are members of the Heath Family (Ericaceae): leatherleaf (*Chamaedaphne calyculata*), bog-rosemary (*Andromeda polifolia*), Labrador-tea (*Rhododendron groenlandicum*) and small cranberry (*Vaccinium oxycoccos*). Sphagnum mosses form a nearly continuous carpet across the landscape.

Open Water Communities

These plant communities occur in shallow to deep water of lakes, ponds, rivers and streams. Plants are free-floating, or submergent (anchored to the bottom and with underwater and/or floating leaves). A controlling factor of these communities is that water levels remain deep enough so that emergent vegetation typical of marshes is unable to establish.

Marshes

Marshes are dominated by emergent plants growing in permanent or nearly permanent shallow to deep water. Shallow water marshes are covered with up to about 15 cm of water for all or most of the growing season, but water levels may sometimes drop to the soil surface. Deep marshes have standing water between 15 and 100 cm or more deep during most of the growing season. A mix of emergent species and aquatic species of the adjoining open water are present.

Sedge Meadows

Sedge meadows are dominated by members of the Sedge Family (Cyperaceae), with grasses (Poaceae) and rushes (*Juncus*) often present. Other herbaceous species are usually present only as scattered individuals. Because of differences in their species composition, sedge meadows can be subdivided into those occurring in northern portions of the region and those occurring southward. Soils are typically organic deposits (peat or muck), and saturated throughout the growing season. Common tussock sedge (*Carex stricta*) is a major component of sedge meadows, forming large hummocks composed of a mass of roots and rhizomes. Periodic fires may help maintain the dominance of the sedges by killing invading shrubs and trees.

Wet Meadows

Wet meadows occur on saturated soils, and are dominated by grasses (especially *Calamagrostis canadensis*) and other types of perennial herbaceous plants. Soils may be inundated for brief periods (1–2 weeks) in the spring following snowmelt or during floods. Shrubs occur only as scattered plants.

Low Prairie

Low prairies are dominated by grasses and grasslike plants. These communities are similar to wet meadows but are dominated by native grasses and other herbaceous species characteristic of the prairie. Low prairie communities primarily occur in southern and western portions of the Upper Midwest region, but are occasional in northern areas on sandy plains or in wet swales.

Calcareous Fens

Calcareous fens are an uncommon wetland type in the Upper Midwest region. They typically develop on sites that are sloping and with a steady flow of groundwater rich in calcium and magnesium bicarbonates (Curtis 1971). The calcium and magnesium bicarbonates precipitate out at the ground surface, leading to a highly alkaline soil. Such conditions are tolerated by a fairly small group of plants termed "calciphiles." Sphagnum mosses are absent or sparse. Due to their uniqueness, calcareous fens often support a large number of uncommon plant species.

Open Bogs

Open bogs are found most commonly in northern portions of the Upper Midwest region, but also occur as small relict features in the south. Technically, true bogs are extremely nutrient-poor, receiving nutrients only from precipitation. Many of the region's bogs may better be termed "poor fens," but given the widespread use of the term among scientists and general public alike, "bog" is used throughout this book. Bogs have a characteristic, nearly continuous carpet of sphagnum moss, and are dominated by shrubs of the Heath Family (Ericaceae), especially leatherleaf (*Chamaedaphne caly-culata*), members of the Sedge Family (Cyperaceae), and scattered herbs such as pitcher plant (*Sarracenia purpurea*) and sundew (*Drosera*). Stunted trees of black spruce (*Picea mariana*) and tamarack (*Larix laricina*) may be present; this type of bog is sometimes termed "muskeg." Soils are composed entirely of saturated organic peat.

Conifer Bogs
Conifer bogs are similar to open bogs except that trees of black spruce and/or tamarack are predominant. Sphagnum moss carpets the ground surface. Shrubs of the Heath Family (Ericaceae), sedges, and a small number of other species occur. Soils are organic, saturated peats.

Shrub-Carrs
Shrub-carrs are dominated by tall shrubs. Soils are organic peat or muck, or alluvial soils of floodplains, and are often saturated throughout the growing season, and sometimes inundated during floods. Willows (*Salix*) and dogwoods (*Cornus*) are especially characteristic.

Alder Thickets
Alder thickets are a tall, deciduous shrub community similar to shrub-carrs except that speckled or tag alder (*Alnus incana*) is dominant, and often forming dense colonies. Alder thickets are common in northern portions of the Upper Midwest region along rivers and streams, or along the wet margins of marshes, sedge meadows and bogs.

Hardwood Swamps
Hardwood swamps are dominated by deciduous trees such as black ash (*Fraxinus nigra*), red maple (*Acer rubrum*), and yellow birch (*Betula alleghaniensis*). Southward, silver maple (*Acer saccharinum*) increases in importance; northward, northern white cedar (*Thuja occidentalis*) may be present. American elm (*Ulmus americana*) may be present, but in greatly reduced numbers than formerly due to losses from Dutch elm disease. Soils are saturated during much of the growing season, and may be covered with shallow water in spring.

Conifer Swamps
Conifer swamps are forested wetlands dominated primarily by tamarack (Larix laricina) and black spruce (*Picea mariana*). In areas where the water is not as stagnant and in areas underlain by limestone, northern white cedar (*Thuja occidentalis*) may form dense stands. Conifer swamps occur almost entirely in northern portions of the Upper Midwest region. Soils are usually peat or muck, and vary in reaction from acidic and nutrient-poor, to neutral or alkaline and relatively fertile. Tamarack is most abundant on acid soils, while northern white cedar is most common on neutral or alkaline soils. Soils are typically saturated much of the growing season, and sometimes covered with shallow water. Sphagnum moss may be present, but does not usually form a continuous mat.

Typical form of sphagnum moss; worldwide there are an estimated 380 species.

Floodplain Forests

Floodplain forests are dominated by deciduous hardwood trees growing on alluvial soils along rivers. These forests often have standing water during the spring, with water levels dropping to the surface (or below) in summer and fall. Typical trees include silver maple (*Acer saccharinum*), green ash (*Fraxinus pennsylvanica*), black willow (*Salix nigra*), eastern cottonwood (*Populus deltoides*), and river birch (*Betula nigra*). The diversity of the undergrowth is high, as flood waters remove and deposit alluvium, creating micro-habitats suitable for many species.

Seasonally Flooded Basins

Seasonally flooded basins are poorly drained, shallow depressions in glacial deposits (kettles), low spots in outwash plains, or depressions in floodplains. The basins may have standing water for a few weeks in the spring or for short periods following heavy rains. Soils are typically dry, however, for much of the growing season. South and west in the Upper Midwest region, attempts may be made to cultivate the basins, and in combination with the fluctuating water level, these areas are often dominated by a wide array of annual or weedy plant species. In far western portions of our region, these shallow basins are termed prairie potholes, and are important habitat for waterfowl. Prairie potholes may support communities of aquatic plants and are often ringed by emergent wetland plants.

Conifer Swamp. Northern white-cedar (*Thuja occidentalis*) dominates the overstory although other trees such as balsam-fir, black spruce, and white spruce are often present. Shrubs include alder (*Alnus incana*), red-osier dogwood (*Cornus sericea*), Labrador tea (*Rhododendron groenlandicum*), blueberry (*Vaccinium angustifolium*), bilberry (*V. myrtilloides*), leatherleaf (*Chamaedaphne calyculata*) and creeping snowberry (*Gaultheria hispidula*). Many herbaceous species are present due to the varied micro-topography. Ferns are common and include northern lady fern (*Athyrium angustum*), cinnamon-fern (*Osmundastrum cinnamomea*) and horsetails (*Equisetum*). Typical sedges are *Carex gynocrates, C. leptalea, C. disperma, C. trisperma,* and *C. interior.* Mosses other than sphagnum are important.

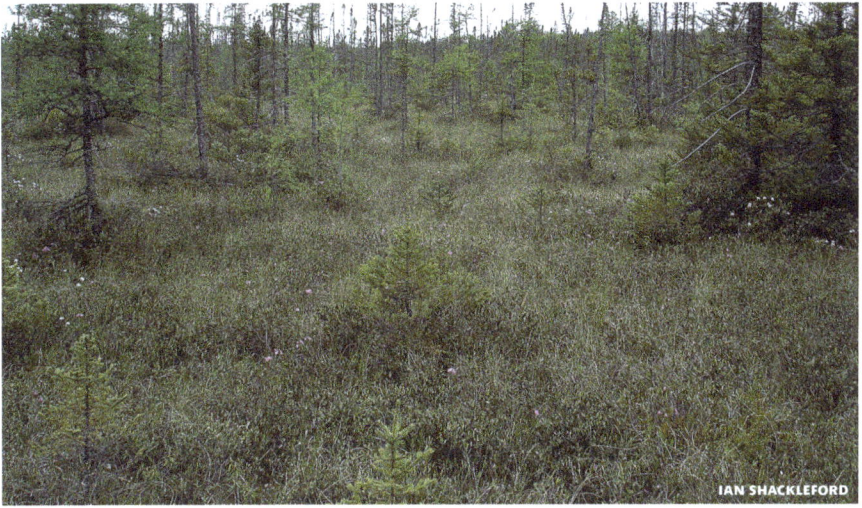

IAN SHACKLEFORD

Open Bog or "muskeg," late spring. Tamarack (*Larix laricina*) and scattered black spruce (*Picea mariana*) are present but stunted in size due to saturated, anaerobic soil conditions. Common shrubs include leatherleaf (*Chamaedaphne calyculata*), bog-laurel (*Kalmia polifolia*), and Labrador-tea (*Rhododendron groenlandicum*). Dominant herbaceous species include running bog sedge (*Carex oligosperma*), tussock cotton-grass (*Eriophorum vaginatum*), and other members of the Sedge Family (Cyperaceae). Sphagnum mosses form a nearly continuous cover across the organic soil.

STEVE CHADDE

Open Water, **Marsh**, and **Sedge Meadow** communities, summer. Many aquatic and wetland plant species are present in this productive pond located in Michigan's Upper Peninsula, including white water-lily (*Nymphaea odorata*), water-shield (*Brasenia schreberi*), buckbean (*Menyanthes trifoliata*), various species of pondweed (*Potamogeton*), softstem bulrush (*Schoenoplectus tabernaemontani*), three-way sedge (*Dulichium arundinaceum*), and common cat-tail (*Typha latifolia*). Adjacent to the pond is a sedge meadow dominated by common tussock sedge (*Carex stricta*), common lakeshore sedge (*Carex lacustris*), and bluejoint (*Calamagrostis canadensis*); sweet gale (*Myrica gale*) is a common shrub at the water's edge.

Hardwood Swamp, early spring. Black ash (*Fraxinus nigra*) is the dominant tree; red maple (*Acer rubrum*) is also present. Marsh-marigold (*Caltha palustris*) is common in the undergrowth, with scattered plants of several species of violet (*Viola*), three-seeded bog sedge (*Carex trisperma*), rough sedge (*Carex scabrata*), sensitive fern (*Onoclea sensibilis*), and orange touch-me-not or jewelweed (*Impatiens capensis*).

Sedge Meadow community, summer. Typical growth-form of common tussock sedge (*Carex stricta*), a dominant component of sedge meadows. The large hummocks, up to one meter tall, are composed of plant roots and the previous years' leaves.

Abbreviations

♦ illustrated species (color photo)

Botanical
subsp. subspecies
var. variety

Conservation Status
END Endangered (state level)
THR Threatened (state level)

Measurements
mm millimeter
cm centimeter
dm decimeter
m meter
x times
± more or less

Geographical
c central
e east
n north
s south
w west

States and Provinces
Ala Alabama
Ariz Arizona
Ark Arkansas
BC British Columbia
c Amer central America
Calif California
Can Canada
Colo Colorado
Conn Connecticut
DC District of Columbia
Del Delaware
Fla Florida
Ga Georgia
Ill Illinois
Ind Indiana
Kans Kansas
Ky Kentucky
La Louisiana
Lab Labrador

LP Lower Peninsula of Michigan
Man Manitoba
Mass Massachusetts
Md Maryland
Mex Mexico
Mich Michigan
Minn Minnesota
Miss Mississippi
Mo Missouri
Mont Montana
N Amer North America
NB New Brunswick
NC North Carolina
ND North Dakota
Neb Nebraska
New Eng New England
Nev Nevada
Nfld Newfoundland
NH New Hampshire
NJ New Jersey
NM New Mexico
NS Nova Scotia
NY New York
Okla Oklahoma
Ont Ontario
Ore Oregon
Pa Pennsylvania
PEI Prince Edward Island
Que Quebec
RI Rhode Island
S Amer South America
Sask Saskatchewan
SC South Carolina
SD South Dakota
Tenn Tennessee
Tex Texas
UP Upper Peninsula of Michigan
USA United States
Va Virginia
Vt Vermont
Wash Washington
Wisc Wisconsin
WVa West Virginia
Wyo Wyoming

Ferns and Fern Allies
Polypodiopsida, Lycopodiopsida

THE FERNS AND THEIR ALLIES in the Upper Midwest region can be classified into two major classes: Polypodiopsida (true ferns) and Lycopodiopsida (clubmosses). Horsetails (Equisetaceae), formerly separated from "true ferns," are now typically included within Polypodiopsida. Parts of a typical true fern are illustrated below.

The Polypodiopsida are represented by ten families in wetlands of the Upper Midwest region (Athyriaceae, Cystopteridaceae, Dryopteridaceae, Equisetaceae, Marsileaceae, Onocleaceae, Ophioglossaceae, Osmundaceae, Salviniaceae, Thelypteridaceae). Three families are included in the Lycopodiopsida (Lycopodiaceae, Selaginellaceae, Isoetaceae).

All ferns and fern-allies reproduce by spores rather than seeds. The spore germinates and produces a tiny flattened thallus termed the gametophyte. The larger, leafy, spore-bearing plant (sporophyte) results from the fertilized egg produced by the pistillate part of the gametophyte. The uncurling of the young leaf or frond is characteristic. The Lycopodiopsida (clubmosses) bear their spores on specialized leaves (sporophylls). The Equisetaceae (horsetails) bear spores in terminal cone-like structures, either on typical green stems, or on specialized fertile stems which lack chlorophyll.

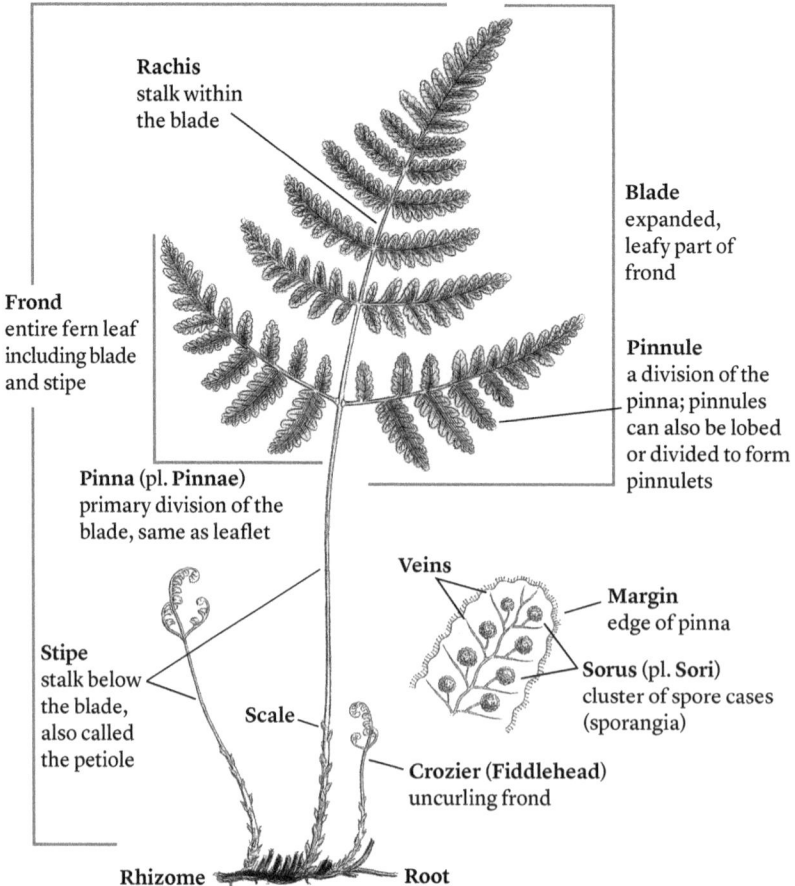

Rachis
stalk within the blade

Blade
expanded, leafy part of frond

Frond
entire fern leaf including blade and stipe

Pinnule
a division of the pinna; pinnules can also be lobed or divided to form pinnulets

Pinna (pl. Pinnae)
primary division of the blade, same as leaflet

Veins

Margin
edge of pinna

Stipe
stalk below the blade, also called the petiole

Scale

Sorus (pl. Sori)
cluster of spore cases (sporangia)

Crozier (Fiddlehead)
uncurling frond

Rhizome

Root

Athyriaceae
Lady Fern Family

MEDIUM TO LARGE FERNS. *Deparia,* formerly considered as a member of *Athyrium,* is now treated as separate species. The degree of blade division helps separate our 2 species (see key), and in contrast to *Deparia, Athyrium* has a more deeply grooved rachis, which is continuous from rachis to costa (vs. discontinuous in *Deparia).*

Athyriaceae LADY FERN FAMILY

1   Blades 2-pinnate (with the pinnae again divided), the pinnules also sometimes deeply lobed . . . . *Athyrium angustum*
1   Blades with deeply lobed pinnae (1-pinnate-pinnatifid) . . . . *Deparia acrostichoides*

Athyrium LADY FERN

Athyrium angustum (Willd.) K. Presl
♦ NORTHERN LADY FERN

DESCRIPTION Native clumped fern, rhizomes short and ascending. Leaves deciduous, sterile and fertile leaves similar; petioles with brown, linear scales; blades elliptic, 2-pinnate, broadest at middle or slightly below middle; pinnae short-stalked or stalkless. Sori generally somewhat curved to hook-shaped, less often straight. **SYNONYM** *Athyrium filix-femina.*
HABITAT moist deciduous woods, shrub thickets, streambanks, hummocks in swamps, wetland margins, and shaded rock outcrops; a wide-ranging circumboreal species.
STATUS MW-FAC | NCNE-FAC | GP-FAC
NOTE Our plants formerly treated as *A. filix-femina* but now typically divided into *Athyrium angustum* (which occurs in ne North America, s to NC, Mo and Neb), and *Athyrium asplenioides* (southern lady-fern) with a more southerly distribution, and with broader fronds, especially at their base.

Deparia GLADE FERN

Deparia acrostichioides (Swartz) M. Kato
♦ SILVERY GLADE FERN

DESCRIPTION Large native fern from creeping rhizomes. Leaves deciduous, 50–100 cm long, sterile and fertile leaves alike; petioles straw-colored (but dark red-brown at base), with brown lance-shaped scales; blades lance-shaped to oblong in outline, tapered at tip and distinctly narrowed toward base; deeply lobed, the segments blunt to somewhat tapered at their tip, margins entire to slightly lobed. Sori crowded, elongate, straight or sometimes curved, the indusia silvery and shiny when young.
SYNONYM *Athyrium thelypterioides.*

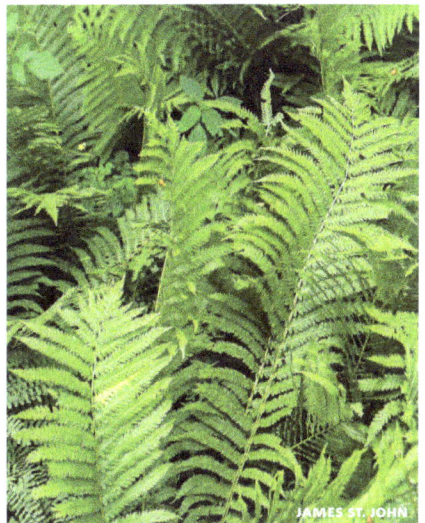

Athyrium angustum

Deparia acrostichioides

HABITAT moist, rich deciduous woods, especially in wetter swales and depressions; streambanks; damp shady slopes.

STATUS MW-FAC | NCNE-FAC

Cystopteridaceae
Bladder Fern Family

Cystopteris BLADDER FERN

Cystopteris bulbifera (L.) Bernh.
◆BULBLET FERN, BLADDER FERN

DESCRIPTION Native clumped fern, rhizomes short and thick. **Leaves** deciduous, 30–100 cm long, sterile and fertile leaves similar but sterile blades usually shorter than fertile; petioles much shorter than blades; blades lance-shaped, 6–15 cm wide at base, long tapered to tip, with 20–30 pairs of pinnae; the veins ending in a notch (sinus). Sori round, on a small vein; indusia hoodlike and attached at its base, covered with scattered, short-stalked glands. Green bulblets, 4–5 mm wide, are produced on lower side of rachis (main stem of leaf) toward upper end of blade, these falling and forming new plants.

HABITAT rocky streambanks, ravines, and moist, shaded, often calcium-rich rocks and cliffs.

STATUS MW-FACW | NCNE-FACW | GP-FACW

NOTE distinguished from **fragile fern** (*Cystopteris fragilis*), a common fern of moist

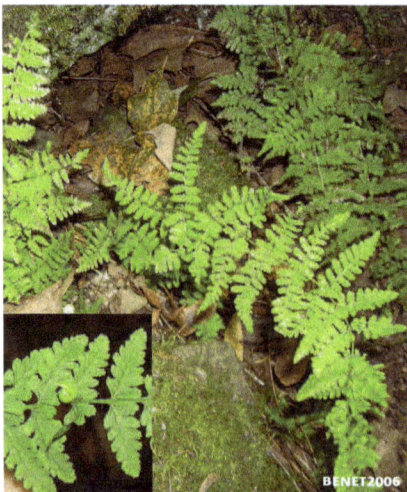

woods, by the blade broadest at base, most veins ending in a notch, and the small bulblets on underside of rachis. In fragile fern, the blade is broadest above its base, most veins end in a tooth, and bulblets absent.

Dryopteridaceae
Wood-Fern Family

Dryopteris WOOD-FERN

Medium to large ferns; rhizomes short, stout and scaly, often covered with old petiole bases. **Leaves** dark green, sometimes evergreen; petioles shorter than blades, straw-colored or green, with chaffy scales near base. Sterile and fertile leaves alike or slightly different; sterile leaves sometimes persisting over winter; blades 1–3 pinnate, the smallest segments commonly toothed or lobed, veins simple to 1- or 2-branched. **Sori** round, on underside veins of pinnae; indusia round to kidney-shaped.

Dryopteris **WOOD-FERN**

1 Blades narrowly oblong, the lowest pair of pinna triangular in outline; sterile and fertile leaves somewhat different, sterile leaves smaller than fertile leaves . . . . *D. cristata*

1 Blades broader; sterile and fertile leaves alike **2**

2 Blade more dissected, 2–3-pinnate **3**

2 Blade less dissected, 1–2-pinnate **4**

3 Leaves deciduous late in season; lowermost inner pinnule longer than next outer one and longer and wider than opposite upper pinnule . . *D. carthusiana*

3 Leaves evergreen; lowermost inner pinnule shorter or equal to adjacent lower pinnule, not distinctly longer and wider than opposite upper pinnule . . . *D. intermedia*

4 Blades large, abruptly narrowed to a short, pointed tip; lowest pair of pinna with wavy outline . . . . *D. goldieana*

4 Blades gradually reduced to long, tapering tips **5**

5 Leaf blades mostly 2–3 dm wide, lowest pinna-pair narrowed at base; sw Mich . . . . *D. celsa*

5 Leaf blades mostly less than 2 dm wide, lowest pinna-pair broadest at or near base; s Wisc and s Mich . . . . *D. clintoniana*

BENET2006

Cystopteris bulbifera

Dryopteris carthusiana (Villars) H.P.Fuchs
♦ SPINULOSE WOOD-FERN

DESCRIPTION Native clumped fern, rhizomes short-creeping. Leaves all alike, deciduous, smooth except for chaffy, pale brown scales near base of petioles; blades 2- to nearly 3-pinnate, 2–6 dm long and 1–4 dm wide, tapered to tip, slightly narrowed at base; pinnae usually 10–15 pairs, alternate to nearly opposite, narrowly lance-shaped; pinnules toothed to deeply lobed, mostly 5–40 mm long and 3–10 mm wide, the teeth tipped with a small spine; innermost lower pinnule longer than next outer one and 2–3x longer than opposite upper pinnule. Sori halfway between midvein and margin; indusia 1 mm wide, without stalked glands.

SYNONYM *Dryopteris spinulosa*.

HABITAT Moist to wet woods, hummocks in swamps, thickets; also drier sand dunes and ridges.

STATUS MW-FACW | NCNE-FACW | GP-FACW

NOTE Similar to **fancy wood-fern** (*Dryopteris intermedia*) but the innermost lower pinnule of the basal pinnae is usually longer than the next outer one, and the leaves are more yellow-green in color than the blue-green, ± evergreen leaves of *Dryopteris intermedia*.

Dryopteris celsa (W. Palmer) Small
LOG-FERN

DESCRIPTION Native fern; rhizomes shallow, creeping, their tips covered by shiny brown or black scales. Leaves deciduous, to 12 dm long, all alike; blades 4–7 dm long and 2–3 dm wide, slightly narrowed at base and tapered gradually to tip; lower pinnae narrowed at base, upper pinnae oblong, uniformly long-tapering; petioles light-green, longer than blade, with scales having dark centers and paler margins. Sori midway between midvein and leaf margin; indusia smooth, 1 mm across.

SYNONYM *Dryopteris goldieana* subsp. *celsa*.

HABITAT Typically on rotting logs and in humus-rich soils of wet woods, swamps, and seepage areas; local and uncommon throughout its range.

STATUS MW-OBL | NCNE-OBL; Mich (THR).

Dryopteris clintoniana (D. C. Eat.) P.
Dowel ♦ CLINTON'S WOOD-FERN

DESCRIPTION Native fern; rhizomes short-creeping. Leaves all ± alike, evergreen (or fertile leaves dying back in winter), 5–10 dm long; blades 3–8 dm long and 1–2 dm wide, slightly narrowed below and long-tapered to tip; pinnae long-triangular, gradually tapered to tip, with sharp-toothed segments; basal pinnae with

Dryopteris carthusiana

Dryopteris clintoniana

a broad base, 1.5–2.5x longer than wide; petioles shorter than blades, scaly at base, scales brown or with darker brown center. Sori near midveins; indusia smooth, 1–2 mm wide.
HABITAT swamps and wet woods.
STATUS MW-FACW | NCNE-FACW

Dryopteris cristata (L.) A. Gray
♦ CRESTED WOOD-FERN

DESCRIPTION Native clumped fern, rhizomes short-creeping with ascending tips. Sterile and fertile Leaves somewhat different, the outer sterile leaves waxy, persistent and smaller than inner fertile leaves; fertile leaves deciduous, 3–8 dm long. Blades 1-pinnate to nearly 2-pinnate, narrowly lance-shaped, 2–6 dm long and 7–15 cm wide, tapered to tip, narrowed at base; pinnae 5–9 cm long and to 4 cm wide, typically twisted to a nearly horizontal position, giving a "venetian blind" appearance to blades; pinnae segments to 20 mm long and 8 mm wide, with small spine-tipped teeth; petioles with sparse, pale brown, long-tapered scales. Sori round, midway between midvein and margin; indusia smooth, 1 mm wide.
HABITAT swamps, thickets, open bogs, fens and seeps.
STATUS MW-OBL | NCNE-OBL | GP-OBL

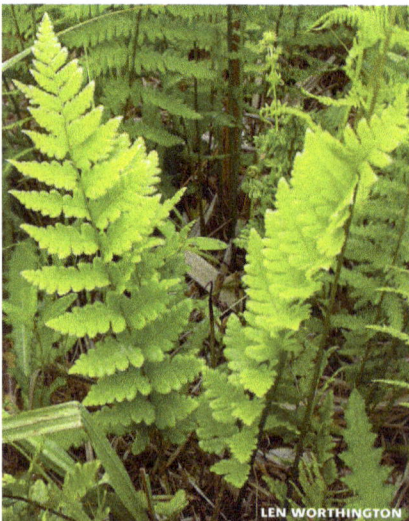

Dryopteris goldieana (Hook.) A. Gray
♦ GOLDIE'S WOOD-FERN

DESCRIPTION Native clumped fern; rhizomes short-creeping, to 1 cm thick, densely scaly. Leaves to 1 m long; blades 30–60 cm long and 20–40 cm wide, deciduous late in season, the upper part abruptly narrowed to a small, tapered tip, the tip often mottled with white; pinnae with small, often rounded teeth; petioles brown, slightly shorter than blades, with narrow, pale brown scales 1–2 cm long, lower scales with a dark midstripe. Sori close to midveins, with a smooth indusia 1–2 mm across.
HABITAT moist hardwood forests, shaded streambanks, talus slopes; soils rich in humus and usually neutral.
STATUS MW-FAC | NCNE-FAC
NOTE Previously spelled as 'goldiana', that spelling now considered incorrect.

Dryopteris intermedia (Muhl.) A. Gray
♦ FANCY WOOD-FERN

DESCRIPTION Native clumped fern, rhizomes ascending. Leaves in an open vaselike cluster of evergreen leaves; blades broadest just above base and abruptly tapered near tip, 2–5 dm long and 1–2 dm wide, 2-pinnate; pinnae at right angles to stem, lowermost inner pinnule usually short-

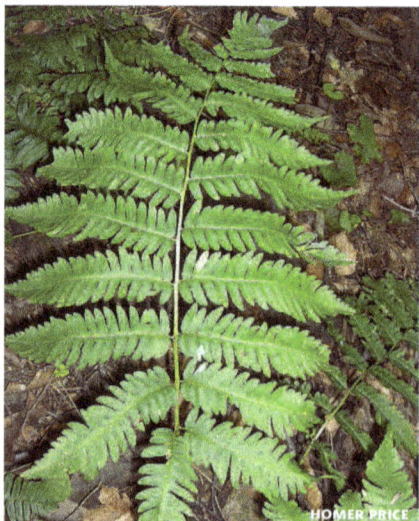

Dryopteris cristata

Dryopteris goldieana

er than next outer pinnule, pinnules toothed and tipped with small spines; petioles 1/3 as long as blade, with pale brown scales with a darker center, petioles and stems with small, gland-tipped hairs. Sori midway between midvein and margin, the indusia 1 mm wide, covered with stalked glands.

HABITAT moist hardwood and mixed hardwood-conifer forests, hummocks in swamps; soils rich in humus and slightly acid to neutral.

STATUS MW-FAC | NCNE-FAC | GP-FACW

Equisetaceae
Horsetail Family

Equisetum
HORSETAIL, SCOURING-RUSH

Rushlike herbs with dark rhizomes. Stems annual or perennial, grooved, usually with large central cavity and smaller outer cavities, unbranched or with whorls of branches at nodes. Leaves reduced to scales, united into a sheath at each node; top of sheath divided into dark-colored teeth. Spores in cones at tips of green or brown fertile stems.

NOTE Stem cross-sections are illustrated for each species; these are useful for identifying *Equisetum* in the field.

Dryopteris intermedia

DOUG MCGRADY

Equisetum **HORSETAIL, SCOURING-RUSH**

1 Stems evergreen (except in *E. laevigatum*); unbranched or with a few scattered branches, branches not in regular whorls (**scouring rushes**) 2

1 Stems annual; usually with regular whorls of branches, sometimes unbranched (**horse-tails**) 5

2 Stems solid (central cavity absent); stems small, slender and sprawling *E. scirpoides*

2 Stems hollow (central cavity present); stems larger, usually upright............ 3

3 Stems 1–3 dm tall, with 5–12 ridges, central cavity to 1/3 diameter of stem............
.... *E. variegatum*

3 Stems usually taller, with 16–50 ridges, central cavity more than half diameter of stem 4

4 Cones with a distinct, small sharp tip; stem sheaths with a black band at tip and base .
.... *E. hyemale*

4 Cones blunt-tipped, sheaths with black band at tip only .... *E. laevigatum*

5 Stems unbranched..................... 6

5 Stems with regular whorls of branches . 10

6 Stems green 7

6 Stems brown or flesh-colored.......... 8

7 Stems with 9–25 shallow ridges; central cavity more than half diameter of stem; sheath teeth entirely black or with narrow white margins .... *E. fluviatile*

7 Stems with 5–10 strongly angled ridges; central cavity less than 1/3 diameter of stem; sheath teeth with white margins and dark centers .... *E. palustre*

8 Sheath teeth papery and red-brown, teeth joined and forming several broad lobes...
.... *E. sylvaticum*

8 Sheath teeth black or brown, not papery, separate or joined in more than 4 small groups 9

9 Stems withering after spores mature, remaining unbranched .... *E. arvense*

9 Stems persistent, becoming branched and green .... *E. pratense*

10 First internode of each branch shorter than the subtending sheath of the main stem 11

10 First internode of each branch equal or longer than the subtending sheath of the main stem 12

11 Stems with 9–25 shallow ridges; central cavity more than half diameter of stem; sheath teeth more than 12, entirely black or with narrow white margins .. *E. fluviatile*

11 Stems with 5-10 strongly angled ridges; central cavity about same size as outer cavities; sheath teeth 5-6, with white margins and dark centers. . . . . *E. palustre*

12 Stem branches themselves branched; sheath teeth papery and red-brown; teeth joined and forming several broad lobes. . . . . *E. sylvaticum*

12 Stem branches unbranched; sheath teeth black or brown, not papery, separate or joined in more than 4 small groups **13**

13 Stem branches ascending; teeth of branch sheaths gradually tapering to a slender tip . . . . *E. arvense*

13 Stem branches spreading; teeth of branch sheaths broadly triangular . . . *E. pratense*

Equisetum arvense L.

◆ COMMON *or* FIELD HORSETAIL

DESCRIPTION Native; stems annual, upright from creeping, branched, tuber-bearing rhizomes covered with dark hairs. Sterile and fertile stems unalike. Sterile stems appearing in spring as fertile wither, green, regularly branched, 1-6 dm tall and 2-5 mm wide, with 10-14 shallow ridges, the ridges usually rough-to-touch; central cavity 1/3-2/3 stem diameter; sheaths with 6-14 persistent, black-brown teeth 1-2 mm long; branches numer-ous in dense whorls, usually without branch-lets, upright or spreading, 3-5-angled, solid. Fertile stems flesh-colored, shorter than sterile stems and with larger sheaths, maturing in early spring and soon withering, unbranched, to 3 dm tall and 8 mm wide; sheaths with 8-12 dark brown teeth. Cones blunt-tipped, long-stalked at end of stem, 0.5-3 cm long.

HABITAT streambanks, meadows, moist woods, ditches, roadsides and along rail-roads; calcareous fens in s part of region.

STATUS MW-FAC | NCNE-FAC | GP-FAC

Equisetum fluviatile L.

◆ WATER-HORSETAIL

DESCRIPTION Native; stems annual, fertile and sterile stems alike, to 1 m or more tall, from smooth, shiny, light brown, creeping rhizomes. Stems with 9-25 shallow, smooth ridges; central cavity large, about 4/5 stem diameter; stem sheaths green, 6-10 mm long; teeth 12-24, persistent, 2-3 mm long, dark brown to black, sometimes with narrow white margins; branches none or few, to many and regularly whorled from middle nodes, spreading, without branchlets, 4-6-angled, hollow. Cones 1-2 cm long at tips of stems, long-stalked, blunt-tipped, deciduous, maturing in summer.

Equisetum arvense

Equisetum fluviatile

HABITAT in standing water of marshes, ponds, peatlands, ditches and swales.
STATUS MW-OBL | NCNE-OBL | GP-OBL

Equisetum hyemale L.
♦ COMMON SCOURING-RUSH

DESCRIPTION Native; stems evergreen, persisting more than 1 year, fertile and sterile stems alike, from black, slender rhizomes. Stems mostly unbranched or with few, short, upright branches from upper nodes, to 15 dm tall but usually shorter, 4–14 mm wide, with 14–50 rounded, very rough ridges; central cavity at least 3/4 stem diameter; stem sheaths 5–15 mm long, with a dark band at tip and usually also at base, the teeth dark brown to black with chaffy margin, 2–4 mm long, deciduous or persistent. Cones stalkless or short-stalked at tips of stems, sharp-pointed, eventually deciduous, 1–2.5 cm long, maturing in summer, or old stems sometimes developing branches with cones in the following spring. HABITAT often forming dense colonies in seeps, wet to moist meadows, shores and streambanks, ditches, roadsides and along railroads; usually where sandy or gravelly.
STATUS MW-FACW | NCNE-FAC | GP-FACW

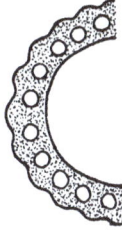

Equisetum laevigatum A. Braun
♦ SMOOTH SCOURING-RUSH

DESCRIPTION Native; stems mostly annual, fertile and sterile stems alike, from brown or black rhizomes. Stems mostly unbranched or with a few upright branches, 3–10 dm tall and 3–8 mm wide, smooth and rather soft, with 10–32 ridges; central cavity 2/3–3/4 stem diameter; stem sheaths with a single dark band at tip, or rarely lowest sheaths with a dark band at base or entirely black; teeth dark brown or black with chaffy margins, free or partly joined in pairs, 1–4 mm long, soon deciduous. Cones short-stalked at tips of stems, rounded with a small sharp point, maturing in early summer and eventually deciduous. HABITAT wet meadows, low prairie, streambanks, floodplains, seeps, and ditches, often where sandy or gravelly.
STATUS MW-FACW | NCNE-FACW | GP-FAC

Equisetum palustre L.
♦ MARSH-HORSETAIL

DESCRIPTION Native; stems annual, fertile and sterile stems alike, from creeping, branched, shiny black rhizomes. Stems 2–8 dm tall, with 5–10 pronounced ridges, the ridges mostly smooth; central cavity small, 1/6–1/3 stem diameter;

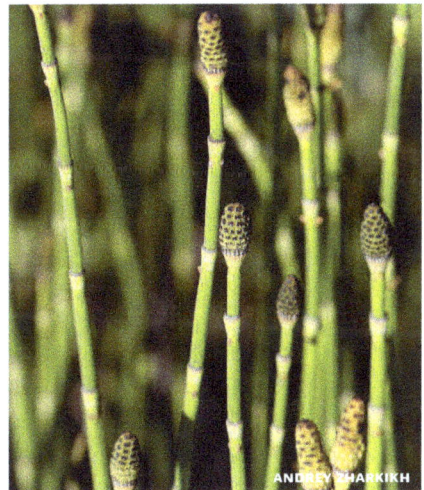

Equisetum hyemale

Equisetum laevigatum

sheaths green, loose and flared upward; teeth 5–6, free or partly joined, persistent, 3–7 mm long, brown to black, with pale, translucent margins; branches few and irregular, to many and whorled at upper nodes, upright, without branchlets, 5–6-angled, hollow. Cones long-stalked at tips of stems, 1–3 cm long, blunt-tipped, maturing in summer, deciduous.

HABITAT in shallow water, along wetland margins, streambanks and fens.

STATUS MW-FACW | NCNE-FACW | GP-FACW

Equisetum pratense Ehrh.
♦MEADOW-HORSETAIL

DESCRIPTION Native; stems annual and erect, sterile and fertile stems unalike, from creeping, dull black rhizomes.

Sterile stems regularly branched, 2–5 dm tall and 1–3 mm wide; 8–18-ridged, the ridges roughened by silica on middle and upper stem; central cavity 1/3–1/2 stem diameter; main stem sheaths 2–6 mm long, the teeth persistent, 1–2 mm long, free or partly joined in pairs, brown with white margins and a dark midstripe; branches slender, many in regular whorls from middle and upper nodes, without branchlets, horizontal or drooping, mostly 3-angled, solid. **Fertile**

stems uncommon, appearing in early spring before sterile stems and persisting, at first unbranched, fleshy and brown (without chlorophyll), later becoming green at nodes and producing many small green branches, mostly 1–3 dm tall; sheaths and teeth about twice as long as on sterile stems.

Cones long-stalked at tips of stems, to 2.5 cm long, blunt-tipped, deciduous.

HABITAT moist woods, meadows, streambanks.

STATUS MW-FACW | NCNE-FACW | GP-FACW

Equisetum scirpoides Michx.
DWARF SCOURING-RUSH

DESCRIPTION Native; stems evergreen, very slender, fertile and sterile stems alike, from widely branching rhizomes. **Stems** 5–30 cm long and only 0.5–1 mm wide, in dense clusters, usually unbranched and zigzagged, upright or trailing; central cavity absent, 3 small outer cavities present; sheaths green with broad black band at tip, loose and flared above, with 3–4 teeth; teeth with white, chaffy margin, ± persistent, but tips usually soon deciduous.

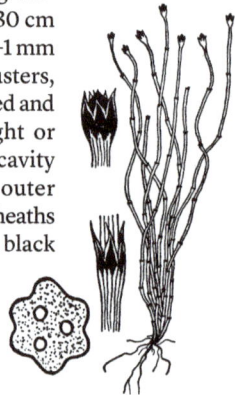

KRISTIAN PETERS

Equisetum palustre

ROBERT FLOGAUS-FAUST

Equisetum pratense

Cones black, small, 3–5 mm long, sharp-tipped.

HABITAT mossy places and moist, shaded woods, the stems often partly buried in humus.

STATUS MW-FAC | NCNE-FAC | GP-FAC

Equisetum sylvaticum L.
♦WOODLAND-HORSETAIL

DESCRIPTION Native; stems annual, erect, sterile and fertile stems unalike, from creeping, shiny light brown rhizomes, tubers occasionally present. **Sterile stems** green, 3–7 dm tall and 1.5–3 mm wide, with 10–18 ridges, rough-to-touch with sharp, hooked silica spines; central cavity 1/2–2/3 stem diameter; sheaths green at base, red-brown and flaring at tip; teeth brown, 3–5 mm long, joined in 3–5 broad lobes. Stems densely branched in regular whorls from the nodes, the branches themselves branched, often curving downward, 4–5-angled, solid. **Fertile stems** at first pink-brown (without chlorophyll), fleshy, unbranched, becoming green and branched as in sterile stems; sheaths and teeth larger than in sterile stems. **Cones** 1.5–3 cm long, stalked, blunt-tipped, deciduous.

HABITAT wet or swampy woods, thickets, usually in partial shade.

STATUS MW-FACW | NCNE-FACW | GP-FACW

Equisetum variegatum Schleicher
♦VARIEGATED SCOURING RUSH

DESCRIPTION Native; stems evergreen, fertile and sterile stems alike, from creeping, much-branched, smooth rhizomes. **Stems** 1–3 dm tall and 1–2.5 mm wide, with 5–12 shallow, rough ridges, branched near base and otherwise usually unbranched; central cavity 1/4–1/3 stem diameter, smaller outer cavities present; sheaths green at base with a broad black band above; teeth persistent, with a dark brown or black midstripe and wide white margins, abruptly narrowed to a hairlike, deciduous tip 0.5–1 mm long. **Cones** to 1 cm long, strongly sharp-tipped, maturing in summer or persisting unopened until following spring.

HABITAT lakeshores, streambanks, wet woods, moist meadows, fens, and ditches, moist sandy soil, frequent near Lake Michigan.

STATUS MW-FACW | NCNE-FACW | GP-FACW

Isoetaceae
Quillwort Family

Isoetes QUILLWORT

Perennial aquatic or emergent herbs. Leaves simple, entire, linear, from a 2–3

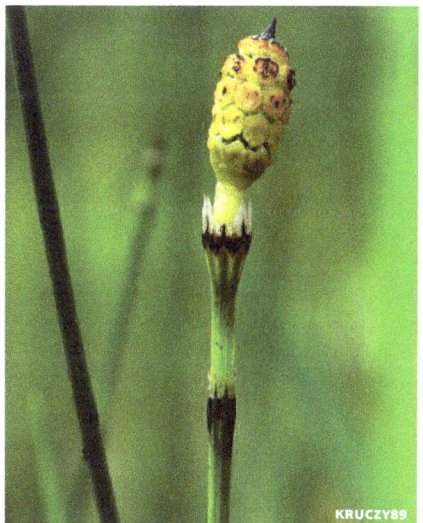

Equisetum sylvaticum

Equisetum variegatum

lobed rhizome (corm). Outermost and innermost leaves typically sterile. Outer fertile leaves have a pocketlike structure (sporangium) bearing large whitish spores (megaspores; illustrated); inner fertile leaves have numerous small microspores.

Isoetes QUILLWORT

1 Megaspores conspicuously covered with small spines.............*I. echinospora*
1 Megaspores not spiny..................**2**
2 Plants normally underwater; leaves without outer fibrous strands . *I. lacustris*
2 Plants underwater or on exposed, drying shores; leaves with 4 or more fibrous strands **3**
3 Plants underwater or emergent; megaspores with a raised, netlike surface; sw LP of Mich............*I. engelmannii*
3 Plants emergent and on drying shores; megaspores with numerous small bumps; sw Minn ....*I. melanopoda*

Isoetes echinospora Durieu
◆SPINY-SPORED QUILLWORT

DESCRIPTION Small, native aquatic herb. Leaves linear, 7–25 or more, 5–15 cm long and 0.5–1.5 mm wide, usually erect, soft, bright green to yellow-green, tapered from base to a very long, slender tip, without peripheral strands from

Isoetes echinospora

base; corm 2-lobed. Sporangium 4–8 mm long, usually brown-spotted when mature, half or more covered by a membranous flap (velum). Megaspores round, white, 0.3–0.6 mm wide, covered with short, sharp to blunt spines.

SYNONYMS *Isoetes braunii, Isoetes muricata*.
HABITAT shallow water (to 1 m deep) of lakes, ponds and slow-moving rivers; plants rooted in mud, sand, or gravel.
STATUS MW-OBL | NCNE-OBL | GP-OBL
NOTE A circumboreal species, and the region's most common quillwort in acidic lakes and ponds.

Isoetes engelmannii A. Braun
ENGELMANN'S QUILLWORT

DESCRIPTION Small, native aquatic herb. Leaves linear, 6–50 cm long and 0.5–2 mm wide, upright, not twisted, usually with 4 peripheral strands from base. Sporangium 6–13 mm long, pale, 1/3–2/3 covered by membranous flap (velum). Megaspores 0.4–0.6 mm wide, with a raised, netlike surface.

HABITAT shallow water of ponds and streams; plants usually underwater in spring, partly emergent later.
STATUS MW-OBL | NCNE-OBL; Mich (END).

Isoetes lacustris L.
LAKE QUILLWORT

DESCRIPTION Small, native aquatic herb. Leaves several to many, 5–20 cm long and 1–2 mm wide, stiff and erect or with leaf tips curved downward, dark green, fleshy and twisted, peripheral strands from base usually absent; corm 2-lobed. Sporangium to 5 mm long, usually not spotted; membranous flap (velum) covering up to half of sporangium. Megaspores round, white, 0.6–0.8 mm wide, with ridges forming an irregular netlike pattern.

SYNONYM *Isoetes macrospora*.
HABITAT underwater in shallow to deep water of cold lakes, ponds and streams.
STATUS MW-OBL | NCNE-OBL

Isoetes melanopoda Gay & Durieu
BLACK-FOOT QUILLWORT

DESCRIPTION Small, native aquatic herb. Leaves 10-50 cm long and 0.5-2 mm wide, black at base with a pale line down middle of inner side, 4 peripheral strands from base usually present. Sporangium 5-20 mm long, brown-spotted when mature, up to 2/3 covered by membranous flap (velum). Megaspores 0.3-0.5 mm wide, covered with short, low ridges.

HABITAT underwater to emergent in temporary ponds, wet streambanks, ditches and swales.

STATUS MW-OBL | NCNE-OBL | GP-OBL; Minn (END).

Lycopodiaceae
Clubmoss Family

LOW, TRAILING, evergreen, perennial herbs resembling large mosses. Leaves needlelike or scalelike, alternate or opposite on stem. Spore-bearing leaves (sporophylls) similar to vegetative leaves or in conelike clusters at tips of upright stems.

Lycopodiaceae CLUBMOSS FAMILY
1 Leafy horizontal stems absent; upright stems in clusters .... *Huperzia*
1 Leafy horizontal stems present and creeping on ground surface; upright stems borne singly along horizontal stems .....
.... *Lycopodiella inundata*

Huperzia FIR-MOSS
Shoots erect; leaves spreading or appressed and upright. Spores borne at base of upper leaves.

Huperzia FIR-MOSS
1 Leaves narrowly obovate with 1-8 irregular teeth; common plants of moist to wet conifer woods .... *H. lucidula*
1 Leaves lance-shaped to oblong lance-shaped, margins entire or with 1-3 small teeth; uncommon on wetland margins ...
.... *H. selago*

Huperzia lucidula (Michaux) Trev.
♦ SHINING FIR-MOSS

DESCRIPTION Native; stems light green, creeping and rooting, upcurving stems forked several times, to 25 cm high and 2 mm wide (stem only), crowded with shiny dark green leaves which persist for more than one season. Leaves in mostly 6 rows, spreading or curved downward, in alternating groups of longer sterile and shorter fertile leaves, giving shoots a ragged look. Sterile leaves 6-12 mm long, toothed and broadest above middle; sporophylls barely widened and with small teeth or entire at tip. Small two-lobed buds (gemmae) produced in some upper leaf-axils; these may sprout into new plants after falling onto moist humus.

SYNONYM *Lycopodium lucidulum.*

HABITAT moist to wet conifer and hardwood forests.

STATUS MW-FACW | NCNE-FAC | GP-FACW

Huperzia selago L.
NORTHERN FIR-MOSS

DESCRIPTION Native; horizontal stems short; upright stems forked from base, 2-3 mm wide (stem only). Leaves persistent, yellow-green, in 8-10 rows, 3-6 mm long and to 1 mm wide, swollen and concave at base, gradually tapered to

Huperzia lucidula

tip, mostly without teeth, uniform in length; leaves appressed to stem, giving stems a smooth, cylindric outline. Sporophylls similar to vegetative leaves; sporangia produced early in season in leaf axils, followed later by sterile leaves. Upper axils with small, 2-lobed reproductive buds (gemmae).

SYNONYM *Lycopodium selago.*

HABITAT an arctic tundra species of thickets, streambanks, cold woods and bog margins.

STATUS MW-FACU | NCNE-FACU

NOTE a hybrid between *Huperzia selago* and *H. lucidula* (*Huperzia xbuttersii*) is reported from ne Minn and nw Wisc.

Lycopodiella BOG CLUBMOSS

Lycopodiella inundata L. Holub
♦NORTHERN BOG CLUBMOSS

DESCRIPTION Native; stems elongate and trailing, deciduous but with evergreen buds at tips, rooting throughout. Leaves in 8-10 rows, those on underside of trailing stems twisted upward, margins ± entire. Fertile branches few, erect, to 1 dm high, with spreading leaves. Cones 1.5-5 cm long and 6-12 mm wide; sporophylls green, base widened and with a pair of teeth.

SYNONYM *Lycopodium inundatum.*

HABITAT Acid, open sphagnum bogs, wet sandy shores and streambanks; disturbed wetlands.

STATUS MW-OBL | NCNE-OBL

Marsileaceae
Water-Clover Family

Marsilea WATERCLOVER
Aquatic perennial herbs, with slender rhizomes rooted in mud, sending up rows of leaves with long petioles and floating blades. Blades divided into 4 pinnae resembling a 4-leaf clover. Staminate and pistillate spores in dark, hair-covered capsules (sporocarps) borne on stalks near base of petioles.

Marsilea WATER-CLOVER

1 Rhizomes rooting along their entire length; leaves ± hairless; sporocarps 2-3 on long, branched stalks; Mich . . . . . *M.quadrifolia*

1 Rhizomes rooting only at nodes; leaves with appressed hairs; sporocarps single on short, unbranched stalks; sw Minn (more common westward) . . . . *M. vestita*

Marsilea quadrifolia L.
♦EUROPEAN WATER-CLOVER

DESCRIPTION Introduced, creeping, perennial fern. Stem a rhizome, rooting at nodes and with 1-3 roots between nodes; petioles 5-20 cm long, sparsely hairy to smooth; blades 5 to 20 mm long and about as wide, usually without hairs. Sporangia in 2-3 oval sporocarps 4-6 mm long, the sporocarps hairy when young, less

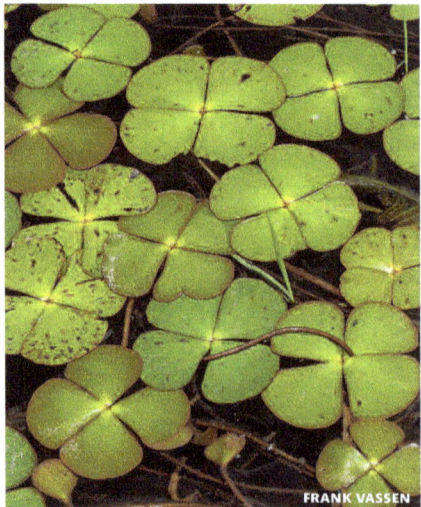

Lycopodiella inundata

Marsilea quadrifolia

so with age, on stalks to 2 cm long, stalks attached from 1–12 mm above base of petiole.

HABITAT introduced, occasionally escaping to ponds and slow-moving streams.

STATUS MW-OBL | NCNE-OBL

Marsilea vestita Hook. & Grev.
HAIRY WATER-CLOVER

DESCRIPTION Native, creeping, perennial fern. Stem a slender rhizome, rooting only at nodes; petioles 2–20 cm long, sparsely hairy; blades floating or emergent, 5–20 mm long and 5–15 mm wide, usually hairy on both sides. Sporangia in a single oval sporocarp borne on a short stalk at or near base of petiole, the sporocarp brown, 4–7 mm long, covered with stiff, flat hairs; sori in 2 rows inside the sporocarp.

SYNONYM *Marsilea mucronata.*

HABITAT shallow water or mud of temporary ponds, floodplains and ditches.

STATUS MW-OBL | GP-OBL; Minn (END).

Onocleaceae
Sensitive Fern Family

LARGE COARSE FERNS; sterile and fertile fronds strongly different, the sterile fronds deciduous, pinnatifid to 1-pinnate-pinnatifid; fertile fronds persistent. Sori enclosed under recurved margin of pinna segment (outer false indusium) and a tiny true inner indusium (membranous or of hairs).

Onocleaceae SENSITIVE FERN FAMILY

1 Sterile blades solitary from creeping rhizomes, deeply divided into lobes (or the lowermost divisions pinnae)...........
....................*Onoclea sensibilis*
1 Sterile blades in a circle from a thick crown; pinnate with lobed pinnules ....
.............*Matteuccia struthiopteris*

Matteuccia OSTRICH FERN

Matteuccia struthiopteris (L.) Todaro
◆ OSTRICH FERN

DESCRIPTION Large, colony-forming native fern; rhizomes deep and long-creeping, black, scaly, producing erect leafy crowns. Sterile leaves upright, 1-pinnate, to 2 m tall and 15–50 cm wide; blades much longer than petioles, abruptly narrowed to tip, gradually tapered to base, stems ± hairy; each pinnae deeply divided into 20 or more pairs of pinnules, these 3–6 mm wide at base and rounded at tip; veins not netlike. Fertile leaves stiff and erect within a circle of sterile leaves, green at first, turning brown or black, much shorter than sterile leaves (to 6 dm tall), produced in mid- to late-summer and often persisting into following year; fertile blades 1-pinnate, pinnae upright or appressed, 2–6 cm long and 2–4 mm wide, the margins inrolled and covering the sori; indusia with a jagged margin.

HABITAT wet and swampy woods, streambanks, seeps, and ditches.

STATUS MW-FACW | NCNE-FAC | GP-FACW

Onoclea SENSITIVE FERN

Onoclea sensibilis L.
◆ SENSITIVE FERN

DESCRIPTION Medium-sized native fern, in clumps of several leaves, spreading by branching rhizomes and forming large patches. Leaves upright, with petioles about as long as blades. Sterile leaves deciduous, 1-pinnate at base, deeply

Matteuccia struthiopteris

cleft upward; the stem broader-winged toward the tip; blades 15–40 cm long and 15–35 cm wide, with 8–12 pairs of opposite pinnae, these deeply wavy-margined or coarsely toothed, 1–5 cm wide, with scattered white hairs on underside veins, the veins joined and netlike. **Fertile leaves** produced in late summer and persisting over winter, shorter than sterile leaves; fertile blades 1-pinnate, pinnae upright, divided into beadlike pinnules with inrolled margins covering the sori; veins not joined. **Sori** round and covered by a hoodlike indusia, becoming dry and hard.

HABITAT Swampy woods and low places in forests, wet meadows, calcareous fens, roadside ditches, wet or moist wheel ruts; sometimes weedy.

STATUS MW-FACW | NCNE-FACW | GP-FACW

Ophioglossaceae
Adder's-Tongue Family

PERENNIAL HERBS from short, erect rhizomes having several fleshy roots. Plants produce one leaf each year on a single stalk (stipe), with bud for next year's leaf at base of stipe. Leaves divided into a fertile segment (sporophyll) and a sterile expanded blade. Sterile blades entire (*Ophioglossum*), or lobed or 1–3x pinnately divided (*Botrychium, Botrypus*). Spores in numerous round sporangia borne on simple or branched fertile blades.

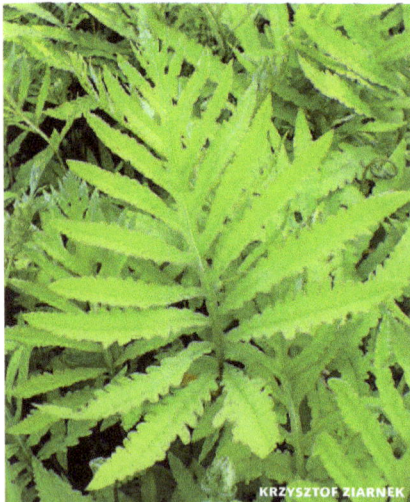

Onoclea sensibilis

Ophioglossaceae **ADDER'S-TONGUE FAMILY**

1 Sterile blades simple, entire; veins netlike; sporangia embedded in rachis of spike . . . . ***Ophioglossum pusillum***

1 Sterile blades pinnately lobed or dissected; veins forked; sporangia exposed, often on a branched structure **2**

2 Plants 50 cm or more tall; blade large and broadly triangular; fertile portion erect, appearing to be a continuation of stipe . . . . ***Botrypus virginianus***

2 Plants smaller, less than 25 cm tall; fertile portion of blade upright or spreading. . . . . ***Botrychium***

Botrychium
GRAPE-FERN, MOONWORT

Mostly small plants with 1 leaf, the blade divided into sterile and fertile segments. Sterile portion of blade pinnately divided or lobed, fertile portion branched to form a panicle bearing the sporangia.

Botrychium **GRAPE-FERN, MOONWORT**

2 Blade broadly triangular in outline, dissected into narrow toothed segments (pinnae) . . . . ***B. lanceolatum***

2 Blade divided into 3–6 fan-shaped pairs . . . . ***B. neolunaria***

Botrychium lanceolatum (S.G. Gmelin)
Angström ♦ LANCE-LEAVED GRAPE-FERN

DESCRIPTION Native; plants 6–30 cm tall, dark green, smooth, appearing in early sum-

Botrychium lanceolatum

mer and persisting to fall. Stems about 5x longer than blades; blades triangular in outline, 1–8 cm long and 1–5 cm wide, stalkless or on a short stalk to 6 mm long, divided into 2–5 pairs of sharp-pointed, toothed pinnae, lowermost pair the largest. Fertile segment 2–9 cm long, mostly twice pinnate, on a stalk 1–3 cm long.

HABITAT moist humus-rich woods, hummocks in swamps, streambanks.

STATUS MW-FACW | NCNE-FACW | GP-FACW; Minn (THR).

NOTE Ours are var. *angustisegmentum* (now sometimes treated as *Botrychium angustisegmentum*), found from Nfld and Ontario to Minnesota, becoming increasingly rare southward.

Botrychium neolunaria Stensvold & Farrar COMMON MOONWORT

DESCRIPTION Native; plants 3–20 cm tall, rubbery-textured, appearing in late spring and withering in summer. Leaf blades 1.5–7 cm long and 1–3 cm wide, stalkless or on a short stalk to 5 mm long; pinnately divided into 3–6 pairs of stalkless pinnae, the pinnae fan-shaped, wider than long and without a midrib; petioles 1.5–3 cm long. Fertile segments 0.5–7 cm long, on stalks about as long as the segments.

HABITAT grassy meadows, sandy or gravelly lakeshores and streambanks, rock ledges and mossy talus; most common on ± neutral soils.

STATUS MW-FACW | NCNE-FACW | GP-FAC; Minn (THR), Wisc (END).

NOTE *B. neolunaria* was formerly treated as *B. lunaria* (L.) Swartz, that species now apparently confined to Maine in the USA and across Canada (Stensvold and Farrar, 2017). Other small species of *Botrychium* have been described including *B. campestre, B. minganense,* and *B. mormo.* For a complete discussion, see *Flora of North America,* Vol. 2 (1993).

Botrypus
RATTLESNAKE FERN

Botrypus virginianus (L.) Holub
♦RATTLESNAKE FERN

DESCRIPTION Native; plants 40–75 cm tall, appearing in spring, withering in autumn, and not overwintering. Leaf blade (trophophore) triangular, sessile, to 25 cm long and to 1.5x as wide, 3–4x pinnate, thin and herbaceous. Pinnae to 12 pairs, usually somewhat overlapping and slightly ascending; pinnules lance-shaped and deeply lobed, the lobes linear, sharply toothed and pointed at tip Spore-bearing portion (sporophore) 2-pinnate, 0.5–1.5x length of trophophore.

SYNONYM *Botrychium virginianum.*

HABITAT Occasional in swamps of cedar and black spruce; more common in moist deciduous woods.

STATUS MW-FACU | NCNE-FACU | GP-FACU

NOTE Rattlesnake fern is a widespread species in North America, found across Canada and nearly all of the USA.

Ophioglossum
ADDER'S-TONGUE

Ophioglossum pusillum Raf.
♦NORTHERN ADDER'S-TONGUE

DESCRIPTION Native; plants erect, 7–30 cm

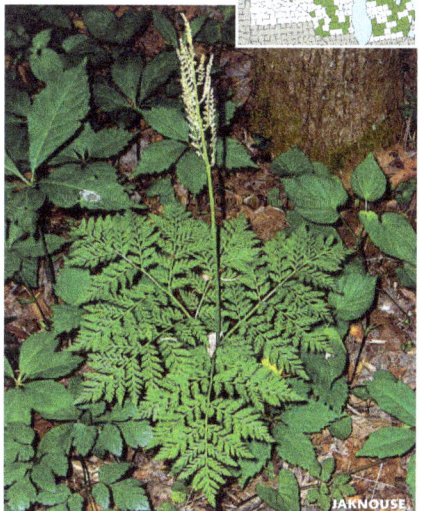

Botrypus virginianus

tall, from slender rhizomes. Leaves 1, entire, on a stalk 3–15 cm long; blades upright, oval to ovate, rounded to acute at tip, 3–8 cm long and 1–4 cm wide, conspicuously net-veined. Sporangia in 2 rows in a terminal, unbranched fertile segment, 1–5 cm long and 2–4 mm wide, on a stalk 6–15 cm long. HABITAT wet sandy meadows and prairies, moist depressions, fens, wetland margins, wet forests. STATUS MW-FACW | NCNE-FACW | GP-FACW NOTE *Ophioglossum vulgatum* L. reported from se Mich (THR), becoming more common in se USA; blades dark green, somewhat shiny, in contrast to pale green and dull blades of *Ophioglossum pusillum*.

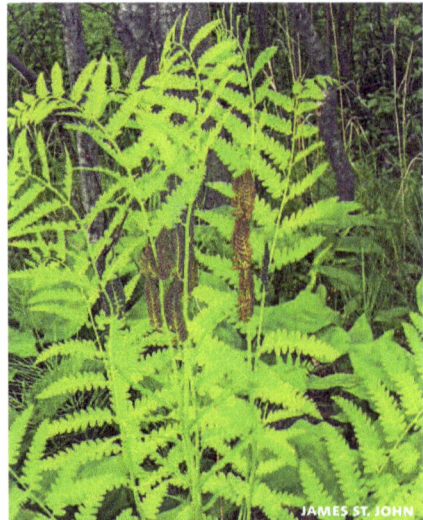

Osmundaceae
Royal Fern Family

Osmundaceae **ROYAL FERN FAMILY**

1   Fertile leaves with one type of leaflet, these on a blade separate from the vegetative leaves; base of leaflets on vegetative leaves bearing a tuft of whitish hairs . . . . *Osmundastrum*

1   Fertile leaves with 2 types of leaflets (some leaflets bearing sporangia and absent on others); base of vegetative leaflets without a tuft of hairs . . . . *Osmunda*

Osmunda ROYAL FERN

Perennial ferns with large rootstocks and exposed crowns covered with old roots and stalks, sending up tufts of coarse leaves. Leaves 1–2-pinnate, differentiated into sterile and fertile segments. Sporangia in round clusters, spores green.

Osmunda **ROYAL FERN**

1   Leaves 2-pinnate, pinnae ± entire; sporangia on upper half of fertile leaves . . . . *O. regalis*

1   Leaves 1-pinnate, sterile pinnae deeply cleft; sporangia only near middle of fertile leaves . . . . *O. claytoniana*

Osmunda claytoniana L.
♦ INTERRUPTED FERN

DESCRIPTION Clumped, native fern to 1 m or more tall; often forming large colonies. Outer leaves usually sterile, with blades 4–10 dm long and 15–30 cm wide; pinnae stalkless and deeply cut into segments, with smooth or slightly hairy margins. Petioles covered with tufts of woolly hairs when young, becoming smooth or sparsely hairy with age, the hairs not forming tufts at pinna-bases (as in *Osmunda cinnamomea*). Inner leaves larger and with 2–5 pairs of fertile pinnae in middle of blade; fertile segments to 6 cm long and 2 cm wide and much smaller than vegetative segments

Ophioglossum pusillum

Osmunda claytoniana

above and below them; sporangia clusters at first green-black, turning dark brown and withering.

SYNONYM *Osmundastrum claytonianum*.

HABITAT moist or seasonally wet depressions in forests, hummocks in swamps, low prairie, wet roadsides; often in drier places than *Osmunda cinnamomea* or *O. regalis*.

STATUS MW-FAC | NCNE-FAC | GP-FAC

Osmunda regalis L.

♦ ROYAL FERN

DESCRIPTION Large, clumped, native fern to 1 m or more tall. Leaf blades broadly ovate in outline, 4–8 dm long and to 3–5 dm wide, 2-pinnate into ± opposite divisions (pinnules), these well-spaced, oblong, rounded at tips, with entire or finely toothed margins. Petioles smooth, green or red-green, to 3/4 length of blade. Fertile leaves with uppermost several pinnae replaced by sporangia clusters.

SYNONYM *Osmunda spectabilis*.

HABITAT bogs, swamps, alder thickets and shallow pools; soils usually acidic.

STATUS MW-OBL | NCNE-OBL | GP-OBL

Osmundastrum

CINNAMON FERN

Osmundastrum cinnamomea L.

♦ CINNAMON-FERN

DESCRIPTION Large, clumped, native fern, to 1 m or more tall. Sterile leaf blades to 30 cm wide, gradually tapered to tip, 1-pinnate, with conspicuous tuft of white or brown woolly hairs at base of each pinna, pinnae stalkless and deeply cleft into segments, with fringe of short hairs on margins; petioles densely hairy when young. Fertile leaves at center of crown, surrounded by taller sterile leaves, without leafy tissue, arising in spring or early summer and turning cinnamon brown, withering and inconspicuous by midsummer.

SYNONYM *Osmunda cinnamomea*.

HABITAT swamps, bog-margins, wooded streambanks, and low wet places; soils acid.

STATUS MW-FACW | NCNE-FACW | GP-FACW

Salviniaceae
Water Fern Family

Azolla MOSQUITO-FERN

Azolla cristata Kaulfuss

♦ MEXICAN MOSQUITO-FERN

DESCRIPTION Small annual aquatic fern, native; plants free-floating or forming floating mats several cm thick, sometimes stranded on mud; roots few and unbranched. Stems lying flat, 1–1.5 cm long, green or red, covered with small, alternate,

Osmunda regalis

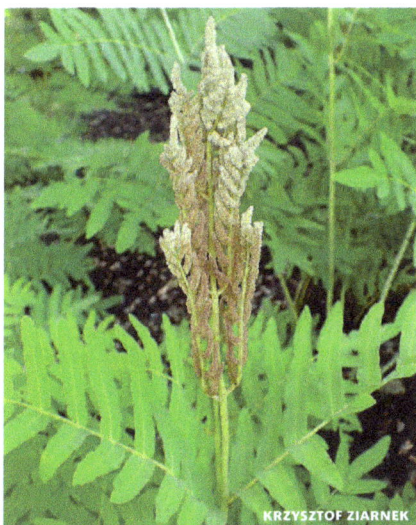

KRZYSZTOF ZIARNEK

Osmundastrum cinnamomea

NICHOLAS T

overlapping leaves in 2 rows. Leaves 2-lobed, the upper lobe to 1 mm long, emergent; lower lobe underwater and larger than the upper. Sporangia of 2 kinds; larger pistillate megaspores (to 0.6 mm long) and tiny staminate spores (micro-sporangia), and borne in separate sporocarps. Sporocarps usually paired on underwater lobes of some leaves. **SYNONYM** *Azolla caroliniana, Azolla mexicana.*

HABITAT quiet water of rivers, ponds, streams and marshes.

STATUS MW-OBL | NCNE-OBL | GP-OBL

NOTE Mosquito-fern is an important genus in rice-producing regions where it is utilized as a natural fertilizer because of its symbiosis with nitrogen-fixing bacteria. The genus is sometimes placed in the family Azollaceae.

Selaginellaceae
Spikemoss Family

Selaginella SPIKEMOSS
Trailing, evergreen herbs with branched, leafy stems, rooting at branching points. Leaves small and overlapping. Spore-bearing leaves similar to vegetative leaves and clustered in cones at ends of branches. Megaspores 4 in each sporangium, yellow or white; micro-spores numerous and very small, red or yellow, covered with small spines.

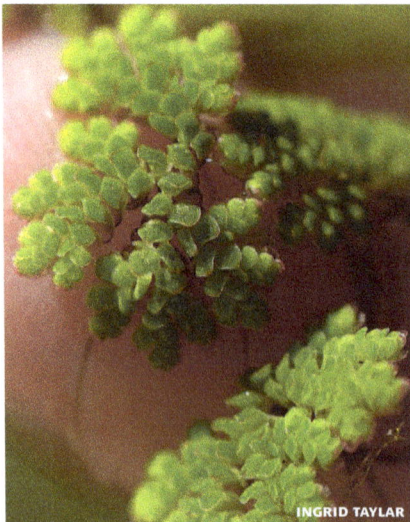

Selaginella SPIKEMOSS

1 Leaves in 4 rows, of 2 kinds: large and spreading, and small and appressed to stem, margins smooth; stems trailing with upright tips; cones 0.5–1 cm long . . . . *S. eclipes*

1 Leaves in many rows, all alike, with hairs on margins; fertile branches upright, cones 2–4 cm long . . . . *S. selaginoides*

Selaginella eclipes Buck
♦ MEADOW SPIKEMOSS

DESCRIPTION Native; plants, forming large, yellow-green mats of branching, trailing stems with upright tips. Stems slender, to 0.4 mm wide. Leaves scalelike, in 4 rows, of 2 types, the larger leaves spreading, 1–2 mm long and 1 mm wide; the smaller leaves appressed to stem, up to 1 mm long and 0.5 mm wide. Cones 0.5–2 cm long, cylindric in 4 rows, the sporophylls similar to lateral leaves and slightly larger. Megaspores white, with a netlike surface.

SYNONYM *Selaginella apoda.*

HABITAT swamps, wet meadows, streambanks and springs, especially where calcium-rich.

STATUS MW-FACW | NCNE-FACW

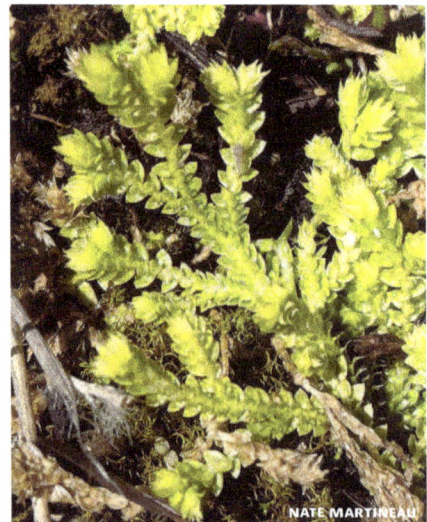

Azolla cristata

Selaginella eclipes

Selaginella selaginoides (L.) Link
♦ NORTHERN SPIKEMOSS

DESCRIPTION Native; plants forming small mats. Sterile stems prostrate, 2–5 cm long; fertile stems upright, deciduous, 5–10 cm high and 0.5 mm wide (stem only), changing upward into broader sporophylls. Leaves in multiple spiral rows, all alike, 2–4 mm long and 1 mm wide, with sharp tips and sparsely hairy margins. Cones ± cylindric but with 4 rounded angles, 1.5–3 cm long and to 5 mm wide. Megaspores yellow-white, with low rounded projections on the 3 flat surfaces.

HABITAT streambanks and lakeshores; cool, mossy talus slopes; conifer bogs and swamps; moist dunes.

STATUS NCNE-FACW; Minn (END), Wisc (END).

Thelypteridaceae
Marsh Fern Family

FERNS WITH CREEPING STEMS, plants from creeping stems and not forming vase-like clumps; blades 1-pinnate. Sori round to oblong; indusia absent in *Phegopteris*; present and tan-colored in *Thelypteris*.

Thelypteridaceae MARSH FERN FAMILY
1 Leaf blades broadly triangular in outline, broadest at base, lowermost pinnae directed downward; indusia absent . . . . *Phegopteris connectilis*
1 Blades lance-shaped in outline, broadest above base; indusia present . . *Thelypteris*

Phegopteris BEECH-FERN
Phegopteris connectilis (Michaux) Watt
♦ NORTHERN BEECH-FERN

DESCRIPTION Native fern; rhizomes long, scaly and densely hairy. Leaves triangular, 15–25 cm long and 6–15 cm wide; blades 1-pinnate, the pinnalike divisions joined by a wing along rachis, except for lowermost pair which are free and angled downward; pinnules oblong, rounded at tip, and usually hairy; petioles longer than blades, hairy, with narrow, brown scales.

SYNONYMS *Dryopteris phegopteris, Thelypteris phegopteris.*

HABITAT cool moist woods, streambanks, sphagnum moss hummocks, shaded rock crevices.

STATUS MW-FACU | NCNE-FACU | GP-FACU

Thelypteris MARSH-FERN
Small to medium ferns from slender rhizomes. Leaves ± hairy; sori small and round; indusia kidney- or horseshoe-shaped.

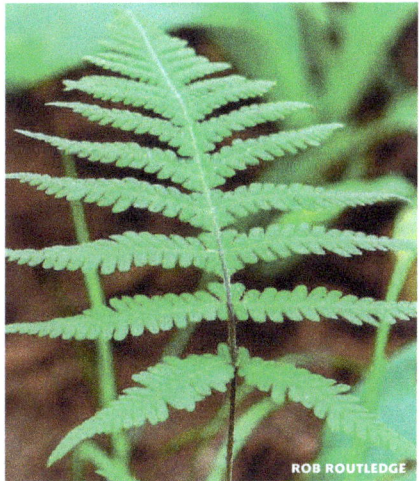

Selaginella selaginoides

Phegopteris connectilis

Thelypteris **MARSH-FERN**

1 Veins of pinnae mostly forked; indusia fringed with hairs; widespread species ... . *T. palustris*

1 Veins not forked; indusia with glands on margin; disjunct in driftless area of sw Wisc .... *T. simulata*

Thelypteris palustris Schott
⬧ MARSH-FERN

DESCRIPTION Native fern; rhizomes spreading and branching. **Leaves** deciduous, erect, 20–60 cm long and to 15 cm wide; blades broadly lance-shaped, short-hairy on rachis and midveins, tapered to tip and only slightly narrowed at base; 1-pinnate, pinnae in 10–25 pairs, mostly alternate, narrowly lance-shaped, to 2 cm wide. Sterile and fertile leaves only slightly different; sterile leaves thin and delicate, pinnules blunt-tipped, 3–5 mm wide, veins once-forked. **Fertile leaves** longer than sterile leaves; pinnules oblong, 2–4 mm wide, the margins rolled under, veins mostly 1-forked; petioles longer than blades, black at base, hairless and without scales. **Sori** round, located halfway between midvein and margin, sometimes partly covered by the rolled under margin; indusia irregular in shape, usually with a fringe of hairs.

SYNONYM *Dryopteris thelypteris.*

HABITAT swamps, low areas in forests, sedge meadows, forest depressions, open bogs, calcareous fens, marshes.

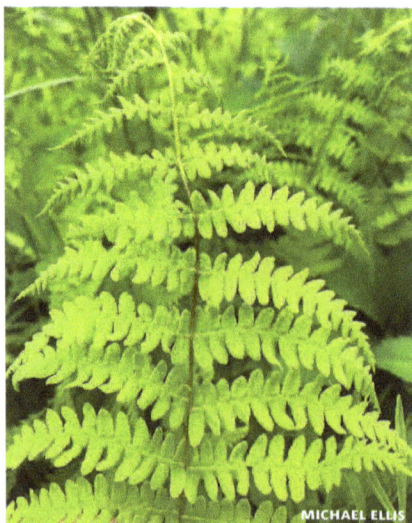

STATUS MW-OBL | NCNE-FACW | GP-OBL

Thelypteris simulata (Davenp.) Nieuwl.
⬧ MASSACHUSETTS FERN

DESCRIPTION Native fern; rhizomes creeping, 2–3 mm wide. **Leaves** fertile leaves taller than sterile leaves; blades 20–40 cm long and 7–15 cm wide, tapered only slightly toward base, hairy on underside; pinnae cleft into oblong lobes which tend to fold together and upward in dry weather; petioles longer than blades, straw-colored, sparsely scaly. **Sori** round; indusia and lower surface of pinnules with red-orange glandular dots.

SYNONYMS *Aspidium simulata, Coryphoteris simulata, Dryopteris simulata, Parathelypteris simulata.*

HABITAT wet, peat-rich meadows; usually in sphagnum moss; soils acidic; uncommon in driftless area of sw Wisc; disjunct from main range of NS south to Va and WVa.

STATUS MW-FACW | NCNE-FACW

NOTE *Thelypteris simulata* is similar to **marsh-fern** (*Thelypteris palustris*) but lower pinnae in Massachusetts fern are narrowed at base next to rachis (only slightly narrowed at base in marsh-fern); veins in both sterile and fertile leaves of *Thelypteris simulata* are unbranched (veins in sterile leaves of *Thelypteris palustris* are mostly forked.

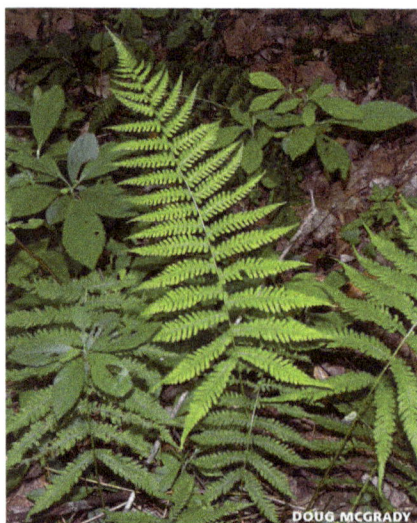

Thelypteris palustris

Thelypteris simulata

CONIFERS
Gymnosperms

CONIFERS are seed plants within the Spermatopsida, subdivision Gynosperm (Latin *gymn-* "naked," Greek *sperma,* "seed"; also termed the Conifers or Gymnospermae). In contrast to the Angiosperms, the seed is not enclosed in an ovary (which develops into a fruit), but are exposed and attached to a cone or other structure. Two families are represented in Upper Midwest wetlands: **Cupressaceae** (*Taxodium distichum, Thuja occidentalis*), and **Pinaceae** (*Abies, Larix, Pinus, Picea*). In *Thuja,* the leaves are small, scalelike and appressed to the branches. The pistillate cones are small, becoming brown with age. In Pinaceae, the leaves are needlelike, and either separate or grouped into bundles of two or more. The staminate and pistillate flowers are borne separately on the same tree. The staminate flowers are borne in a herbaceous cone; the pistillate flowers are in a woody cone with woody, overlapping scales.

Cupressaceae
Cypress Family

Thuja ARBOR-VITAE

Thuja occidentalis L.
⬥NORTHERN WHITE CEDAR

DESCRIPTION Native shade-tolerant tree to 20 m tall, cone-shaped with widely spreading branches, sometimes layered at base, trunk to 1 m wide or more. Bark reddish or gray-brown, in long shreddy strips. Twigs flattened, in fanlike sprays. Leaves scalelike and overlapping, 3–6 mm long and 1–2 mm wide, yellow-green, aromatic, persisting for 1–2 years. Seed cones small, brown, 1 cm long, maturing in fall and persisting over winter.

HABITAT cold, poorly drained swamps where *Thuja* may form dense stands; soils neutral or basic, usually highly organic, water not stagnant. Also along streams, on gravelly and sandy shores of Great Lakes, and dry soils over limestone.

STATUS MW-FACW | NCNE-FACW | GP-FACW

NOTE Northern white cedar is an important winter food for deer, and where deer are abundant, stands of northern white cedar may have conspicuous browse lines. Deer browsing may also limit successful regeneration of cedar seedlings.

ADDITIONAL SPECIES Bald cypress (*Taxodium distichum* (L.) L.C. Rich.), a member of the Cupressaceae, was collected in Michigan in 2010 in wet places and along rivers in Kalamazoo County. This tree is more common in swamps from s Ill southward to the Gulf Coast.

Pinaceae
Pine Family

RESINOUS TREES with evergreen or deciduous, needlelike leaves. Staminate and pistillate cones separate but borne on same tree. Staminate cones small and soft, falling after pollen is shed. Female cones larger, with woody scales arranged in a spiral. Seeds on upper surface of scales.

Although usually on well-drained soils in our area, **jack pine** (***Pinus banksiana***) and **eastern white pine** (***Pinus strobus***) are occasionally found in boggy habitats. When present they occur as stunted trees atop mossy hummocks. North of our region, jack pine is more typically a wetland species. **Eastern hemlock** (*Tsuga*

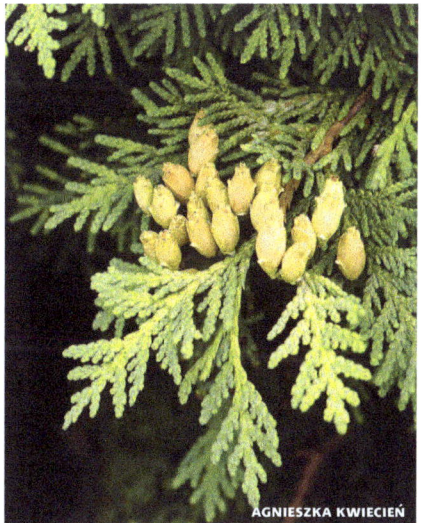

AGNIESZKA KWIECIEŃ

Thuja occidentalis

canadensis) sometimes occurs on swamp margins and in seasonally wet forest depressions.

Pinaceae **PINE FAMILY**

1 Leaves in clusters of 10–20; deciduous ...
   ....*Larix laricina*
1 Leaves single and alternate on branches; persistent **2**
2 Leaves flattened in cross-section, soft; cones upright ....*Abies balsamea*
2 Leaves 4-sided in cross-section, stiff; cones drooping.........................*Picea*

Abies FIR

Abies balsamea (L.) Miller
◊ BALSAM FIR

DESCRIPTION Native shade-tolerant tree to 25 m tall, crown spire-like, trunk to 6 dm wide; lower branches often drooping. **Bark** thin, smooth and gray, becoming brown and scaly with age.
Twigs sparsely short-hairy. **Leaves** evergreen, linear, 12–25 mm long and 1–2 mm wide, blunt or with a small notch at tip, flat in cross-section, twisted at base and arranged in one plane (especially on lower branches), or spiraled on twigs. **Seed cones** 5–10 cm long and 1.5–3 cm wide, with broadly rounded scales.

HABITAT cold boreal forests, swamps, and moist forests in n portion of the region; southward, mostly restricted to fens.
STATUS MW-FACW | NCNE-FAC | GP-FAC

Larix LARCH

Larix laricina (Duroi) K. Koch
◊ TAMARACK, EASTERN LARCH

DESCRIPTION Native shade-intolerant tree to 20 m tall, crown narrow, trunk to 6 dm wide. **Bark** smooth and gray when young, becoming scaly and red-brown. **Twigs** yellow-brown, ± horizontal or with upright tips. **Leaves** deciduous, in clusters of 10–20, linear, 1–2.5 cm long and less than 1 mm wide, soft, blunt-tipped, bright green, turning yellow in fall. **seed cones** 1–2 cm long and 0.5–1 cm wide, ripening in fall and persisting on trees for 1 year.

HABITAT cold, poorly drained swamps, bogs and wet lakeshores; southward in Midwest region, confined to wet depressions.
STATUS MW-FACW | NCNE-FACW | GP-FACW

Picea SPRUCE

Evergreen trees; bark thin and scaly, resin blisters common in white spruce (*Picea glauca*). Leaves linear, square in cross-section, stiff, spreading in all directions around twig. Cones borne on last year's

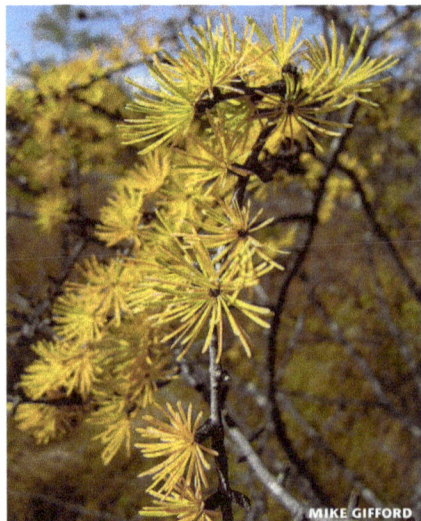

Abies balsamea

Larix laricina

branches, drooping. Seeds wing-margined.

Picea **SPRUCE**

1   Leaves mostly 6–15 mm long; cones 1.5–3 cm long; twigs with fine hairs *P. mariana*
1   Leaves mostly 15–20 cm long; cones 2.5–6 cm long; twigs mostly without hairs . . . . . .
. . . . *P. glauca*

Picea glauca (Moench) Voss
◆ WHITE SPRUCE

DESCRIPTION Moderately shade-tolerant native tree to 30 m tall (often smaller), crown conelike, trunk to 60 cm or more wide; branches slightly drooping, hairless. **Bark** thin, gray-brown. **Leaves** evergreen, linear, 1.5–2 cm long, 4-angled in cross-section, stiff, waxy blue-green, sharp-tipped. **Seed cones** 2.5–6 cm long, scales fan-shaped, rounded at tip, the tip entire.
HABITAT moist to sometimes wet forests; absent from wetlands where water is stagnant.
STATUS MW-FACU | NCNE-FACU | GP-FACU

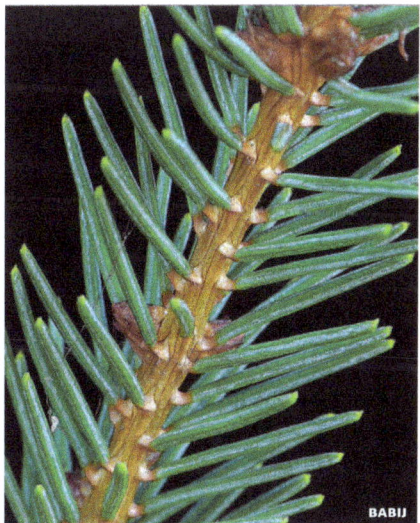

Picea mariana (Miller) BSP.
◆ BLACK SPRUCE

DESCRIPTION Moderately shade-tolerant native tree to 25 m tall (often smaller), crown narrow, often clublike at top, trunk to 25 cm wide; branches short and drooping, often layered at base. **Bark** thin, scaly, gray-brown. **Leaves** evergreen, linear, 6–18 mm long, 4-angled in cross-section, stiff, waxy blue-green, mostly blunt-tipped. **seed cones** 1.5–3 cm long, scales irregularly toothed, persisting for many years.
HABITAT cold, acid, sphagnum bogs, swamps, and lakeshores; often where water is slow-moving and low in oxygen; less common in calcium-rich, well-aerated swamps dominated by northern white cedar (*Thuja occidentalis*).
STATUS MW-FACW | NCNE-FACW | GP-FACW
NOTE distinguished from **white spruce** (*Picea glauca*) by its shorter needles, the branches with fine, white to red-brown hairs, the smaller, rounded seed cones with toothed scale margins, and its occurrence in generally wetter (and sometimes stagnant) habitats.

DICOTS
Dicotyledoneae

ANGIOSPERMS, (angion- "vessel," sperm "seed"), part of division Spermatopsida, form the world's largest group of vascular plants. Rather than cones as in the Gymnosperms, the reproductive structures are flowers, and seeds are enclosed in fruits that typically develop from the

Picea glauca

Picea mariana

ovary. Angiosperms are divided into two classes, the Dicotyledoneae (sometimes termed the Magnoliopsida and often shortened to "Dicots") and the Monocotyledoneae (or Liliopsida, the "Monocots"). The names are derived from the presence of either one or two "seed leaves" or cotyledons. The Dicots are a large and diverse group of plants, and include a variety of trees, shrubs, and herbaceous species (these often simply called "wildflowers"). Monocots include familiar families such as the orchids (Orchidaceae), grasses (Poaceae), and sedges (Cyperaceae). Although not always clear-cut, Dicots and Monocots may be distinguished from one another by a combination of the following characters:

DICOTS
- Embryo with two cotyledons
- Flower parts in multiples of 4 or 5
- Major leaf veins netlike
- Stem vascular bundles in a ring
- Roots develop from radicle
- Secondary growth often present

MONOCOTS
- Embryo with single cotyledon
- Flower parts in multiples of 3
- Major leaf veins parallel
- Stem vacular bundles scattered
- Roots adventitious
- Secondary growth (wood) absent

Acanthaceae
Acanthus Family

Justicia WATER-WILLOW

Justicia americana (L.) M. Vahl
♦ AMERICAN WATER-WILLOW

DESCRIPTION Native perennial herb, spreading by rhizomes and sometimes forming large colonies. **Stems** usually unbranched, smooth, 5–10 dm long. **Leaves** opposite, linear to lance-shaped, entire, 8–16 cm long and 10–25 mm wide, tapered to a tip, long-tapered to base, ± stalkless. **Flowers** in upright spikes on long stalks from upper leaf axils; spikes 1–3 cm long, crowded with flowers; sepals 5, green; petals

5, joined below into a tube, pale violet to purple, marked with darker purple on lower lip, 8–12 mm wide; stamens 2. **Fruit** a cylindric capsule 1–2 cm long; seeds 3 mm wide, covered with wartlike bumps. **Flowering** June–Aug.

SYNONYM *Dianthera americana.*

HABITAT shallow water, muddy pond and lakeshores, mud bars.

STATUS MW-OBL | NCNE-OBL; Mich (THR, mostly along Huron and Raisin Rivers).

ADDITIONAL SPECIES another member of this family, **smooth ruellia (*Ruellia strepens*)** is known from moist to wet forests in Lenawee County, Mich (END); its flowers are tubular and blue to violet in color.

Adoxaceae
Muskroot Family

Our 2 woody genera, *Sambucus* and *Viburnum,* were previously included in Caprifoliaceae; some current works now place these genera in Viburnaceae.

Adoxaceae **MUSKROOT FAMILY**
1   Leaves pinnately compound .. *Sambucus*
1   Leaves simple............... *Viburnum*

Sambucus ELDER
Shrubs or small trees. Stems pithy, the bark with wartlike lenticels. Leaves pinnately divided. Flowers small, white, per-

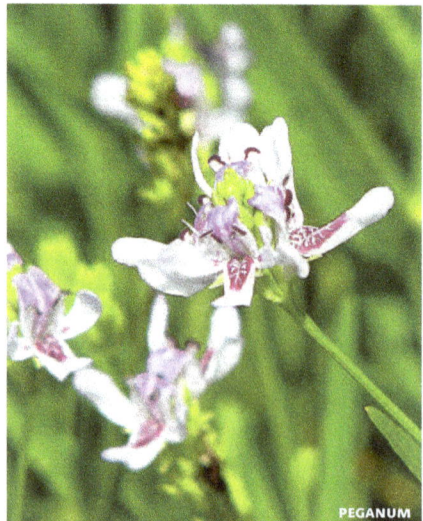

PEGANUM

Justicia american

fect, 5-lobed; in large, rounded clusters at ends of stems. Fruit a red or dark purple, berrylike drupe.

Sambucus ELDER

1 Flowers opening in summer after leaves developed, in broad, ± flat clusters; fruit purple-black, edible; leaflets usually 7 . . . . *S. nigra*

1 Flowers opening in late spring with unfolding leaves, in pyramid-shaped or rounded clusters; fruit red, inedible; leaflets usually 5 . . . . *S. racemosa*

Sambucus nigra L.
◊ COMMON ELDER

DESCRIPTION Native shrub to 3 m tall; roots spreading underground and forming thickets. **Stems** young stems soft or barely woody, smooth; older stems with warty gray-brown bark; inner pith white. **Leaves** large, opposite, pinnately divided into 5–11 (usually 7) leaflets, the lower pair of leaflets sometimes divided into 2–3 segments; leaflets lance-shaped to oval, tapered to a long sharp tip, base often asymmetrical, smooth or hairy on underside, especially along veins; margins with sharp, forward-pointing teeth. **Flowers** small, white, 5-parted, 3–5 mm wide, numerous, in flat or slightly rounded clusters 10–15 cm wide at ends of stems. **Fruit** a round, purple-black, berrylike drupe, edible. **Flowering** July–Aug (blooming when fruit of Sambucus racemosa is about ripe).

SYNONYM *Sambucus canadensis.*

HABITAT floodplain forests, swamps, wet forest depressions, thickets, shores, meadows, roadsides, fencerows.

STATUS MW-FACW | NCNE-FACW | GP-FAC

Sambucus racemosa L.
◊ RED-BERRIED ELDER

DESCRIPTION Native shrub to 3 m tall. **Stems** soft or barely woody; twigs yellow-brown and hairy, branches with warty gray-brown bark; inner pith red-brown. **Leaves** large, opposite, pinnately divided into 5–7 (usually 5) leaflets, the leaflets lance-shaped to ovate, tapered to a long sharp tip, smooth or hairy on underside; margins with small, sharp, forward-pointing teeth. **Flowers** small, white, 5-parted, 3–4 mm wide, many, in elongate, pyramidal clusters at ends of stems, the clusters 5–12 cm long and usually longer than wide. **Fruit** a round, red, berrylike drupe, inedible. **Flowering** May–June (flowers opening with developing leaves).

SYNONYM *Sambucus pubens.*

HABITAT occasional in swamps and thickets; more common in moist deciduous forests, roadsides and fencerows.

STATUS MW-FACU | NCNE-FACU | GP-FACU

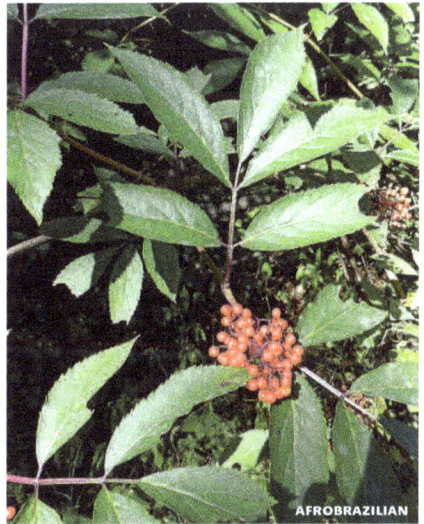

Sambucus nigra

Sambucus racemosa

Viburnum VIBURNUM

Shrubs or small trees. Leaves simple, entire or toothed, often palmately lobed. Flowers white or pink, in rounded clusters at ends of stems, sometimes outer florets larger and sterile. Fruit a drupe with a large seed; white, yellow, pink, or orange at first, maturing to orange, red, or blue-black.

Viburnum VIBURNUM

1   Leaves not lobed; pinnately veined.......
     .... *V. nudum*
1   Leaves 3-lobed; palmately veined from base of leaf...........................**2**
2   Flowers on leafy shoots at ends of stems, of 2 types; outer flowers large and showy, inner flowers smaller .... *V. opulus*
2   Flowers few, all alike, on short branches from lateral buds............... *V. edule*

Viburnum edule (Michx.) Raf.
SQUASHBERRY, LOW-BUSH CRANBERRY

DESCRIPTION Native shrub 1–2 m tall. **Stems** upright or spreading; twigs brown-purple, smooth, often angled or ridged. **Leaves** opposite, mostly shallowly 3-lobed and palmately veined (leaves at ends of stems often unlobed), 5–12 cm long and 3–12 cm wide, tapered to a sharp tip; underside veins hairy; margins coarsely toothed; petioles 1–3 cm long. **Flowers** creamy-white, small, in few-flowered, stalked clusters 1–3 cm wide, on short, 2-leaved branches from lateral buds on last year's shoots. **Fruit** a round, berrylike drupe 6–10 mm long, yellow at first, becoming orange or red. **Flowering** June–July.
SYNONYM *Viburnum eradiatum, V. pauciflorum.*
HABITAT moist conifer forests, thickets, forest openings, talus slopes.
STATUS MW-FACW | NCNE-FACW | GP-FACW; Wisc (END), Mich (THR, Isle Royale and Keweenaw County).

Viburnum nudum L.
♦ WITHE-ROD, WILD RAISIN

DESCRIPTION Native shrub to 4 m tall. **Young stems** brown-scurfy at first, becoming smooth; winter buds gold-brown. **Leaves** opposite, oval to oblong, 5–10 cm long and 3–5 cm wide, tapered to an abrupt, blunt tip, main vein on underside brown-hairy; margins entire or shallowly toothed; petioles grooved, 5–15 mm long. **Flowers** creamy-white, unpleasantly scented, all-alike, in ± flat-topped clusters 5–10 cm wide at ends of stems. **Fruit** a round to oval drupe, 6–10 mm long, yellow-white at first, then pink, ripening to blue or blue-black and covered with a waxy bloom. **Flowering** May–July.
SYNONYM *Viburnum cassinoides.*
HABITAT cedar swamps, open bogs, fens, floodplain forests, wetland margins; occasional in drier woods.
STATUS MW-FACW | NCNE-FACU | GP-FACW

Viburnum opulus L.
♦ HIGH-BUSH CRANBERRY

DESCRIPTION Native shrub, 3–4 m tall; young stems smooth. **Leaves** opposite, maple-like, sharply 3-lobed and palmately veined, 5–10 cm long and about as wide, the lobes tapered to sharp tips; smooth or hairy beneath, especially on the veins; margins entire or coarsely toothed, petioles grooved, 1–3 cm long, with several club-shaped glands present near base of blade. **Flowers** white, in large, flat-topped clusters 5–15 cm wide at ends of stems; outer flowers sterile with large petals, surrounding the inner, smaller fertile flowers. **Fruit** an

TED BODNER

Viburnum nudum

orange to red, round or oval drupe, 10–15 mm long. June.

HABITAT swamps, fens, streambanks, shores, ditches.

STATUS MW-FAC | NCNE-FACW | GP-FAC

NOTE Our plants var. *americanum*, sometimes considered a separate species, *Viburnum trilobum*.

Amaranthaceae
Amaranth Family

ANNUAL OR PERENNIAL HERBS, often in alkaline soil. Stems often angled or jointed, succulent in *Salicornia*. Leaves simple, alternate, or occasionally opposite (*Salicornia*), sometimes covered with thin, flaky scales giving a mealy appearance. Flowers 1 to many, small, green or red-tinged, clustered in leaf axils or at ends of stems; perfect, or staminate and pistillate flowers separate; sepals usually 5; petals absent; ovary superior, 1-chambered. Fruit a 1-seeded utricle.

Amaranthaceae **AMARANTH FAMILY**

1   Leaves reduced to scales, opposite; stems succulent; w Minn only   *Salicornia rubra*

1   Leaves not scalelike, mostly alternate; stems not succulent; more widespread in Upper Midwest region **2**

2   Leaves without petioles, linear, round in cross-section . . . . *Suaeda calceoliformis*

2   Leaves mostly with petioles, blades broader **3**

3   Flowers perfect (with both staminate and pistillate parts), in spikes which have small leafy bracts throughout; fruit surrounded by the persistent sepals and petals . . . . *Chenopodium*

3   Flowers unisexual (plants monoecious or dioecious); tepals and bracts acute, scarious or fruit in most if not all flowers enveloped by a pair of bracteoles (perianth absent) **4**

4   Bracts and tepals all acute, scarious (staminate flowers), absent in pistillate flowers . . . . . . *Amaranthus tuberculatus*

4   Bracts beneath pistillate flowers broad and usually tuberculate and toothed, herbaceous in texture . . . . *Atriplex patula*

Amaranthus AMARANTH

Amaranthus tuberculatus (Moq.) Sauer
◆ WATER HEMP

DESCRIPTION Native annual. Stems erect to spreading, usually much-branched, 2–15 dm tall, usually hairless.

Leaves alternate, ovate to lance-shaped, variable in size, larger leaves 4–10 cm long, smaller leaves 1–4 cm long. Flowers either staminate or pistillate and on different plants, in spikes from leaf axils and at ends of stems; staminate flowers with 5 sepals, 2–3 mm long and 5 stamens; pistillate flowers without

Viburnum opulus

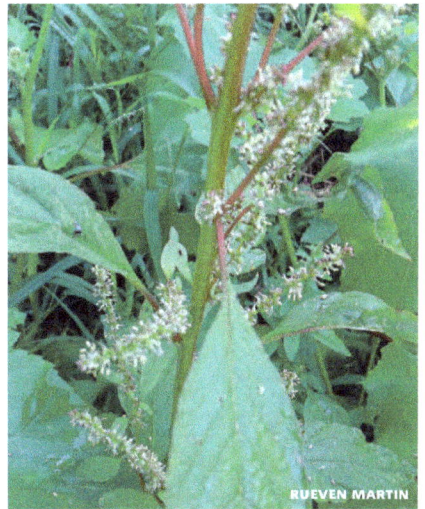

Amaranthus tuberculatus

sepals or petals (rarely with 1–2 small sepals). Fruit a utricle 1–2 mm long; seeds red-brown, 1 mm wide. Flowering July–Sept.

SYNONYM *Acnida altissima.*

HABITAT exposed muddy shores, streambanks, swamps, wet meadows, ditches.

STATUS MW-OBL | NCNE-OBL | GP-OBL

Atriplex SALTBUSH

Atriplex patula L.
♦ HALBERD-LEAF SALTBUSH, SPEARSCALE

DESCRIPTION Taprooted annual herb; introduced. Stems erect to sprawling, usually branched, 2–10 dm long. Leaves alternate (or the lowest opposite), lance-shaped or triangular, 2–8 cm long and 1–6 cm wide, with outward pointing basal lobes, gray and mealy when young, becoming dull green and smooth with age; petioles present, or absent on upper leaves. Flowers tiny, green; either staminate or pistillate but on same plant, usually intermixed in crowded spikes from leaf axils and at ends of stems, the spikes simple or branched, without bracts or with a few small bracts near base of spikes; staminate flowers with a 5-lobed group of sepals, stamens 5; pistillate flowers without sepals or petals, surrounded by 2 sepal-like, small bracts, these expanding and enclosing fruit when mature. Fruit lens-shaped, dark

brown to black, 1–3 mm wide. Flowering Aug–Sept.

SYNONYMS *Atriplex acadiensis, Atriplex hastata.*

HABITAT shores, streambanks and mud flats, usually where brackish; disturbed places.

STATUS MW-FACW | NCNE-FACW | GP-FACW

Chenopodium GOOSEFOOT

Taprooted annual herbs. Stems erect to spreading. Leaves alternate, mostly lance-shaped to broadly triangular, somewhat fleshy and often mealy on lower surface. Flowers perfect, small and numerous, green or red-tinged, in dense spikelike clusters from leaf axils or at ends of stems, the spikes with small leafy bracts; sepals often curved over the fruit; petals absent; stamens 1–5; styles 2–3. Fruit a 1-seeded utricle.

Chenopodium GOOSEFOOT

1 Leaves persistently white-mealy on underside, dull green above . *C. glaucum*

1 Leaves not white-mealy when mature, green on upper and lower sides, often red-tinged . . . . *C. rubrum*

Chenopodium glaucum L.
♦ OAK-LEAVED GOOSEFOOT

DESCRIPTION Annual herb. Stems upright to sprawling, 1–6 dm long, usually branched from

Atriplex patula

Chenopodium glaucum

base, sometimes red-tinged. Leaves lance-shaped to ovate, 1–4 cm long and to 2 cm wide, dull green above, densely white-mealy on underside (especially when young); margins entire, wavy, or with few rounded teeth; petioles slender, shorter on upper leaves. Flowers in small, often branched, spikelike clusters from leaf axils, the spikes often shorter than leaves; sepals mostly 3; petals absent. Seeds dark brown, shiny, 1 mm wide. Flowering Aug–Oct.

SYNONYMS *Chenopodium salinum, Oxybasis glauca.*

HABITAT shores, streambanks, and disturbed areas such as railroad ballast and barnyards; soils often brackish.

STATUS MW-FACW | NCNE-FACW | GP-FAC. In e USA, considered introduced; native in Minn and w USA.

Chenopodium rubrum L.
♦ ALKALI-BLITE

DESCRIPTION Annual herb. Stems usually erect, sometimes sprawling, 1–8 dm long, often branched from base. Leaves lance-shaped to broadly triangular, 2–10 cm long and 1–8 cm wide, green and often red-tinged on both surfaces, smooth, not mealy; margins wavy-toothed or lobed; petioles present. Flowers small, in upright, branched spikes from leaf axils and at ends of stems, the spikes often longer than leaves; sepals 3, red; petals absent.

Seeds dark brown, shiny, to 1 mm wide. Flowering Aug–Oct.

SYNONYM *Oxybasis rubra.*

HABITAT lakeshores, streambanks, disturbed areas.

STATUS MW-OBL | NCNE-OBL | GP-OBL; considered native in Minn, adventive further east.

Salicornia GLASSWORT

Salicornia rubra A. Nels.
♦ WESTERN GLASSWORT

DESCRIPTION Taprooted annual herb; plants succulent, green to bright red. Stems 0.5–2 dm long; branches opposite, fleshy, jointed at nodes, breaking apart when plants are trampled. Leaves opposite, small and scalelike, 1–2 mm long. Flowers in spikes 1–5 cm long at ends of stems; perfect, or some flowers pistillate only; sepals enclosing flower except for a small opening from which stamens and style branches protrude; petals absent. Seeds 1 mm long. Flowering Aug–Oct.

HABITAT shores, seeps and ditches; soils brackish.

STATUS MW-OBL | NCNE-OBL | GP-OBL; apart from Minn, considered introduced in the Upper Midwest region from w USA.

ADDITIONAL SPECIES **Slender glasswort** *(Salicornia europaea)* has been collected from several salty, disturbed locations in s Wisc and s Mich. This species is more common in salt marshes along the Atlantic coast;

Chenopodium rubrum

Salicornia rubra

distinguished from western glasswort by having the joints of the spike mostly longer than wide (in western glasswort, the joints of the spike are about as wide or wider than long).

Suaeda SEA-BLITE

Suaeda calceoliformis (Hook.) Moq.
◆ PLAINS SEA-BLITE

DESCRIPTION Annual taprooted herb. Stems upright to sprawling, usually branched, 0.5–6 dm long. Leaves alternate, linear, flat on 1 side and convex on other, green, succulent, 5–30 mm long and 1 mm wide, reduced to wider bracts 1–5 mm long in the head; petioles absent. Flowers small, perfect, or staminate or pistillate flowers separate, green or sometimes red-tinged, in dense clusters of 3–7 flowers in bract axils; sepals joined, deeply 5-lobed, the lobes unequal and hooded; stamens 5. Fruit a utricle enclosed by the sepals; seeds black, shiny, about 1 mm wide. Flowering July–Sept.
SYNONYM *Suaeda depressa*.
HABITAT brackish wetlands, and along salted highways.
STATUS MW-FACW | NCNE-FACW | GP-FACW; considered introduced in Wisc and of Mich.

Anacardiaceae
Sumac Family
Toxicodendron
POISON-SUMAC
Shrubs or vines. Leaves alternate, divided into 3 or more leaflets. Flowers in branched inflorescence; petals 5, green-white to yellowish. Fruit a white to yellowish drupe.

Toxicodendron **POISON-SUMAC**
1 Leaflets 3; margins entire or toothed . . . . .
. . . . *T. radicans*
1 Leaflets 7–13; margins entire . . . . *T. vernix*

Toxicodendron radicans (L.) Kuntze
◆ COMMON POISON-IVY

DESCRIPTION Native vine or low shrub. Leaves alternate, divided into 3 dull to shiny ovate to elliptic leaflets, 5–15 cm long; margins entire or with a few coarse teeth. Flowers green-white or yellowish, 25 or more in a branched inflorescence from leaf axil; petals 5. Fruit a white drupe, 3–5 mm wide, these often drooping. Flowering May–July.
SYNONYM *Rhus radicans*.
HABITAT in northern portion of region, occasional in swamps and moist rocky forests, more common along roadsides, clearings,

Suaeda calceoliformis

Toxicodendron radicans

and on sand dunes. Southward, found in floodplain forests, swamps and drier upland woods.

STATUS MW-FAC | NCNE-FAC | GP-FACU

CAUTION Common poison-ivy is a skin-irritant.

Toxicodendron vernix (L.) Kuntze
◆ POISON-SUMAC

DESCRIPTION Native shrub or small tree to 5 m tall, often branched from base. Leaves alternate, divided into 7–13 leaflets, the leaflets oblong to oval, 4–6 cm long, tapered to a pointed tip; margins entire, smooth; the leaves turning bright red in the fall. Flowers small, white or green, in panicles to 2 dm long; sepals 5, joined at base; petals 5, not joined; stamens 5. Fruit a round, gray-white drupe, 4–5 mm wide. Flowering June–July.

SYNONYM *Rhus vernix*.

HABITAT tamarack swamps, thickets, floating bog mats and bog margins, often in partial shade.

STATUS MW-OBL | NCNE-OBL | GP-OBL

CAUTION Poison-sumac is a skin-irritant.

Apiaceae
Carrot Family

BIENNIAL OR PERENNIAL aromatic herbs with hollow stems. Leaves alternate and

JAMES MILLER & TED BODNER

Toxicodendron vernix

sometimes also from base of plant, mostly compound; petioles sheathing stems. Flowers small, perfect (with both staminate and pistillate parts), regular, in flat-topped or rounded umbrella-like clusters (umbels); sepals 5 or absent; petals 5, white or greenish (yellow in *Pastinaca sativa* and *Zizia aurea*). Fruit 2-chambered, separating into 2, 1-seeded fruits when mature.

Apiaceae **CARROT FAMILY**

1 Leaf margins and inflorescence bracts with stiff spines; flowers in dense-bracted heads............ *Eryngium yuccifolium*

1 Leaf margins and bracts without spines; flowers in umbels...................... **2**

2 Flowers yellow **3**

2 Flowers white or green-white.......... **4**

3 Fruit winged; leaves once compound .... *Pastinaca sativa*

3 Fruit with prominent ribs but not winged; leaves 2–3x compound ...... *Zizia aurea*

4 Leaves simple .... *Hydrocotyle*

4 Leaves divided or compound **5**

5 Stems covered with woolly hairs; fruits and ovaries hairy ..... *Heracleum maximum*

5 Stems not covered with woolly hairs; fruits and ovaries hairless................... **6**

6 Lateral veins of the leaflets end in the notches (lobes or sinuses) between the teeth; or bulblets in upper leaf axils .... *Cicuta*

6 Lateral veins of the leaflets end in the teeth; bulblets absent **7**

7 Leaves finely divided, the smallest divisions deeply pinnately lobed; fruits winged, longer than wide .... *Conioselinum chinense*

7 Smallest divisions of leaves toothed; fruits various............................... **8**

8 Divisions of stem leaves linear to narrowly lance-shaped, more than 4x longer than wide **9**

8 Divisions of stem leaves broader, less than 4x longer than wide................... **10**

9 Leaflets entire or with several coarse, irregularly spaced teeth *Oxypolis rigidior*

9 Leaflets finely toothed....... *Sium suave*

10 Petiole sheaths large and inflated .... *Angelica atropurpurea*

10 Petiole sheaths not inflated **11**

11 Large plants more than 1 m tall; stems with many purple spots; fruit with prominent ribs .... *Conium maculatum*

11 Smaller plants, less than 1 m tall, often partially underwater; stems not purple-spotted; fruit with indistinct ribs ....
.... *Berula erecta*

Angelica ANGELICA

Angelica atropurpurea L.
◆ PURPLESTEM-ANGELICA

DESCRIPTION Native perennial herb; toxic. Stems stout, 2–3 m tall, ± smooth, often streaked with purple and green. Leaves alternate, lower leaves 3-parted, 1–3 dm long, on long petioles; upper leaves smaller, less compound, on shorter petioles, or reduced to bladeless sheaths; leaflets ovate to lance-shaped, smooth, 4–10 cm long; margins sharp toothed. Flowers in rounded small clusters (umbelets), these grouped into large rounded umbels 1–2 dm wide; petals white to green-white. Fruit oval, 4–6 mm long, winged. Flowering May–July.

HABITAT springs, seeps, calcareous fens, streambanks, shores, marshes, sedge meadows, wet depressions in forests; often where calcium-rich.

STATUS MW-OBL | NCNE-OBL | GP-OBL

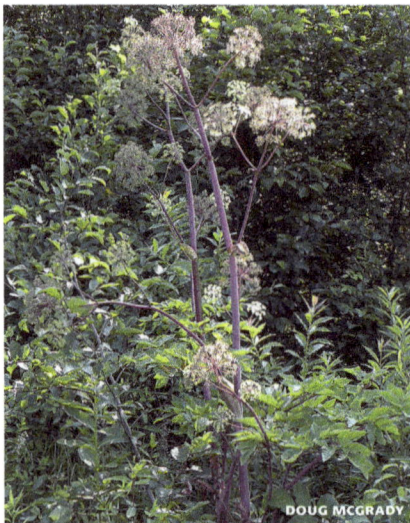

Berula WATER PARSNIP

Berula erecta (Huds.) Cov.
◆ CUT-LEAF WATER PARSNIP

DESCRIPTION Native perennial herb. Stems erect to trailing, sparsely branched, 4–8 dm long, often rooting along trailing portion. Leaves alternate, 1-pinnate, basal leaves larger and less dissected than stem leaves, oblong, 5–20 cm long and 2–10 cm wide; leaflets lance-shaped to ovate; margins toothed or lobed. Flowers grouped into 5–15 small clusters (umbelets) 1 cm wide, these grouped into umbels 3–6 cm across; flowers white, 1–2 mm wide; sepals small or absent. Fruit oval or round, slightly flattened, 1–2 mm long, but seldom maturing. Flowering July–Sept.

SYNONYM *Berula pusilla.*

HABITAT shallow water, springs, spring-fed streams, marshes, swamps, often where calcium-rich.

STATUS MW-OBL | NCNE-OBL | GP-OBL; Mich (THR).

Cicuta WATER-HEMLOCK

Biennial or perennial toxic herbs. Leaves alternate, 2–3-pinnate; leaflets narrow or lance-shaped, entire or toothed; leaf veins ending in the lobes (sinuses) and not at teeth as in other members of family. Flowers white or green, in few to many umbels. Fruit oval or round, ribbed.

DOUG MCGRADY

Angelica atropurpurea

STAN SHEBS

Berula erecta

Cicuta WATER-HEMLOCK

1 Upper leaflet axils usually with bulblets; leaflets narrow, to 5 mm wide *C. bulbifera*
1 Bulblets absent; leaflets usually much more than 5 mm wide .... *C. maculata*

Cicuta bulbifera L.
♦ BULBLET-BEARING WATER-HEMLOCK

DESCRIPTION Native biennial or perennial herb; toxic; fibrous-rooted or with a few thickened, tuberlike roots. **Stems** slender, upright, 3–10 dm tall, not thickened at base. **Leaves** alternate along stem, to 15 cm long and 10 cm wide, pinnately divided; leaflets mostly linear, 1–5 mm wide, margins sparsely toothed to entire; upper leaves reduced in size, undivided or with few segments, with 1 to several bulblets 1–3 mm long, in axils. **Flowers** white, in umbels 2–4 cm wide. **Fruit** round, 1–2 mm wide, but rarely maturing. Aug–Sept.
HABITAT streambanks, lake and pond shores, marshes, swamps, open bogs, thickets, springs and ditches.
STATUS MW-OBL | NCNE-OBL | GP-OBL
CAUTION toxic.

Cicuta maculata L.
♦ COMMON WATER-HEMLOCK

DESCRIPTION Native biennial or perennial herb; toxic. **Stems** single or several together, often branched, 1–2 m long, distinctly hollow above the chambered and tuberous-thickened base. **Leaves** from base of plant and alternate on stem, mostly 10–30 cm long and 5–20 cm wide; basal leaves larger and longer stalked than stem leaves; leaflets linear to lance-shaped, 3–10 cm long and 5–35 mm wide; margins toothed. **Flowers** white, in several to many umbels, these 6–12 cm wide in fruit, on stout stalks 5–15 cm long. **Fruit** round to ovate, 2–4 mm long, with prominent ribs. **Flowering** June–Sept.
SYNONYM *Cicuta douglasii*.
HABITAT wet meadows, marshes, swamps, moist to wet forests, thickets, shores, streambanks, springs.
STATUS MW-OBL | NCNE-OBL | GP-OBL
CAUTION *Cicuta maculata* considered the most toxic plant in North America; the tuberous roots, chambered stem base and young shoots are especially dangerous if eaten.

Conioselinum
HEMLOCK-PARSLEY

Conioselinum chinense (L.) BSP.
HEMLOCK-PARSLEY

DESCRIPTION Native perennial herb. **Stems** erect, slender to stout, 5–15 dm tall, smooth or often with short, rough hairs in flower head. **Leaves** alternate, triangular in outline, 1–3x pinnate, on short, winged stalks; leaflets lance-shaped, deeply

ROB ROUTLEDGE
Cicuta bulbifera

WENDELL SMITH
Cicuta maculata

lobed, 2–4 cm long. Flowers white, in long-stalked umbels 3–12 cm wide. Fruit oval or oblong, 2–5 mm long. Flowering Aug–Sept. **HABITAT** tamarack swamps, forest seeps, streambanks, fens. **STATUS** MW-FACW | NCNE-FACW; historically known from se Wisc.

Conium POISON HEMLOCK

Conium maculatum L.
♦ POISON HEMLOCK

DESCRIPTION Introduced biennial herb. Stems stout, branched, purple-spotted, 1–2 m long. Leaves alternate, 2–4 dm long, 3–4x pinnately divided, the leaflets toothed or sharply lobed. Flowers white, in many umbelets, these grouped in umbels to 6 cm wide. Fruit ovate, ribbed, 3 mm long. Flowering June–July.
HABITAT Weedy on shores, streambanks, waste ground and roadsides, especially on moist, fertile soil; introduced; now found throughout most of s Canada and USA.
STATUS MW-FACW | NCNE-FACW | GP-FACW
CAUTION Poison hemlock is very toxic; plants also have a strong, unpleasant odor.

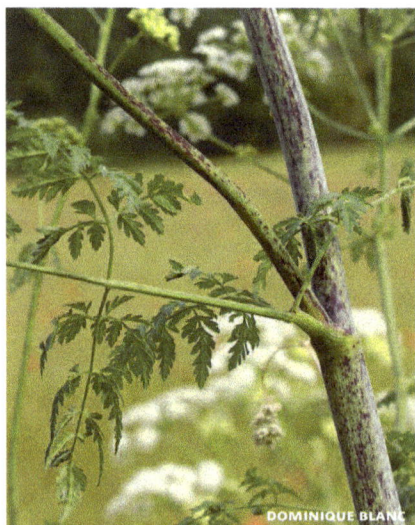

Eryngium ERYNGIUM

Eryngium yuccifolium Michx.
♦ RATTLESNAKE-MASTER

DESCRIPTION Erect, smooth, native perennial herb. Stems stout, unbranched except near the infloresence, to 1 m or more long. Leaves alternate, blue-green, mostly near base of plant and often downward-curved; narrowly lance-shaped, to 1 m long and 1–3 cm wide, parallel-veined, succulent, clasping stem; margins with small prickles or spines. Flowers in spherical, thistle-like umbels 2–3 cm wide, the flowers numerous, petals 5, green-white; each flower subtended by a bract. Flowering July–Sept.
HABITAT wet to dry sandy prairie and oak barrens, margins of swamps and marshes, sedge and grass-dominated portions of prairie fens, thickets along streams; plants sometimes found in small populations along power lines and railroad rights-of-way.
STATUS MW-FAC | NCNE-FAC | GP-FACW; Mich (THR).

Heracleum COW-PARSNIP

Heracleum maximum Bartr.
♦ COW-PARSNIP

DESCRIPTION Large native perennial herb. Stems stout, hairy, 1–2 m long. Leaves alter-

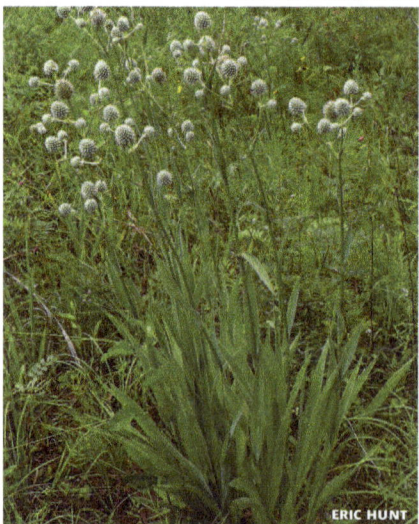

Conium maculatum

Eryngium yuccifolium

nate, nearly round in outline, divided into 3 leaflets; leaflets 1–4 dm long and as wide, margins coarsely toothed. Flowers white, in large umbels, the terminal umbel 1–2 dm wide. Fruit obovate, 8–12 mm long and nearly as wide, often hairy. Flowering May–July.

SYNONYM *Heracleum lanatum, Heracleum sphondylium* subsp. *montanum.*

HABITAT streambanks, thickets, wet meadows, moist forest openings and disturbed areas.

STATUS MW-FACW | NCNE-FACW | GP-FAC

CAUTION skin irritant.

Hydrocotyle PENNYWORT

Small perennial herbs. Stems prostrate, often rooting at nodes. Leaves round or kidney-shaped, with shallowly lobed margins and long petioles. Flowers small, white, in stalked or stalkless umbels. Fruit of 2 compressed carpels, ± round in outline.

Hydrocotyle **PENNYWORT**

1 Leaf petiole attached to base of blade; flowers stalkless from leaf nodes; e Minn, Wisc, Mich . . . . *H. americana*

1 Leaf petiole attached to center of blade; flowers at ends of long stalks; Mich LP . . . . *H. umbellata*

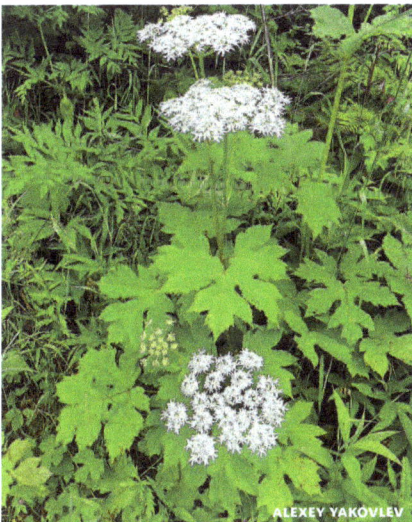

Heracleum maximum

Hydrocotyle americana L.
⬧ MARSH PENNYWORT

DESCRIPTION Small native perennial herb. Stems slender and creeping, 10–20 cm long. Leaves round to kidney-shaped, 1–5 cm wide; margins with 7–12 shallow lobes. Flowers white, in ± stalkless umbels from nodes; umbels 2–7-flowered. Fruit 1–2 mm wide, ribbed. Flowering June–Sept.

HABITAT conifer swamps, streambanks, shores, wet forest depressions.

STATUS MW-OBL | NCNE-OBL

ADDITIONAL SPECIES Buttercup pennywort *(Hydrocotyle ranunculoides)*, more common s of our region, is reported from s Minn. It differs from marsh pennywort *(Hydrocotyle americana)* by its long-stalked umbels and deeply lobed leaves (vs. stalkless umbels and only shallowly lobed leaves in *Hydrocotyle americana.*

Hydrocotyle umbellata L.
WATER PENNYWORT

DESCRIPTION Smooth native perennial herb. Stems creeping or floating in shallow water. Leaves round, 3–6 cm wide, margins with rounded teeth or shallow lobes; petioles 3–12 cm long, longer than leaves and attached to center of blades. Flowers white, in umbels of 15–35 flowers, umbels on stalks to 15 cm long. Fruit 2–3 mm wide, notched at both ends. Flowering Aug–Sept.

HABITAT streambanks, sandy shores, tamarack swamps; often in shallow water.

STATUS MW-OBL | NCNE-OBL

Hydrocotyle americana

Oxypolis WATER-DROPWORT

Oxypolis rigidior (L.) Raf.
COMMON WATER-DROPWORT

DESCRIPTION Native perennial herb. **Stems** stout or slender, to 1.5 m long, with few branches and leaves. **Leaves** 1-pinnate; leaflets 5–9, linear to oblong, 5–15 cm long and 5–40 mm wide; margins entire or with scattered coarse teeth. **Flowers** white, on stalks 5–20 mm long, in loose umbels to 15 cm wide. **Fruit** rounded at ends, 4–6 mm long and 3–4 mm wide. **Flowering** July–Sept.

SYNONYM *Oxypolis turgida.*

HABITAT swamps, thickets, marshes, moist or wet prairie, calcareous fens.

STATUS MW-OBL | NCNE-OBL | GP-OBL

NOTE Similar to **water-parsnip** (*Sium suave*) but differs in having entire to irregularly toothed leaves and a slightly grooved stem; Sium has finely toothed leaf margins and a more deeply grooved stem.

Pastinaca WILD PARSNIP

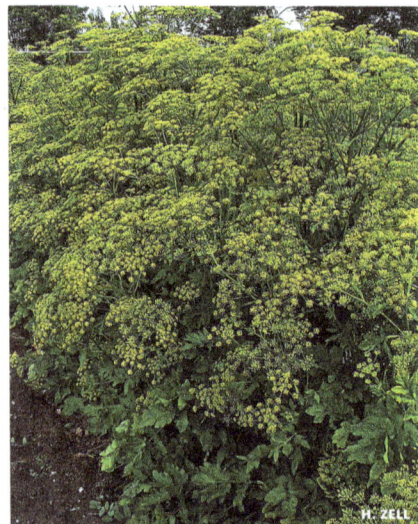

Pastinaca sativa L. ♦ WILD PARSNIP

DESCRIPTION Introduced, biennial taprooted herb; often invasive. **Stems** erect, 0.5–2 m long, branched, ± smooth to finely hairy, conspicuously angled and grooved. **Leaves** oblong to ovate in outline, 1-pinnate, the leaflets 5–11, coarsely toothed and lobed or divided. **Flowers** yellow, in flat-topped compound umbels. **Fruit** 4–6 mm wide, oblong to round. Jul–Aug.

HABITAT wet-mesic prairie, calcareous fens; also on roasides and in disturbed areas.

STATUS [no rating]; native of Eurasia, sporadic across much of USA.

CAUTION avoid contact with wild parsnip as plants are skin irritants.

Sium WATER-PARSNIP

Sium suave Walter ♦ WATER-PARSNIP

DESCRIPTION Native, perennial emergent herb. **Stems** single, smooth, 5–20 dm long, strongly ribbed upward; stem base thickened and hollow with cross-partitions. **Leaves** 1-pinnate, on long, hollow stalks (shorter stalked above); leaflets 7–17 per leaf, linear to lance-shaped, 5–10 cm long and 3–15 mm wide; margins with fine, sharp, forward-pointing teeth; finely dissected underwater leaves often present from spring to midsummer. **Flowers** white or green-white, 1–2 mm wide, in stalked umbels 4–12 cm wide at ends of stems and from side branches. **Fruit** oval, 2–3 mm long, with prominent ribs. **Flowering** July–Sept.

HABITAT wet forest depressions, marshes, swamps, streambanks, pond and lake margins, ditches; usually in shallow water.

Pastinaca sativa

Sium suave

STATUS MW-OBL | NCNE-OBL | GP-OBL

NOTE distinguished from the toxic **water-hemlock** (*Cicuta maculata*) by its leaves only once compound; water-hemlock has leaves 2–3 times pinnately compound.

Zizia ZIZIA

Zizia aurea (L.) W. D. J. Koch
◆ GOLDEN ALEXANDERS

DESCRIPTION Native perennial herb, from thickened roots. Stems erect, branched or un-branched, smooth, to 1 m long. Leaves alternate, 2–3x compound; leaflets ovate to ovate lance-shaped, to 4 cm long and 2.5 cm wide; margins sharp-toothed; on petioles to 10 cm long. Flowers in compound umbels at ends of stems; petals 5, yellow, 2 mm long, incurved at tip; stamens 5, alternating with petals. Fruit a flattened, ribbed achene, to 4 mm long. Flowering May–July.

HABITAT sedge meadows, thickets, river-banks, wet forests; often with poison sumac.

STATUS MW-FAC | NCNE-FAC | GP-FAC

NOTE the inwardly folded yellow flower petals are distinctive.

Apocynaceae
Dogbane Family

Asclepias MILKWEED

Asclepias incarnata L.
◆ SWAMP-MILKWEED

DESCRIPTION Native perennial herb, from thick rhizomes; plants with milky juice. Stems stout, to 1.5 m long, branched above, smooth except for short, appressed hairs on upper stem. Leaves opposite, simple, mostly lance-shaped, 6–15 cm long and 1–5 cm wide, tapered to a sharp tip, margins entire, petioles short. Flowers pink to purple-red, numerous in umbels at ends of stems and from upper leaf axils, perfect, regular; sepals 5, spreading; petals 5, 4–6 mm long and curved downward; stamens 5; flowers with 5 petal-like "hoods," each with an awl-shaped "horn" projecting from the opening. Fruit a follicle (1-chambered and opening on one side only) with many seeds, the seeds having tufts of white hairs. Flowering June–Aug.

HABITAT openings in conifer swamps, marshes, beaver ponds, streambanks, ditches, open bogs and fens; plants often in shallow water.

STATUS MW-OBL | NCNE-OBL | GP-FACW

NOTE Swamp-milkweed is an important food for monarch butterfly caterpillars which feed on the leaves; the flowers also produce large amounts of nectar.

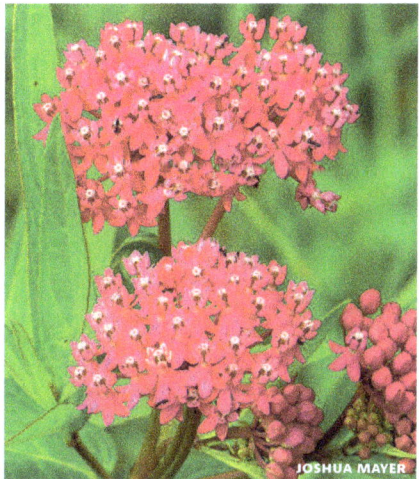

Zizia aurea

Asclepias incarnata

Aquifoliaceae
Holly Family

SHRUBS with alternate, simple leaves. Flowers from leaf axils, 4–8-parted, usually either staminate or pistillate, sometimes perfect, on same (*Ilex verticillata*) or different (*Ilex mucronata*) plants. Fruit a berrylike drupe.

Ilex HOLLY

1 Leaves tipped with a short, sharp point; margins mostly entire or with a few scattered teeth; petals linear . . . .*I. mucronata*

1 Leaves not tipped with a short, sharp point; margins toothed; petals oblong . . . .*I. verticillata*

Ilex HOLLY

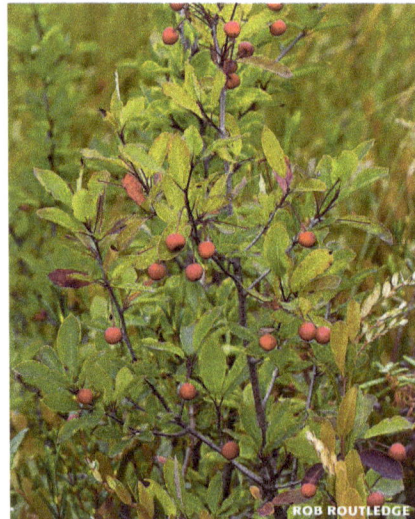

Ilex mucronata (L.) Powell, Savol. & Andrews

⧫MOUNTAIN-HOLLY, CATBERRY

DESCRIPTION Much-branched native shrub to 3 m tall; young twigs purple-tinged. **Leaves** deciduous, alternate, oval or ovate, 3–6 cm long and 2–3 cm wide, bright green above, dull and paler below, tip of leaf with a small, sharp point; margins entire or with small scattered teeth, on purple-red stalks 1 cm long. **Flowers** very small, yellow-white, on threadlike stalks from leaf axils; staminate flowers usually in small groups, pistillate flowers single. **Fruit** a purple-red berrylike drupe, 5–6 mm wide. **Flowering** May–June.

SYNONYM *Nemopanthus mucronatus*.

HABITAT open bogs (especially near outer moat), swamps, thickets, wet depressions in forests, lakeshores.

STATUS MW-OBL | NCNE-OBL

Ilex verticillata (L.) A. Gray

⧫WINTERBERRY

DESCRIPTION Native shrub to 5 m tall; twigs smooth, finely ridged. **Leaves** deciduous, alternate, obovate to oval, tapered to a tip, dull green above, paler below; margins with incurved teeth. **Flowers** small, green-white, on short stalks from leaf axils, opening before leaves fully expanded in spring; staminate flowers in crowded clusters, pistillate flowers 1 or several in a group. **Fruit** a berrylike drupe, orange or red, 5–6 mm wide and persisting into winter. June.

HABITAT swamps, open bogs, thickets, shores and streambanks.

STATUS MW-FACW | NCNE-FACW

Ilex mucronata

Ilex verticillata

Araliaceae
Ginseng Family

Oplopanax DEVIL'S CLUB

Oplopanax horridus (J. E. Smith) Miq.
◆DEVIL'S CLUB

DESCRIPTION Native spiny shrub, 1–2 m tall, forming small colonies. Stems with dense covering of very sharp spines to 1 cm long. Leaves deciduous, alternate, 2–4 dm wide, palmately lobed, dark green and smooth above, paler below with spines on veins; petioles spine-covered. Flowers small, green-white, in a dense raceme 1–2 dm long; petals 5. Fruit a bright red, berrylike drupe, 4–6 mm long. Flowering July.

HABITAT moist ravines and mixed woods.
STATUS NCNE-FACW; in Upper Midwest region, known only from several islands in Lake Superior: Isle Royale and adjacent islands (Mich, THR) and Porphyry Island (Ontario). Disjunct from main range of Montana and Oregon north to Alaska.

Asteraceae
Aster Family

ANNUAL, BIENNIAL, OR PERENNIAL HERBS. Leaves simple or compound, opposite, alternate, or whorled. Flowers perfect (with both staminate and pistillate parts) or single-sexed (sometimes sterile) and of 2 types: ray (or ligulate) and disk (or tubular). Ray flowers joined at base and with a long, flat, segment above (the ray); disk flowers tube-shaped with 5 lobes or teeth at tip. Flowers clustered in 1 of 3 types of heads which resemble a single flower and are attached to a common surface (the receptacle): ray flowers only (as in dandelion, *Taraxacum*); disk flowers only (discoid, as in tansy, *Tanacetum*); and heads with both ray and disk flowers (radiate), the ray flowers surrounding the disk flowers (as in sunflower, *Helianthus*). In addition to flowers, the receptacle may also have scales called chaff; if no scales are present, the receptacle is termed naked. Each head is surrounded by involucral bracts (sometimes called phyllaries); collectively, the bracts are termed the involucre, comparable to the group of sepals (calyx) subtending an individual flower. Fertile flowers have 1 pistil tipped by a 2-cleft style (undivided in sterile flowers); 5 stamens; the ovary (and achene) often topped by a pappus composed of several to many scales, awns or hairs. Fruit a seedlike achene.

Asteraceae **ASTER FAMILY**

1 Plants aquatic, underwater leaves whorled and dissected into narrow segments . . . . *Bidens beckii*
1 Plants not aquatic (sometimes emergent); leaves not as above **2**
2 Plants with white milky juice; heads with ray flowers only. **3**
2 Plants with watery juice; heads with both ray and disk flowers, or disk flowers only **4**
3 Ray flowers yellow; leaves in a basal rosette . . . . *Crepis runcinata*
3 Ray flowers pink, purple, or rarely white; stems leafy . . . . *Nabalus racemosus*
4 Leaves with sharp spines. . . . . *Cirsium*
4 Leaves without spines. **5**
5 Leaves opposite or whorled. **6**
5 Leaves at least in part alternate, or the leaves mostly all at base of plant **12**
6 Receptacle (disk to which flowers are attached) not chaffy **7**
6 Receptacle chaffy. **9**
7 Leaves whorled; heads pale pink to reddish or purple (very rarely albino) *Eutrochium*
7 Leaves opposite (sometimes a few upper ones alternate); heads white or blue. **8**
8 Leaf blades petiolate, ovate, less than twice as long as broad (rarely 2.5 times). . . . . *Ageratina*
8 Leaf blades sessile, or if petiolate, lanceolate and ca. 2–4 times as long as broad. *Eupatorium*

Oplopanax horridus

9 Pappus (bristles or scales atop ovary/achene) of bristly hairs or bristle-tipped scales **10**

9 Pappus absent. **11**

10 Involucral bracts in 2 series, the outer often leaflike and much larger than the inner; pappus of 2-4 awl-shaped bristles, the bristles with sharp, usually downward-pointing barbs . . . . *Bidens*

10 Involucral bracts in several series and overlapping, ± equal in size; pappus of 2 lance-shaped, unbarbed awns *Helianthus*

11 Rays yellow, 1 cm or more long; large perennial herb, 1-3 m tall. . . . . . . *Silphium*

11 Rays small, white; annual herb to 1 m tall . . . . *Eclipta prostrata*

12 Heads with both disk and ray flowers; the rays yellow **13**

12 Heads with disk flowers only; or with both disk and ray flowers, the rays not yellow **20**

13 Pappus of 2-several awns or scales. **14**

13 Pappus of many hairlike bristles **17**

14 Receptacle not chaffy; leaves tapered to a stalkless base, continuing downward as wings on the stem . *Helenium autumnale*

14 Receptacle chaffy; leaves short-stalked, not continuing downward on the stem. **15**

15 Leaves all alternate . . . . *Rudbeckia*

15 Lower leaves opposite, upper leaves alternate. **15**

16 Stem not wing-margined; achenes of disk flowers compressed but not strongly flattened . . . . *Helianthus*

16 At least upper stem wing-margined; achenes of disk flowers strongly flattened . . . . *Verbesina alternifolia*

17 Involucral bracts in 1 series, of similar length and not overlapping in rows. **18**

17 Involucral bracts in several series, of different lengths and overlapping **19**

18 Upper half of unbranched portion of stem less leafy than the lower half (i.e., leaves sparser and distinctly smaller); plants perennial, often with crowded basal leaves. . . . . *Packera*

18 Upper half of unbranched portion of stem nearly or quite as leafy as the lower half (i.e., leaves ± equally numerous and of similar size and shape, or only gradually reduced upwards); plants without basal rosette of persistent leaves, plants annual. . . . . *Tephroseris palustris*

19 Flowers in a corymblike (± flat-topped) head; leaves narrow, 2-10 mm wide, entire, dotted with glands. . . . . *Euthamia*

19 Flowers in a paniclelike head; leaves wider, 1-4 cm wide, toothed, not gland-dotted . . . . *Solidago*

20 Heads either staminate or pistillate and of different shapes, the staminate flowers in small heads above the larger pistillate heads; involucral bracts of the pistillate heads with hooked spines, enclosing the flowers to form a bur . . . . *Xanthium strumarium*

20 Heads with both staminate and pistillate flowers, or rarely either staminate or pistillate; involucral bracts neither spiny nor bur-like **21**

21 Main leaves large and at base of plant, arrowhead-shaped or palmately lobed, white-woolly at least on underside; plants flowering in spring to early summer before leaves develop . . . . *Petasites*

21 Main leaves along stem, neither arrowhead-shaped nor palmately lobed; plants flowering late-summer or fall (except the earlier flowering *Erigeron philadelphicus*). **22**

22 Heads with disk flowers only **23**

22 Heads with both disk and ray flowers (the rays narrow and ± equal to the involucral bracts in *Erigeron lonchophyllus,* and therefore inconspicuous) **29**

23 Leaves pinnately dissected; pappus absent . . . . *Artemisia biennis*

23 Leaves simple, entire or toothed; pappus of many hairlike bristles. **24**

24 Plants annual; leaves linear **25**

24 Plants perennial (sometimes biennial in *Erigeron lonchophyllus*); leaves wider . . . **26**

25 Stems with wool-like hairs. . . . . *Gnaphalium uliginosum*

25 Stems without hairs . . . . *Symphyotrichum ciliatum*

26 Flowers white. **27**

26 Flowers purple or pink-purple **28**

27 Leaves entire or with a few teeth; involucral bracts 10-15 *Arnoglossum plantagineum*

27 Leaves sharply toothed; involucral bracts 5 . . . . *Senecio suaveolens*

28 Flowers in a long, spikelike head; leaves linear. . . . . *Liatris spicata*

28 Flowers in an open corymb; leaves lance-shaped. . . . . *Vernonia fasciculata*

29 Pappus of 2 bristles about as long as achene and several very small bristles . . . . *Boltonia asteroides*

29 Pappus of many hairlike bristles **30**

30 Rays very narrow, to 0.5 mm wide . . . .
. . . . *Erigeron*

30 Rays wider than 0.5 mm **31**

31 Heads in elongate panicles; midvein of involucral bracts expanded to a ± diamond-shape above middle of bract . . . .
. . . . *Symphyotrichum*

31 Heads in a ± flat-topped cluster; midrib of involucral bract ± same width its entire length **32**

32 Leaves linear, less than 7 mm wide; rays pink. . . . . *Oclemena nemoralis*

32 Leaves broader, more than 9 mm wide; rays white. . . . . *Doellingeria umbellata*

Ageratina SNAKEROOT

Ageratina altissima (L.) King & H.E. Robins.
◆ WHITE SNAKEROOT

DESCRIPTION Native perennial herb. Stems 1–3, 3–15 dm long, smooth or with short hairs. Leaves opposite, ovate, 5–16 cm long and 3–12 cm wide, smooth or hairy, especially on underside veins, margins coarsely sharp-toothed; petioles 1–3 cm long. Flower heads of bright white disk flowers only, in a flat-topped or rounded inflorescence; involucre 3–5 mm high, smooth or short-hairy, the involucral bracts linear, nearly equal, in 1–2 series. Fruit a ± smooth achene; pappus of long, slender bristles. Flowering July–Oct.

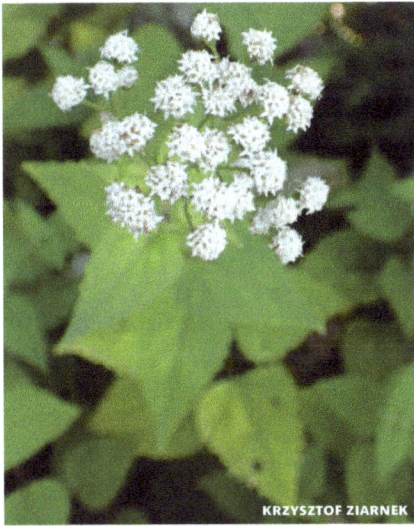

SYNONYMS *Ageratina altissima, Eupatorium rugosum.*

HABITAT floodplain forests, cedar swamps, thickets, streambanks, wooded ravines.

STATUS MW-FACU | NCNE-FACU

Arnoglossum
INDIAN PLANTAIN

Arnoglossum plantagineum Raf.
◆ TUBEROUS INDIAN PLANTAIN

DESCRIPTION Large, native, perennial herb, from a tuberous base and fleshy roots. Stems stout, smooth, 5–18 dm long, winged. Leaves alternate, mostly at base and lower stem of plant, thick, oval to ovate, strongly 5–9-veined, 5–20 cm long and 2–10 cm wide; margins entire or slightly wavy; petioles long on lower leaves, becoming short or absent on upper leaves. Flower heads many, of white disk flowers only, in a branched inflorescence at end of stem; involucral bracts 5, of equal length. Fruit a smooth achene; pappus of many, rough white bristles. Flowering June–Aug.

SYNONYMS *Cacalia plantaginea, Cacalia tuberosa, Mesadenia tuberosa.*

HABITAT marshes, low prairie, fens, sedge meadows, calcareous shores.

STATUS MW-FAC | NCNE-FAC | GP-FAC; Minn (THR), Wisc (THR).

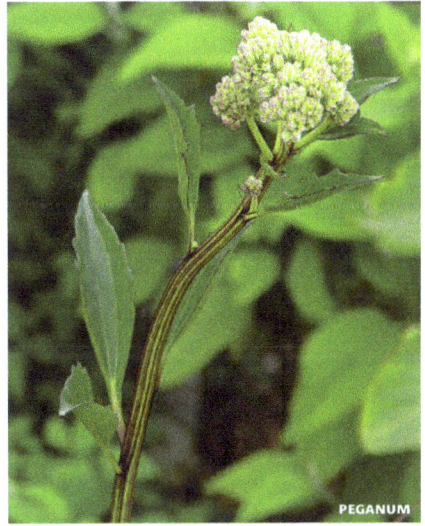

Ageratina altissima

Arnoglossum plantagineum

Artemisia
WORMWOOD, SAGE

Artemisia biennis Willd.
♦BIENNIAL WORMWOOD

DESCRIPTION Taprooted, annual or biennial herb. Stems erect, to 1 m or more long, often branched, smooth, only faintly scented. Leaves alternate, pinnately dissected nearly to middle, 5–12 cm long and 2–5 cm wide, the segments linear and toothed. Flowers in stalkless heads from upper leaf axils; the heads composed of many small green disk flowers, grouped into spikelike inflorescences, with leafy bracts much longer than the clusters of heads; pappus absent. Fruit a small oblong achene. Aug–Sept.

HABITAT Sandy lakeshores, streambanks, ditches, mud flats, disturbed areas; often where seasonally flooded.

STATUS MW-FACW | NCNE-FACW | GP-FACU; introduced; found across ne, nc and w USA.

Bidens BEGGAR-TICKS

Weedy annual or biennial herbs, or an aquatic perennial (*Bidens beckii*). Leaves opposite, simple, lobed, or pinnately divided. Flower heads with both disk and ray flowers, or with disk flowers only; ray flowers often about 8, yellow; involucral bracts in 2 series, the outer row leaflike and spreading, the inner row much shorter and erect; receptacle ± flat and chaffy. Fruit a flattened achene; pappus of 2–5 barbed awns which persist atop the achene; the body of achene barbed or with stiff hairs (at least on the angles), the "stick-tights" facilitating dispersal of seed by animals.

***Bidens* BEGGAR-TICKS**

1　Plants aquatic, underwater leaves whorled and dissected into narrow segments . . . .*B. beckii*

1　Plants not aquatic (sometimes emergent); leaves not as above **2**

2　Leaves simple and toothed, or rarely lobed; achenes 3–4-awned **3**

2　Leaves all (or mostly) pinnately divided or compound; achenes 2-awned **5**

3　Leaves with a petiole 1–4 cm long . . . .*B. tripartita*

3　Leaves mostly stalkless **4**

4　Heads nodding when mature; outer involucral bracts widely spreading . . . .*B. cernua*

4　Heads mostly upright; outer involucral bracts ± erect . . . .*B. tripartita*

5　Heads with disk flowers only, or with short rays less than 5 mm long **6**

5　Heads with both disk and ray flowers, the rays over 10 mm long **8**

6　Outer involucral bracts 2–5 (usually 4), not fringed with hairs . . . .*B. discoidea*

6　Outer involucral bracts 6 or more, fringed with hairs (at least near base) **7**

7　Disk flowers orange; outer involucral bracts mostly 6–8 . . . .*B. frondosa*

7　Disk flowers yellow; outer involucral bracts 10 or more . . . .*B. vulgata*

8　Achenes broad and ovate, mostly more than 3 mm wide . . . .*B. aristosa*

8　Achenes narrow and ± straight-sided, less than 3 mm wide . . . .*B. trichosperma*

Bidens aristosa (Michx.) Brit.
BEARDED BEGGAR-TICKS

DESCRIPTION Native annual or biennial herb. Stems 3–12 dm long, much-branched, ± smooth. Leaves pinnately divided, 5–15 cm long, the segments narrowly lance-shaped, hairy on underside; margins with coarse, forward-pointing teeth

Artemisia biennis

or shallow lobes; petioles 1–3 cm long. Flower heads 2–5 cm wide, numerous on leafless stalks; rays yellow, usually 8, 1–2.5 cm long; the outer involucral bracts 8–10, 5–10 mm long, margins smooth or fringed with hairs. Fruit a flattened achene, 5–7 mm long; pappus of 2 (rarely 4) barbed awns, or sometimes absent. Flowering Aug–Sept.

HABITAT marshy areas, ditches, disturbed wetlands.
STATUS MW-FACW | NCNE-FACW | GP-FACW

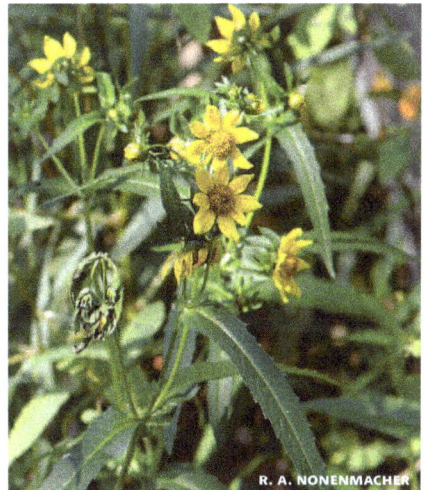

Bidens beckii Torr. ex Spreng.
WATER BEGGAR-TICKS

DESCRIPTION Native perennial aquatic herb. Stems 0.4–2 m long, little-branched. Underwater leaves opposite or whorled, dissected into threadlike segments; emersed leaves simple, opposite, lance-shaped to ovate, margins with forward-pointing teeth, petioles absent. Flower heads single or few at ends of stems; rays 6–10, gold-yellow, 1–1.5 cm long, notched at tip; involucral bracts smooth. Fruit an achene, ± round in section, 10–15 mm long; pappus of 3–6 slender awns, longer than achenes, the upper portion of awn with downward-pointing barbs. Flowering June–Sept.

SYNONYM *Megalodonta beckii*.

HABITAT quiet, shallow to deep water of lakes, ponds, rivers and streams.
STATUS MW-OBL | NCNE-OBL | GP-OBL

Bidens cernua L.
♦ NODDING BEGGAR-TICKS

DESCRIPTION Native annual herb. Stems often branched, to 1 m long, smooth or with spreading hairs. Leaves opposite, smooth, lance-shaped to oblong lance-shaped, 3–16 cm long and 0.5–5 cm wide; margins with sharp, forward-pointing teeth and often rough-to-touch; petioles absent, the leaves usually clasping at base. Flower heads many, globe-shaped, 1.5–3 cm wide, usually nodding after flowering; rays yellow, 6–8, to 1.5 cm long, or absent; outer involucral bracts 4–8, unequal in length, the margins often fringed with hairs. Fruit a ± straight-sided achene, 5–7 mm long, with downward-pointing barbs on margins; pappus with 4 (sometimes 2) awns, the awns with downward-pointing barbs. Flowering July–Oct.

HABITAT exposed sandy or muddy shores, streambanks, marshes, forest depressions, wet meadows, ditches and other wet places.
STATUS MW-OBL | NCNE-OBL | GP-OBL

Bidens discoidea (T. & G.) Britton
FEW-BRACTED BEGGAR-TICKS

DESCRIPTION Native annual herb. Stems smooth, 3–10 dm long. Leaves opposite, smooth, divided into 3-leaflets, the leaflets lance-shaped, the terminal leaflet largest, to 10 cm long and 4 cm wide; margins with coarse, forward-pointing teeth; petioles slender, 1–6 cm long. Flower heads many on slender stalks, the disk to 1 cm wide; rays absent; outer involucral bracts usually 4, leaflike, much longer than disk. Fruit a flattened achene, 3–6 mm long; pappus of 2 awns to 2 mm long, with short, upward pointing bristles. Flowering Aug–Sept.

HABITAT hummocks or logs in swamps, open muddy shores; usually where shaded.
STATUS MW-FACW | NCNE-FACW | GP-FACW

R. A. NONENMACHER

Bidens cernua

Bidens frondosa L.
♦DEVIL'S BEGGAR-TICKS

DESCRIPTION Native annual herb. **Stems** erect, 2–10 dm tall, branched, ± smooth, purple-tinged. **Leaves** pinnately divided into 3–5 segments, the segments lance-shaped, to 10 cm long and 3 cm wide, underside sometimes with short hairs; margins with coarse, forward-pointing teeth; petioles slender, 1–6 cm long. **Flower heads** many on long stalks; disk flowers orange, the disk to 1 cm wide; rays absent or very small; the outer involucral bracts usually 8, green and leaflike, longer than disk, fringed with hairs on margins. **Fruit** a flattened, nearly black achene, 5–10 mm long; pappus of 2 slender awns with downward-pointing barbs. **Flowering** July–Oct.

HABITAT wet, sandy or gravelly shores, forest depressions, streambanks, pond margins; weedy in wet disturbed areas.

STATUS MW-FACW | NCNE-FACW | GP-FACW

Bidens trichosperma (Michx.) Britt.
CROWNED BEGGARTICKS

DESCRIPTION Native annual or biennial herb. **Stems** branched, 3–15 dm tall, smooth, often purple. **Leaves** opposite, smooth, to 15 cm long, pinnately divided into 3–7 narrow leaflets; margins coarsely toothed or deeply lobed to sometimes entire;

petioles 3–15 mm long. **Flower heads** with both disk and ray flowers, large and numerous on slender stalks; rays about 8, gold-yellow, 1–2.5 cm long; outer involucral bracts 6–10, to 1 cm long, short-hairy on margins, inner bracts shorter. **Fruit** a flattened achene, 5–9 mm long, with long, stiff hairs on margins; pappus of 2 short, scalelike awns, 1–2 mm long. **Flowering** July–Oct.

SYNONYM *Bidens coronata*.

HABITAT open bogs, fens, tamarack swamps, shores, streambanks, marshes, sand bars.

STATUS MW-OBL | NCNE-OBL | GP-OBL

Bidens tripartita L.
♦THREE-LOBE BEGGARTICKS

DESCRIPTION Native annual herb. **Stems** green-purple to yellow, to 2 m long, usually branched, smooth. **Leaves** opposite, smooth, the lower leaves sometimes deeply lobed, 3–15 cm long and 1–4 cm wide; margins with coarse, forward-pointing teeth; petioles present on lower leaves, upper leaves short-petioled or stalkless. **Flower heads** several to many, 1–2 cm wide, upright; disk flowers orange-yellow; rays absent, or few and 3–4 mm long; outer involucral bracts 4–9, usually not much longer than the head. **Fruit** an achene, 3–7 mm long; pappus of 2–4 downwardly barbed awns, about half as long as achene. **Flowering** Aug–Oct.

SYNONYMS *Bidens comosa, Bidens connata*.

HABITAT exposed muddy shores, streambanks, marshes, pond, forest depressions, wet meadows, ditches and other wet places.

STATUS MW-OBL | NCNE-FACW | GP-FACW

ANDREY ZHARKIKH

Bidens frondosa

ENRICO BLASUTTO

Bidens tripartita

Bidens vulgata Greene
♦TALL BEGGAR-TICKS

DESCRIPTION Native annual herb. **Stems** to 2 m tall, smooth or upper stem and leaves short-hairy. **Leaves** opposite, pinnately divided into 3–5 segments, the segments lance-shaped, to 15 cm long and 5 cm wide, with prominent veins; margins with sharp, forward-pointing teeth; petioles present. **Flower heads** on stout, leafless stalks, disk flowers yellow; ray flowers usually present, small, yellow; outer involucral bracts about 13, leaflike. **Fruit** a flattened, olive-green or brown achene, 10–12 mm long; pappus of 2 awns with downward-pointing barbs. **Flowering** Aug–Oct.
SYNONYM *Bidens puberula.*
HABITAT streambanks, wet meadows, wet forests; weedy in moist disturbed areas.
STATUS MW-FACW | NCNE-FAC | GP-FAC

Boltonia BOLTONIA

Boltonia asteroides (L.) L'Her.
♦WHITE BOLTONIA

DESCRIPTION Native perennial herb, fibrous-rooted, sometimes with shallow rhizomes. **Stems** stout, erect, 3–15 dm long, smooth. **Leaves** alternate, lance-shaped or oval, 5–16 cm long and 0.5–2 cm wide, becoming smaller in the head, narrowed to stalkless or slightly clasping base; margins entire but rough-to-touch. **Flower heads** many, with both disk and ray flowers, 1.5–2.5 cm wide; disk flowers yellow; rays white, pink, or lavender, 5–15 mm long; involucres 2.5–5 mm high, the bracts overlapping, chaffy on outer margins, with a green midvein. **Fruit** a flattened, obovate achene, 1–2 mm long, with a winged margin; pappus of 2 awns and 2–4 shorter bristles. **Flowering** Aug–Sept.
SYNONYM *Boltonia latisquama.*
HABITAT seasonally flooded muddy shores, wet meadows, marshes, low prairie.
STATUS MW-OBL | NCNE-FACW | GP-FACW

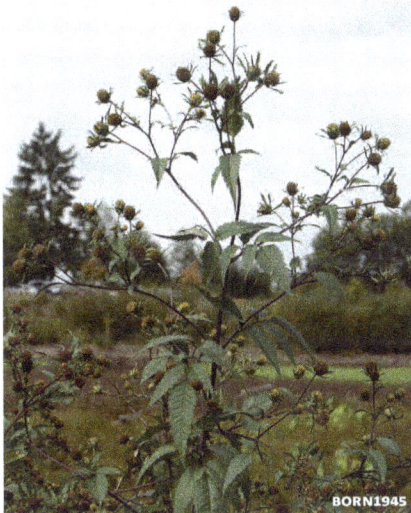

Cirsium THISTLE

Biennial or perennial herbs. Stems and leaves often spiny. Leaves from base of plant or alternate on stem. Flower heads of pink to purple disk flowers only; involucral bracts tipped with spines. Fruit a smooth achene; pappus of many slender bristles.

ADDITIONAL SPECIES Colonies of **Canada thistle** (*Cirsium arvense,* invasive) are sometimes found on wetland margins, especially where the soil is disturbed. It is distinguished from other thistles by its rhizomatous habit and lack of a rosette of basal leaves.

Bidens vulgata

Boltonia asteroides

Cirsium THISTLE

1 Stem not winged; leaf undersides and involucral bracts often with cobwebby hairs; involucral bracts 2–3.5 cm long, flowers deep rose-purple; native and not weedy . . . . *C. muticum*

1 Stem with spiny wings; leaf undersides and involucral bracts usually without cobwebby hairs; involucral bracts 1–2 cm long, flowers pale pink-purple; introduced and weedy . . . . *C. palustre*

Cirsium muticum Michx.
⬧SWAMP-THISTLE

DESCRIPTION Stout, biennial native herb. Stems 0.5–2 m long, branched in head, with long, soft hairs when young, becoming ± smooth. Leaves deeply lobed into pinnate segments, 1–2 dm long, underside often with matted, cobwebby hairs, becoming ± smooth with age; margins toothed and often tipped with spines; petioles present on lower leaves, stem leaves stalkless. Flower heads of purple or pink disk flowers only, single on leafless stalks over 1 cm long at ends of stems; involucre 2–3.5 cm high; the involucral bracts overlapping, densely hairy with cottony hairs (especially on margins), sometimes tipped with a short spine 0.5 mm long. Fruit an achene, 5 mm long; pappus of long, slender bristles. Flowering Aug–Oct.

HABITAT swamps, thickets, calcareous fens, sedge meadows, streambanks, shores.
STATUS MW-OBL | NCNE-OBL | GP-FACW

Cirsium palustre (L.) Scop.
⬧MARSH THISTLE,
EUROPEAN SWAMP-THISTLE

DESCRIPTION Introduced biennial herb; often invasive. Stems 0.5–2 m tall, spiny. Leaves to 20 cm long, deeply lobed into pinnate segments, covered with loosely matted hairs or ± smooth, tapered at base and continued downward on stem as spiny wings; margin teeth spine-tipped. Flower heads of purple disk flowers only, on short stalks mostly less than 1 cm long; involucre 1–2 cm high; the involucral bracts overlapping, not spine-tipped. Fruit an achene, 3 mm long; pappus of slender bristles to 1 cm long. Flowering June–Aug.

HABITAT roadside ditches and adjacent wetlands, including swamps, thickets and fens, especially where disturbed; resembling the native *Cirsium muticum* in these habitats.
STATUS NCNE-FACW

Crepis HAWK'S BEARD

Crepis runcinata (James) T. & G.
⬧DANDELION HAWK'S BEARD

DESCRIPTION Native perennial herb with milky juice. Stems 2–6 dm long, smooth or

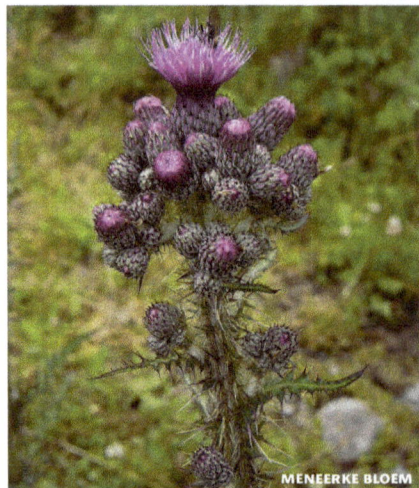

Cirsium muticum

Cirsium palustre

sparsely hairy, the stem leaves small and bractlike. Leaves in a rosette at base of plant, oblong lance-shaped to oval, 5–20 cm long and 1–4 cm wide, rounded at tip, tapered to a petiolelike base, margins entire or with widely spaced teeth. Flower heads 1–10, 1–2 cm wide, of yellow ray flowers only; involucre 8–15 mm high, with gland-tipped hairs, the involucral bracts in 2 series, the outer bracts shorter than inner. Fruit a brown achene, round in section, 4–5 mm long; pappus of many white slender bristles. Flowering June–July.

SYNONYM *Crepis glaucella*.

HABITAT wet meadows, low prairie, shores and swales, especially where alkaline.

STATUS MW-FACW | NCNE-FACW | GP-FAC

Doellingeria WHITE-TOP

Doellingeria umbellata (P. Mill.) Nees
♦ PARASOL WHITE-TOP

DESCRIPTION Native perennial herb, from thick rhizomes. Stems 0.5–2 m long, upper stem with appressed, short hairs. Leaves alternate, lance-shaped to oblong lance-shaped, 4–15 cm long and 1–4 cm wide, rough-to-touch above, densely short-hairy below; margins entire; petioles short, or absent on upper leaves. Flower heads usually many, 1–1.5 cm wide, in a ± flat-topped inflorescence; involucre 3–5 mm high, the involucral bracts short-hairy and overlapping; rays 5–10, white, 5–8 mm long. Fruit a nerved achene; pappus nearly white. Flowering July–Sept.

SYNONYM *Aster umbellatus*.

HABITAT openings in swamps and moist forests, thickets, streambanks, sedge meadows, calcareous fens, roadside ditches, rocky shores of Lake Superior.

STATUS MW-FACW | NCNE-FACW | GP-OBL

Eclipta YERBA-DE-TAJO

Eclipta prostrata (L.) L.
♦ YERBA-DE-TAJO

DESCRIPTION Native annual herb. Stems spreading, branched, 5–8 dm long, with rough, appressed hairs, often rooting at the nodes. Leaves opposite, lance-shaped, 2–10 cm long and 0.5–2.5 cm wide, margins with shallow teeth; petioles absent, or short on lower leaves. Flower heads with both disk and ray flowers, in clusters of 1–3 at ends of stems or from leaf axils, on stalks or nearly stalkless; the disk 4–6 mm wide; rays short, nearly white. Fruit a flat-topped achene, 2–3 mm long; pappus a crown of very short bristles. Flowering July–Oct.

SYNONYMS *Eclipta alba*, *Verbesina alba*.

HABITAT mud flats, muddy stream banks and ditches, where somewhat weedy.

STATUS MW-FACW | NCNE-FACW | GP-FACW

Crepis runcinata

Doellingeria umbellata

Eclipta prostrata

Erigeron DAISY, FLEABANE

Biennial or perennial herbs with simple, alternate leaves. Flower heads with both disk and ray flowers; disk flowers yellow; rays white to pink, very narrow, only to about 0.5 mm wide; involucral bracts in 1–2 series, linear, about equal in length, green in middle and at base, translucent at tip and on upper margins. Fruit a flattened achene; pappus of 20–30 slender, rough bristles.

Erigeron **DAISY, FLEABANE**

1 Leaves clasping stem; common and widespread ....*E. philadelphicus*

1 Leaves not clasping stem; uncommon plants of w Minn and Mich UP..........**2**

2 Leaves all ± same size; uncommon plant of Mich UP................*E. hyssopifolius*

2 Leaves becoming smaller upward on stem; uncommon plant of n Minn..............
..........................*E. lonchophyllus*

Erigeron hyssopifolius Michx.
HYSSOP-DAISY

DESCRIPTION Native perennial herb. Stems usually many, 1–4 dm long, smooth or sparsely hairy. Leaves thin, ± linear, 1–3 cm long and to 5 mm wide, smooth, lowest leaves scalelike; margins entire; petioles short. Flower heads 1–5, with both disk and ray flowers, on long, nearly leafless stalks; involucre 4–6 mm high; disk 5–12 mm wide; rays 20–50, white or sometimes pink-purple, to 8 mm long. Fruit an achene; pappus of long bristles, sometimes also with a ring of very short bristles. Flowering July–Aug.
HABITAT calcareous fens and cedar swamps; also rocky Lake Superior shoreline in Ont.
STATUS NCNE-FACW; Mich UP (THR); main range northward across Canada.

Erigeron lonchophyllus Hook.
LOW MEADOW FLEABANE

DESCRIPTION Native biennial or short-lived perennial herb. Stems 1–4 dm tall, with spreading hairs. Leaves alternate, lower leaves oblong lance-shaped, 5–15 cm long and 1–5 mm wide, tapered to a short petiolelike base, upper leaves linear and stalkless, not clasping, margins entire, fringed with hairs. Flower heads several to many, 1–1.5 cm wide; involucre 5–10 mm high, the involucral bracts coarsely hairy, the outer bracts shorter than inner; rays many, white, turning brown at tip, only to 0.2 mm wide. Fruit an achene; pappus of slender, rough bristles. Flowering July–Sept.
HABITAT wet meadows, low prairie, seeps.
STATUS MW-FACW | NCNE-FACW | GP-FACW

Erigeron philadelphicus L.
♦ PHILADELPHIA DAISY

DESCRIPTION Native biennial or short-lived perennial herb. Stems 1 to several, branched in head, 2–7 dm long, usually long-hairy. Leaves alternate, lower leaves spatula-shaped, 5–15 cm long and 1–4 cm wide, tapered to a short petiole; upper leaves smaller, lance-shaped, clasping at base, hairy to nearly smooth, rounded at tip; margins entire or with rounded teeth. Flower heads few to many, with both disk and ray flowers, 1.5–2.5 cm wide; involucre 3–6 mm high, the involucral bracts hairy, of ± equal length; rays many, white to deep pink, 5–10 mm long and to 0.5 mm wide. Fruit a short-hairy achene; pappus of long rough bristles. Flowering May–Aug.
HABITAT wet meadows, shores, streambanks, wet woods, floodplains, springs; also weedy in open disturbed areas and lawns.
STATUS MW-FACW | NCNE-FAC | GP-FAC

D. GORDON ROBERTSON

Erigeron philadelphicus

Eupatorium JOE-PYE-WEED

Eupatorium perfoliatum L.

◆BONESET

DESCRIPTION Native perennial herb, from a thick rhizome. Stems stout, 3–15 dm tall, with long, spreading hairs. Leaves opposite, mostly joined at the broad base and perforated by the stem (upper leaves sometimes separate), lance-shaped, 6–20 cm long and 1.5–5 cm wide, upper surface sparsely hairy, underside hairy, both sides dotted with yellow glands; margins finely toothed and rough-to-touch; petioles absent. Flower heads of dull white disk flowers only, in a flat-topped inflorescence; involucre 3–6 mm high, the involucral bracts green with white margins, hairy, overlapping in 3 series. Fruit a black achene, 1–2 mm long; pappus of long slender bristles. Flowering July–Sept.

HABITAT marshes, wet meadows, low prairie, shores, streambanks, ditches, cedar swamps, thickets, calcareous fens. Often occurring with **spotted joe-pye-weed** (*Eupatorium maculatum*).

STATUS MW-OBL | NCNE-FACW | GP-FACW

Euthamia

FLAT-TOPPED GOLDENROD

Perennial herbs, spreading by rhizomes. Stems leafy. Leaves alternate, covered with resinous dots; margins entire; petioles absent or very short. Flower heads small, of yellow disk and ray flowers, in a ± flat-topped cluster at ends of stems; involucre somewhat sticky. Fruit an achene; pappus of slender white bristles.

Euthamia **FLAT-TOPPED GOLDENROD**

1 Largest stem leaves 4 mm or more wide, with 3 conspicuous longitudinal veins . . . . *E. graminifolia*

1 Largest stem leaves less than 3 mm wide, with single longitudinal vein (midrib) and sometimes with faint pair of longitudinal veins **2**

2 Involucre 3.5–4.5 mm long; Mich . . . . *E. caroliniana*

2 Involucre 5-6.5 mm long; Minn, Wisc . . . . *E. gymnospermoides*

Euthamia caroliniana (L.) Greene ex Porter & Britt

◆LAKES FLAT-TOPPED GOLDENROD

Native perennial herb, spreading by rhizomes. Stems smooth, 3–8 dm long, branched in head. Leaves alternate, linear, 2–8 cm long and 2–3 mm wide, often with clusters of smaller leaves in axils of main leaves, 1-veined or with another pair of fainter veins, with glandular dots; margins entire, sometimes rough-to-touch; petioles absent. Flower heads of yellow disk and ray flowers, in flat-topped clusters at ends of stems; involucre sticky, 3–5 mm long; the

R. A. NONENMACHER

Eupatorium perfoliatum

Euthamia caroliniana

involucral bracts overlapping. Fruit a short-hairy achene; pappus of slender white bristles.

SYNONYM *Euthamia remota.*

HABITAT Sandy or mucky shores (especially on recently exposed lakeshores), interdunal swales; sometimes in dry sandy openings.

STATUS MW-FACW | NCNE-FAC; main range Atlantic coast.

Euthamia graminifolia (L.) Nutt.
♦COMMON FLAT-TOPPED GOLDENROD

DESCRIPTION Native perennial herb, spreading by rhizomes. Stems erect, 5–15 dm tall, smooth to hairy, usually branched in head. Leaves alternate, linear to narrowly lance-shaped or oval, 3–15 cm long and 3–10 mm wide, 3-veined, with small glandular dots; margins entire, smooth or rough-to-touch; petioles absent or very short. Flower heads small, in flat-topped clusters at ends of stems; with yellow disk and ray flowers, the rays small, to 1 mm long; involucre 3–5 mm high, somewhat sticky, the involucral bracts overlapping in several series, yellow or green-tipped. Fruit a finely hairy achene, 1 mm long; pappus of many white, slender bristles. Flowering Aug–Sept.

SYNONYM *Solidago graminifolia.*

HABITAT shores, wet meadows, low prairie, springs, fens, swamps, interdunal wetlands, streambanks, often where sandy or gravelly; also weedy in abandoned fields.

STATUS MW-FACW | NCNE-FAC | GP-FACW

Euthamia gymnospermoides Greene
TEXAS GOLDENTOP

DESCRIPTION Native perennial from branched, creeping rhizomes. Stems 4–10 dm

Euthamia graminifolia

tall, glabrous. Leaves glabrous except for the slightly scabrous margins, densely and strongly glandular-punctate, obscurely to sometimes evidently 3-nerved, without any additional lateral nerves, the basal and lower cauline ones soon deciduous, the others numerous, not much reduced upwards, linear, mostly 4–9 cm long and 1.5–5 mm wide. Flower heads terminal, corymbiform, the heads sometimes sessile in small glomerules, but not infrequently more pedunculate; involucre 4.5–6.5 mm high, glutinous; heads mostly 14–20-flowered, the short rays 10–14, the disk-flowers 4–6. Achenes hairy; pappus of bristles. Flowering Aug.–Oct.

SYNONYM *Solidago gymnospermoides.*

HABITAT open, often sandy places.

STATUS MW FACW | NCNE FACW | GP-FAC; historical report for Mich.

Eutrochium JOE-PYE-WEED

Perennial herbs from a thick rhizome. Stems stout, erect. Leaves whorled, lower leaves smaller; margins toothed. Flower heads of pink, purple or white disk flowers only, usually many in a more or less flat-topped head at ends of stems; involucral bracts overlapping or nearly equal length. Fruit an angled achene; pappus of many slender bristles.

ADDITIONAL SPECIES Green-stemmed Joe-pye-weed (*Eutrochium purpureum*) is present on usually drier sites in mostly southern portions of the Upper Midwest region. Its stem is purple only near the nodes, and with leaves in whorls of only 3–5.

Eutrochium JOE-PYE-WEED

1 Stems waxy-coated; heads with less than 8 flowers; uncommon plant of s portion of region . . . . *E. fistulosum*

1 Stems not waxy-coated; heads with more than 8 flowers; common and widespread . . . . *E. maculatum*

Eutrochium fistulosum (Barrat) E. Lamont
HOLLOW-STEMMED JOE-PYE-WEED

DESCRIPTION Native perennial herb. Stems waxy, purple, hollow, to 3 m long. Leaves narrowly oval, in whorls of

4–7, 10–30 cm long and 3–15 cm wide; margins with small, rounded teeth; petioles short. Flower heads of pink-purple disk flowers only; involucre to 1 cm long, the involucral bracts overlapping. Fruit an achene; pappus of slender bristles. Flowering July-Sept.

SYNONYMS *Eupatoriadelphus fistulosus, Eupatorium fistulosum.*

HABITAT marshes, wet swamp openings, wet areas between dunes.

STATUS MW-OBL | NCNE-FACW | GP-FACW; Mich (THR).

Eutrochium maculatum (L.) E. Lamont
◊ SPOTTED JOE-PYE-WEED

DESCRIPTION Native perennial herb. Stems 5–20 dm long, spotted or tinged with purple, short-hairy above, especially on branches of head. Leaves in whorls of mostly 4–5, lance-shaped to ovate, 5–20 cm long and 2–7 cm wide, upper surface with sparse short hairs, underside often densely short-hairy; margins with sharp, forward-pointing teeth; petioles to 2 cm long. Flower heads of light pink to purple disk flowers only, the inflorescence ± flat-topped; involucres 6–9 mm high, purple-tinged, the involucral bracts overlapping. Fruit a black, angled achene, 2–4 mm long; pappus of long, slender bristles. Flowering July–Sept.

SYNONYMS *Eupatoriadelphus maculatus, Eupatorium maculatum.*

HABITAT wet meadows, marshes, low prairie, shores, streambanks, ditches, cedar swamps, open bogs, calcareous fens.

STATUS MW-OBL | NCNE-OBL | GP-OBL

Gnaphalium CUDWEED
Gnaphalium uliginosum L.
◊ MARSH CUDWEED

DESCRIPTION Introduced annual herb. Stem branched and spreading, 10–25 cm tall, covered with white wool-like hairs. Leaves alternate, entire, linear or oblong lance-shaped, 2–4 cm long and to 5 mm wide, with sparse woolly hairs. Flower heads of whitish disk flowers only, the rays absent, in numerous clusters from leaf axils and at end of stem-branches, shorter than the subtending leaves. Fruit a smooth or bump-covered achene.

SYNONYM *Filaginella uliginosa.*

HABITAT Introduced from Europe and now weedy, especially on streambanks and in wet to dry disturbed areas.

STATUS MW-FAC | NCNE-FAC | GP-FAC

Helenium SNEEZEWEED
Helenium autumnale L.
◊ COMMON SNEEZEWEED

DESCRIPTION Native perennial herb. Stems single or clustered, erect, 3–13 dm tall,

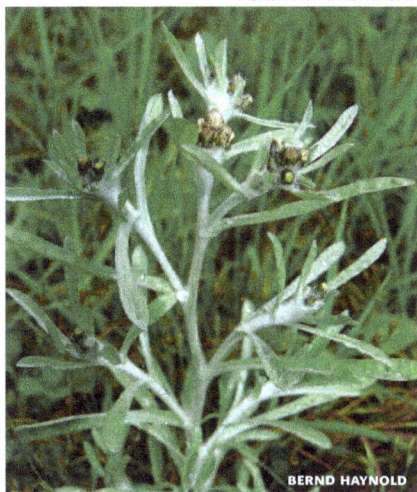

Eutrochium maculatum
D. GORDON ROBERTSON

Gnaphalium uliginosum
BERND HAYNOLD

smooth or finely hairy, branched in head. Leaves alternate, bright green, lance-shaped to oval, 4–12 cm long and 0.5–3.5 cm wide, usually short-hairy, glandular-dotted; margins entire to shallowly toothed; petioles absent, the blades tapered to a narrow base extending downward as wings on stem. Flower heads ± round, 1.5–4 cm wide; few to many on slender stalks in a leafy inflorescence, with both disk and ray flowers, the disk flowers yellow to brown, the rays yellow and 3-lobed, 1.5–2.5 cm long; involucral bracts in 2–3 series, linear, short-hairy, bent downward with age. Fruit a finely hairy, 4–5-angled achene, 1–2 mm long; pappus of several translucent, awn-tipped scales. Flowering July–Sept.

HABITAT wet meadows, shores, streambanks, marshes, swamps.

STATUS MW-FACW | NCNE-FACW | GP-FACW

Helianthus SUNFLOWER

Large perennial herbs (those included here), with fibrous or fleshy roots and short to long rhizomes. Stems unbranched or branched above. Leaves usually opposite on lower part of stem and alternate above, lance-shaped, margins entire or with forward-pointing teeth; petioles present. Flower heads large, mostly 1 to several (rarely many), at ends of stems and branches, with yellow disk and ray flowers, the rays large and showy;

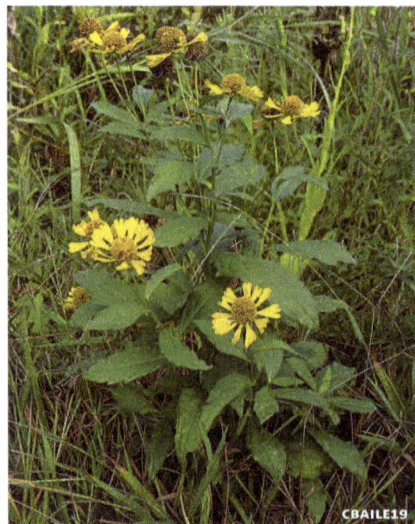

involucre of several series of narrow, overlapping bracts; receptacle chaffy. Fruit a flattened achene; pappus of 2 deciduous, awn-tipped scales.

Helianthus SUNFLOWER

1 Upper side of leaf rough-to-touch; stems hairy . . . . *H. giganteus*
1 Upper side of leaf not (or only slightly) rough-to-touch; stems smooth, often waxy-coated, or sometimes with sparse hairs on upper stem . . . *H. grosseserratus*

Helianthus giganteus L.
SWAMP-SUNFLOWER

DESCRIPTION Native perennial herb, with short rhizomes and thick, fleshy roots. Stems 1–3 m long, often purple, with coarse hairs or sometimes nearly smooth, often branched in head. Upper leaves generally alternate, lower leaves opposite; lance-shaped, 6–20 cm long and 1–4 cm wide, base with 3 main veins, upper surface very rough-to-touch, underside with short, stiff hairs; margins toothed to ± entire; petiole short or absent. Flower heads 3–6 cm wide, several to many, on long stalks in an open inflorescence; with yellow disk and ray flowers, the rays 1.5–3 cm long; involucral bracts narrow, awl-shaped, green or dark near base, hairy or margins fringed with hairs. Fruit a smooth achene; pappus of 2 awl-shaped scales. Flowering July–Sept.

HABITAT wet meadows, low prairie, sedge meadows, fens, floodplain forests, streambanks.

STATUS MW-FACW | NCNE-FACW | GP-FAC

Helianthus grosseserratus Martens
♦ SAWTOOTH SUNFLOWER

DESCRIPTION Native perennial herb, with fleshy roots, spreading by rhizomes and forming colonies. Stems 1–3 m tall, short-hairy in head, smooth and often waxy below, purple or blue-green. Leaves upper leaves alternate, lower leaves opposite; lance-shaped, 10–20 cm long and 2–5 cm wide, rough-to-touch on both sides, also densely short hairy on the paler underside; margins with coarse, forward-pointing teeth, upper leaves often entire; petioles 1–

Helenium autumnale

4 cm long. Flower heads 3–8 cm wide, several to many at ends of stems and branches; with yellow disk flowers and deep yellow ray flowers, the rays 2.5–4 cm long; involucral bracts narrowly lance-shaped, fringed with hairs and sometimes hairy on back. Fruit a smooth achene, 3–4 mm long; pappus of 2 lance-shaped scales. Flowering July–Oct.
HABITAT wet meadows, low prairie, streambanks, swamps, ditches, roadsides.
STATUS MW-FACW | NCNE-FACW | GP-FACW

Liatris BLAZING STAR

Liatris spicata (L.) Willd.
♦ MARSH BLAZING STAR, GAYFEATHER
DESCRIPTION Native perennial herb. Stems smooth, upright, 5–20 dm tall. Leaves many, lower leaves linear, often ± blunt-tipped, 10–30 cm long; upper leaves smaller, linear to awl-shaped. Flower heads stalkless, of blue-purple (rarely white) disk flowers only, numerous in an elongated spike; involucral bracts appressed, overlapping in series of 4–6; pappus of barbed hairs. Fruit an achene.
HABITAT low prairie,

SIXFLASHPHOTO

calcareous fens, wet meadows, wet sandy flats, conifer swamps.
STATUS MW-FAC | NCNE-FAC
ADDITIONAL SPECIES Prairie blazing star (*Liatris pycnostachya*) occurs in moist prairies and calcareous fens of Minn, s to Miss and westward, occasionally as an escape from gardens in the east. Its involucral bracts spread outward at tip in contrast to the appressed bracts of *Liatris spicata*.

Nabalus RATTLESNAKE-ROOT

Nabalus racemosus (Michx). DC.
♦ GLAUCOUS RATTLESNAKE-ROOT
DESCRIPTION Native perennial herb. Stems slender, erect, ridged, 4–18 dm tall, smooth and somewhat waxy, hairy in the head. Leaves thick, smooth and waxy; lower leaves oval to obovate, 10–20 cm long and 2–10 cm wide; margins shallowly toothed; petioles long and winged; stem leaves becoming smaller upwards, stalkless and partly clasping the stem. Flower heads many in a narrow, elongate inflorescence, of ray flowers only, pink or purplish; involucre 9–14 mm high, purple-black, long-hairy. Fruit a linear achene; pappus of straw-colored bristles. Flowering Aug–Sept.
SYNONYM *Prenanthes racemosa*.
HABITAT Sandy or gravelly shores, streambanks, wet meadows, low prairie, fens, rocky shores of Lake Superior.

Helianthus grosseserratus

Nabalus racemosus

STATUS MW-FACW | NCNE-FACW | GP-FACU

ADDITIONAL SPECIES **White rattlesnake-root** (*Nabalus albus*), typically found in deciduous woods, sometimes occurs in wet woods or along streams in the Upper Midwest region. It differs from *Nabalus racemosa* by its usually deeply lobed leaves and hairless involucre.

Oclemena
BOG NODDING-ASTER

Oclemena nemoralis (Aiton) Greene
LEAFY BOG-ASTER

DESCRIPTION Native perennial herb, spreading by rhizomes. Stems 1-6 dm long, finely hairy, the hairs slightly sticky. Leaves alternate, all from the stem, firm, linear to oval or oblong, 1-5 cm long and to 1 cm wide, upper surface rough-to-touch, finely hairy below (at least along veins); margins often rolled under; petioles absent. Flower heads 1 to several, on slender stalks; involucre 5-7 mm high, the involucral bracts awl-shaped, overlapping, green-purple; ray flowers 15-25, pink or lilac-purple, 10-15 mm long. Fruit a glandular-hairy achene; pappus white. Aug-Sept.

SYNONYM *Aster nemoralis*.

HABITAT fens and patterned peatlands (especially where sphagnum mosses are sparse), occasional on sandy shores; often with **sweet gale** (*Myrica gale*) and **twig-rush** (*Cladium mariscoides*).

STATUS NCNE-OBL; UP of Mich, where at western edge of range.

Packera GROUNDSEL
Erect perennial or annual herbs. Leaves alternate or from base of plant, stalked near base, stalkless and usually smaller upward. Flower heads with both disk and ray flowers, few to many in clusters at ends of stems; disk flowers perfect and yellow, the rays yellow; involucral bracts in 1 series and not overlapping, of equal lengths; receptacle flat or convex, not chaffy. Fruit an achene, nearly round in section; pappus of slender bristles.

Packera GROUNDSEL
1 Heads with disk flowers only . *P. indecora*
1 Heads with both ray and disk flowers ... 2
2 Basal leaves heart-shaped at base ....
   ....*P. aurea*
2 Basal leaves ovate ....*P. pseudaurea*

Packera aurea (L.) A. & D.Löve
♦ HEART-LEAVED GROUNDSEL

DESCRIPTION Native perennial herb, from a spreading crown or rhizome. Stems single or clumped, 3-8 dm long, slightly hairy when young, soon becoming smooth. Basal leaves heart-shaped, 5-10 cm long and to as wide, often purple-tinged, on long petioles, the margins with rounded teeth. Stem leaves much smaller and ± pinnately lobed, becoming stalkless. Flower heads several to many, the disk 5-10 mm wide, rays gold-yellow, 6-13 mm long involucre 5-8 mm high, the involucral bracts often purple-tipped. Fruit a smooth achene; pappus of slender white bristles. Flowering May-July.

SYNONYMS *Senecio aureus, Senecio gracilis*.

HABITAT floodplain forests, wet forest depressions, swamp openings and hummocks, sedge meadows, thickets, fens, ditches.

STATUS MW-FACW | NCNE-FACW | GP-FACW

NICHOLAS A. TONELLI

Packera aurea

Packera indecora (Greene) A. & D.Löve
TALLER DISCOID GROUNDSEL

DESCRIPTION Native perennial herb. **Stems** 3–8 dm long, smooth apart from woolly hairs in leaf axils. **Basal leaves** ovate, 3–6 cm long and 2–4 cm wide, margins with coarse, forward-pointing teeth, the petioles longer than blades. **Stem leaves** few, much smaller, deeply cleft, ± stalkless. **Flower heads** few to several on slender stalks and forming a rounded cluster; with yellow disk flowers only; involucre 6–10 mm high, the involucral bracts often purple-tipped. **Fruit** a smooth achene; pappus of long slender bristles. **Flowering** July–Aug. **SYNONYM** *Senecio discoideus, Senecio indecorus.*

HABITAT cedar swamps, moist conifer or mixed conifer and deciduous forests, rocky Lake Superior shores, streambanks. **STATUS** NCNE-FACW | GP-FACW; Wisc (THR), Mich (THR).

Packera pseudaurea Rydb.
WESTERN HEART-LEAVED GROUNDSEL

DESCRIPTION Native perennial herb, from a crown or short rhizome. **Stems** single or few, solid, 2–5 dm long, smooth or with tufts of woolly hairs in leaf axils when young. **Basal leaves** ovate to oval, 2–4 cm long and 1–2 cm wide, underside often purple, margins with rounded teeth, petioles long and slender. **Stem leaves** 2–6 cm long and 0.5–2 cm wide, pinnately cleft at least near base, stalkless and often clasping. **Flower heads** 1–1.5 cm wide, few to many in a single cluster; with both disk and ray flowers, the rays pale yellow, 6–10 mm long; involucre 4–7 mm high, the involucral bracts green. **Fruit** a smooth achene, 1–2 mm long; pappus of white bristles. **Flowering** May–July. **SYNONYM** *Senecio pseudaureus.*

HABITAT wet meadows, low prairie, fens. **STATUS** MW-FACW | NCNE-FACW | GP-FACW

Petasites SWEET COLTSFOOT

Perennial herbs, spreading by rhizomes. Leaves mostly from base of plant on long petioles, arrowhead-shaped or palmately lobed, with white woolly hairs on underside; stem leaves alternate, reduced to bracts. Flowering before or as leaves expand in spring, the heads white, the flowers mostly either staminate and pistillate and on different plants; or the heads sometimes with both staminate and pistillate flowers, the staminate heads usually with disk flowers only, the pistillate heads with all disk flowers or sometimes with short rays; involucral bracts in a single series; receptacle not chaffy. Fruit a linear, ribbed achene; pappus of many white, slender bristles.

Petasites SWEET COLTSFOOT
1 Leaf blades palmately lobed; northern portions of region ....*P. frigidus*
1 Leaf blades arrowhead-shaped, toothed and not lobed; uncommon in Upper Midwest region ....*P. sagittatus*

Petasites frigidus (L.) Fries.
♦ NORTHERN SWEET COLTSFOOT

DESCRIPTION Native perennial herb, spreading by rhizomes. **Stems** 1–6 dm long, smooth or short-hairy in the head.

Leaves mostly from base of plant, triangular to nearly round in outline, palmately lobed, 5–30 cm wide, upper surface green and smooth, underside densely white-hairy, sometimes becoming smooth with age; mar-

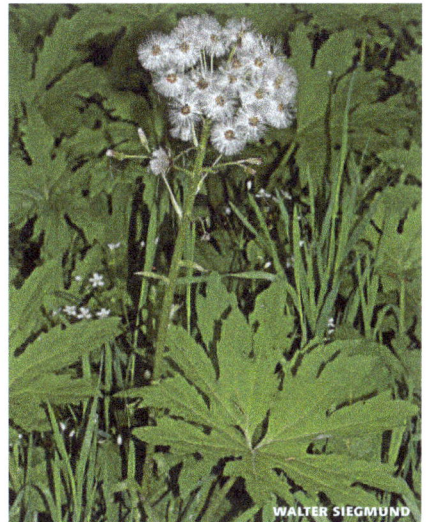

WALTER SIEGMUND

Petasites frigidus

gins coarsely toothed; petioles of basal leaves 1-3 dm long; stem leaves small and bractlike, 2-6 cm long. Flower heads nearly white, staminate and pistillate flowers mostly on separate plants; rays of pistillate heads to 7 mm long, involucre 4-9 mm high. Fruit a narrow achene; pappus of many slender bristles. Flowering May-June.

SYNONYM *Petasites palmatus.*

HABITAT wet conifer forests and swamps, wet trails and clearings, aspen woods.

STATUS MW-FACW | NCNE-FACW | GP-FAC

Petasites sagittatus (Banks) A. Gray
♦ARROWHEAD SWEET COLTSFOOT

DESCRIPTION Native perennial herb. Stems 3-6 dm tall, sparsely covered with woolly white hairs. Leaves mostly from base of plant, arrowhead-shaped, 10-40 cm long and 3-30 cm wide, upper surface smooth to sparsely hairy, densely white hairy below; margins wavy with outward-pointing teeth; petioles 1-3 dm long; the stem leaves reduced in size. Flower heads ± white; rays of pistillate heads 8-9 mm long. Fruit a linear achene; pappus of slender bristles. Flowering May-June.

SYNONYM *Petasites frigidus* var. *sagittatus.*

HABITAT wet meadows, marshes, sedge meadows, open swamps.

STATUS MW-FACW | NCNE-FACW | GP-FAC; Wisc (THR), Mich (THR).

NOTE *Petasites sagittatus* sometimes treated

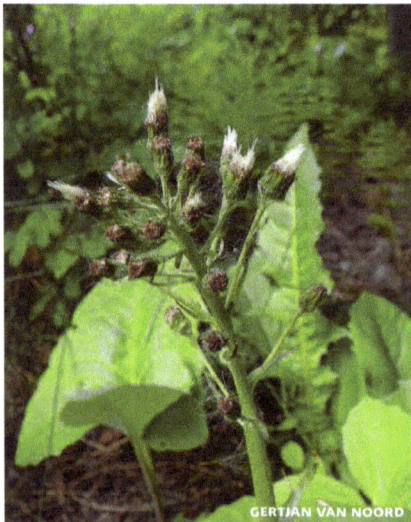

as *Petasites frigidus* var. *sagittatus* (Pursh) Cherniawsky.

Rudbeckia CONEFLOWER

Perennial herbs. Stems and leaves rough-hairy. Leaves alternate, entire to deeply lobed. Flower heads with both disk and ray flowers, the rays yellow to orange; involucral bracts green, overlapping; receptacle rounded, chaffy. Fruit a smooth, 4-angled achene; pappus absent or a short crown.

Rudbeckia CONEFLOWER

1 Leaves simple, not lobed ...... *R. fulgida*
1 Main leaves 3-7 lobed 2
2 Largest leaves 3 lobed; disk dark purple .. .. *R. triloba*
2 Largest leaves 5-7 lobed; disk green-yellow .... *R. laciniata*

Rudbeckia fulgida Aiton
♦EASTERN CONEFLOWER

DESCRIPTION Native perennial herb, often spreading by stolons. Stems to 1 m long, hairy. Leaves alternate, the lower leaves lance-shaped to oblong, 5-10 cm long and 2-4 cm wide, hairy; margins entire or sparsely toothed; petioles long and winged; upper leaves smaller, stalkless. Flower heads several, 2-4 cm wide, at ends

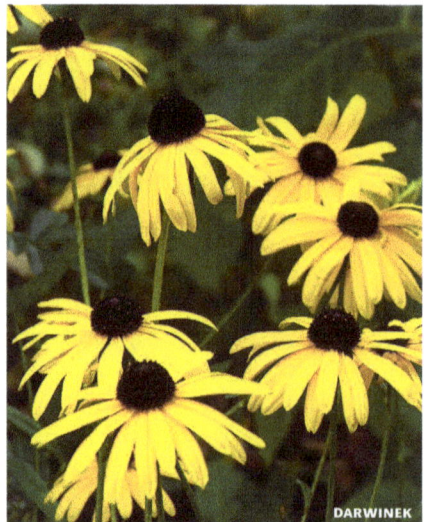

Petasites sagittatus

Rudbeckia fulgida

of long stalks, with both disk and ray flowers, the disk round, dark purple or brown; the rays yellow to orange, notched at tip; involucre 1–2 cm high. Fruit an achene. Aug–Sept.

SYNONYM *Rudbeckia speciosa* var. *sullivantii.*

HABITAT sedge meadows, calcareous fens, streambanks, low prairie.

STATUS MW-OBL | NCNE-OBL | GP-FAC

Rudbeckia laciniata L.
♦ TALL *or* CUTLEAF CONEFLOWER

DESCRIPTION Native perennial herb, from a woody base. Stems branched, 0.5–3 m long, smooth and often waxy. Leaves alternate, to 30 cm wide, deeply lobed, nearly smooth to hairy on underside; margins coarsely toothed as well as lobed, or entire on upper leaves; petioles long on lower leaves, becoming short above. Flower heads several to many at ends of stems, with both disk and ray flowers, disk flowers green-yellow, rays lemon-yellow, drooping, 3–6 cm long; involucral bracts of unequal lengths; receptacle round at first, becoming cylindric. Fruit a 4-angled achene. Flowering July–Sept.

HABITAT floodplain forests, swamps, streambanks, thickets, ditches; usually in partial or full shade.

STATUS MW-FACW | NCNE-FACW | GP-FAC

Rudbeckia triloba L.
BROWN-EYED SUSAN

DESCRIPTION Native perennial herb. Stems 0.5–1.5 m long, with coarse spreading hairs or sometimes nearly smooth. Leaves alternate, thin, coarsely hairy (or sometimes nearly hairless), the basal leaves broadly ovate to heart-shaped, with long petioles; the stem leaves narrower, with short petioles or the petioles absent, usually with some leaves deeply trilobed; margins sharp-toothed or nearly entire. Flower heads several to many at ends of stems, with both disk and ray flowers, the disk dark purple, rays 6–12, yellow but usually orange at base, 1–3 cm long; involucral bracts of unequal lengths, green and leaflike. Fruit a 4-angled achene. Flowering July–Sept.

HABITAT edges of wet forests and marshes, wet praire, shorelines.

STATUS MW-FACU | NCNE-FACU | GP-FACU

Senecio RAGWORT

Senecio suaveolens (L.) Pojark.
♦ SWEET-SCENTED INDIAN PLANTAIN

DESCRIPTION Native perennial herb, from fleshy roots. Stems ± smooth, grooved, 1–2.5 m tall, leafy to the inflorescence. Leaves alternate, smooth; lower

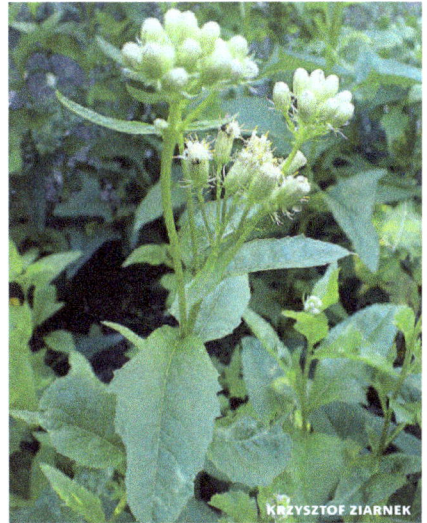

Rudbeckia laciniata

Senecio suaveolens

leaves triangular with a pair of outward-pointing lobes at base, 5–20 cm long and nearly as wide; upper leaves smaller and often not lobed; margins sharply and irregularly toothed; petioles winged. Flower heads of disk flowers only, in a ± flat-topped inflorescence, the disk about 1 cm wide; disk flowers white or light pink; involucre 1 cm long, the main involucral bracts 10–15. Fruit an achene; pappus of many soft, white bristles. Flowering July–Sept.
SYNONYMS *Cacalia suaveolens, Hasteola suaveolens, Synosma suaveolens.*
HABITAT riverbanks, shores, calcareous fens, wet low areas.
STATUS MW-FACW | NCNE-FACW; Minn (END).

Silphium
CUP-PLANT, ROSIN-WEED
Tall perennial herbs, with resinous juice. Leaves opposite or all from base of plant, broadly ovate. Flower heads with yellow disk and ray flowers, in clusters at ends of stems; involucral bracts overlapping, receptacle ± flat, chaffy. fruit an achene; pappus absent or of 2 small scales from top of achene.

Silphium **CUP-PLANT, ROSIN-WEED**
1 Stems leafy and 4-angled. *S. perfoliatum*
1 Leaves all at base of plant; stems round. . .
. . . . *S. terebinthinaceum*

Silphium perfoliatum L.
◆ CUP-PLANT, INDIAN-CUP
DESCRIPTION Native perennial herb, spreading by rhizomes. Stems erect, 4-angled, smooth, 1–2.5 m long. Leaves opposite, broadly ovate, 8–30 cm long and 4–15 cm wide, rough-to-touch, margins coarsely toothed, the lower leaves often short-stalked and joined by wings on the petioles; upper leaves joined at base, forming a cup around stem. Flower heads several to many in an open inflorescence, with both disk and ray flowers, the disk 1.5–2.5 cm wide, the rays yellow, 1.5–2.5 cm long; involucre 1–2.5 cm high, the involucral bracts ovate, nearly equal, fringed with hairs on margins; receptacle flat, chaffy. Fruit a flat, obovate achene, 8–10 mm long and 5–

6 mm wide, the margins narrowly winged; pappus absent. Flowering July–Sept.
HABITAT floodplain forests, streambanks, springs.
STATUS MW-FACW | NCNE-FACW | GP-FAC; Mich (THR).

Silphium terebinthinaceum Jacq.
◆ ROSIN-WEED, PRAIRIE-DOCK
DESCRIPTION Taprooted, native perennial herb. Stems 1–3 m long, branched above, ± leafless, smooth. Leaves main leaves all from base of plant, ovate, leathery, 1–5 dm long and 1–3 dm wide, usually heart-shaped at base, usually rough-to-touch, margins sharply toothed, petioles long; stem leaves few and reduced to large bracts. Flower heads many in an open inflorescence, with yellow disk and ray flowers, the disk 1.5–3 cm wide, the rays 2–3 cm long; involucre 1–2.5 cm high, the involucral bracts smooth, loose and overlapping. Fruit an obovate, narrowly winged achene; pappus reduced to 2 small teeth at top of achene. Flowering

FRANK MAYFIELD

Silphium perfoliatum

July–Sept.
HABITAT low prairie, fens; especially where calcium-rich.
STATUS MW-FAC | NCNE-FAC

Solidago GOLDENROD

Erect perennials, spreading by rhizomes or from a crown. Leaves alternate, margins entire or toothed. Flower heads small, many, in flat-topped (corymblike), rounded (paniclelike) or spikelike clusters at ends of stems; the flowers sometimes mostly on 1 side of inflorescence branches (secund) in species with paniclelike heads; the heads with yellow disk and ray flowers; involucral bracts in several overlapping series, papery at base and tipped with green; receptacle flat or convex, not chaffy. Fruit an achene, angled or nearly round in cross-section; pappus of many slender white bristles.

Solidago **GOLDENROD**

1   Heads in a ± flat-topped cluster at end of stems **2**

1   Heads in an elongate or pyramid-shaped cluster **4**

2   Rare species along Lakes Michigan and Huron shoreline in Michigan; rays 3–5 mm long . . . . *S. houghtonii*

2   Widespread species; rays 1–3 mm long . . **3**

3   Flower stalks rough-hairy; upper leaves folded inward along midrib, with 3 or more veins at base . . . . *S. riddellii*

3   Flower stalks ± smooth; leaves flat, not 3-veined at base . . . . *S. ohioensis*

4   Flower heads spiraled around branches of inflorescence . . . . *S. uliginosa*

4   Flower heads mostly on upper side of branches of inflorescence **5**

5   Stem leaves with 3 prominent veins . . . . *S. gigantea*

5   Stem leaves with strong midvein only, not 3-veined **6**

6   Leaves smooth, lowest stem leaves clasping stem . . . . *Solidago uliginosa*

6   Upper leaf surface rough-to-touch, lowest stem leaves not clasping stem . . . . *Solidago patula*

Solidago gigantea Aiton
♦SMOOTH GOLDENROD

DESCRIPTION Native perennial herb, from stout rhizomes, often forming colonies. Stems 0.5–2 m long, mostly smooth, sometimes waxy, short-hairy on upper branches. Leaves alternate, lance-shaped to oval, 6–15 cm long and 1–4 cm wide, prominently 3-veined, tapered to an stalkless or short, petiolelike base, smooth, or sparsely hairy on underside veins; margins with sharp, forward-pointing teeth. Flower heads many, in large paniclelike clusters, on 1 side of the spreading branches, with yellow disk and ray flowers, the rays 2–3 mm long; involucre 2–5 mm high, the involucral bracts linear. Fruit a short-hairy achene, 1–2 mm long; pappus of slender white bristles. Flowering July–Sept.

HABITAT wet meadows, streambanks, swamps, floodplain forests, thickets, marshes, calcareous fens, ditches; also in moist to dry open woods and roadsides.

STATUS MW-FACW | NCNE-FACW | GP-FAC

NOTE Canada goldenrod (*Solidago canadensis*) is a common species that sometimes occurs in moist to wet open areas (as well as drier places). It is similar to smooth goldenrod but plants generally smaller and densely short-hairy on leaf undersides and upper stems.

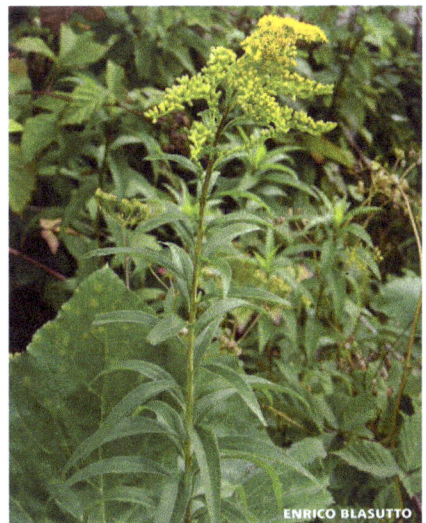

ENRICO BLASUTTO

Solidago gigantea

Solidago houghtonii T. & G.
HOUGHTON'S or GREAT LAKES GOLDENROD

DESCRIPTION Native perennial herb, from a crown. Stems slender, 2-5 dm long, smooth below, short-hairy in the inflorescence. Leaves alternate, largest at base of plant, becoming smaller upward, linear, to 20 cm long and 1-2 cm wide; margins entire but rough-to-touch; lower leaves tapered to a petiolelike base, petioles absent on upper leaves. Flower heads often many, crowded in flat-topped to rounded clusters at ends of stems, with yellow disk and ray flowers; involucre 5-8 mm high, the involucral bracts rounded at tip. Fruit a smooth, angled achene; pappus of slender, sometimes feathery hairs. Flowering July-Sept.

SYNONYM *Oligoneuron houghtonii.*

HABITAT moist flats and depressions between dunes near n shores of Lakes Huron and Michigan, also in nearby fens, limestone pavements on Drummond Island, Mich; usually where calcium-rich.

STATUS NCNE-OBL; Mich (THR), w NY, and s Ont is total range for this species (federally listed as threatened).

Solidago ohioensis Riddell
♦ OHIO GOLDENROD

DESCRIPTION Native perennial herb, from a crown. Stems 5-10 dm long, smooth. Leaves alternate, largest at base of plant and becoming smaller upward, lance-shaped to oblong lance-shaped, to 2 dm long and 1-5 cm wide, pinnately-veined, margins entire or slightly toothed near tip, rough-to-touch; tapered to a long petiole on lower leaves, upper leaves stalkless. Flower heads many in a branched, flat-topped to rounded inflorescence at ends of stems, with yellow disk and ray flowers; involucre smooth, 4-5 mm high, the involucral bracts rounded at tip. Fruit a smooth, angled achene; pappus of slender white bristles. Flowering July-Sept.

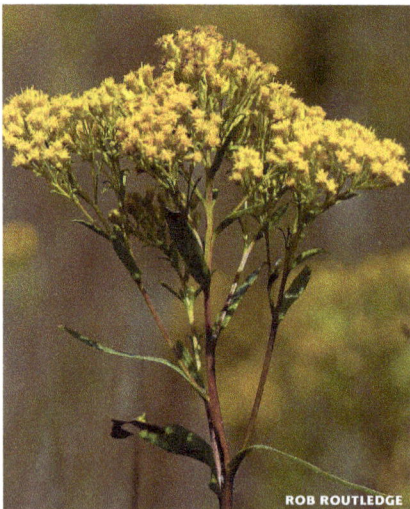

SYNONYM *Oligoneuron ohioense.*

HABITAT wet, sandy or gravelly shores, streambanks, sedge meadows, calcareous fens, low prairie; soils often calcium-rich.

STATUS MW-OBL | NCNE-OBL

Solidago patula Muhl.
♦ ROUGH-LEAVED GOLDENROD

DESCRIPTION Native perennial herb, from a crown. Stems 5-20 dm long, lower stem smooth, strongly angled, upper stem short-hairy. Leaves alternate, largest at base of plant, 1-3 dm long and 4-10 cm wide, becoming smaller upward, oval to ovate, pinnately veined, upper surface rough-to-touch, underside smooth; margins with large, forward-pointing teeth; petioles long and winged on lower leaves, upper leaves stalkless. Flower heads in a paniclelike head, the branches spreading and curved downward at tip, flower heads mostly on 1 side of branches, with yellow disk and ray flowers; involucre 3-4 mm high, the involucral bracts tapered to a sharp or rounded tip. Fruit a sparsely hairy achene; pappus of slender white bristles. Flowering

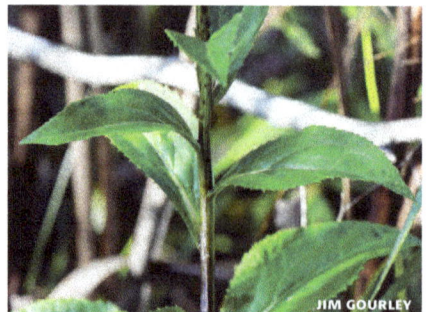

Solidago ohioensis

Solidago patula

Aug–Sept.

HABITAT swamps, thickets, calcareous fens, sedge meadows.

STATUS MW-OBL | NCNE-OBL | GP-OBL

Solidago riddellii Frank
RIDDELL'S GOLDENROD

Native perennial herb, from a crown and sometimes also with rhizomes. **Stems** 2–10 dm long, smooth but sometimes sparsely hairy in head. **Leaves** alternate, smooth, largest at base of plant, these often early-deciduous, lance-shaped to linear, 10–20 cm long and 5–30 mm wide, becoming smaller upward, the upper leaves sickle-shaped and folded along midrib; margins entire; petioles of lower leaves long and winged, upper leaves stalkless and clasping stem. **Flower heads** many, crowded in a branched, rounded to flat-topped inflorescence, the heads not confined to 1 side of the branches, with yellow disk and ray flowers, the rays 1–2 mm long; involucre 5–6 mm high, the involucral bracts rounded at tip. **Fruit** a smooth achene, 1–2 mm long; pappus of slender bristles. **Flowering** Aug–Oct.

SYNONYM *Oligoneuron riddellii*.

HABITAT wet meadows, calcareous fens, low prairie, lakeshores, streambanks.

STATUS MW-OBL | NCNE-OBL | GP-OBL

Solidago uliginosa Nutt.
♦ NORTHERN BOG GOLDENROD

DESCRIPTION Native perennial herb, from a branched crown. **Stems** stout, 5–15 dm long, smooth but finely hairy in the head. **Leaves** alternate, largest at base of plant, 5–35 cm long and 1–5 cm wide, becoming smaller upward, lance-shaped to oblong lance-shaped, smooth, margins finely toothed, or entire on upper leaves, rough-to-touch; lower leaves tapered to long petioles, somewhat clasping stem, upper leaves stalkless. **Flower heads** in a long, crowded spikelike inflorescence, the branches ascending, straight or curved downward at tip, the heads sometimes mostly on 1 side of branches, with yellow disk and ray flowers; involucre 3–5 mm high, the inner involucral bracts rounded at tip, the outer often acute. **Fruit** a ± smooth achene; pappus of slender bristles. **Flowering** Aug–Sept.

HABITAT conifer swamps, fens, open bogs, low prairie, wet meadows, interdunal wetlands, Lake Superior rocky shore.

STATUS MW-OBL | NCNE-OBL | GP-OBL

Symphyotrichum
WILD ASTER

Mostly perennial herbs. Leaves simple, alternate. Flower heads with both ray and disk flowers (disk flowers only in *S. ciliatum*); ray flowers white, pink, blue or purple, usually more than 0.5 mm wide (in contrast to the fleabanes, *Erigeron*); disk flowers red, purple or yellow; involucral bracts in 2 or more series, usually over-lapping; receptacle naked (not chaffy), ± flat; pappus of numerous hairlike bristles.

NOTE genus *Symphyotrichum* includes most species formerly included in *Aster*.

Symphyotrichum **WILD ASTER**

1 Upper stem leaves stalkless, the base of leaf clasping stem; involucral bracts sometimes with gland-tipped hairs **2**

1 Upper stem leaves not clasping; involucral bracts without gland-tipped hairs **3**

2 Leaf margins entire; involucral bracts and flower stalks with glands ....
.... *S. novae-angliae*

2 Leaf margins usually with at least some teeth; involucral bracts without glands ...
.... *S. puniceum*

3 Plants annual; rays absent or less than 2 mm long **4**

3 Plants perennial; rays present and larger **5**

4 Rays absent; leaves less than 4 mm wide, with small, rough prickles on margins ....
.... *S. ciliatum*

RYAN HODNETT

Solidago uliginosa

4 Rays present but very small; leaves more than 6 mm wide; leaf margins smooth.... .... *S. subulatum*

5 Involucral bracts covered with hairs...... .... *S. ontarionis*

5 Involucral bracts smooth, or only fringed with hairs on margins **6**

6 Involucral bracts tipped with a short sharp spine .... *S. pilosum*

6 Involucral bracts not spine-tipped **7**

7 Leaf undersides net-veined, the veins dark green, the spaces enclosed by the veins paler and not elongate; rays blue .... *S. praealtum*

7 Leaf undersides not distinctly net-veined with paler spaces, or if somewhat net-veined, the spaces elongate; rays mostly white to pink.......................... **8**

8 Leaves hairy, at least on underside midvein; disk flowers deeply lobed (to half or more the length of expanded portion of corolla above the tube) **9**

8 Leaf undersides ± smooth; disk flowers shallowly lobed (to 1/3 or less the length of expanded portion of corolla above tube) **10**

9 Leaf underside smooth except for hairs on midvein; stems in a clump from a crown.. .... *S. lateriflorum*

9 Leaf underside covered with short hairs; stems single from slender rhizomes .... *S. ontarionis*

10 Flower heads less than 1.5 cm wide, on stalks 1–5 cm long .... *S. dumosum*

10 Flower heads 1.5–3 cm wide, on stalks less than 1 cm long....................... **11**

11 Stems less than 2 mm wide; leaves less than 6 mm wide, margins often rolled under; flowers in clusters of 1–15 heads, the flower stalks spreading.............. *S. borealis*

11 Stem 3 mm or more wide; leaves more than 6 mm wide, margins flat; flowers in clusters of 20 heads or more, the flower stalks usually ascending . *S. lanceolatum*

Symphyotrichum boreale (Torr. & Gray)
A. & D. Löve ◆ NORTHERN BOG-ASTER

DESCRIPTION Native perennial herb, from rhizomes 1–2 mm wide. Stems erect, slender, 3–8 dm tall and to 2 mm wide, unbranched below, usually branched in the head; smooth except for lines of short, appressed hairs below base of upper leaves.

Leaves alternate, linear, 4–12 cm long and 2–6 mm wide, sometimes slightly clasping at base, margins rough-to-touch, petioles absent. Flower heads usually few to rarely many, in an open, broad inflorescence; the heads 1.5–2 cm wide; involucre 5–7 mm high, the involucral bracts overlapping, often purple at tips and on margins; ray flowers 20–50, white to light blue or lavender, 1–1.5 cm long. Fruit an achene; pappus of pale hairs. Flowering Aug–Sept.

SYNONYMS *Aster borealis*, *Aster junciformis*.
HABITAT conifer swamps, calcareous fens, open bogs, wet meadows, shores, seeps.
STATUS MW-OBL | NCNE-OBL | GP-OBL

Symphyotrichum ciliatum (Ledeb.)
Nesom RAYLESS ASTER

DESCRIPTION Taproot-ed, introduced annual herb. Stems un-branched and erect, to branched and spreading, 2–6 dm long, smooth. Leaves alternate, linear, 2–10 cm long and mostly 2–5 mm wide, margins fringed with scattered hairs, petioles absent. Flower heads several to many, in an open inflorescence which forms much of plant; flower heads bell-shaped, 1–2 cm wide, involucre 5–10 mm high, the involucral bracts mostly green, linear, of equal length or slightly overlapping; ray flowers absent. Fruit a flattened achene, 1–2 mm long; pappus of many long, soft hairs. Flowering Aug–Sept.

SYNONYMS *Aster brachyactis*, *Brachyactis angusta*, *Brachyactis ciliata*.

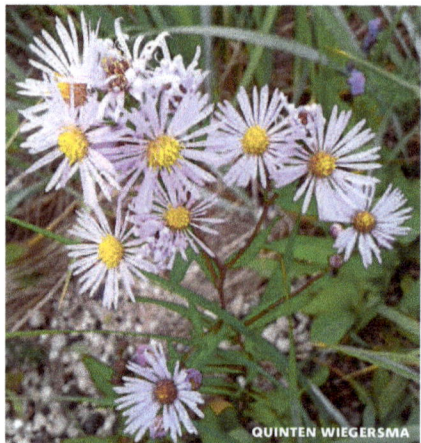

Symphyotrichum boreale

HABITAT shores (including along Great Lakes), streambanks, wet meadows, roadside ditches, usually where brackish. **STATUS** MW-FAC | NCNE-FAC | GP-FACW; considered native in Minn; introduced from western North America in Wisc and Mich.

Symphyotrichum dumosum (L.)
Nesom BUSHY ASTER

DESCRIPTION Native perennial herb, spreading by long or short rhizomes. **Stems** 3–10 dm long, branched above, smooth or with fine hairs on upper stem. **Leaves** alternate, all from stem, linear to narrowly lance-shaped, 3–10 cm long and to 1 cm wide, much smaller and bractlike on branches, upper surface usually rough-to-touch, underside smooth, margins rough-to-touch, petioles absent. **Flower heads** many in an open, branched inflorescence, on stalks 2 cm or more long; involucre 4–6 mm long, smooth, the involucral bracts broad and green at tip, overlapping; ray flowers 15–30, usually white, rarely blue to pale violet, 5–10 mm long. **Fruit** a hairy achene; pappus white. **Flowering** Aug–Oct.
SYNONYM *Aster dumosus*.
HABITAT moist to wet sandy or mucky shores, interdunal swales, sedge meadows, sometimes where calcium-rich; also in drier oak and jack pine woods.
STATUS MW-FAC | NCNE-FAC | GP-FAC

Symphyotrichum lanceolatum
(Willd.) Nesom ♦ EASTERN LINED ASTER

DESCRIPTION Native perennial herb, forming colonies from long rhizomes. **Stems** 0.5–1.5 m long, upper stem with lines of hairs. **Leaves** alternate, all on stem, lance-shaped to linear, 8–15 cm long and 3–30 mm wide, upper surface smooth or slightly rough-to-touch, margins toothed or sometimes entire; petioles absent or blades tapered to petiolelike base, sometimes slightly clasping stem. **Flower heads** many in an elongate leafy inflorescence; the involucre 3–6 mm high, the involucral bracts tapered to a green tip, smooth or margins fringed with hairs, strongly overlapping; ray flowers 20–40, usually white, sometimes lavender or blue, 4–12 mm long. **Fruit** an achene;

pappus white. **Flowering** Aug–Oct.
SYNONYMS *Aster hesperius, Aster interior, Aster lanceolatus, Aster paniculatus, Aster simplex.*
HABITAT marshes, wet meadows, fens, swamp openings, low prairie, streambanks and shores.
STATUS MW-FAC | NCNE-FACW | GP-FACW
NOTE one of the region's most common asters, found in a wide variety of habitats.

Symphyotrichum lateriflorum (L.)
A.& D. Löve
♦ CALICO ASTER, GOBLET ASTER

DESCRIPTION Native perennial herb, from a branched base or short rhizome. **Stems** 3–14 dm long, smooth to finely hairy. **Leaves** from base of plant and along stem; basal leaves ovate, with a long petiole; stem leaves alternate, lance-shaped, 5–15 cm long and 0.5–3 cm wide, branch leaves much smaller; upper surface rough-to-touch or smooth, underside smooth except for finely hairy midvein; margins with forward-pointing teeth or sometimes entire; petioles short or absent; basal and lower stem leaves usually soon deciduous. **Flower heads** many, mostly on 1 side of inflorescence branches; involucre smooth, 4–6 mm long, the involucral bracts overlapping in 3–4 series, with a green-purple tip; ray flowers 9–15, white or pale purple, 4–7 mm long. **Flowering** Aug–Oct.

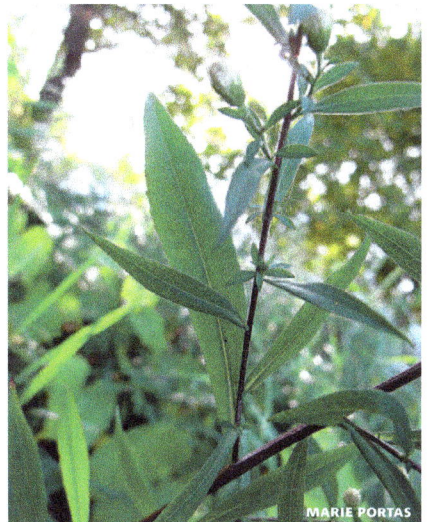

MARIE PORTAS

Symphyotrichum lanceolatum

SYNONYMS *Aster agrostifolius, Aster hirsuti-caulis, Aster lateriflorus.*

HABITAT swamps, thickets, floodplain forests, shores, streambanks, roadsides, moist hardwood and mixed conifer-hardwood forests; often where shaded.

STATUS MW-FACW | NCNE-FAC | GP-FACW

Symphyotrichum novae-angliae
(L.) Nesom ♦ NEW ENGLAND ASTER

DESCRIPTION Native perennial herb, from a short rhizome or crown. Stems stout, erect, 4–10 dm long, with stiff, spreading, sometimes gland-tipped hairs. Leaves alternate, lance-shaped, 3–7 cm long and 1–2.5 cm wide, upper surface rough-to-touch or with short hairs, underside soft hairy, base of leaf strongly clasping stem, margins entire, petioles absent. Flower heads several to many, in clusters at ends of branches, 1.5–3 cm wide; involucre 7–12 mm high, the involucral bracts awl-shaped, glandular-hairy, sometimes purple; ray flowers 40 or more, blue-violet to less often red or pink, 1–2 cm long. Fruit a hairy achene; pappus red-tinged. Flowering Aug–Oct.

SYNONYM *Aster novae-angliae.*

HABITAT wet meadows, low prairie, shores, thickets, calcareous fens, roadsides; usually in moist or wet open areas.

STATUS MW-FACW | NCNE-FACW | GP-FACW

Symphyotrichum ontarionis (Wieg.)
Nesom LAKE ONTARIO ASTER

DESCRIPTION Native perennial herb, from long creeping rhizomes. Stems branched, 3–8 dm long, upper stems with short spreading hairs. Leaves alternate, thin, oblong lance-shaped, 5–10 cm long and 1–3 cm wide (upper leaves smaller), upper surface rough-hairy to nearly smooth, underside finely to densely hairy; margins with sharp, forward-pointing teeth above middle of blade; petioles absent. Flower heads 1–2 cm wide, on short stalks from short leafy branches; involucre smooth to finely hairy, 5–7 mm high, the involucral bracts overlapping; ray flowers white, 9 or more. Fruit an achene; pappus white. Flowering Sept–Oct.

SYNONYM *Aster ontarionis.*

HABITAT floodplain forests, river terraces, thickets.

STATUS MW-FAC | NCNE-FAC | GP-FAC

Symphyotrichum pilosum (Willd.)
Nesom FROST ASTER

DESCRIPTION Native perennial herb, from a large crown. Stems to 1.5 m long, ± smooth or stems and leaves with spreading hairs. Leaves alternate, lower leaves oblong lance-shaped, 5–10 cm long and 1–2 cm wide, stalked; upper leaves small-

Symphyotrichum lateriflorum

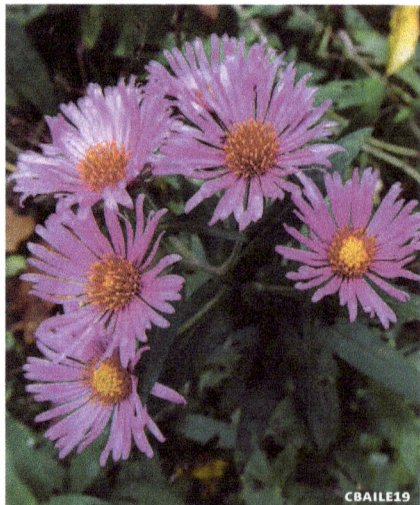

Symphyotrichum novae-angliae

er, linear, stalkless; margins entire or slightly toothed; petioles fringed with hairs; basal leaves and lower stem leaves soon deciduous (or basal leaves persistent). Flower heads at ends of small branches, forming an open inflorescence; involucre urn-shaped, narrowed near middle and flared upward, 3–5 mm high, smooth, involucral bracts overlapping to nearly equal in length, green-tipped; ray flowers 15–35, white. Fruit an achene; pappus white. Flowering Aug–Sept. SYNONYM *Aster pilosus*.

HABITAT sandy and gravelly shores, interdunal swales, wet meadows; often where calcium-rich; sometimes weedy in disturbed fields and roadsides.

STATUS MW-FACU | NCNE-FACU | GP-FACU

Symphyotrichum praealtum (Poir.) Nesom VEINY LINED ASTER

DESCRIPTION Native perennial herb, spreading by rhizomes and forming colonies. Stems to 1 m long, with lines of hairs, especially in upper stem. Leaves alternate, firm, lance-shaped, 6–12 cm long and 1–2 cm wide, upper surface rough-to-touch to nearly smooth, underside smooth or finely hairy, with conspicuous netlike veins surrounding lighter colored areas (areole); margins ± entire; petioles absent, the base of leaves often slightly clasping. Flower heads at ends of short leafy branches; involucre 5–7 mm high, the involucral bracts overlapping; rays 5–15 mm long, blue-purple (rarely white). Fruit an achene; pappus white. Flowering Sept–Oct.

SYNONYMS *Aster nebraskensis, Aster praealtus, Aster woldeni*.

HABITAT wet meadows, low prairie, moist fields, thickets.

STATUS MW-FACW | NCNE-FACW | GP-FACW

Symphyotrichum puniceum (L.) A.& D. Löve ◆PURPLE-STEM ASTER, SHINING ASTER

DESCRIPTION Native perennial herb, from short to long creeping rhizomes. Stems stout, 0.5–2 m long, unbranched, or branched in head, lower stem mostly smooth or sparsely hairy, upper stem and branches with lines of fine hairs. Leaves

alternate, crowded, especially on upper part of stem, often shiny, lance-shaped to oblong, 5–15 cm long and 1–4 cm wide, clasping at base, margins ± entire, petioles absent. Flower heads few to many in a leafy inflorescence; involucre 5–12 mm high, the involucral bracts tapered to a tip; ray flowers 20–60, pale blue or lavender (sometimes white), to 2 cm long. Fruit a smooth or finely hairy achene; pappus nearly white. Flowering Aug–Sept.

SYNONYMS *Aster firmus, Aster lucidulus, Aster puniceus*.

HABITAT swamps, streambanks, shores, sedge meadows, calcareous fens, marshes, thickets, and roadside ditches; sometimes forming large colonies.

STATUS MW-OBL | NCNE-OBL | GP-OBL

Symphyotrichum subulatum (Michx.) Nesom ANNUAL SALTMARSH ASTER

DESCRIPTION Introduced annual herb, from a short taproot; plants smooth and fleshy. Stems to 1 m long. Leaves alternate, linear to lance-shaped, to 20 cm long and 1 cm wide; margins ± entire; petioles absent, base sometimes slightly clasping; bracts in head much smaller and awl-shaped. Flower heads usually many in an open inflorescence; involucre 5–8 mm high, the involucral bracts overlapping in 3–4 series, often purple-tinged at tips and margins; rays 20–30, small, blue-purple.

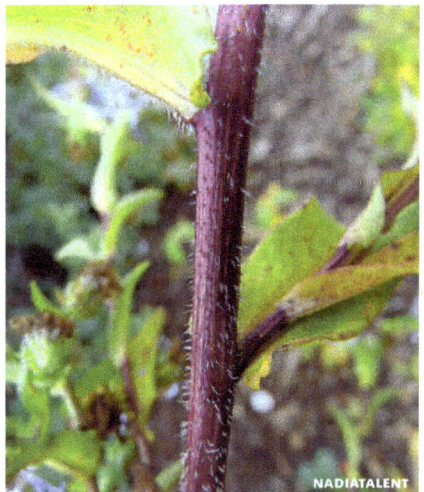

Symphyotrichum puniceum

Fruit a compressed achene; pappus white. **Flowering** Sept–Oct. **SYNONYM** *Aster subulatus*. **HABITAT** ditches along salted highways. **STATUS** MW-OBL | NCNE-FACW | GP-OBL; local in se Mich where considered introduced; main range along Atlantic coast in s USA where in salt marshes.

Tephroseris SQUAW-WEED

Tephroseris palustris (L.) Reichenb.
♦CLUSTERED MARSH SQUAW-WEED

DESCRIPTION Native annual or biennial herb. **Stems** stout, single, 2–10 dm long, hollow near base, sparsely to densely hairy. **Leaves** lance-shaped to oblong or the lower spatula-shaped, 5–20 cm long and 0.5–5 cm wide, smooth or hairy, rounded at tip; margins entire to coarsely toothed or cleft; lower leaves stalked, upper leaves stalkless and clasping at base. **Flower heads** 1–1.5 cm wide, usually many in crowded clusters; with both disk and ray flowers, the rays pale yellow, 4–9 mm long; involucre 4–8 mm high, the involucral bracts chaffy near tip. **Fruit** a smooth achene, 1–3 mm long; pappus bristles white, very slender and numerous, lengthening after flowering. **Flowering** May–Aug. **SYNONYM** *Senecio congestus*. **HABITAT** Shores and mud flats.

Tephroseris palustris

STATUS MW-FACW | NCNE-FACW | GP-FACW; historically known from Wisc and Mich.

Verbesina WINGSTEM

Verbesina alternifolia (L.) Britt. ex Kearney WINGSTEM

DESCRIPTION Large native perennial herb. **Stems** usually winged, sometimes round, 1–3 m long, finely hairy to nearly smooth. **Leaves** alternate, lance-shaped or oblong lance-shaped, 10–25 cm long and 2–8 cm wide, becoming smaller above, rough-to-touch; margins with forward-pointing teeth or nearly entire; petioles short. **Flower heads** in open clusters at ends of stems and branches, with both disk and ray flowers, the disk globe-shaped, 1–1.5 cm wide; rays 2–10, yellow, 1–3 cm long; involucral bracts narrow, loosely spreading. **Fruit** a winged (or sometimes wingless) achene; pappus of 2 spreading awns. **Flowering** Aug–Oct. **SYNONYMS** *Actinomeris alternifolia, Ridan alternifolia*. **HABITAT** floodplain forests, thickets, fens, streambanks. **STATUS** MW-FACW | NCNE-FACW | GP-FAC

Vernonia IRONWEED

Vernonia fasciculata Michx.
♦SMOOTH IRONWEED

DESCRIPTION Stout, native perennial herb, from a thick rootstock. **Stems** erect, single or clumped, 5–12 dm long, red or purple, smooth but short-hairy on branches of the head. **Leaves** alternate, lance-shaped, 5–15 cm long and 1–4 cm wide, smooth above, underside finely pitted, margins sharp-toothed, petioles short. **Flower heads** usually many, crowded in flat-topped clusters to 10 cm wide, with purple disk flowers only; involucre 6–9 mm high, the involucral bracts overlapping, green with purple tips; receptacle flat, not chaffy. **Fruit** a ribbed achene, 3–4 mm long; pappus of purple to brown, slender bristles. **Flowering** July–Sept. **HABITAT** marshes, low prairie, streambanks. **STATUS** MW-FACW | NCNE-FACW | GP-FAC

Xanthium COCKLEBUR

Xanthium strumarium L.
◆COMMON COCKLEBUR

DESCRIPTION Weedy, taprooted, native annual herb; plants variable in size and habit, rough-to-touch or sometimes nearly smooth. Stems 2–15 dm long, often brown-spotted. Leaves alternate, ovate to nearly round, sometimes with 3–5 shallow lobes, 3–15 cm long and 2–20 cm wide, margins with blunt teeth; petioles 3–10 cm long. Flower heads either staminate or pistillate, the staminate flowers brown, in clusters of small round heads at ends of stems above the larger pistillate heads; pistillate heads in several to many clusters from leaf axils, each head with 2 flowers, with a spiny involucre enclosing the head; petals absent. Fruit a brown bur formed by the involucre, 1.5–3 cm long, covered with hooked prickles; achenes thick, 1 in each of the 2 chambers of the bur. Flowering Aug–Sept.

HABITAT shores, streambanks, wet meadows, sand bars, dried depressions, often where disturbed; also in cultivated and abandoned fields, roadsides and waste places.

STATUS MW-FAC | NCNE-FAC | GP-FAC; widespread and weedy across most of North America.

Balsaminaceae
Touch-Me-Not Family

Impatiens TOUCH-ME-NOT
Smooth annual herbs with hollow, succulent stems and shallow, weak roots. Leaves simple, alternate, the blades shallowly toothed. Flowers with both staminate and pistillate parts, irregular, yellow to orange-yellow, pouchlike and spurred, hanging from the petioles in few-flowered racemes from upper leaf axils; sepals 3, petal-like; petals 3; stamens 5. Fruit a 5-valved capsule; the mature capsules splitting when jarred or touched, scattering the seeds away from parent plants (however, small cleistogamous, or self-fertile, flowers lacking petals are sometimes produced in summer and are often the only flowers found on plants growing in shaded situations).

Impatiens TOUCH-ME-NOT

1　Flowers orange-yellow, usually with red-brown spots.................*I. capensis*
1　Flowers pale yellow, spots faint or absent.
　.............................*I. pallida*

Impatiens capensis Meerb.
◆ORANGE TOUCH-ME-NOT, JEWEL-WEED

DESCRIPTION Native annual herb. Stems 3–10 dm long, usually branched above. Leaves ovate to oval, 3–9 cm long and 1.5–4 cm wide, tapered to tip or rounded and tipped with a short slender point, margins shallowly and irregularly toothed; petioles longest on lower leaves, shorter upward, 0.5–5 cm long. Flowers orange-yellow, 1.5–3 cm long, usually mottled with red-brown spots, with a spur recurved parallel to the sac and to half its length. Fruit a capsule about 2 cm long, splitting when mature to forcefully eject the seeds. Flowering July–Sept.

Vernonia fasciculata

Impatiens capensis

HABITAT swamps, low areas in woods, floodplain forests, thickets, streambanks, shores, marshes, fens, springs; often where disturbed.
STATUS MW-FACW | NCNE-FACW | GP-FAC

Impatiens pallida Nutt.
YELLOW TOUCH-ME-NOT

DESCRIPTION Native annual herb, similar to **orange touch-me-not** (*I. capensis*) but much less common. *Impatiens pallida* is typically larger, the leaves to 12 cm long and 8 cm wide, and more finely toothed than those of orange touch-me-not. **Flowers** pale yellow, unspotted or with faint redbrown spots, 2–4 cm long, the spur recurved at a right angle to sac, and to 1/4 length of sac. **Flowering** July–Sept.
HABITAT floodplain forests, low spots in woods, swamps, stream-banks, shores; often where somewhat disturbed.
STATUS MW-FACW | NCNE-FACW | GP-FACW

Betulaceae
Birch Family

TREES OR SHRUBS. Leaves deciduous, simple, alternate, with toothed margins and pinnate veins. Flowers small, staminate and pistillate flowers separate on same plant, crowded into catkins (aments) that open in spring before leaves fully open; staminate catkins hang downward; cone-like pistillate catkins erect or drooping. Fruit a small, 1-seeded, winged nutlet.

Betulaceae BIRCH FAMILY
1 Female catkins in loose clusters of several catkins; scales of pistillate catkins persistent, becoming hard and stiff *Alnus*
1 Female catkins single; scales of pistillate catkins soon deciduous . . . . *Betula*

Alnus ALDER
Thicket-forming shrubs, or an introduced tree (*Alnus glutinosa*). Leaves ovate, toothed on margins. Staminate flowers in long, drooping catkins which fall after shedding pollen; pistillate flowers in short, persistent conelike clusters. Fruit a flattened achene with winged margins.

Alnus ALDER
1 Introduced tree mostly in s Upper Midwest region; leaves broadly rounded, tip rounded to blunt or notched . *A. glutinosa*
1 Shrubs; leaves ovate to oval, tapered to a sharp tip; widespread species 2
2 Twigs and young leaves sticky, leaves with small, sharp teeth; catkins on long stalks; winter buds not stalked; fruit broadly winged, northern portion of region only . . . . *A. viridis*
2 Twigs and young leaves not sticky, leaves unevenly double-toothed; catkins stalkless or on short stalks; winter bud stalked, blunt at tip; fruit narrowly winged; widespread in northern and central areas of region . . . . *A. incana*

Alnus glutinosa (L.) Gaertner
BLACK ALDER

DESCRIPTION Introduced tree to 20 m tall; twigs, young leaves and fruits sticky-to-touch. **Leaves** oval to nearly round, tip rounded or with a small notch, veins 5–8 on each side of midvein, dark green and ± shiny above, paler below, margins finely toothed. Staminate and pistillate flowers separate on same tree; pistillate catkins drooping from leaf axils, 1.5–2.5 cm long and 10–12 mm wide. **Flowering** April–May.
SYNONYMS *Alnus alnus, Alnus vulgaris.*
HABITAT floodplain forests, riverbanks; also in drier places; introduced from Eurasia and planted as an ornamental; occasionally escaping and naturalized.
STATUS MW-FACW | NCNE-FACW

Alnus incana (L.) Moench
♦SPECKLED ALDER, TAG ALDER

DESCRIPTION Thicket-forming native shrub to 5 m tall; twigs red-brown, waxy, with conspicuous pale lenticels. **Leaves** ovate to oval, broadest near or below middle, 6–14 cm long and 4–7 cm wide, dark green and smooth above, paler and hairy below; margins sharply toothed and shallowly lobed; petioles 1–2.5 cm long. **Flowers** in catkins clustered at ends of branches; staminate catkins developing in late summer, short-stalked, elongate, 4–9 cm long; pistillate

catkins appear in late summer, stalkless, rounded, 1–2 cm long and to 1 cm wide, the scales unlobed, becoming conelike, persistent. Fruit a flat nutlet, narrowly winged on margin, 2–4 mm long. Flowering April–June.

SYNONYM *Alnus rugosa*.

HABITAT swamps, thickets, bog margins, shores and streambanks.

STATUS MW-FACW | NCNE-FACW | GP-FACW

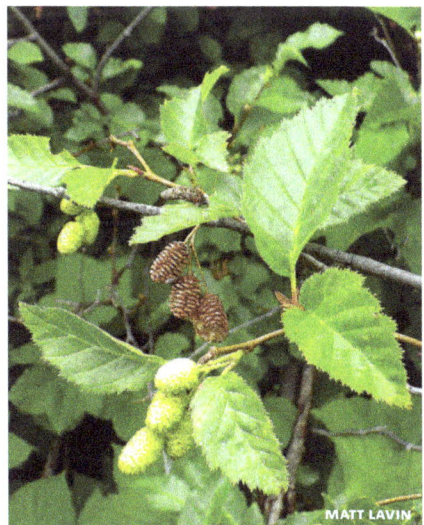

Alnus viridis (Villars) Lam.
♦ GREEN or MOUNTAIN ALDER

DESCRIPTION Thicket-forming native shrub to 4 m tall; bark red-brown to gray; twigs brown, sticky, somewhat hairy, lenticels pale and scattered. Leaves round-oval, bright green above, slightly paler and shiny below, sticky when young, margins wavy with small, sharp teeth; petioles 6–12 mm long. Flowers in catkins; staminate catkins stalked, slender, developing in late summer and expanding in spring; pistillate catkins appear in spring, becoming long-stalked, blunt and conelike, persistent, 1–2 cm long. Fruit a nutlet, 2–3 mm long, with a pale, thin wing.

SYNONYMS *Alnus alnobetula*, *Alnus crispa*, *Alnus mollis*.

HABITAT lakeshores, wet depressions in woods, rock outcrops, beaches along Lake Superior.

STATUS MW-FAC | NCNE-FAC | GP-FAC

Betula BIRCH

Trees or shrubs (often with many stems from base); bark sometimes peeling in thin layers. Leaves deciduous, alternate, sharply toothed. Staminate and pistillate flowers separate on same plant, catkins appearing in fall, opening the following spring, staminate flowers in drooping slender catkins; pistillate flowers in erect conelike catkins. Fruit an achene with a winged margin.

ADDITIONAL SPECIES Paper birch (*Betula papyrifera*) a common tree across most of the Upper Midwest region, is sometimes found on hummocks in swamps and on wetland margins. It is identified by its characteristic white papery bark marked by horizontal lenticels.

Betula BIRCH

1 Shrub to 2 m tall; bark not shredding; leaves to 5 cm long . . . .*B. pumila*
1 Small to large trees; bark shredding with age 2
2 Bark yellow-gray; leaves rounded at base, margins not wavy-toothed; widespread . . . .*B. alleghaniensis*
2 Bark red-brown; leaves wedge-shaped at base, margins wavy-toothed; absent from Mich . . . .*B. nigra*

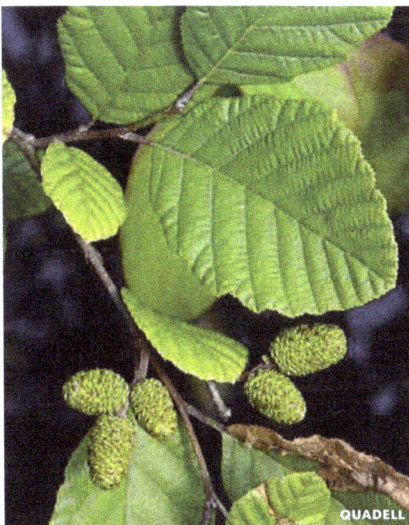

Alnus incana

Alnus viridis

Betula alleghaniensis Britton
♦ YELLOW BIRCH

DESCRIPTION Medium to large native tree, to 25 m. Bark on young trees thin and smooth with conspicuous horizontal lenticels, becoming yellow-gray and shredding into thin, shaggy horizontal strips; bark of old trees breaking into large plates. Twigs hairy when young, becoming smooth and shiny, wintergreen-scented when crushed. Leaves alternate, simple, ovate, tapered to a short, sharp tip, dark green above, paler yellow-green below, 6–12 cm long, margins coarsely double-toothed, petioles grooved and hairy. Staminate and pistillate flowers in catkins, separate on same tree, appearing before leaves in spring; staminate catkins drooping, yellow-purple, 7–10 cm long; pistillate catkins erect, green, 2–4 cm long, ± stalkless. Fruit a winged nutlet, 3–5 mm wide. Flowering April–May.

HABITAT north in our region, occasional in swamps, thickets, streambanks and forest depressions with red maple (*Acer rubrum*), black ash (*Fraxinus nigra*), black spruce (*Picea mariana*), eastern hemlock (*Tsuga canadensis*) and speckled alder (*Alnus incana*); more common in moist forests with sugar maple (*Acer saccharum*) and American beech (*Fagus grandifolia*). Southward, mostly confined to deciduous swamps with red maple, black ash and tamarack (*Larix laricina*).

STATUS MW-FAC | NCNE-FAC | GP-FACU

Betula nigra L. RIVER BIRCH

DESCRIPTION Small or medium native tree to 20 m tall, trunk to 6 dm wide, crown rounded. Bark red-brown, shredding and curly; twigs slender, red-brown; buds pointed, hairy. Leaves alternate, simple, ovate, 4–8 cm long, upper surface smooth, lower surface paler and densely hairy; margins coarsely double-toothed, except untoothed near base; petioles with woolly hairs. Flowers staminate and pistillate flowers small, separate but on same tree; staminate flowers in slender drooping clusters; pistillate flowers in short, woolly clusters. Fruit a small hairy nutlet with a 3-lobed, winged margin, crowded in a cylindrical cone 1.5–3 cm long. Flowering May.

HABITAT floodplain forests, riverbanks, swamps; along Miss and Minn Rivers in Minn, Wisc.

STATUS MW-FACW | NCNE-FACW | GP-FACW

Betula pumila L. ♦ BOG BIRCH

DESCRIPTION Native shrub 1–3 m tall. Bark dull gray or brown. Twigs gray, short-hairy and dotted with resin glands, becoming red-brown and waxy with age. Leaves leathery, rounded to obovate, 2–4 cm long and 1–3 cm wide, dark green above, paler and often waxy below; margins coarsely toothed, the teeth blunt

Betula alleghaniensis

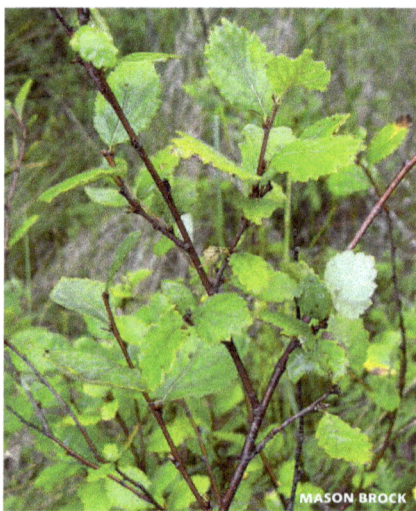

Betula pumila

or sharp; petioles 3–6 mm long. Flowers in catkins; staminate catkins stalkless, cylindric, 15–20 mm long and 2–3 mm wide; pistillate catkins stalked, cylindric, 1–2 cm long and 5 mm wide; scales 3-lobed. Fruit a flat, winged, rounded nutlet, 2–3 mm long and 2–4 mm wide. Flowering May.

SYNONYM *Betula glandulosa* var. *glandulifera*.

HABITAT swamps, bogs, fens, seeps; often where calcium-rich.

STATUS MW-OBL | NCNE-OBL | GP-OBL

Boraginaceae
Borage Family

ANNUAL OR PERENNIAL HERBS with usually bristly stems and alternate, bristly leaves. Flowers typically in a spirally coiled, spikelike head that uncurls as flowers mature; flowers perfect (with both staminate and pistillate parts), with 5 petals, 4–5 sepals, and 5 stamens. Fruit a dry capsule with 4 nutlets.

Boraginaceae **BORAGE FAMILY**

1 Flowers tubelike, the petal lobes erect or slightly spreading ....*Mertensia*

1 Flowers tubelike, but petal lobes abruptly flared and flattened ....*Myosotis*

Mertensia BLUEBELL

Perennial herbs; plants smooth or hairy. Leaves alternate and entire. Flowers usually blue (pink in bud), tube-, funnel- or bell-shaped, petals widened and shallowly lobed at tip; in small clusters at ends of stems and branches. Fruit a smooth or wrinkled nutlet.

Mertensia **BLUEBELL**

1 Leaves and sepals hairy ..*M. paniculata*

1 Leaves and sepals without hairs ....
.........................*M. virginica*

Mertensia paniculata (Aiton) G. Don.
NORTHERN BLUEBELLS

DESCRIPTION Native perennial herb. Stems erect, 3–10 dm long, branched above, smooth or with sparse hairs. Basal leaves ovate, rounded at base; stem leaves lance-shaped to ovate, 5–15 cm long, tapered to a tip, hairy, entire; petioles short on lower leaves, upper leaves ± stalkless. Flowers blue-purple, narrowly bell-shaped, 10–15 mm long, on slender stalks, in few-flowered racemes at ends of stems and branches; sepal lobes lance-shaped, 3–6 mm long, with dense, short hairs. Fruit a nutlet. Flowering June–July.

HABITAT conifer swamps, streambanks, seeps.

STATUS MW-FAC | NCNE-FAC | GP-FAC; historical records from Mich UP.

Mertensia virginica (L.) Pers.
♦EASTERN BLUEBELLS

DESCRIPTION Native perennial herb; plants smooth. Stems upright, 3–7 dm long. Leaves oval to obovate, entire, 5–15 cm long, rounded or blunt at the tip; upper leaves stalkless, lower leaves with winged petioles. Flowers showy, blue-purple, trumpet-shaped, 5-lobed at tip, 2–3 cm long, stalked, in a cluster at end of stem; sepals rounded at tip, 3 mm long. Fruit a nutlet. Flowering April–May.

HABITAT floodplain forests, moist deciduous forests, streambanks; sometimes escaping from gardens where grown as an ornamental.

STATUS MW-FACW | NCNE-FAC; Mich (END).

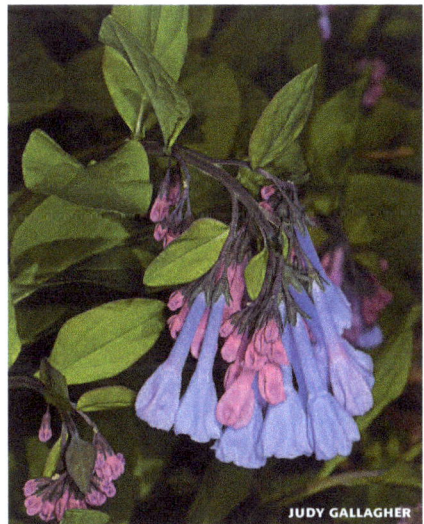

JUDY GALLAGHER

Mertensia virginica

Myosotis
FORGET-ME-NOT,
SCORPION-GRASS

Perennial (sometimes annual) herbs; plants with short, appressed hairs. Leaves alternate and entire. Flowers blue, tube-shaped and abruptly flared outward at tip, in a 1-sided raceme. Fruit a nutlet.

Myosotis
FORGET-ME-NOT, SCORPION-GRASS

1 Plants without stolons; lobes of sepals as long or longer than corolla tube; flowers up to 6 mm wide; nutlets longer than style . . . . *M. laxa*
1 Plants creeping and spreading by stolons; lobes of sepals shorter than corolla tube; flowers mostly 6 mm or more wide; nutlets shorter than style . . . . *M. scorpioides*

Myosotis laxa Lehm.
SMALLER FORGET-ME-NOT

DESCRIPTION Short-lived native perennial (sometimes annual) herb. **Stems** slender, 1–4 dm long, often lying on ground at base, but not creeping, with fine, short, appressed hairs. **Leaves** oblong or spatula-shaped, 2–6 cm long. **Flowers** blue, on stalks usually much longer than the flower, in 1-sided clusters at ends of stems; sepals covered with short hairs, sepal lobes shorter than the tube; petal lobes shorter or slightly longer than the tube. **Fruit** a nutlet distinctly longer than the style. **Flowering** June–Sept.

HABITAT Cedar swamps, wet shores and streambanks.

STATUS MW-OBL | NCNE-OBL | GP-OBL

Myosotis scorpioides L.
♦WATER SCORPION-GRASS

DESCRIPTION Colony-forming, introduced perennial herb. **Stems** 2-6 dm long, with short, appressed hairs, often creeping at base and producing stolons. **Leaves** 3-8 cm long and 0.5-2 cm wide, lower leaves oblong lance-shaped, upper leaves oblong or oval; stalkless or the lower leaves on short petioles. **Flowers** blue with a yellow center, tube-shaped, abruptly flared

at tip, in a 1-sided raceme at ends of stems; flower stalks spreading in fruit; sepals with short, appressed hairs, sepal lobes equal or shorter than the tube. **Fruit** a nutlet shorter than the style. **Flowering** May–Sept.

SYNONYM *Myosotis palustris.*

HABITAT streambanks, shores, ditches, swamps, wet depressions in forests; introduced and naturalized throughout much of USA and s Canada.

STATUS MW-OBL | NCNE-OBL | GP-OBL

Brassicaceae
Mustard Family

ANNUAL, BIENNIAL, OR PERENNIAL HERBS. Leaves simple or compound, alternate on stems or basal, smooth or hairy, some species with branched hairs. Flowers in terminal or lateral clusters (racemes), the lower portion often fruiting while tip in flower, the stalks elongating in fruit. Flowers perfect, cross-shaped, with 4 sepals and 4 yellow, white, pink or purple petals; stamens 6, the outer 2 stamens shorter than the inner 4; pistil 1, style 1, ovary superior. Fruit a cylindrical (silique) or round (silicle) pod with 2 chambers and 1 to many seeds in 1 or 2 rows in each chamber.

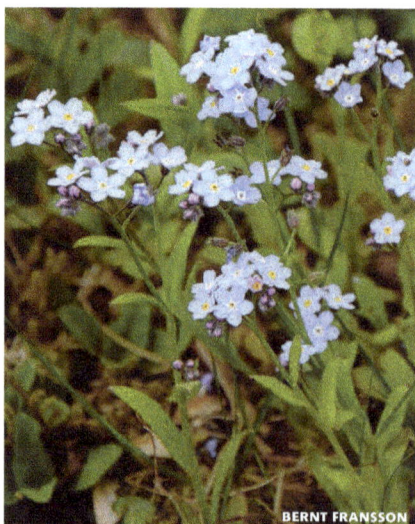

BERNT FRANSSON

Myosotis scorpioides

Brassicaceae **MUSTARD FAMILY**

1 Flowers yellow **2**
1 Flowers white, pink, or purple **3**
2 Leaf segments rounded or broadly oval in outline, the terminal segment much larger than lateral segments; plants without hairs .... *Barbarea orthoceras*
2 Leaf segments tapered to a tip, the terminal segments about same size as lateral segments; plants hairy or smooth *Rorippa*
3 Petals pink or purple **4**
3 Petals white or green **5**
4 Plants smooth throughout.............. .... *Iodanthus pinnatifidus*
4 Plants hairy, at least on lower stems or leaves .... *Cardamine*
5 Leaves all basal, linear and entire, plants usually flowering and fruiting underwater; uncommon aquatic plant of ne Minn, Isle Royale and UP of Mich .... *Subularia aquatica*
5 Leaves from base of plant and on stems, pinnately divided; widespread.......... **6**
6 Plants aquatic; underwater leaves divided into many linear, threadlike segments, the segments easily detaching from stem .... *Rorippa aquatica*
6 Plants on land or in water; leaves broader, not threadlike **7**
7 Plants often in shallow water, rooting along underwater nodes of stem; basal rosette absent; fruit often curved, on spreading stalks; seeds netlike on surface .... *Nasturtium officinale*
7 Plants of moist sites; rooting only at base, basal rosette usually present on young plants; fruit straight and upright **8**
8 Plants strongly onion- or garlic-scented .. .. *Alliaria petiolata*
8 Plants not strongly scented .. *Cardamine*

Alliaria ALLIARIA

Alliaria petiolata (M. Bieb.) Cavara & Grande
◆ GARLIC-MUSTARD

DESCRIPTION Intro-duced, invasive bienni-al herb, strongly onion-or garlic-scented when crushed (especially in spring and early summer); root slender, white, "s"-shaped at top. First-year plants a rosette of 3-4 round, scallop-margined leaves; leaves remain green over winter.

Second-year plants with 1-2 flowering stems. Stems to 1 m long. Leaves narrowly oblong to ovate; petioles long on lower leaves; stem leaves alternate, triangular-shaped, margins with large teeth; leaves at base long-petioled; petioles on stem leaves becoming shorter. Flowers many; petals white, 4. Fruit a slender capsule (silique) 3-6 cm long, with a single row of black seeds. Flowering May.
HABITAT moist, shaded forest openings, streambanks, roadsides, floodplain forests; soils not highly acidic.
STATUS MW-FAC | NCNE-FACU | GP-FACU

Barbarea WINTER-CRESS

Barbarea orthoceras Ledeb.
NORTHERN WINTER-CRESS

DESCRIPTION Native bi-ennial herb; plants smooth or with sparse covering of unbranched hairs. Stems 3-8 dm long, unbranched, or branched above. Leaves simple or with 1-4 pairs of lateral lobes, the middle and upper leaves deeply lobed. Flow-ers in racemes; on short stalks to 1 mm long, the stalks clublike at tip; petals yellow, 3-5 mm long. Fruit upright, 2-4 cm long, with a beak 0.5-2 mm long. Flowering June-July.
HABITAT rocky shores, swamps and wet woods.
STATUS MW-OBL | NCNE-OBL | GP-OBL

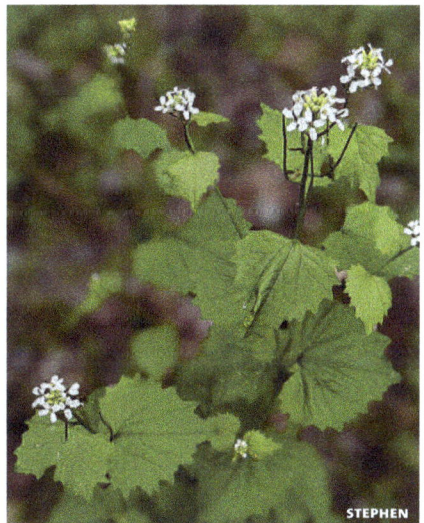

Alliaria petiolata

Cardamine
BITTER-CRESS, TOOTHWORT

Annual, biennial, or perennial herbs, smooth or with short hairs near base of stem. Leaves simple to pinnately divided, the basal leaves often different in shape than stem leaves. Flowers in racemes or umbel-like clusters; sepals green to yellow, early deciduous; petals white. Fruit a 2-chambered, linear pod (silique), the seeds in a single row in each chamber.

Cardamine **BITTER-CRESS, TOOTHWORT**

1 Stem leaves simple or with 1-2 small lobes only. **2**
1 Stem leaves pinnately dissected, with 2 or more deep lobes **3**
2 Petals pink to purple (rarely white); sepals purple, turning brown with age . . . . *C. douglassii*
2 Petals white (rarely pink); sepals green, turning yellow with age . . . . . . *C. bulbosa*
3 Plants annual or biennial; petals 2-4 mm long. . . . . *C. pensylvanica*
3 Plants perennial; petals 5 mm or more long . . . . *C. pratensis*

Cardamine bulbosa (Schreb. ex Muhl.)
Britton, Sterns & Poggenb. ◆SPRING-CRESS

DESCRIPTION Native perennial herb. **Stems** 1 to several from a short thick tuber, unbranched or with a few branches above, 2-6 dm long, smooth or with short hairs on lower stems. **Leaves** simple, sparsely to densely covered with short hairs; basal leaves round or heart-shaped, on long petioles, withering before plants in full flower; stem leaves 4-8, oblong to oval, 2-7 cm long and 0.5-2.5 cm wide; petioles shorter upward on stem. **Flowers** in racemes; sepals green, turning yellow after flowering, 2-4 mm long; petals white (rarely pink), 6-15 mm long. **Fruit** a silique, 1-2.5 cm long and 1-2 mm wide, with a style beak 2-4 mm long, on spreading stalks 1-3 cm long, the pod often falling before mature. **Flowering** May-June. **SYNONYM** *Cardamine rhomboidea*. **HABITAT** wet forest depressions, floodplain forests, streambanks, wet meadows, swamps, calcareous fens. **STATUS** MW-OBL | NCNE-OBL | GP-OBL

Cardamine douglassii Britton
◆PINK SPRING-CRESS

DESCRIPTION Native perennial herb, spreading by shallow rhizomes; plants with dense to sparse hairs. **Stems** to 6 dm long. **Leaves** simple, basal leaves round or heart-shaped, deciduous before plants in full flower; stem leaves 3-5, narrowly oblong to ovate; petioles long on lower leaves, becoming shorter above. **Flowers** in a raceme; sepals purple, turning brown with age; petals pink, purple, or rarely white. **Fruit** an upright silique, 2-3 cm long and 2-3 mm wide, on a stalk 1-2 cm long. **Flowering** April-May. **HABITAT** floodplain forests and low deciduous woods, often in shade. **STATUS** MW-FACW | NCNE-FACW

Cardamine pensylvanica Muhl.
PENNSYLVANIA BITTER-CRESS

DESCRIPTION Native annual or biennial herb. **Stems** erect or spreading, to 6 dm long, usually hairy on lower stem. **Leaves** pinnately divided into 2-5 pairs of lateral leaflets and a single terminal segment, 4-8 cm long and 1-4 cm wide, the leaflets entire or with a few teeth or lobes, the terminal leaflet largest, 1-4 cm long and 1-2 cm wide; petioles shorter than blades,

Cardamine bulbosa

becoming shorter upward. Flowers in a raceme; sepals 1–2 mm long; petals white, 2–4 mm long. Fruit an upright silique, 2–3 cm long and to 1 mm wide, with a style-beak to 2 mm long, on stalks 5–15 mm long. Flowering May–Sept.

SYNONYM *Cardamine parviflora.*

HABITAT streambanks, swamps, and wet forests (often where seasonally flooded); wet, disturbed areas.

STATUS MW-FACW | NCNE-FACW | GP-FACW

Cardamine pratensis L.
CUCKOO-FLOWER

DESCRIPTION Native perennial upright herb. Stems 2–5 dm long. Leaves basal leaves on long petioles, divided into 3–8 broad leaflets, 5–20 mm long, the terminal segment largest and ± entire; lower stem leaves similar to basal ones, becoming shorter and with shorter petioles upward on stem; stem leaves with 7–17 oval to linear leaflets. Flowers in a crowded raceme; petals white, 8–15 mm long. Fruit an upright silique, 2.5–4 cm long, with a style-beak 1–2 mm long, on stalks 8–15 mm long. Flowering May–June.

SYNONYM *Cardamine palustris.*

HABITAT peatlands, tamarack and cedar swamps, wet depressions in forests.

STATUS [not rated]

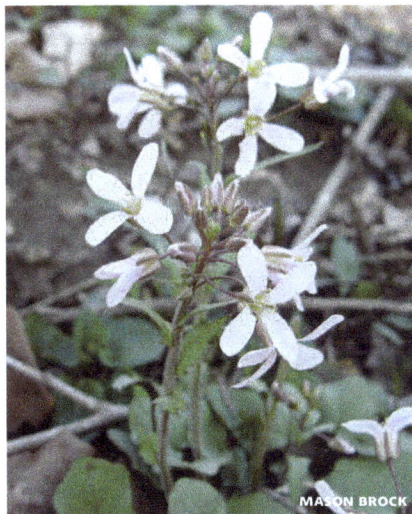

Cardamine douglassii

Iodanthus PURPLE ROCKET

Iodanthus pinnatifidus (Michx.) Steudel
PURPLE ROCKET

DESCRIPTION Native perennial herb; plants smooth. Stems to 1 m long, unbranched except in head. Leaves lance-shaped to oval or oblong, leaf base often with lobes which clasp stem; lower leaves often divided at base into 1–4 pairs of small segments; margins deep-toothed; petioles short. Flowers in a branched raceme, on stalks 5–10 mm long, pale violet to white; sepals rounded at tip, 3–5 mm long; petals 10–13 mm long. Fruit a linear, cylindric silique, 2–4 cm long and 1–2 mm wide, on spreading stalks. Flowering June–July.

HABITAT wet or moist floodplain forests.

STATUS MW-FACW | NCNE-FACW | GP-FACW; Minn (END).

Nasturtium WATERCRESS

Nasturtium officinale R.Br.
♦WATERCRESS

DESCRIPTION Introduced perennial herb; plants smooth. Stems underwater, floating, or trailing on mud; rooting from lower nodes. Leaves 4–12 cm long and 2–5 cm wide, pinnately divided into 3–9 segments, the lateral segments round to ovate in outline, the terminal segment largest; margins entire or with a few shallow rounded teeth; petioles present. Flowers in 1 to several racemes per stem, flat-topped and elongating in fruit; flowers 5 mm wide, sepals green-

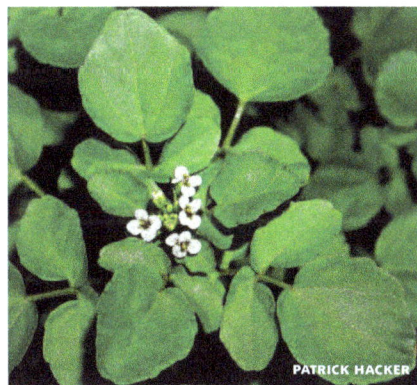

Nasturtium officinale

white, oblong, 1–3 mm long; petals white, sometimes purple-tinged, obovate, 4–5 mm long. Fruit a linear, often curved pod (silique), 1–2.5 cm long and 2 mm wide, tipped with a short style beak to 1 mm long. Flowering May–Sept.

SYNONYM *Rorippa nasturtium-officinale.*

HABITAT springs, slow-moving streams, ditches.

STATUS MW-OBL | NCNE-OBL | GP-OBL; introduced and naturalized across North America, sometimes used as an edible green.

Rorippa YELLOW-CRESS

Annual, biennial or perennial herbs; plants smooth or with unbranched hairs. Leaves sometimes in a basal rosette in young plants, toothed to pinnately divided, petioles short or absent. Flowers small, in racemes at ends of stems or from lateral branches; sepals green to yellow, deciduous by fruiting time; petals yellow or white, shorter to longer than sepals. Fruit a short-cylindric to linear pod (silique), mostly 2-chambered, the seeds in 2 rows.

Rorippa YELLOW-CRESS

1 Plant truly aquatic, the submersed leaves dissected in a bipinnate pattern into filiform segments (midvein present, the lateral segments again dissected), frequently detaching readily from the stem; petals white . . . . *R. aquatica*

1 Plant terrestrial or aquatic but even if in water the leaves with definite flat lobes (not bipinnately dissected) and not falling from the stem; petals yellow **2**

2 Plants annual or biennial, taprooted; petals shorter or equal to sepals **3**

2 Plants perennial, roots creeping; petals longer than sepals **4**

3 Stalks of fruit 3 mm or more long; fruit to 1.5x longer than its stalk . . . . . *R. palustris*

3 Stalks of fruit 1–2 mm long; fruit more than 1.5x longer than its stalk . . . *R. sessiliflora*

4 Stems sprawling or spreading; lateral leaf segments entire or with few shallow teeth; beak of fruit 1–2 mm long . . . . . *R. sinuata*

4 Stems ± erect; lateral leaf segments with sharp teeth; beak of fruit 0.5–1 mm long . . . . *R. sylvestris*

Rorippa aquatica (Eat.) Palmer & Steyermark
LAKE-CRESS

DESCRIPTION Native perennial, fibrous-rooted herb; stems and leaves smooth, usually underwater. Underwater leaves pinnately dissected into many threadlike segments; emersed leaves, if present, lance-shaped, 3–7 cm long, coarsely toothed. Flowers on spreading stalks to 1 cm long; sepals turning upright; petals white, 6–8 mm long. Fruit oval, 5–8 mm long, 1-chambered, tipped by a persistent slender style 2–4 mm long, but apparently rarely maturing. Flowering June–Aug.

SYNONYM *Armoracia aquatica, Armoracia lacustris, Neobeckia aquatica.*

HABITAT quiet water in lakes, rivers and streams, muddy shores.

STATUS MW-OBL | NCNE-OBL | GP-OBL; Wisc (END), Mich (THR).

Rorippa palustris (L.) Besser
♦COMMON YELLOW-CRESS

DESCRIPTION Native annual or biennial herb. Stems erect, usually 1, to 1 m long, unbranched or branched upward. Leaves lance-shaped to obovate, mostly pinnately divided; the blades oblong to oblong lance-shaped, 5–30 cm long and 2–6 cm wide, middle stem leaves usually with basal

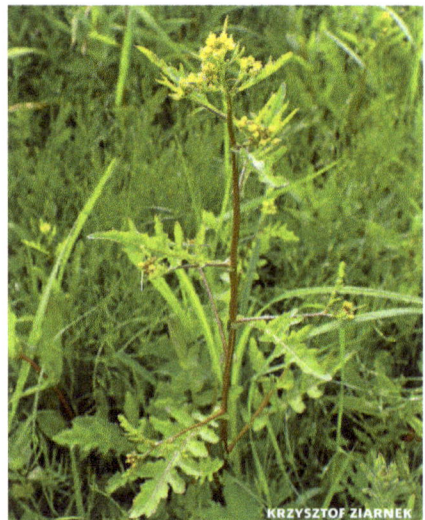

KRZYSZTOF ZIARNEK

Rorippa palustris

lobes and clasping stem, smooth to densely hairy on lower surface; margins deeply lobed and slightly wavy; petioles short or absent. Flowers in racemes at ends of stems and from leaf axils, the terminal raceme flowering and fruiting first, the oldest siliques on lowest portions of raceme; sepals green, 1–3 mm long, early deciduous; petals yellow, drying white, 2–3 mm long. Fruit a round to short-cylindric pod, 3–10 mm long and 1–3 mm wide, straight-sided or slightly tapered to tip, on stalks 3–10 mm long. Flowering June–Sept.

SYNONYM *Rorippa islandica.*

HABITAT marshes, wet meadows, shores, streambanks, ditches and other wet places.

STATUS MW-OBL | NCNE-OBL | GP-OBL

Rorippa sessiliflora (Nutt.) A. S. Hitchc.
SOUTHERN YELLOW-CRESS

DESCRIPTION Native annual (to biennial) herb. Stems erect, 2–4 dm long, branched, smooth. Lower leaves oblong, coarsely round-toothed, lower part of blade usually deeply cleft; upper leaves smaller, ovate, entire or toothed. Flowers in racemes from ends of branches and upper leaf axils; sepals yellow, petals absent; stamens 3–6. Fruit a pod (silique), 6–10 mm long and 3–4 mm wide, often somewhat sickle-shaped, on short, spreading or ascending stalks 1–2 mm long, the style beak very short. Flowering June–July.

SYNONYM *Radicula sessiliflora.*

HABITAT muddy shores and streambanks.

STATUS MW-OBL | NCNE-OBL | GP-OBL; historical records from Minn.

Rorippa sinuata (Nutt.) A. S. Hitchc.
WESTERN YELLOW-CRESS

DESCRIPTION Native perennial herb, spreading by rhizomes. Stems usually several, sprawling, 1–4 dm long, sparsely to densely covered with blunt-tipped hairs. Leaves all from stem (basal leaves absent), 2–8 cm long and 0.5–2 cm wide, oblong, pinnately divided into 5–7 pairs, sometimes with basal lobes clasping stem, margins entire or with a few teeth. Flowers in racemes at ends of stems and from upper leaf axils, all flowering at about same time

or flowers from axils first; sepals yellow-green, 3–5 mm long, early deciduous; petals yellow, 4–6 mm long, longer than sepals. Fruit a linear pod (silique), 5–12 mm long and 1–2 mm wide, tapered to the style beak, on upright to spreading stalks, 4–10 mm long. Flowering June–Aug.

SYNONYM *Radicula sinuata.*

HABITAT stream and riverbanks, ditches, and other low places, especially where sandy.

STATUS MW-FACW | NCNE-FACW | GP-FACW

Rorippa sylvestris (L.) Besser
CREEPING YELLOW-CRESS

DESCRIPTION Introduced perennial herb, spreading by rhizomes and sometimes stolons. Stems erect, branched above, 2–6 dm long, smooth or sparsely hairy on lower stem; basal rosettes present on young plants. Stem Lleaves pinnately divided, oblong in outline, 3–15 cm long and 2.5 cm wide, gradually reduced in size upward on stem, margins usually toothed; petioles present on lower leaves, petioles absent on upper leaves. Flowers in racemes at ends of stems and from upper leaf axils, all flowering at about same time or the oldest siliques on lower portion of terminal racemes; sepals yellow-green, 2–3 mm long; petals yellow, 3–5 mm long, to 2 mm longer than the sepals. Fruit a linear pod (silique), 4–10 mm long and to 1 mm wide, usually upright on spreading stalks 5–10 mm long. Flowering June–Aug.

SYNONYM *Radicula sylvestris.*

HABITAT wet forests, lakeshores, muddy streambanks and ditches; sometimes weedy; introduced to North America from Europe, now found across much of Canada and USA.

STATUS MW-OBL | NCNE-OBL | GP-FACW

Subularia AWLWORT
Subularia aquatica L.
◆ WATER AWLWORT

DESCRIPTION Small, native annual aquatic herb; plants underwater or sometimes on muddy shores. Stems 3–10 cm long. Leaves all basal, awl-shaped or linear, 1–5 cm long. Flowers small, 2–10, widely separated in a raceme; sepals per-

sistent, petals white. Fruit a short, oval or oblong pod (silicle), 2–4 mm long. Flowering June–Aug.

HABITAT cold lakes in shallow water to 1 m deep.

STATUS NCNE-OBL | GP-OBL; Minn (THR), and Isle Royale and UP of Mich (END). Circumboreal, s to New Eng, Mich, Minn, Colo, and Calif.

Cabombaceae
Water-Shield Family

AQUATIC PERENNIAL HERBS with floating and/or underwater leaves. Flowers perfect (with both staminate and pistillate parts), white (*Cabomba*), or purple (*Brasenia*), emergent on long stalks from upper nodes, sepals 3–4, petal-like, petals 3–4. Fruit a capsule of 3 to many segments.

Cabombaceae WATER-SHIELD FAMILY

1 Leaves floating, alternate and entire; flowers purple, stamens 12–18 . . . . *Brasenia schreberi*

1 Leaves mostly underwater, opposite and dissected; flowers white, stamens 6 . . . . *Cabomba caroliniana*

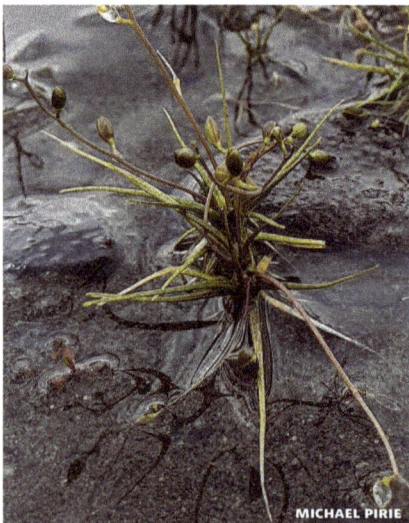

Subularia aquatica

Brasenia WATER-SHIELD

Brasenia schreberi J. F. Gmelin
♦WATER-SHIELD

DESCRIPTION Native perennial aquatic herb; underwater portions of plant with a slippery jelly-like coating. Stems to 2 m long. Leaf blades floating, oval, 4–12 cm long and half as wide; petiole attached to center of blade underside. Flowers dull-purple, on stalks to 15 cm long from leaf axils; sepals 3, petals 3, 12–15 mm long. Fruit oblong, 3–5 mm long. July.

HABITAT quiet ponds and lakes; water usually acidic.

STATUS MW-OBL | NCNE-OBL | GP-OBL

Cabomba FANWORT

Cabomba caroliniana A. Gray
FANWORT

DESCRIPTION Introduced perennial aquatic herb; our only aquatic species with opposite, highly dissected leaves on distinct petioles. Stems branched, to 2 m long. Leaves all underwater or with a few small floating leaves; underwater leaves opposite, 2–5 cm wide, palmately dissected into narrow segments, on petioles 1–3 cm long; floating leaves small, oblong, 6–20 mm long, often lobed at ends, petiole attached at cen-

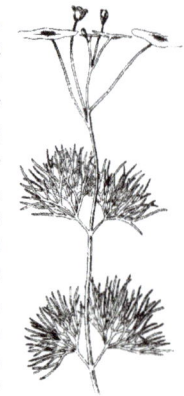

Brasenia schreberi

ter of blade underside. Flowers white with yellow base, on stalks 3–10 cm long from upper leaf axils; sepals and petals 3 each, 6–12 mm long, obovate stamens 6. Fruit with 3 segments.

HABITAT lakes and streams.

STATUS MW-OBL | NCNE-OBL | GP-OBL

NOTE Popular as an aquarium plant, fanwort is considered introduced in our region from the s USA; the species is also potentially invasive.

Campanulaceae
Bellflower Family

PERENNIAL HERBS. Stems usually with milky sap. Leaves simple, alternate. Flowers in racemes at ends of stems or single from upper leaf axils, perfect (with both staminate and pistillate parts), 5-parted, regular and funnel-shaped *(Campanula)* or irregular *(Lobelia)*; petals blue, white or scarlet; stamens separate or joined into a tube around style. Fruit a many-seeded capsule.

Campanulaceae **BELLFLOWER FAMILY**

1   Flowers regular, stamens separate; plants weak and reclining on surrounding plants .... *Campanula aparinoides*

1   Flowers irregular, stamens united to form a tube around the style; plants with upright stems........................... *Lobelia*

Campanula BELLFLOWER
Campanula aparinoides Pursh
◆ MARSH-BELLFLOWER

DESCRIPTION Native perennial herb, spreading by slender rhizomes. Stems slender, weak, usually reclining on other plants, 2–6 dm long, 3-angled, rough-to-touch. Leaves linear or narrowly lance-shaped, larger below and smaller upward on stem, 2–8 cm long and 2–8 mm wide, tapered to a sharp tip; margins and midvein on leaf underside often rough; petioles absent. Flowers single on long slender stalks from upper leaf axils, funnel-shaped, sepals triangular to lance-shaped, 2–5 mm long; petals pale blue to white, 5–12 mm

long. Fruit a capsule, opening near its base to release seeds. Flowering July–Sept.

HABITAT sedge meadows, marshes, calcareous fens, conifer swamps (cedar, tamarack), thickets, open bogs; soils often calcium-rich.

STATUS MW-OBL | NCNE-OBL | GP-OBL

Lobelia LOBELIA

Perennial herbs. Stems single, usually with milky juice. Leaves alternate. Flowers irregular, in racemes at ends of stems; white, bright red, or pale to dark blue, often with white or yellow markings; 2-lipped, the 3 lobes of lower lip spreading, the 2 lobes of upper lip erect or pointing forward, divided to base, the anthers projecting through the split; stamens 5, joined to form a tube around style, the lower 2 anthers hairy at tip and shorter than other 3. Fruit a capsule.

Lobelia **LOBELIA**

1   Stem leaves narrow, to 4 mm wide, margins entire or with a few small teeth; or leaves all from base of plant **2**

1   Stem leaves broader, 1–5 cm wide, margins toothed **3**

2   Leaves all from base of plant, hollow, round in cross-section; plants usually underwater; n portions of region....... *L. dortmanna*

2   Leaves all from stem, flat and linear; wetland habitats; widespread in region... .... *L. kalmii*

3   Flowers small, to 1.5 cm long **4**

3   Flowers larger, 2–4 cm long............. **5**

4   Inflorescence usually branched; hypanthium equalling the corolla or nearly so, inflated in fruit................ *L. inflata*

4   Inflorescence unbranched; hypanthium shorter than corolla, not much inflated in fruit......................... *L. spicata*

5   Flowers bright red (rarely white), 3 cm or more long .... *L. cardinalis*

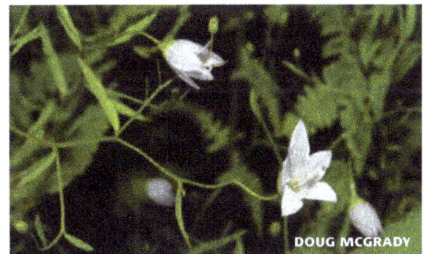

Campanula aparinoides
DOUG MCGRADY

5 Flowers blue with white stripes on lower lip, less than 2.5 cm long . . . . *L. siphilitica*

Lobelia cardinalis L.
♦CARDINAL-FLOWER

Native perennial herb. **Stems** erect, usually unbranched, 5–15 dm long, hairy to smooth. **Leaves** lance-shaped to oblong, 10–15 cm long and 3–5 cm wide, tapered to a point, margins toothed; lower leaves on short petioles, upper leaves ± stalkless. **Flowers** bright scarlet (rarely white), in racemes 1–4 dm long, the racemes with small, leafy, linear bracts; flowers 2–4 cm long, on hairy stalks 5–15 mm long. **Flowering** July–Sept.

HABITAT floodplain forests, swamps, thickets, streambanks, shores and ditches; sometimes in shallow water.

STATUS MW-OBL | NCNE-OBL | GP-FACW

Lobelia dortmanna L.
WATER-LOBELIA

DESCRIPTION Native perennial herb; plants usually underwater or sometimes on exposed sandy shores. **Stems** upright, hollow, smooth, with milky juice. **Leaves** in dense rosettes at base of plants, fleshy, hollow, linear, 3–8 cm long, rounded at tip; stem leaves tiny. **Flowers** pale blue or white, 1–2 cm long, in a few-flowered

raceme; sepals 2 mm long. **Flowering** July–Sept.

HABITAT shallow water of acidic lakes and ponds; wet, sandy shores.

STATUS MW-OBL | NCNE-OBL

Lobelia inflata L.
♦INDIAN-TOBACCO

DESCRIPTION Annual herb; the long, irregular hairs at base of stem are distinctive. **Stems** erect, usually branched, villous, to 1 m tall. **Leaves** sessile or subsessile, obovate, 5–8 cm long, 1.5–3.5 cm wide, more or less serrate, usually pubescent. **Flowers** in racemes terminating the branches, 1–2 dm long; lower bracts foliaceous, the upper gradually reduced; pedicels 3–8 mm long, glabrous or puberulent, bracteolate at the base. Flowers 7–10 mm long; sepals linear, 3–5 mm long; corolla blue or white, the lower lip pubescent; hypanthium much inflated in fruit. **Flowering** July–Oct.

HABITAT edges of swamps and rivers; also in dry to moist open woods; disturbed places such as roadsides, ditches, borrow pits, utility line clearings.

STATUS MW-FACU | NCNE-FACU | GP-FAC

NOTE long used in herbal medicine.

Lobelia kalmii L. ♦BROOK LOBELIA

DESCRIPTION Small native perennial herb. **Stems** erect, smooth, 1–4 dm long, un-

Lobelia cardinalis

Lobelia inflata

branched or with a few branches above, sometimes with a rosette of small, obovate leaves at base of plant. Stem Leaves linear, 1–5 cm long and 1–5 mm wide, blunt to sharp-tipped, margins with a few small teeth. Flowers blue with a white center, 6–10 mm long, in an open raceme, the flowers on stalks 4–10 mm long. Flowering July–Oct.

HABITAT wet, sandy or gravelly shores, wet meadows, interdunal wetlands, conifer swamps (cedar, tamarack), rock ledges and crevices; usually where calcium-rich.

STATUS MW-OBL | NCNE-OBL | GP-OBL

Lobelia siphilitica L.
♦ GREAT BLUE LOBELIA

DESCRIPTION Native perennial herb. Stems stout, erect, 3–12 dm long. Leaves oblong or oval, smaller upward, 6–12 cm long and 1–3 cm wide, tip sharp or blunt, margins irregularly toothed, petioles absent. Flowers dark blue, in crowded racemes 1–3 dm long; the lower lip blue and white-striped, 1.5–2.5 cm long, on ascending stalks 4–10 mm long; sepals triangular to lance-shaped, 5–20 mm long, usually with narrow lobes near base; petals absent. Flowering Aug–Sept.

HABITAT swamps, floodplain forests, thickets, streambanks, calcareous fens, wet meadows.

STATUS MW-OBL | NCNE-FACW | GP-OBL

Lobelia spicata Lam.
SPIKED LOBELIA

DESCRIPTION Native perennial herb. Stems unbranched, 3–10 dm long, hairy toward base. Leaves obovate to lance-shaped, 5–10 cm long, hairy, becoming smaller above. Flowers pale blue to white, 6–10 mm long, on stalks 2–4 mm long, in a slender, crowded raceme; base of sepals often with distinct, curved lobes (auricles), 1–2 mm long. Flowering May–Aug.

HABITAT moist to wet prairie (sometimes where disturbed), wet meadows, swamp margins.

STATUS MW-FAC | NCNE-FAC | GP-FAC

Caprifoliaceae
Honeysuckle Family

SHRUBS, VINES, OR HERBS with opposite, mostly simple leaves. Flowers perfect (with both staminate and pistillate parts), mostly 5-parted. Fruit a fleshy berry or dry capsule.

Caprifoliaceae HONEYSUCKLE FAMILY

1 Flowers numerous, in rather dense terminal inflorescences (at ends of stem and branches) .... *Valeriana*

1 Flowers axillary or on paired pedicels on a peduncle **2**

Lobelia kalmii

Lobelia siphilitica

2 Plants small, creeping, evergreen; flowers paired and nodding at tips of slender stalks
.... *Linnaea borealis*

2 Plants larger shrubs, upright, deciduous ...
.... *Lonicera*

Linnaea TWINFLOWER

Linnaea borealis L. ♦ TWINFLOWER

DESCRIPTION Evergreen, trailing native vine. **Stems** slightly woody, to 1–2 m long, with numerous short, erect, leafy branches to 10 cm long; branches green to red-brown, finely hairy; older stems woody, 2–4 mm wide. **Leaves** opposite, simple, evergreen, oval to round, 1–2 cm long, blunt at tip, upper surface and margins with short, straight hairs; margins rolled under, with a few rounded teeth near tip; petiole short, short-hairy. **Flowers** small, pink to white, bell-shaped, shallowly 5-lobed, slightly fragrant, in nodding pairs atop a Y-shaped stalk to 10 cm long, the stalk with gland-tipped hairs and 2 small bracts at the fork and a pair of smaller bracts at base of each flower. **Fruit** a small, dry, 1-seeded capsule. **Flowering** June–Aug.

SYNONYM *Linnaea americana*.

HABITAT hummocks in cedar swamps and thickets, moist conifer woods, on rotten logs and mossy boulders.

STATUS MW-FAC | NCNE-FAC | GP-FACU; common

in northern portions of our region, becoming local southward; circumpolar.

Lonicera HONEYSUCKLE

Shrubs (those included here) or vines. Leaves opposite, simple, entire. Flowers long and tubular or funnel-shaped, in pairs from leaf axils. Fruit a few-seeded, blue or red berry.

Lonicera **HONEYSUCKLE**

1 Flowers and fruits on short stalks (1 cm or less), fruit blue; bark red-brown and peeling in papery layers .... *L. villosa*

1 Flowers and fruits on longer stalks (1 cm or more), fruit red; bark gray-brown and shredding .... *L. oblongifolia*

Lonicera oblongifolia (Goldie) Hook.
♦ SWAMP FLY-HONEYSUCKLE

DESCRIPTION Thicket-forming native shrub 1–1.5 m tall. **Branches** upright, with shredding bark and solid pith. **Twigs** green to purple, smooth. **Leaves** opposite, oblong or oval, 3–8 cm long and 1–4 cm wide, rounded or blunt at tip, underside hairy when young, becoming smooth; margins entire, not fringed with hairs; petioles absent or to 1–2 mm long. **Flowers** yellow-white, tube-shaped with 2 spreading lips, 10–15 mm long, in pairs at ends of slender stalks up to 4 cm long from leaf axils. **Fruit** an orange-red to red (or sometimes purple), few-seeded berry composed of the 2 joined ovaries. **Flowering** May–June.

SYNONYM *Xylosteon oblongifolia*.

HABITAT cedar and tamarack swamps, fens, open bogs, wet streambanks and shores; often over limestone.

STATUS MW-OBL | NCNE-OBL | GP-OBL

Linnaea borealis

Lonicera oblongifolia

Lonicera villosa (Michx.) J.A. Schultes
♦ WATERBERRY

DESCRIPTION Native shrub to 1 m. Branches upright, red-brown to gray, outer thin layers soon peeling to expose red-brown inner layers. Twigs purple-red, with long, soft hairs. Leaves opposite, oval to oblong, 2–6 cm long and 1–3 cm wide, blunt or rounded at tip, upper surface dark green, underside paler and hairy, especially on veins; margins fringed with hairs and often rolled under; petioles absent or to 1–2 mm long. Flowers yellow, tubular to funnel-shaped, 10–15 mm long, in pairs on short hairy stalks from axils of lower leaves. Fruit an edible dark blue berry consisting of the 2 joined ovaries. Flowering May–July.

SYNONYM *Lonicera caerulea*.

HABITAT cedar and tamarack swamps, thickets, fens, shores.

STATUS MW-FACW | NCNE-FACW | GP-FACW

Valeriana VALERIAN

Perennial, strongly scented herbs. Leaves from base of plant and opposite along stem, simple to pinnately divided. Flowers somewhat irregular, in branched heads at ends of stems; calyx inrolled when young, later expanding and spreading; petals joined into a tube-shaped, 5-lobed corolla; stamens 3. Fruit a 1-chambered achene.

Valeriana **VALERIAN**
1   Leaves thick, parallel-veined; plants from a stout, carrotlike taproot......... *V. edulis*
1   Leaves thin, net-veined; plants from a creeping or ascending rhizome ....
.... *V. uliginosa*

Valeriana edulis Nutt.
♦ COMMON VALERIAN, TOBACCO-ROOT

DESCRIPTION Native perennial herb, from a stout taproot. Stems smooth, 3–12 dm long. Leaves thick, ± parallel-veined, often hairy when young, becoming smooth or with the margins fringed with hairs when mature; basal leaves oblong lance-shaped, 1–3 dm long and 1–2 cm wide, margins entire or with several lobes, tapered to a winged petiole; stem leaves stalkless, pinnately divided into lance-shaped segments. Flowers both perfect and single-sexed, the different types often on different plants; perfect and staminate flowers 2–4 mm wide, pistillate flowers to 1 mm wide, in widely branched panicles at ends of stems; corolla 5-lobed, yellow-white. Fruit an ovate achene, 3–4 mm long. Flowering May–June.

SYNONYM *Valeriana ciliata*.

HABITAT wet meadows, calcareous fens, low prairie.

STATUS MW-FACW | NCNE-FACW | GP-FAC; s Minn (THR), s Mich (THR).

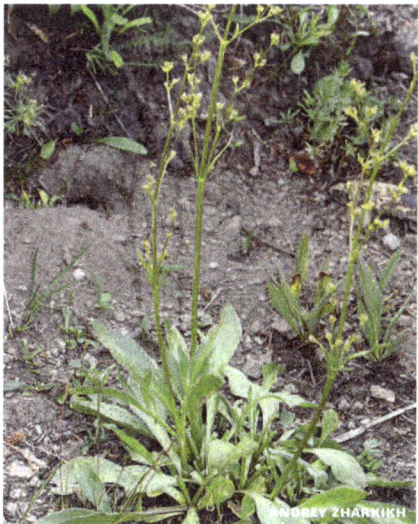

Lonicera villosa

Valeriana edulis

Valeriana uliginosa (T. & G.) Rydb.
♦BOG VALERIAN

DESCRIPTION Native perennial herb, from a stout rhizome or crown. Stems 3–8 dm long, ± smooth. Leaves thin, net-veined; basal leaves obovate, 6–14 cm long and 1–3 cm wide, tapered to a long petiole; margins entire or with several lobes; stem leaves 2–6 pairs, pinnately divided into 3–15 ovate segments, margins often fringed with fine hairs; petioles present. Flowers all perfect, in clusters grouped into panicles at ends of stems; corolla 5-lobed, pale pink, 5–7 mm long. Fruit a smooth, ovate achene, 3–4 mm long. Flowering May–July.

SYNONYMS *Valeriana septentrionalis* var. *uliginosa*, *Valeriana sitchensis* ssp. *uliginosa*.

HABITAT conifer swamps (especially of cedar and tamarack), marshes, calcareous fens, wet meadows; soils often alkaline.

STATUS [not rated]

Caryophyllaceae
Pink Family

ANNUAL OR PERENNIAL HERBS. Leaves simple, entire, mostly opposite but sometimes alternate or whorled. Stems often swollen at nodes. Flowers perfect (with both staminate and pistillate parts) or imperfect, in open or compact heads at ends of stems or from leaf axils; sepals usually 5, separate or joined into a tube; petals 5 (sometimes 4), separate, often lobed or toothed, sometimes absent; stamens 3–10, anthers often distinctly colored. Fruit a few- to many-seeded capsule.

Caryophyllaceae **PINK FAMILY**

1 Leaves succulent; stipules present . . . . *Spergularia*

Valeriana uliginosa

1 Leaves not succulent; stipules absent . . . **2**
2 Sepals joined to form a toothed or lobed tube . . . . *Silene nivea*
2 Sepals free or joined only at base *Stellaria*

Silene CATCHFLY, CAMPION

Silene nivea (Nutt.) Otth
SNOWY CAMPION

DESCRIPTION Native perennial herb, spreading by rhizomes; plants smooth or with a few short hairs. Stems 2–3 dm long. Leaves mostly on stem, opposite, lance-shaped or oblong, 5–10 cm long and 1–3 cm wide, stalkless or on short petioles. Flowers few, mostly in leaf axils; sepals joined to form a tubelike flower 1.5 cm long; petals white, stamens 10, styles 3. Fruit a 1-chambered capsule. Flowering June–July.

SYNONYM *Silene alba*.

HABITAT streambanks, wooded ravines, calcareous fens.

STATUS MW-FACW | NCNE-FACW | GP-FACW; rare in Minn (THR), Wisc (THR), Mich (THR).

Spergularia SAND-SPURREY

Low, succulent herbs. Leaves opposite, linear or reduced to bristles. Flowers in branched clusters at ends of stems; sepals and petals each 5. Fruit a capsule.

Spergularia **SAND-SPURREY**

1 Sepals less than 4 mm long, stamens 2–5, seeds wingless . . . . *S. marina*
1 Sepals more than 4 mm long, stamens 9–10, seeds thin-winged . . . . *S. media*

Spergularia marina (L.) Griseb.
♦LESSER SALT SPURREY

DESCRIPTION Native annual herb. Stems upright, or sprawling, to 30 cm long, smooth or with gland-tipped hairs.

Leaves fleshy, 5–40 mm long and to 1.5 mm wide, tipped with a short spine. Flowers pink or white; stamens 2–5; seeds less than 1 mm wide, usually without a wing. Flowering Summer.
SYNONYM *Spergularia salina* var. *salina*.
HABITAT highway ditches where salted.
STATUS MW-OBL | NCNE-FACW | GP-OBL

Spergularia media (L.) C. Presl.
SALT SPURREY

DESCRIPTION Introduced annual or perennial herb; plants smooth or with a few gland-tipped hairs. Stems upright or sprawling, to 40 cm long. Leaves fleshy, 1–5 cm long and to 2 mm wide, usually tipped with a short spine. Flowers white; stamens 9–10; seeds about 1 mm wide, with a wing 0.1–0.3 mm wide. Flowering Summer.
SYNONYM *Spergularia maritima*.
HABITAT highway ditches where salted.
STATUS MW-FACU | NCNE-FACU | GP-FACU

Stellaria STITCHWORT

Low, spreading or erect perennials (ours), mostly without hairs. Stems slender, 4-angled. Flowers single in forks of stems or in few-flowered clusters at ends of stems; sepals green with translucent margins; petals white, lobed or deeply cleft (sometimes absent in *Stellaria borealis*); stamens 10 or less; styles 3 Fruit an ovate or oblong capsule.

Stellaria STITCHWORT

1 Plants large, stems to 8 dm long; styles 5 . . . . *S. aquatica*

Spergularia marina

1 Plants smaller; styles 3–4 . . . . **2**
2 Flowers single in forks of stems, not subtended by membranous bracts **3**
2 Flowers in branched terminal clusters, subtended by small membranous bracts **4**
3 Stems 25 cm or more long; seeds smooth . . . . *S. borealis*
3 Stems to 20 cm long; seeds rough . . . . *S. crassifolia*
4 Petals shorter than sepals **5**
4 Petals much longer than sepals **6**
5 Flowers mostly in clusters from leaf axils; seeds covered with small bumps; se Minnesota only . . . . *S. alsine*
5 Flowers in clusters at ends of stems; seeds smooth; northern portions of region . . . . *S. borealis*
6 Heads open and widely branched, the stalks spreading; leaves spreading to ascending, widest at or above middle . . . . *S. longifolia*
6 Head narrow, the stalks erect to ascending; leaves upright, widest near base . . . . *S. longipes*

Stellaria alsine Grimm
BOG-STITCHWORT

DESCRIPTION Introduced annual herb. Stems sprawling, smooth, angled, rooting at nodes. Leaves oval to oblong, 1.5–3 cm long and to 1 cm wide; lower leaves on petioles, upper leaves stalkless. Flowers in few-flowered clusters from leaf axils, sepals 3–4 mm long, petals white, shorter than sepals. Fruit a capsule, longer than the sepals; seeds 0.5 mm long, covered with small bumps. Flowering May–Aug.
SYNONYM *Alsine uliginosa*.
HABITAT marshes, streambanks, seeps.
STATUS MW-OBL | NCNE-OBL; considered introduced in our region from main range of ne USA; known from se Minnesota near Mississippi River.

Stellaria aquatica (L.) Scop.
◆ GIANT CHICKWEED

DESCRIPTION Introduced perennial herb, spreading by rhizomes. Stems sprawling and matted, to 8 dm long, rooting at nodes, covered with gland-tipped hairs. Leaves ovate to lance-shaped, 2–8 cm

long and 1–4 cm wide, petioles short or absent. Flowers in open, leafy clusters at ends of stems; sepals 5–9 mm long; petals white, much longer than sepals. Fruit a capsule; seeds 0.8 mm long, covered with small bumps. Flowering June–Oct.

SYNONYMS *Alsine aquatica, Myosoton aquaticum.*

HABITAT streambanks, ponds, wet or moist disturbed areas, often in partial shade.

STATUS MW-FACW | NCNE-FAC | GP-FAC

Stellaria borealis Bigelow
NORTHERN STITCHWORT

DESCRIPTION Native perennial herb, spreading by rhizomes. Stems sprawling, to 5 dm long, branched, angled. Leaves lance-shaped, narrowed at base, 1–5 cm long and 2–8 mm wide, margins hairy. Flowers in clusters at ends of stems; sepals 2–4 mm long; petals usually absent. Fruit a dark capsule, longer than sepals; seeds to 1 mm long, nearly smooth. Flowering June–Aug.

SYNONYMS *Alsine borealis, Stellaria calycantha.*

HABITAT openings and hollows in conifer forests, margins of ponds and marshes.

STATUS MW-OBL | NCNE-FACW | GP-FACW

Stellaria crassifolia Ehrh.
◆FLESHY STITCHWORT

DESCRIPTION Native perennial. Stems sprawling and matted to erect, freely branched, 8–30 cm long, fleshy, smooth. Leaves soft, oval to lance-shaped, narrowed at base, 1–3 cm long and 1–3 mm wide. Flowers single in forks of stem, nodding on stalks 1–3 cm long;

sepals 2–4 mm long; petals longer than sepals. Fruit an ovate capsule, to 5 mm long and longer than the sepals; seeds red-brown, to 1 mm long. Flowering June–July.

HABITAT streambanks and wet shores.

STATUS MW-OBL | NCNE-FACW | GP-OB; Mich (END).

Stellaria longifolia Muhl.
LONG-LEAVED STITCHWORT

DESCRIPTION Native perennial herb. Stems sprawling, prominently 4-angled, usually freely branched, 1–5 dm long.

Leaves spreading to ascending, linear to lance-shaped, 2–5 cm long and 1–6 mm wide, widest at or above middle, tapered at both ends. Flowers in branched clusters at ends of stems; sepals 3–5 mm long; petals longer than sepals. Fruit a green-yellow to brown capsule, usually longer than the sepals; seeds light brown, about 1 mm long. Flowering May–July.

HABITAT wet meadows, marshes, shrub thickets, swamps, streambanks, pond margins.

STATUS MW-FACW | NCNE-FACW | GP-FACW

Stellaria longipes Goldie
LONG-STALKED STITCHWORT

DESCRIPTION Native perennial herb, spreading by rhizomes. Stems erect, or sprawling and matted, 5–30 cm long.

Leaves upright, stiff and shiny-waxy, linear or lance-shaped, 1–4 cm long and 1–4 mm wide, widest near base, tapered to tip. Flowers in branched clusters at ends of stems or appearing lateral from stem; sepals 4–5 mm long; petals slightly longer than sepals. Fruit a straw-colored to shiny purple capsule, longer than the sepals; seeds red-brown, oblong to oval, about 1 mm long. Flowering May–July.

HABITAT wet meadows, ditches and thickets;

Stellaria aquatica

Stellaria crassifolia

in Michigan, sand dunes near Lakes Michigan and Superior.

STATUS MW-OBL | NCNE-FACU | GP-OBL

Celastraceae
Bittersweet Family

Parnassia
GRASS-OF-PARNASSUS

Smooth perennial herbs. Leaves all from base of plant but often with 1 stalkless leaf near middle of stalk, margins entire; petioles present. Flowers large, white, single at ends of stalks; calyx 5-lobed; petals white, veined, spreading; fertile stamens 5, alternating with petals; staminodes (infertile stamens) attached to base of petals and divided into threadlike segments tipped with glandular knobs; stigmas 4. Fruit a 4-chambered capsule with numerous seeds.

NOTE sometimes placed in own family: Parnassiaceae.

ADDITIONAL SPECIES *Parnassia parviflora* DC. known from Door County, Wisc (END), along Lake Michigan shoreline in cracks in wet limestone rocks, or on open, moist, sandy beaches and dunes; petals white with green veins; flowering July–Aug.

Parnassia GRASS-OF-PARNASSUS

1 Sepals with narrow translucent margins; staminodes (sterile stamens) 3-parted, not widened at base .... *P. glauca*

1 Sepal margins green; staminodes 5 to many-parted .... *P. palustris*

Parnassia glauca

Parnassia glauca Raf.
⬩AMERICAN GRASS-OF-PARNASSUS

DESCRIPTION Smooth native perennial herb. Leaves from base of plant and usually with one ± stalkless stem leaf; broadly ovate to nearly round, 2–7 cm long and 1–5 cm wide, rounded to a blunt or somewhat pointed tip; margins entire; petioles long. Flowers single atop a stalk 1–4 dm long; sepals ovate, 2–5 mm long, with a narrow, translucent margin; petals white with green veins, 1–2 cm long; staminodes 3-parted from near base, shorter than to equal to stamens. Fruit a capsule about 1 cm long. Flowering Aug–Sept.

HABITAT calcareous fens and wet meadows.

STATUS MW-OBL | NCNE-OBL | GP-OBL

range Minn, Wisc, Mich.

Parnassia palustris L.
⬩NORTHERN GRASS-OF-PARNASSUS

DESCRIPTION Smooth native perennial herb. Leaves from base of plant and usually with one clasping, heart-shaped leaf below middle of stalk; ovate to nearly round, 1–3 cm long, rounded or blunt at tip; margins entire; petioles long and slender. Flowers single atop a stalk 1.5–4 dm long; sepals lance-shaped, to 1 cm long, green throughout; petals white with green veins, ovate, 1–1.5 cm long, longer than sepals; staminodes many-parted from the

Parnassia palustris

widened tip, 5–9 mm long. Fruit a capsule. Flowering July–Sept.

HABITAT calcareous fens, shores, stream-banks and wet meadows.

STATUS MW-OBL | NCNE-OBL | GP-OBL; n and c Minn; rare in n Wisc (THR), and UP and n LP of Mich (THR).

Ceratophyllaceae
Hornwort Family

Ceratophyllum
COONTAIL, HORNWORT

Aquatic perennial herbs, often forming large patches; roots absent, but plants usually anchored to substrate by pale, modified leaves. Stems slender, branched. Leaves in whorls, with more than 4 leaves per node, whorls crowded at ends of stems (hence the common name of coontail), dissected 2–3x into narrow segments. Flowers small, inconspicuous in leaf axils, staminate and pistillate flowers separate on same plant, staminate usually above pistillate on stems.

Ceratophyllum **HORNWORT FAMILY**

1 Leaves usually stiff, forked 1–2 times, margins coarsely toothed; achenes with 2 spines near base .... *C. demersum*

1 Leaves limp, some larger leaves forked 3–4 times, margins not toothed; achenes with several spines along each margin ....
.... *C. echinatum*

Ceratophyllum demersum

Ceratophyllum demersum L.
♦ COMMON HORNWORT

DESCRIPTION Native aquatic herb. Stems long, branched. Leaves in whorls of 5–12 at each node, stiff, 1–3 cm long, 1-2-forked; leaf segments linear, 0.5–1 mm wide, coarsely toothed. Fruit an oval achene, 4–6 mm long, with 2 spines at base.

HABITAT shallow to deep water of lakes, ponds, backwater areas, ditches, and may form dense masses; water typically neutral or alkaline.

STATUS MW-OBL | NCNE-OBL | GP-OBL

Ceratophyllum echinatum A. Gray
PRICKLY HORNWORT

DESCRIPTION Native aquatic herb. Similar to *C. demersum*, but leaves usually limp, larger leaves usually 3- or sometimes 4-forked, the segments narrower and mostly without teeth. Fruit an achene with 2 spines at base and several unequal spines on achene body.

SYNONYM *Ceratophyllum muricatum.*

HABITAT lakes, ponds, quiet water of rivers and streams; water acidic.

STATUS MW-OBL | NCNE-OBL | GP-OBL

Cornaceae
Dogwood Family

Cornus
BUNCHBERRY, DOGWOOD

Shrubs, or herbaceous shoots from a woody rhizome in bunchberry (*Cornus canadensis*). Leaves opposite or whorled, simple, entire. Flowers perfect, 4-parted, sepals and petals small, in a rounded or flat-topped cluster. Fruit a red or white berrylike drupe with 1–2 hard seeds.

Cornus **BUNCHBERRY, DOGWOOD**

1 Plants herbaceous from a woody rhizome, less than 3 dm tall; leaves whorled ....
.... *C. canadensis*

1 Taller shrubs, 5 dm or more tall; leaves opposite or alternate 2

2 Leaves alternate on stems *C. alternifolia*

2 Leaves opposite 3

3 Leaves with stiff, rough hairs on upper surface; s portion of region . . . . *C. drummondii*

3 Leaves smooth, not rough-hairy above; widespread species **4**

4 Fruit white; young twigs densely short-hairy; se Mich. . . . . *C. amomum*

4 Fruit blue; young twigs ± smooth **5**

5 Twigs gray; leaves with fewer than 5 pairs of lateral veins . . . . *C. racemosa*

5 Twigs red; leaves with 5 or more pairs of lateral veins . . . . *C. sericea*

Cornus alternifolia L. f.

◆ ALTERNATE-LEAVED DOGWOOD, PAGODA DOGWOOD

DESCRIPTION Native shrub, to 5 m tall. Twigs red-green or brown, somewhat shiny, alternate on stems, pith white. Leaves alternate, sometimes crowded and appearing whorled near ends of stems, oval to ovate, 5–12 cm long and 3–7 cm wide, tapered to a sharp tip, underside finely hairy; lateral veins 4–5 pairs, these curving toward tip of blade; margins entire; petioles to 5 cm long. Flowers small, creamy-white, in crowded, flat-topped or rounded clusters at ends of stems. Fruit a round, blue, berrylike drupe, 6 mm wide, atop a red stalk. Flowering May–July.

SYNONYM *Svida alternifolia*.

HABITAT swamps, thickets, streambanks and springs; also in drier deciduous and mixed forests.

STATUS MW-FAC | NCNE-FACU | GP-FACU

Cornus amomum Miller

SILKY DOGWOOD

DESCRIPTION Native shrub, 1–3 m tall. Older branches red and gray-streaked, young twigs gray, finely hairy; pith brown. Leaves opposite, oval to ovate, 5–12 cm long and 2–5 cm wide, usually less than half as wide as long, tapered to a sharp tip, lateral veins 4–6 on each side, underside finely hairy; margins entire; petioles 1–2 cm long, often curved and causing the leaves to droop. Flowers small, creamy-white, in flat-topped or slightly rounded, hairy clusters. Fruit a round, blue or blue-white, berrylike drupe, 8 mm wide, atop a long stalk. Flowering June–July (our latest flowering dogwood).

SYNONYM *Cornus obliqua*.

HABITAT conifer swamps, marshes, open bogs, calcareous fens, lakeshores, streambanks, wet dunes.

STATUS MW-FACW | NCNE-FACW; Mich (END).

Cornus canadensis L.

◆ BUNCHBERRY, DWARF CORNEL

DESCRIPTION Native perennial from horizontal, woody rhizomes, often forming large colonies. Stems erect, green, 1–2 dm tall, with a pair of small bracts on lower stem, topped with a whorl-like cluster of 4–6 leaves. Leaves oval to obovate, 4–7 cm long, tapered at both ends; lateral veins 2–3 pairs, arising from midvein below middle of blade; margins entire; petioles

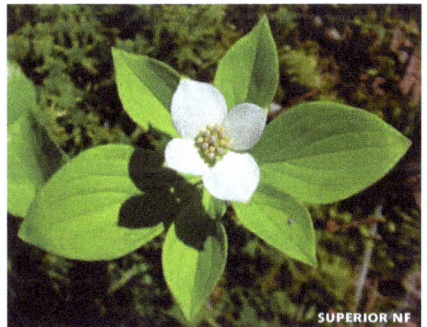

Cornus alternifolia

Cornus canadensis

short or absent. Flowers small, yellow-green or creamy-white in a single cluster at end of a stalk 1–3 cm long; flowers surrounded by 4 white or pinkish, petal-like showy bracts, 1–2 cm long, these soon deciduous. Fruit a cluster of round, bright red berrylike drupes, the drupes 6–8 mm wide. Flowering June–July.

HABITAT cedar swamps, thickets and moist conifer forests, often on hummocks or rotting logs; also in drier, mixed conifer-deciduous forests.

STATUS MW-FAC | NCNE-FAC | GP-FACU

Cornus drummondii C. A. Meyer.
ROUGH-LEAVED DOGWOOD

DESCRIPTION Native shrub, 2–4 m tall, sometimes forming thickets. Twigs red and finely hairy when young, becoming gray-brown and smooth; pith brown. Leaves opposite, lance-shaped to ovate, 5–8 cm long and 3–5 cm wide, abruptly tapered to a rounded tip, rough-to-touch on upper surface, underside finely hairy; lateral veins 3 or 4 on each side of midvein; margins entire; petioles finely hairy, to 2 cm long. Flowers small, creamy-white, many, in loose ± flat-topped clusters. Fruit a white, round, berrylike drupe, 6 mm wide, on a red-purple stalk. Flowering May–June.

HABITAT streambanks, thickets.

STATUS MW-FAC | NCNE-FAC | GP-FAC

Cornus racemosa Lam.
◆NORTHERN SWAMP DOGWOOD

DESCRIPTION Native shrub, 1–3 m tall, often forming dense thickets. Twigs red, becoming gray or light brown;

pith usually brown. Leaves opposite, lance-shaped to oval, 4–9 cm long and 2–4 cm wide, abruptly tapered to a rounded tip, underside with short hairs; lateral veins 3 or 4 on each side of midvein; margins entire; petioles to 1 cm long. Flowers small, creamy-white, ill-scented, in numerous, open, elongated clusters. Fruit a round, berrylike drupe, at first lead-colored, becoming white, 5 mm wide, on red stalks. Flowering June–July.

SYNONYM *Cornus foemina*.

HABITAT lakeshores, streambanks, swamps, thickets, marshes, moist woods, low prairie.

STATUS MW-FAC | NCNE-FAC | GP-FAC

Cornus sericea L.
◆RED-OSIER DOGWOOD

DESCRIPTION Many-stemmed native shrub, 1–3 m tall, forming thickets. Branches upright or prostrate and rooting; twigs and young branches red; pith white. Leaves opposite, green, ovate to oval, mostly 5–15 cm long and 2–7 cm wide, tapered to a tip, soft hairy on underside; margins entire; petioles to 2.5 cm long. Flowers small, white, many in flat-topped or slightly rounded clusters. Fruit a round, white or blue-tinged, berrylike drupe, 6–9 mm wide. Flowering May–Aug.

SYNONYMS *Cornus alba, Cornus stolonifera*.

HABITAT swamps, marshes, shores, streambanks, floodplain forests, shrub thickets, calcareous fens; also on sand dunes. Com-

Cornus racemosa

Cornus sericea

mon, but plants often heavily browsed where reduced in size and abundance where deer numbers are high.

STATUS MW-FACW | NCNE-FACW | GP-FACW

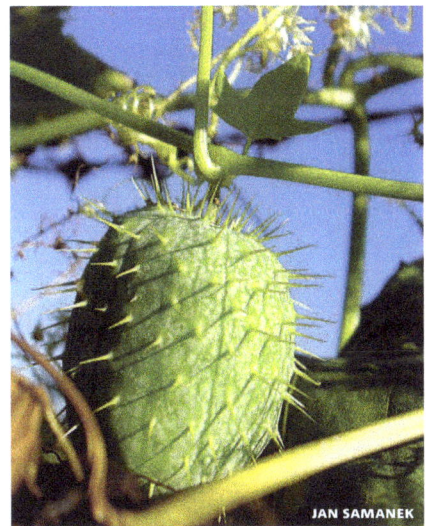

Crassulaceae
Stonecrop Family

Crassula PYGMY-WEED

Crassula aquatica (L.) Schönl
◆ PYGMY-WEED

DESCRIPTION Small native annual herb. **Stems** branched, 2–10 cm long. **Leaves** opposite, linear, succulent, 3–6 mm long, spreading, margins entire, petioles absent. **Flowers** small, 1 mm long, single in leaf axils; petals white or green-white, 4, erect or slightly spreading; stamens 4, alternate with petals; pistils 3–4, with a short style. **Flowering** Aug–Sept.

SYNONYM *Tillaea aquatica.*

HABITAT muddy shores and shallow water.

STATUS MW-OBL | NCNE-OBL | GP-OBL; rare in Minn (THR). Nfld and Que s to Md, La, Tex and w USA; Eurasia.

Curcurbitaceae
Cucumber Family

ANNUAL HERBACEOUS VINES (our species). Flowers green or white, either staminate or pistillate and on same plants (ours). Fruit a dry or fleshy, cucumber- or squash-like fruit (*pepo*).

Curcurbitaceae **GOURD FAMILY**

1 Staminate flowers 6-lobed; fruit inflated, 3–5 cm long, 4-seeded . *Echinocystis lobata*

1 Staminate flowers 5-lobed; fruit not inflated, to 1.5 cm long, 1-seeded . . . .
. . . . *Sicyos angulatus*

Echinocystis
WILD CUCUMBER

Echinocystis lobata (Michx.) T. & G.
◆ WILD CUCUMBER, BALSAM-APPLE

DESCRIPTION Native annual vining herb, to 5 m or more long. **Leaves** round in outline, with 3–7 (usually 5) sharp, triangular lobes; petioles 3–8 cm long. **Flowers** white; staminate flowers 8-10 mm wide, with lance-shaped lobes, in long, upright racemes; pistillate flowers 1 to several on short stalks from leaf axils. **Fruit** green, ovate, inflated, 3–5 cm long, with soft prickles. **Flowering** Aug–Sept.

HABITAT floodplain forests, wet deciduous forests, streambanks, thickets, and waste ground.

STATUS MW-FACW | NCNE-FACW | GP-FAC

CAVAN ALLEN
Crassula aquatica

JAN SAMANEK
Echinocystis lobata

Sicyos BUR-CUCUMBER

Sicyos angulatus L. BUR-CUCUMBER
DESCRIPTION Native annual vining herb, to 2
m long. **Stems** angled,
sticky-hairy, with
branched tendrils.
Leaves round in outline, with 3-5 shallow,
toothed lobes, rough on both sides; petioles
hairy, 3-10 cm long. **Flowers** green or white;
staminate flowers 8-10 mm wide, 5-lobed,
on stalks 10 cm or more long; pistillate flowers on stalks to 8 cm long. **Fruit** yellow,
ovate, 1.5 cm long, hairy and spine-covered.
Flowering Aug-Sept.
HABITAT floodplain forests, wet deciduous
forests, streambanks, thickets and waste
ground.
STATUS MW-FACW | NCNE-FACW | GP-FACW

Droseraceae
Sundew Family

Drosera SUNDEW
Perennial herbs. Leaves all from base of
plant, covered with stalked, sticky glands
that trap and digest insects. Flowers
white, several, on 1 side of erect, leafless
stalks, the stalks nodding at tip; with 5
petals and 5 sepals; stamens mostly 5,
styles 3. Fruit a dry, many-seeded capsule.

Drosera **SUNDEW**
1 Leaves widely spreading, blades round,
wider than long ....*D. rotundifolia*
1 Leaves upright, blades linear or broad at tip
and tapered to base, longer than wide ..**2**
2 Leaf blades linear, 10-20x longer than
wide; young petals pink.......*D. linearis*
2 Leaf blades broad near tip and narrowed to
base, 2-7x as long as wide; young petals
white **3**
3 Blades 2-3x as long as wide, petioles
without hairs; flower stalks from side of
plant base and curving upward ....
.........................*D. intermedia*
3 Blades 5-7x as long as wide, petioles with
some hairs; flower stalks erect from center
of plant base ....*D. anglica*

Drosera anglica Hudson
♦ENGLISH or GREAT SUNDEW
DESCRIPTION Native
perennial insectivorous
herb. **Leaves** leaf
blades obovate to spat-
ula-shaped, 15-35 mm
long and 3-4 mm wide, upper surface covered with gland-tipped hairs; petioles 3-6
cm long, smooth or with few glandular
hairs. **Flowers** 1-9 in a racemelike cluster
atop a stalk 6-25 cm tall; sepals 5-6 mm
long; petals white, 6 mm long, spatula-
shaped. Seeds black, 1 mm long, with fine
lines. **Flowering** June-Aug.
HABITAT floating sphagnum mats, calcareous

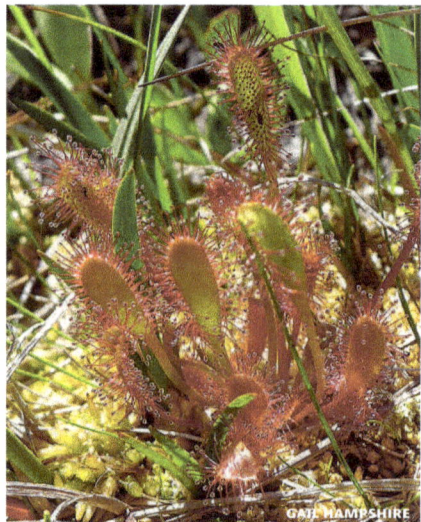

Drosera anglica

Drosera intermedia

fens, wet areas between dunes.

STATUS MW-OBL | NCNE-OBL.

NOTE *Drosera anglica* is similar to spoon-leaved sundew (*Drosera intermedia*) but rarely occurring together. Plants of *Drosera anglica* are generally larger, with shorter petioles (1–3x as long as leaf blades vs. 2.5–3.5x as long in *Drosera intermedia*).

Drosera intermedia Hayne
♦SPOON-LEAVED SUNDEW

DESCRIPTION Native perennial insectivorous herb. Leaves in a basal rosette and also usually along lower stem; spatula-shaped, 2–4 mm wide, upper surface covered with long, gland-tipped hairs; petioles smooth, 2–5 cm long. Flowers on stalks to 20 cm tall; sepals 3–4 mm long; petals white, 4–5 mm long. Seeds red-brown, to 1 mm long, covered with small bumps. Flowering July–Sept.

HABITAT low spots in open bogs, sandy shores, often in shallow water.

STATUS MW-OBL | NCNE-OBL | GP-OBL

Drosera linearis Goldie
♦LINEAR-LEAVED SUNDEW

Native perennial insectivorous herb. Leaves leaf blades linear, 2–5 cm long and 2 mm wide; petioles smooth, 3–7 cm long. Flowers 1–4 atop stalks 6–15 cm tall; flowers 6–8 mm wide; sepals 4–5 mm long; petals obovate, 6 mm long, white. Seeds black, less than 1 mm long, with small craterlike pits on surface. Flowering June–Aug.

HABITAT calcareous fens, wet areas between dunes near Great Lakes; rarely in sphagnum moss.

STATUS MW-OBL | NCNE-OBL; Wisc (THR).

Drosera rotundifolia L.
♦ROUND-LEAVED SUNDEW

DESCRIPTION Native perennial insectivorous herb. Leaves leaf blades ± round, wider than long, 2–10 mm long and as wide or wider, covered with long, red, gland-tipped hairs; abruptly tapered to a petiole longer than blade; petioles 2–5 cm long covered with gland-tipped hairs. Flowers 2–15 in a ± 1-sided, racemelike cluster, on a leafless stalk 10-30 cm tall; flowers 4–7 mm wide, sepals 5, 4–5 mm long; petals white to pink, longer than sepals; stamens 5, shorter than petals. Seeds light brown, shiny and with fine lines, 1–1.5 mm long. Flowering July–Aug.

HABITAT swamps and open bogs, usually in sphagnum; wet sandy shores.

STATUS MW-OBL | NCNE-OBL | GP-OBL

Drosera linearis

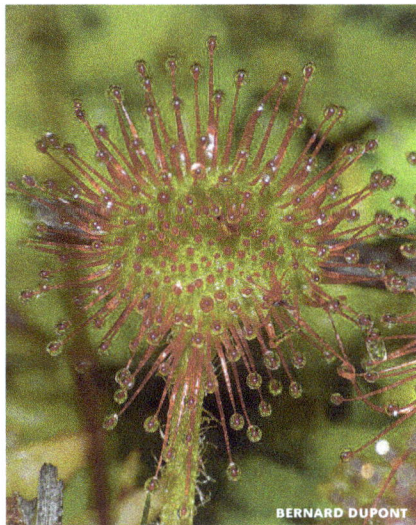

Drosera rotundifolia

Elatinaceae
Waterwort Family

Elatine WATERWORT

Small, branched, annual herbs of shallow water, shores and mud flats. Leaves simple, opposite, entire or toothed, with small membranous stipules. Flowers small, single from leaf axils, perfect; sepals, petals and stamens 2–3 (ours); styles 3; ovary superior, 3–4-chambered. Fruit a capsule with numerous small seeds.

Elatine WATERWORT
1 Flowers with 2 sepals and 2 petals, seeds all at base of fruit .... *E. minima*
1 Flowers with 3 sepals and 3 petals; seeds at differing levels in fruit .... *E. triandra*

Elatine minima (Nutt.) Fischer & C. A. Meyer
SMALL WATERWORT

DESCRIPTION Native annual herb, forming small mosslike mats on mud; plants smooth, with branches to 5 cm long. **Leaves** opposite, oblong to obovate, rounded at tip, to 4 mm long, petioles absent. **Flowers** small, single and stalkless in leaf axils, sepals 2, petals 2. **Fruit** a round capsule; seeds with rows of small, rounded pits.

HABITAT shallow water and wet shores along lakes and ponds, usually where sandy or mucky.

STATUS MW-OBL | NCNE-OBL

NOTE *Elatine minima* is typically a smaller plant than the otherwise similar *Elatine triandra*.

Elatine triandra Schkuh.
♦LONG-STEMMED WATERWORT

DESCRIPTION Introduced annual herb; plants small, matted, to 15 cm long, somewhat fleshy, smooth, branched from base, the branches sprawling or floating, often rooting at nodes. **Leaves** opposite, linear to obovate, 3–10 mm long and 1–3 mm wide, margins entire, petioles absent; stipules very small. **Flowers** small, single and stalkless in leaf axils, 1.5–2 mm wide, sepals 3, petals 3. **Fruit** a round capsule, 1–2 mm wide; seeds 0.5 mm long, ridged and with tiny, angled pits. **Flowering** July–Sept.

HABITAT mud flats or in shallow water of lakes and ponds.

STATUS NCNE-OBL

Elatine triandra

Ericaceae
Heath Family

SHRUBS OR SCARCELY WOODY SHRUBS. Leaves evergreen or deciduous, mostly alternate, simple, with entire or toothed margins. Flowers usually perfect (with staminate and pistillate parts), urn- or vase-shaped, white, pink, or cream-colored; stamens as many (or 2x as many) as petals. Fruit a berry or dry capsule.

Ericaceae HEATH FAMILY
1 Herbaceous plants; leaves all basal...... 2
1 Woody plants; leave alternate or opposite 3
2 Flowers white, single, nodding at end of stalk .... *Moneses uniflora*
2 Flowers pink, several in racemes .... *Pyrola asarifolia*
3 Leaves deciduous............ *Vaccinium* .... (blueberries)
3 Leaves evergreen 4
4 Leaves with shiny, orange-yellow resinous dots (especially on underside) .... *Gaylussacia baccata*

4 Leaves without resinous dots **5**
5 Plants creeping on ground surface **6**
5 Plants upright **7**
6 Stems covered with brown hairs; fruit a white berry . . . . *Gaultheria hispidula*
6 Stems not covered with brown hairs; fruit a red berry . . . . . . *Vaccinium* (cranberries)
7 Leaf margins distinctly rolled under **8**
7 Leaf margins not rolled under **10**
8 Leaf underside densely covered with woolly hairs . . . . *Rhododendron groenlandicum*
8 Leaf underside white; woolly hairs absent **9**
9 Leaves alternate . . *Andromeda polifolia*
9 Leaves opposite . . . . *Kalmia polifolia*
10 Leaves alternate . . . . *Chamaedaphne calyculata*
10 Leaves opposite . . . . *Kalmia angustifolia*

Andromeda BOG-ROSEMARY

Andromeda polifolia L.
♦BOG-ROSEMARY

DESCRIPTION Low, upright or trailing native shrub, 3–6 dm tall. Stems gray to blackish; twigs brown, with hairs in lines running down stems, or sometimes smooth. Leaves evergreen and leathery, often blue-green, linear or narrowly oval, 2–5 cm long and 3–10 mm wide, the tip sharp-pointed and tipped with a small spine, the base tapered to the stem or a short petiole, dark green above and whitened below by short stiff hairs; margins entire and distinctly rolled under. Flowers in drooping clusters at ends of branches, white or often pink, urn-shaped, 5-parted, 5–6 mm long, on curved stalks to 8 mm long. Fruit a rounded capsule to 5 mm wide, the style persistent from indented top of capsule; fruit drooping at first, but erect when mature. Flowering May–June.

SYNONYM *Andromeda glaucophylla*.

HABITAT sphagnum bogs, black spruce and tamarack swamps.

STATUS MW-OBL | NCNE-OBL | GP-OBL

Chamaedaphne
LEATHERLEAF

Chamaedaphne calyculata (L.) Moench
♦LEATHERLEAF

DESCRIPTION Upright native shrub to 1 m tall. Older stems gray, the outer bark shredding to expose the smooth, red inner bark. Twigs brown, with fine hairs and covered with small, round scales. Leaves evergreen and leathery, becoming smaller toward ends of flowering branches, oval, 1–5 cm long and 3–15 mm wide, the tip rounded or pointed, brown-green and smooth above, pale brown with a covering of small, round

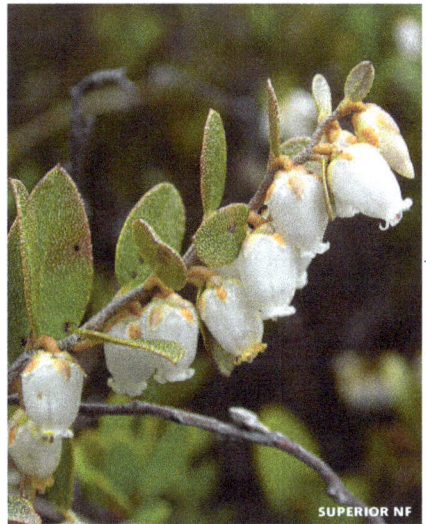

Andromeda polifolia

Chamaedaphne calyculata

scales below; margins entire or with small rounded teeth; petioles short. Flowers white, urn-shaped or cylindric, in 1-sided, leafy racemes, hanging from axils of reduced leaves near ends of branches; 5-parted, 5–7 mm long, on stalks 2–5 mm long. Fruit a brown, rounded capsule to 6 mm wide, the hairlike style persistent from indented top of capsule; capsules persisting on branches for several years. Flowering May–June. HABITAT open bogs, black spruce and tamarack swamps, peaty lakeshores and streambanks, often forming low, dense thickets. STATUS MW-OBL | NCNE-OBL | GP-OBL

Empetrum CROWBERRY

Empetrum nigrum L.
BLACK CROWBERRY

DESCRIPTION Much-branched, low native shrub; sometimes forming mats 1-2 m wide. Leaves evergreen and needlelike, dark green and leathery, linear-oblong, only 4–8 mm long, rounded or blunt at tip, narrowed at base to a short stalk, margins rolled under. Flowers small, pink to purple, single in axils of upper leaves, either staminate or pistillate or perfect. Fruit purple-black to black, berry-like, 4–6 mm wide, somewhat juicy, with 6-9 hard nutlets. Flowering July–Aug. HABITAT cedar and black spruce swamps, rocky shorelines; also in drier, sandy, pine woods. STATUS NCNE-FAC; uncommon in Mich (THR), mostly near Lake Superior.

Gaultheria WINTERGREEN

Gaultheria hispidula (L.) Muhl.
♦CREEPING SNOWBERRY

DESCRIPTION Low, creeping, matted shrub; native. Stems 2–4 dm long, covered with brown hairs. Leaves crowded, evergreen, oval to nearly round, 4–10 mm long and to 5 mm wide, abruptly tapered to tip, green above, underside paler, with brown, bristly hairs; margins rolled under; petioles short. Flowers few, single in leaf axils, white, bell-shaped, 4-parted, 2–4 mm long, on curved stalks 1 mm long. Fruit a translucent, juicy, white berry 5–10 mm wide, with a slight wintergreen-flavor. Flowering May–June. HABITAT open bogs, swamps, wet conifer woods, often in moss on hummocks or downed logs. STATUS MW-FACW | NCNE-FACW | GP-FACW

Gaylussacia HUCKLEBERRY

Gaylussacia baccata (Wangenh.) K. Koch.
♦BLACK HUCKLEBERRY

DESCRIPTION Medium-sized native shrub. Stems upright, much-branched, 3–10 dm long; branches brown, finely hairy when young, dark and smooth with age. Leaves alternate, deciduous, leathery, oval, 2–5 cm long and 1–2.5 cm wide; dark green above, paler below, both sides with shiny, orange-yellow resinous dots; margins entire, often fringed with small hairs; petioles 2–4 mm long. Flowers yellow-orange or red-tinged, cylindric, 5-lobed, 4–6 mm long, in ± 1-sided racemes from lateral branches, the flowers on short, gland-

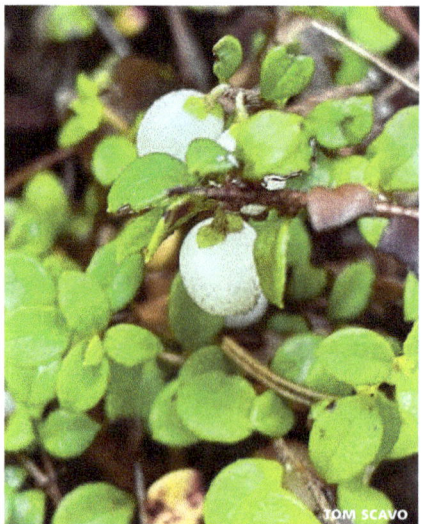

Gaultheria hispidula

dotted stalks 4–5 mm long. Fruit a red-purple to black, berrylike drupe, 6–8 mm long, with 10 nutlets; edible but seedy. Flowering May–June.
HABITAT open bogs, usually with tamarack and leatherleaf (*Chamaedaphne calyculata*); more common in dry, acid, sandy or rocky habitats.
STATUS MW-FACU | NCNE-FACU

Kalmia LAUREL

Evergreen shrubs. Leaves opposite, entire and leathery (our species). Flowers showy, 5-parted, in lateral or terminal clusters. Fruit a rounded capsule.

Kalmia LAUREL
1 Stems round; leaves oval, stalked; flowers in spreading or drooping clusters from leaf axils; n Mich. . . . . *K. angustifolia*
1 Stems flattened; leaves narrower, stalkless; flowers in upright terminal clusters; widespread in n portions of region . . . . *K. polifolia*

Kalmia angustifolia L.

♦ SHEEP-LAUREL

DESCRIPTION Native shrub, 6–10 dm tall. Older stems smooth and gray. Twigs round in section, brown and finely hairy when young. Leaves opposite or in whorls of 3, evergreen and leathery, oval to oval, 2–5 cm long and 5–20 mm wide, tip blunt or rounded; dark green above, paler and smooth or with scattered stalked glands below; margins entire and somewhat rolled under; petioles 3–10 mm long. Flowers showy, deep pink, several to many in lateral clusters from axils of previous year's leaves, saucer-shaped, 5-parted, 9–12 mm wide, on stalks to 2 cm long. Fruit a round capsule to 5 mm wide, the style persistent. Clusters of capsules may persist for several years. Flowering June–July.
HABITAT open bogs and wet conifer forests; sometimes in dry jack pine forests.
STATUS NCNE-FAC

Kalmia polifolia Wangenh.

♦ BOG-LAUREL

DESCRIPTION Low native shrub, to 6 dm tall. Stems older stems dark; twigs swollen at nodes, flattened and 2-edged in section, smooth, pale brown when young. Leaves opposite, evergreen and leathery, linear to narrowly oval, 1–4 cm long and 6–12 mm wide, tip blunt or narrowed to an abrupt point; dark green and smooth above, white below with a covering of short, white hairs, midrib on underside with large purple, stalked glands; margins entire and rolled under; petioles absent. Flowers showy, pale to rose-pink, in terminal clusters at ends of current year's branches, saucer-shaped, 5-parted, 8–11 mm wide, on stalks to 3 cm

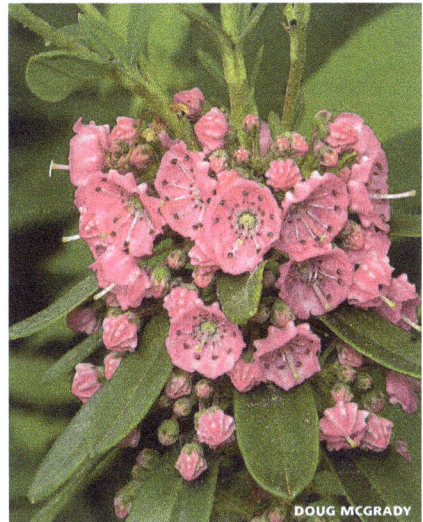

Gaylussacia baccata

Kalmia angustifolia

long. Fruit a round capsule to 6 mm wide, tipped by the persistent style, the capsules in upright clusters. Flowering May–June. **HABITAT** sphagnum peatlands, black spruce and tamarack swamps.
STATUS MW-OBL | NCNE-OBL

Moneses
ONE-FLOWERED SHINLEAF

Moneses uniflora (L.) A. Gray
◆ ONE-FLOWERED SHINLEAF

DESCRIPTION Low perennial herb, roots creeping. Stems to 10 cm long. Leaves decid-uous, mostly at base of plant, opposite or in whorls of 3, nearly round, margins entire or finely toothed. Flowers white, single at end of long stalk, nodding, 1–2 cm wide; petals 5. Fruit a round capsule, opening from top downward. Flow-ering July–Aug.
SYNONYM *Pyrola uniflora.*
HABITAT cedar swamps, wet conifer or mixed conifer and deciduous forests.
STATUS MW-FAC | NCNE-FAC | GP-FAC

Pyrola
SHINLEAF, WINTERGREEN

Pyrola asarifolia Michx.
◆ PINK SHINLEAF
DESCRIPTION Perennial herb, spreading by rhizomes. Stems to 3

dm long. Leaves persisting over winter, all near base of plant, kidney-shaped, 3–4 cm long and 3–5 cm wide, margins shallowly rounded-toothed; flower stalk with 1–3 small, scalelike leaves. Flowers nodding in a raceme; sepals triangular, 2–3 mm long; petals 5, 5–7 mm long, pink to pale purple. Fruit a capsule opening from base upward. Flowering June–Aug.
SYNONYM *Pyrola uliginosa.*
HABITAT cedar swamps, peatlands, marly wetlands, and interdunal wetlands.
STATUS MW-FACW | GP-FACU

Rhododendron
LABRADOR-TEA

Rhododendron groenlandicum
(Oeder) K.A. Kron & Judd. ◆ LABRADOR-TEA
DESCRIPTION Native shrub, to 1 m tall. Stems older stems gray or red-brown; twigs covered with woolly, curly brown hairs. Leaves alternate, evergreen and leathery, fragrant when rubbed, narrowly oval to oblong, 2.5–5 cm long and 5–20 mm wide, rounded at tip; dark green and smooth above, the midvein sunken; underside cov-ered with tan to rust-colored curly hairs; margins entire and rolled under; petioles short. Flowers creamy-white, in rounded clusters at ends of branches, 5-parted, to 1 cm wide, on finely hairy stalks 1–2 cm long.

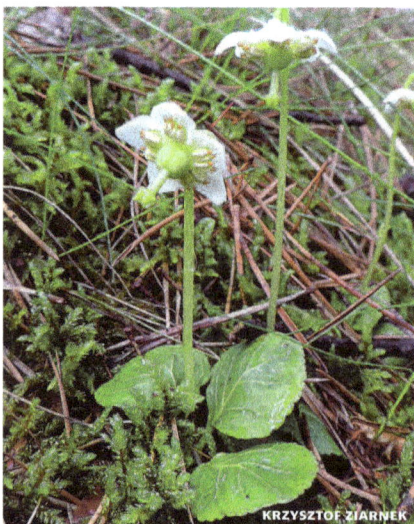

Kalmia polifolia

Moneses uniflora

Fruit a lance-shaped capsule 5–6 mm long, the style persistent and hairlike; capsules splitting at base to release numerous small seeds, the empty capsules persistent on stems for several years. Flowering May–June.

SYNONYM *Ledum groenlandicum.*

HABITAT sphagnum bogs, swamps and wet conifer forests.

STATUS MW-OBL | NCNE-OBL | GP-FACW

Vaccinium
BLUEBERRY, CRANBERRY

Deciduous or evergreen shrubs. Leaves alternate, simple. Flowers 4- or 5-parted, single in leaf axils or in clusters in axils or at ends of branches; ovary inferior. Fruit a many-seeded, red, blue, or black berry. The genus may be divided into subgroups, two of which occur in wetlands of the Upper Midwest region: **blueberries** (*Vaccinium angustifolium, V. corymbosum, V. myrtilloides*) and **cranberries** (*Vaccinium macrocarpon, V. oxycoccos, V. vitis-idaea*).

Vaccinium **BLUEBERRY, CRANBERRY**

1 Plants low and trailing; leaves evergreen, less than 5 mm wide; mature fruit a red berry (**cranberries**) **2**

1 Plants upright and bushy; leaves deciduous, more than 5 mm wide; mature fruit a blue to black berry (**blueberries**) **4**

2 Leaf underside with black, bristly glands; ne Minn, nw Wisc, and Isle Royale, Mich . . . . *V. vitis-idaea*

2 Leaf underside without black glands; widespread, especially in northern portions of region **3**

3 Leaves blunt or rounded at tip (and sometimes notched), pale below; bracts on flower stalk green and leaflike (more than 1 mm wide) . . . . *V. macrocarpon*

3 Leaves tapered to pointed tip, white below; bracts on flower stalk red and narrow (less than 1 mm wide) . . . . *V. oxycoccos*

4 Tall shrubs (usually 1–2 m tall) . . . . *V. corymbosum*

4 Low shrubs (usually less than 0.5 m tall) **5**

5 Leaf underside hairy; margins with small, bristle-tipped teeth . . . . *V. angustifolium*

5 Leaf underside without hairs or only sparsely hairy; margins entire, usually fringed with fine hairs . . . . *V. myrtilloides*

Vaccinium angustifolium Aiton

◆LOWBUSH BLUEBERRY

DESCRIPTION Low native shrub 1–6 dm tall, forming colonies from surface runners. Older stems red-brown to black. Twigs green-brown, with hairs in lines down stems, or sometimes smooth. Leaves deciduous, bright green oval, 2–5 cm long and 5–15 mm wide, smooth on both sides or sparsely hairy on veins; margins finely toothed with bristle-tipped teeth; peti-

Pyrola asarifolia

Rhododendron groenlandicum

oles very short. Flowers in clusters, opening before or with leaves, white or pale pink, narrowly bell-shaped, 5-parted, 4–6 mm long. Fruit blue and wax-covered, 5–12 mm wide, edible and sweet. Flowering April–June Fruit ripening July–Aug.

HABITAT sphagnum peatlands and wetland margins; also in dry, sandy openings and forests.

STATUS MW-FACU | NCNE-FACU | GP-FACU

Vaccinium corymbosum L.
HIGHBUSH BLUEBERRY

DESCRIPTION Native shrub, 1–3 m tall. Older stems red-brown to black. Twigs green-brown, with small hairs in lines down stems. Leaves deciduous, dark green, oval to ovate, 3–8 cm long and 1.5–3 cm wide, tapered to an often bristle-pointed tip; smooth above or sometimes finely hairy along veins, green or paler below, hairy at least on veins; margins entire with a fringe of hairs, or with small, gland-tipped teeth; petioles very short. Flowers in clusters at ends of branches, opening with leaves, white or pink, urn-shaped or cylindric, 5-parted, 6–10 mm long (our largest flowered blueberry). Fruit blue or blue-black, wax-covered, 7–12 mm wide, edible, sweet and juicy. Flowering May–June, Fruit ripening July–Aug.

HABITAT moist, low forests and swamps, shrubby peatlands and wetland margins.

STATUS MW-FACW | NCNE-FACW | GP-FACW

Vaccinium macrocarpon Aiton
⧫CRANBERRY

DESCRIPTION Evergreen trailing shrub; native. Stems slender, to 1 m or more long, with branches to 2 dm tall. Leaves leathery, oblong-oval, 5–15 mm long and 2–5 mm wide, rounded or blunt at tip, pale on underside; margins flat or slightly rolled under; petioles absent or very short. Flowers white to pink, 1 cm wide, 4-lobed, the lobes turned back at tips, single or in clusters of 2–6, on stalks 1–3 cm long, the stalks with 2 bracts, the bracts green, 2–4 mm long and 1–2 mm wide. Fruit red, 1–1.5 cm wide, edible but tart, often persisting over-winter. Flowering June–July Fruit ripening Aug–Sept.

SYNONYM *Oxycoccus macrocarpon*.

HABITAT sphagnum bogs, swamps and peaty pond margins.

STATUS MW-OBL | NCNE-OBL | GP-OBL

NOTE *Vaccinium macrocarpon* is the cultivated cranberry (grown in central and northern Wisc); fruit typically larger than the similar *Vaccinium oxycoccos*.

Vaccinium myrtilloides Michx.
⧫VELVETLEAF-BLUEBERRY

DESCRIPTION Native shrub, often forming colonies. Stems 3–6 dm long, red-brown to black with numerous wartlike lenticels; young twigs green-brown, densely velvety white-hairy. Leaves deciduous, thin and soft, oval, 2–5 cm long and 1–2.5 cm wide, dark green above, paler and soft hairy below, not waxy; margins entire and finely hairy; petioles very short. Flowers in clusters at ends of short, leafy branches, opening with leaves, creamy or green-white, tinged with pink, bell-shaped or short-cylin-

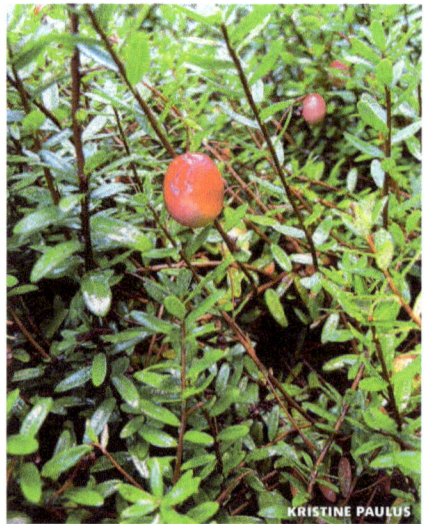

Vaccinium angustifolium

Vaccinium macrocarpon

dric, 5-parted, 4–5 mm long. Fruit blue, wax-covered, 6–9 mm wide; edible but tart. Flowering May–July. Fruit ripening July–Sept.

HABITAT sphagnum bogs and swamps; also in dry to moist woods and clearings.

STATUS MW-FACW | NCNE-FACW | GP-FACW

Vaccinium oxycoccos L.
♦ SMALL CRANBERRY

DESCRIPTION Ever-green trailing shrub; native. Stems slender, 0.5 m or more long, with upright branches 1–2 dm tall. Leaves leathery, ovate to oval or narrowly triangular, 2–10 mm long and 1–3 mm wide, pointed or rounded at tip, strongly whitened on underside; margins flat or strongly rolled under; petioles absent or very short. Flowers pale pink, 1 cm wide, 4-lobed, the lobes turned back at tips, single or in clusters of 2–4, on stalks 1–3 cm long, the stalks with 2 bracts usually at or below middle of stalk, the bracts red, scalelike, to 2 mm long and less than 1 mm wide. Fruit pale and red-speckled when young, becoming red, 6–12 mm wide, edible but tart. Flowering June–July. Fruit ripening Aug–Sept.

SYNONYM *Oxycoccus oxycoccos.*

HABITAT wet, acid, sphagnum bogs.

STATUS MW-OBL | NCNE-OBL | GP-OBL

Vaccinium vitis-idaea L.
♦ MOUNTAIN CRANBERRY

DESCRIPTION Low ever-green, trailing shrub; native. Older stems brown-black with peel-ing bark, branching, the branches upright, slender, 1–2 dm long, often forming mats. Twigs green-brown to red, ± smooth. Leaves alternate, leathery, oval to oval, 0.5–2 cm long and 4–15 mm wide, rounded or slightly indented at tip; upper surface dark green, shiny and smooth, paler and with dark bristly glands below; margins entire and rolled under; petioles hairy, 1–2 mm long. Flowers white to pink, bell-shaped and 4-lobed, style longer than petals, several in 1-sided clusters at ends of branches, the flowers on short glandular stalks, the stalks with 2 small bracts at base. Fruit a dark red berry, to 1 cm wide, persisting over winter, tart but edible, especially the following spring. Flowering June–July.

HABITAT sphagnum bogs; also in drier, sandy or rocky places.

STATUS MW-FAC | NCNE-FAC | GP-FAC; Wisc (END), Mich (END, Keweenaw County and Isle Royale); circumboreal.

NOTE also known as **lingen** or **red whortle-berry**. Mountain cranberry can be distin-guished from the more common cranberries (*V. macrocarpon* and *V. oxycoccos*) by the black, bristly, glandular dots on the leaf under-side.

Vaccinium vitis-idaea

Vaccinium myrtilloides

Vaccinium oxycoccos

Fabaceae
Pea Family

PERENNIAL SHRUBS (*Amorpha*) and herbs. Leaves alternate, pinnately divided, the terminal leaflet sometimes modified as a tendril (*Lathyrus*). Flowers in simple or branched racemes, perfect (with both staminate and pistillate parts), irregular, 5-lobed (only 1 lobe in *Amorpha*), the upper lobe (banner) larger than the other lobes, with 2 outer lateral petals (wings), and 2 inner petals which are partly joined (the keel), and enclosing the 10 stamens and style; pistil 1, ovary 1-chambered, maturing into a pod.

Fabaceae PEA FAMILY

1 Shrub; flowers 1-lobed, only the banner present . . . . *Amorpha fruticosa*
1 Herbs; flowers 5-lobed 2
2 Leaves with tendrils . . *Lathyrus palustris*
2 Leaves without tendrils 3
3 Leaves even-pinnate . . . *Senna hebecarpa*
3 Leaves not even-pinnate 4
4 Plants vinelike and sprawling; widespread . . . . *Apios americana*
4 Plants upright, not vinelike; w and sc Minn. . . . . *Astragalus agrestis*

Amorpha
FALSE INDIGO, LEAD-PLANT

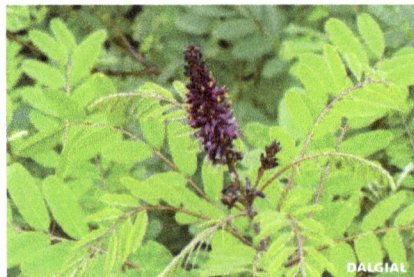

Amorpha fruticosa L.
◆FALSE INDIGO

DESCRIPTION Much-branched native shrub, mostly 1–3 m tall. Twigs tan to gray. Leaves pinnately divided, 5–15 cm long; leaflets 9–27, oval to obovate, 1–4 cm long and 0.5–3 cm wide, upper surface smooth, underside short-hairy, margins entire, petioles 2–5 cm long, stipules absent. Flowers dark purple, in dense spikelike racemes 2–15 cm long at ends of stems; petals 1-lobed, only the banner present, 3–5 mm long, folded to enclose the 10 stamens. Fruit an oblong pod, curved near tip, 5–7 mm long, spotted with glands, with 1–2 seeds. Flowering June–July.
HABITAT wet meadows, streambanks, shores.
STATUS MW-FACW | NCNE-FACW | GP-FACW

Amorpha fruticosa

Apios GROUND-NUT

Apios americana Medic.
◆COMMON GROUND-NUT

DESCRIPTION Perennial herbaceous native vine, rhizomes with a necklace-like series of 2 or more tubers; plants with milky juice. Stems to 1 m long, climbing over other plants. Leaves pinnately divided; main leaves with 5–7 leaflets; leaflets ovate, 4–6 cm long, tapered to a point, smooth to short-hairy beneath, margins entire. Flowers brown-purple, 10–13 mm long, single or paired, in crowded racemes from leaf axils. Fruit a linear pod, 5–10 cm long. Flowering July–Aug.
HABITAT floodplain forests, thickets, shores, wet meadows, low prairie.
STATUS MW-FACW | NCNE-FACW | GP-FAC
NOTE also known as wild-bean and Indian-potato.

Astragalus MILKVETCH

Astragalus agrestis Douglas
FIELD MILKVETCH

DESCRIPTION Perennial native herb, from slender rhizomes. Stems slender, 1–3 dm tall, smooth or sparsely hairy. Leaves pinnately divided; leaflets 13–21, oblong to lance-shaped, sparsely hairy on both sides, margins entire. Flowers purple, upright, 15–20 mm long, in crowded, long-stalked racemes, 2–4 cm long; sepals covered with mix of black and white hairs. Fruit an erect, ovate pod, 7–9 mm long, covered with stiff hairs. Flowering May.
HABITAT wet meadows and low prairie.
STATUS MW-FACW | NCNE-FACW | GP-FACU

ADDITIONAL SPECIES Alpine milk-vetch (*Astragalus alpinus)* is rare on moist, sandy or gravelly shores in n Minn and nw Wisc, disjunct from its main range further n and w (state endangered, Minn and Wisc). The violet flowers open in spring and continue to flower through the summer.

Lathyrus
VETCHLING, WILD PEA

Lathyrus palustris L. MARSH-PEA

DESCRIPTION Perennial vining native herb, spreading by rhizomes. Stems to 1 m long, strongly winged, climbing and clinging to surrounding plants by tendrils. Leaves pinnately divided, with 4–8 leaflets and a terminal leaflet modified into a tendril; leaflets linear to lance-shaped, 2–7 cm long and 3–20 mm wide; stipules prominent, ± arrowhead-shaped, 1–3 cm long; margins entire; petioles absent. Flowers in racemes from leaf axils, 2–6 flowers per raceme, red-purple, drying blue to blue-violet; sepals irregular, 7–10 mm long, the lowest lobe longest; petals 12–20 mm long. Fruit a flat, many-seeded pod, 3–5 cm long. Flowering June–Aug.

HABITAT conifer swamps, thickets, wet meadows, marshes, streambanks, calcareous fens, low prairie.
STATUS MW-FACW | NCNE-FACW | GP-FACW

Senna WILD SENNA

Senna hebecarpa (Fern.) Irwin & Barneby
◆ NORTHERN WILD SENNA

DESCRIPTION Native perennial herb. Stems erect, mostly unbranched, 1–2 m long, smooth or with scattered hairs on upper stem. Leaves compound, divided into 6–10 pairs of leaflets, the leaflets oblong or oval, 2–5 cm long and 1–2 cm wide, rounded at end but with a small, sharp tip; margins fringed with hairs; leaf petioles short. Flowers perfect, yellow, 1–2 cm long, in many-flowered, crowded racemes from upper leaf axils; sepals unequal; petals nearly alike, 10–15 mm long. Fruit a pod, 6–10 cm long and 5–8 mm wide, sparsely hairy; seeds flat, oval, longer than wide. Flowering July–Aug.

SYNONYM *Cassia hebecarpa.*

HABITAT floodplain forests, streambanks, fens.

STATUS MW-FACW | NCNE-FACW

NOTE the similar **Maryland senna (*Senna marilandica*)**, occurs in our region in s Wisc. It is found along streams, in thickets and open woods. In *Senna marilandica* the gland on the leaf petiole is not stalked (vs. stalked in *Senna hebecarpa*).

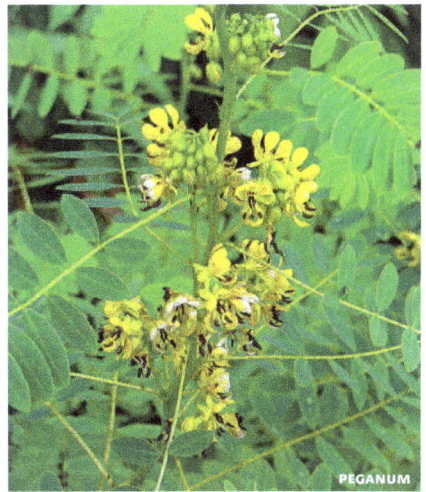

Apios americana

Senna hebecarpa

Fagaceae
Beech Family

Quercus OAK
Deciduous trees (our species). Leaves alternate, simple, lobed, pinnately veined. Staminate and pistillate flowers separate but on same tree. Fruit a nut (acorn) partially enclosed by a cuplike structure (cupule).

Quercus OAK
1 Leaf margins lobed, not bristle-tipped....
   ....*Q. bicolor*
1 Leaf margins tipped with sharp bristles . **2**
2 Acorn 10–13 mm long; acorn cup 10–16 mm wide ....*Q. palustris*
2 Acorn 15–30 mm long; acorn cup 16–25 mm wide.....................*Q. shumardii*

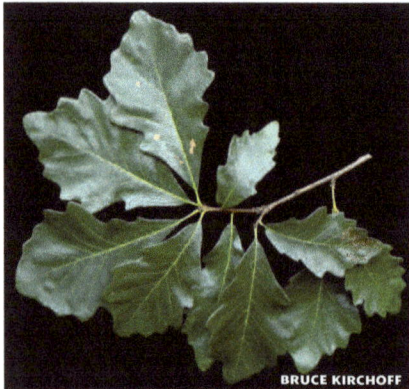

Quercus bicolor Willd.
◆SWAMP WHITE OAK
DESCRIPTION Native tree to 20 m tall; trunk to 1 m wide; crown broad and rounded. **Bark** gray-brown, deeply furrowed, becoming flaky; twigs gray to yellow-brown; buds clustered at branch tips, yellow-brown, smooth or sparsely hairy. **Leaves** alternate, broadest above middle, to 15 cm long and 10 cm wide, smooth or hairy on upper surface, white and soft hairy on underside; margins with coarse, rounded teeth or shallow lobes; petioles 2–3 cm long. **Flowers** staminate and pistillate flowers separate but on same tree, appearing with leaves in spring; staminate flowers in slender, drooping catkins, pistillate flowers in groups of 2–4. **Fruit** a pair of acorns, on stalks 2–3 cm long, the acorns ovate, pale brown, 2.5–4 cm long, the cup rough and hairy, covering about 1/3 of acorn.
HABITAT floodplain forests, low woods and swamps.
STATUS MW-FACW | NCNE-FACW

Quercus palustris Muenchh. ◆PIN-OAK
DESCRIPTION Native tree to 25 m tall; trunk less than 1 m wide; crown narrowly rounded; lower branches drooping. **Bark** light or dark brown, only shallowly furrowed. **Twigs** red-brown to dark gray; buds clustered at branch tips, red-brown to dark gray, smooth. **Leaves** alternate, to 15 cm long and 10 cm wide, divided more than halfway to middle into 5–7 bristle-tipped lobes, upper surface dark green and shiny, lower surface paler and with hairs on veins; petioles 3–4 cm long. **Flowers** staminate and pistillate flowers separate but on same tree, appearing with leaves in spring; staminate flowers in slender, drooping catkins, pistillate flowers in groups of 1–3. **Fruit** acorns in groups of 1–4, with or without stalks, the acorns hemisphere-shaped, pale brown, 10–14 mm long, the cup thin, finely hairy, covering less than 1/4 of acorn.
HABITAT floodplain forests, low wet woods, swamps; tolerant of periodic flooding.
STATUS MW-FACW | NCNE-FACW | GP-FAC

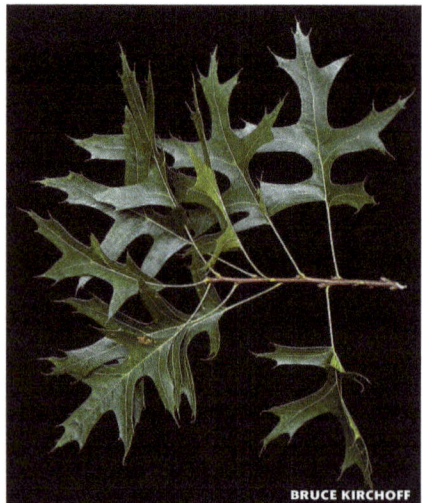

Quercus bicolor

Quercus palustris

Quercus shumardii Buckl.
SHUMARD'S OAK

DESCRIPTION Native tree to 30 m tall; bark gray-brown to dark brown, becoming deeply furrowed. Twigs gray to light brown, soon hairless. Leaves deeply 5–7-lobed, the lobes sometimes with teeth or small lobes; hairless except for tufts of branched hairs in underside vein-axils. Fruit acorn 1.5–3 cm long; the cup covering 1/4–1/3 of the nut.

HABITAT low wet woods, floodplains, moist slopes.

STATUS MW-FACW | NCNE-FACW | GP-FAC

Gentianaceae
Gentian Family

ANNUAL, BIENNIAL, OR PERENNIAL HERBS; plants usually smooth. Leaves simple, entire, opposite or whorled, stem leaves without petioles. Flowers often showy, perfect (with both staminate and pistillate parts), regular, single at end of stems or in clusters; petals 4-5, blue, purple, white or green, joined for at least part of their length; stamens 4 or 5. Fruit a 2-chambered, many-seeded capsule enclosed by the withered, persistent petals.

Gentianaceae GENTIAN FAMILY
1 Leaves reduced to small, narrow scales less than 3 mm long. . . . . *Bartonia*
1 Leaves well-developed, not scalelike. . . . **2**
2 Flowers pink . . . . *Sabatia angularis*
2 Flowers blue, purple-green, or white **3**
3 Petals 4, spurred at base; flowers green, tinged with purple . . . . . . *Halenia deflexa*
3 Petals 4, with fringed lobes; or petals 5 and not spurred; blue, purple or white **4**
4 Petals 4, fringed; flowers on stalks longer than the flowers; seeds covered with small bumps . . . . *Gentianopsis*
4 Petals 5, not fringed; flower stalks short or absent; seeds smooth **5**
5 Flowers 2.5–4 cm long, on short stalks; seeds flattened and winged . . . . *Gentiana*
5 Flowers 1–2 cm long, stalkless; seeds round . . . . *Gentianella*

Bartonia SCREW-STEM

Slender annual or biennial herbs. Stems pale green to yellow or purple. Leaves reduced to small opposite or alternate scales. Flowers small, 4-parted, green-white to green-yellow, bell-shaped, in slender panicles or racemes at ends of stems.

Bartonia SCREW-STEM
1 Leaf-scales mostly alternate, or the lower opposite . . . . *B. paniculata*
1 Leaf-scales mostly opposite. . *B. virginica*

Bartonia paniculata (Michx.) Muhl.
SCREW-STEM

DESCRIPTION Native annual or biennial herb. Stems slender, 2–4 dm long, upright or lax. Leaves small and scale-like, 1–2 mm long, mostly alternate, or the lower leaves opposite. Flowers yellow-white or greenish, 2–4 mm long, in panicles 5–20 cm long; the flowers on slender, arched and spreading stalks; sepals awl-shaped, 2 mm long; petals lance-shaped; anthers yellow. Fruit a capsule. Flowering Aug–Sept.

SYNONYM *Bartonia lanceolata*.

HABITAT tamarack swamps, fens, sphagnum bogs, open wetlands.

STATUS MW-OBL | NCNE-OBL | GP-OBL; Mich (THR); more common in se USA.

Bartonia virginica (L.) BSP.
♦YELLOW SCREW-STEM

DESCRIPTION Native annual or biennial herb. Stems slender, erect, yellow-green, 1–4 dm long. Leaves mostly opposite, small and scalelike, 1–2 mm long. Flowers green-yellow or green-white, 3–4 mm long, in a slender raceme or panicle, the branches and flower stalks opposite and upright; sepals awl-shaped; petals oblong, tapered to a rounded tip. Fruit a capsule 2–3 mm long. Flowering Aug–Sept.

HABITAT swamps (often in sphagnum moss), open bogs, wet woods and depressions, sandy shores and ditches.

STATUS MW-FACW | NCNE-FACW; Minn (END).

Gentiana GENTIAN

Perennial herbs, with thick, fibrous roots. Leaves opposite or whorled, simple, margins entire, petioles absent. Flowers large, blue, green-white or yellow, 5-parted, in clusters near ends of stems; petals forming a tubelike, shallowly lobed flower, the lobes alternating with a folded membrane as long or longer than petal lobes; stamens 5. Fruit a 2-chambered capsule.

Gentiana GENTIAN

1 Leaves and petal lobes fringed with tiny hairs (under 10x magnification); widespread in Upper Midwest region . . . . .
. . . . *G. andrewsii*

1 Leaves and petal lobes smooth; plants of northern portions of region **2**

2 Leaves dark green, linear, less than 1 cm wide; Michigan UP only . . . . *G. linearis*

2 Leaves light green, lance-shaped to ovate, larger leaves more than 1 cm wide; northern portions of region *G. rubricaulis*

Gentiana andrewsii Griseb.
◆ BOTTLE GENTIAN

DESCRIPTION Native perennial herb. **Stems** erect, single or few together, 2–8 dm long, unbranched, smooth.
Leaves opposite, lance-shaped, 4–12 cm long and 1–3 cm wide, margins fringed with hairs. **Flowers** 1 to many, stalkless in upper leaf axils, 3–5 cm long; sepals forming a tube around petals, the sepal lobes unequal, fringed with hairs; petals forming a tubelike flower, usually remaining closed, the folds between petal lobes finely fringed (use hand lens to see this) and longer than the petal lobes. **Fruit** a capsule; seeds winged. Aug-Sept.

HABITAT wet meadows, swamps and wet woods, thickets, low prairie, shores, ditches.

STATUS MW-FACW | NCNE-FACW | GP-FAC

Gentiana linearis Froelich
NARROW-LEAVED GENTIAN

DESCRIPTION Native perennial herb. **Stems** smooth, 2–8 dm long. **Leaves** dark green, linear to narrowly lance-shaped, 4–9 cm long and less than 1 cm wide, margins entire. **Flowers** opening slightly, several in a cluster at end of stem and upper leaf axils; sepal lobes green, linear, 4–10 mm long; petals blue (sometimes white), 3–5 cm long, the lobes ovate, rounded, not fringed. **Fruit** a capsule; seeds winged. Aug-Sept.

HABITAT wet meadows, shores, streambanks.

STATUS NCNE-FACW; Mich (THR); disjunct from main range in ne USA.

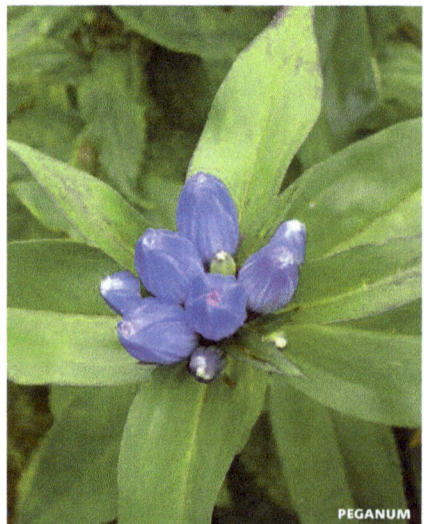

Bartonia virginica

Gentiana andrewsii

Gentiana rubricaulis Schwein.
GREAT LAKES GENTIAN

DESCRIPTION Native perennial herb. **Stems** smooth, 3–7 dm long. **Leaves** pale green, lance-shaped, 4–8 cm long and 2–3 cm wide, margins entire. **Flowers** 3–5 cm long, green-blue below, blue above, narrowly open, in a cluster at end of stem; sepal lobes oblong, 4–12 mm long, chaffy and translucent near base. **Fruit** a capsule; seeds winged.

SYNONYM *Gentiana linearis* var. *latifolia*.

HABITAT wet meadows, peatlands, streambanks, thickets, conifer swamps, Lake Superior rocky shores; soils usually calcium-rich.

STATUS MW-OBL | NCNE-OBL | GP-OBL

Gentianella GENTIAN

Annual or biennial herbs. Leaves opposite, stalkless, margins entire. Flowers 4–5-parted, blue to white, funnel-shaped or tubular; in clusters from ends of stems or upper leaf axils; petals withering and persistent around capsule.

Gentianella GENTIAN

1 Petal lobes with fringe of hairs at base; nw Minn only . . . . *G. amarella*
1 Petal lobes not fringed with hairs at base; s portions of our area . . . . . *G. quinquefolia*

Gentianella amarella (L.) Boerner
NORTHERN GENTIAN; FELWORT

DESCRIPTION Native annual or biennial herb. **Stems** unbranched, 1–6 dm long. **Leaves** lance-shaped, 2–6 cm long; margins entire. **Flowers** blue, 10–15 mm long, 5-parted (rarely 4-parted); single from middle leaf axils, several in a cluster from upper axils, on stalks less than 1 cm long; sepal lobes linear, 3–5 mm long; petals forming a tubelike to flared, funnel-like flower, petal lobes 3–5 mm long, the base of lobe fringed with hairs 2 mm long. **Fruit** a capsule. **Flowering** July–Aug.

SYNONYMS *Gentianella acuta, Gentiana amarella*.

HABITAT low prairie, wet or moist sandy or gravelly soil; nw Minn, more common in w USA.

STATUS MW-OBL | NCNE-OBL | GP-FACW

Gentianella quinquefolia (L.) Small
STIFF GENTIAN

DESCRIPTION Native annual or biennial herb. **Stems** 4-angled, 2–8 dm long, usually branched. **Leaves** at base spatula-shaped, upper leaves lance-ovate, 2–7 cm long; margins entire. **Flowers** blue (rarely white), 15–25 mm long, in clusters of 1–7 flowers at ends of stems or upper leaf axils, on stalks to 1 cm long; sepal lobes lance-shaped; petals forming a narrowly funnel-shaped flower, petal lobes 4–6 mm long, not fringed with hairs at base. **Fruit** a capsule; seeds round. Aug–Sept.

SYNONYMS *Gentianella occidentalis, Amarella occidentalis*.

HABITAT wet meadows, streambanks, moist woods; often where calcium-rich.

STATUS MW-FAC | NCNE-FAC; rare in Mich (THR).

Gentianopsis
FRINGED GENTIAN

Smooth, taprooted, annual or biennial herbs. Leaves opposite, stalkless, margins entire. Flowers 1 to several, showy, blue,

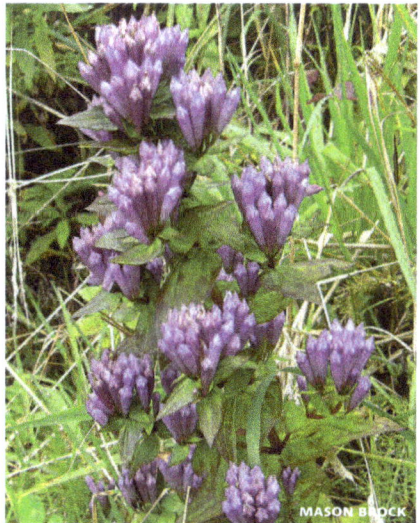

Gentianella quinquefolia

sometimes tinged with white on outside, long-stalked, at ends of stems and branches, 4-parted; sepals oblong cone-shaped; petals deeply lobed, forming a tubular or bell-shaped flower, the lobes ragged or fringed at tips and sometimes on sides, without a folded membrane between the lobes (present in *Gentiana*); stamens 4. Fruit a capsule; seeds covered with bumps.

Gentianopsis **FRINGED GENTIAN**

1 Upper leaves lance-shaped to ovate; petal lobes long-fringed across tip and sides, the fringes 2–5 mm long .... *G. crinita*
1 Upper leaves linear; tips of petal lobes ragged with short, fine teeth, often fringed on sides .... *G. virgata*

Gentianopsis crinita (Froelich) Ma
♦ **FRINGED GENTIAN**

DESCRIPTION Native annual or biennial herb. **Stems** erect, 2–7 dm long, usually branched above. Basal leaves spatula-shaped, smaller than stem leaves; stem leaves ovate, 2–6 cm long and 1–2.5 cm wide, the base usually clasping stem; margins entire. **Flowers** bright blue, 3–6 cm long, 4-parted, single at ends of main stems and branches, on stalks 5–20 cm long; sepals forming a tube, 1–2 cm long; petals joined to form a funnel-like to bell-shaped flower, the petal lobes fringed across tip and part way down sides with linear fringes 2–6 mm long. **Fruit** a capsule, broadest at middle. **Flowering** Aug–Oct.
SYNONYM *Gentiana crinita*.
HABITAT wet meadows, streambanks, ditches, wet woods; soils usually calcium-rich and sandy or gravelly.
STATUS MW-OBL | NCNE-FACW | GP-OBL

Gentianopsis virgata (Raf.) Holub
♦ **LESSER FRINGED GENTIAN**

DESCRIPTION Native annual herb, similar to *G. crinita* but smaller. **Stems** simple or few-branched, 1–5 dm long. Basal leaves spatula-shaped; stem leaves linear to linear lance-shaped, 2–5 cm long and 2–7 mm wide, tapered to a blunt tip, the base not clasping stem; margins entire. **Flowers** bright blue, 2–5 cm long, mostly 4-parted, single on stalks at ends of stems; sepal tube 6–15 mm long; petals forming a tubelike flower, flared toward tip, petal lobes ragged toothed across tips, often fringed on sides. **Fruit** a capsule. **Flowering** Sept–Oct.
SYNONYMS *Gentiana procera*, *Gentianopsis procera*.
HABITAT sandy and gravelly shores, wet meadows, calcareous fens, wetlands between dunes near Great Lakes; soils usually calcium-rich.
STATUS MW-OBL | NCNE-OBL | GP-OBL

Halenia SPURRED GENTIAN

Halenia deflexa (J. E. Smith) Griseb.
♦ **SPURRED GENTIAN**

DESCRIPTION Native annual. **Stems** erect, simple or few-branched, rounded 4-angled, 15–40 cm long. **Leaves** opposite, lower leaves spatula-shaped, narrowed to a petiole; stem leaves lance-shaped to ovate, 2–5 cm long and 1–2.5 cm wide, stalkless; margins entire. **Flowers** green, tinged with purple, 10–12 mm long, 4-parted, on stalks to 4 cm long, in loose clusters of

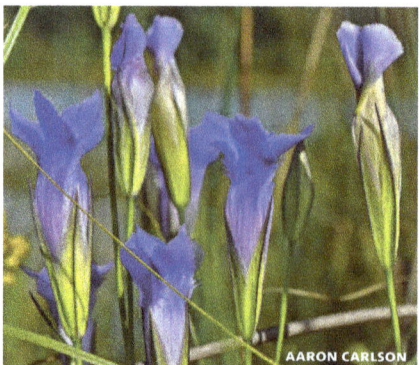

Gentianopsis crinita

Gentianopsis virgata

5–9 flowers at ends of stems; petals lance-shaped, usually with downward-pointing spurs at base, the spurs to 5 mm long. Fruit an oblong capsule. Flowering July–Aug. HABITAT cedar swamps, moist conifer woods (especially along shores), old logging roads. STATUS MW-FAC | NCNE-FAC | GP-FAC

Sabatia
MARSH-PINK, SEA-PINK

Sabatia angularis (L.) Pursh
◆ COMMON MARSH-PINK

DESCRIPTION Native biennial. Stems stout, 2–8 dm tall, sharply 4-angled, branches opposite, or alternate near base. Leaves opposite, ovate, 1–4 cm long, 3–7-nerved, clasping at base, margins entire. Flowers showy, rose-pink with a greenish center, single at ends of numerous branches, on stalks 1–3 cm long; sepal lobes linear, half as long as petals; petal lobes 1–2 cm long, obovate; style 2-parted. Fruit an oblong capsule. Flowering Aug–Sept. HABITAT sandy shores, wet areas between dunes. STATUS MW-FAC | NCNE-FAC | GP-FAC; Mich (THR).

ARX FORTIS

Sabatia angularis

SUPERIOR NF

Halenia deflexa

Grossulariaceae
Currant Family

Ribes GOOSEBERRY, CURRANT
Small to medium shrubs with upright to spreading stems, the stems smooth or with spines at nodes and sometimes also with bristles between the nodes. Leaves alternate, palmately 3–5-lobed; margins toothed. Flowers 1–several in clusters or few to many in racemes; perfect, regular, ovary inferior; sepals 5; petals 5, green to white or yellow, shorter than the sepals; stamens 5, alternate with the petals; styles 2. Fruit a many-seeded berry, usually topped by persistent, dry flower parts.

NOTE *Ribes* are of two types, currants and gooseberries. **Currants** lack spines and bristles (except in *Ribes lacustre*) and the stalk of the berry is jointed at its tip so that the berries detach from the stalk. **Gooseberries** have spines and bristles and the berry stalk is not jointed.

Ribes GOOSEBERRY, CURRANT
1 Stems with spines or bristles, at least at nodes 2
1 Stems without spines 5
2 Spines and bristles persistent; berries with gland-tipped hairs or bristles 3
2 Spines and bristles deciduous during summer; berries without glands or bristles 4
3 Stems bristly between nodes; leaves with disagreeable scent when rubbed; berries purple-black, with gland-tipped hairs . . . . *R. lacustre*
3 Bristles between nodes usually absent; leaves not ill-scented; berries wine-red and spiny . . . . *R. cynosbati*
4 Spines and bristles soft; leaves without glands on underside; bracts below flowers with long hairs . . . . *R. hirtellum*
4 Spines and bristles firm; leaves with scattered glands on underside (at least on veins); bracts below flowers with gland-tipped hairs . . . . *R. oxyacanthoides*
5 Stems upright; leaves dotted with resinous glands (at least on underside); berries black 6
5 Stems spreading and reclining; leaves not resin-dotted; berries red 7

6 Leaves dotted on both sides with yellow to brown resinous glands; flowers in drooping clusters .... *R. americanum*

6 Leaves resin-dotted only on underside; flowers in upright clusters .... *R. hudsonianum*

7 Plants with skunklike odor when rubbed; berries with gland-tipped hairs .... *R. glandulosum*

7 Plants without skunklike odor; berries smooth .... *R. triste*

Ribes americanum Miller
♦ EASTERN BLACK CURRANT

DESCRIPTION Native shrub, 1–1.2 m tall. Stems without spines or bristles, young stems finely hairy; branches upright to spreading; twigs gray-brown and smooth, black with age. Leaves 3–8 cm long and 3–10 cm wide, 3-lobed and usually with 2 additional shallow lobes at base, dotted with shiny, yellow to brown resinous glands, especially on underside, smooth or short-hairy above, hairy below; margins coarsely toothed; petioles hairy and resin-dotted, 3–6 cm long. Flowers creamy-white to yellow, bell-shaped, 8–12 mm long; 6–15 in drooping racemes 3–8 cm long; each flower with a linear bract longer than the flower stalk, the stalks 2–3 mm long; sepals 4–5 mm long, rounded; petals blunt, 2–3 mm long; stamens about equaling petals. Fruit an edible, smooth, black berry, 6–10 mm wide. Flowering April–June.

SYNONYM *Ribes floridanum.*

HABITAT moist to wet forests, swamps, marsh and lake borders, streambanks.

STATUS MW-FACW | NCNE-FACW | GP-FACW

Ribes cynosbati L.
♦ PRICKLY GOOSEBERRY

DESCRIPTION Native shrub to 6–9 dm tall. Branches upright to spreading. Stems and branches with 1–3 spines at nodes, outer bark peeling off, inner bark brown-purple to black; young stems brown-gray, finely hairy. Leaves 3–8 cm long and 3–7 cm wide, 3–5-lobed, the lobes rounded at tips; upper surface dark green, sparsely hairy, underside paler, finely hairy and with gland-tipped hairs along veins; margins with coarse, round teeth; petioles 2.5–4 cm long, finely hairy and with scattered gland-tipped hairs. Flowers green-yellow, bell-shaped, 6–9 mm long, in clusters of 2–3 from spurs on old wood, on stalks with gland-tipped hairs. Fruit a red-purple berry, 8–12 mm wide, covered with stiff, brown spines. Flowering May–June.

HABITAT occasional in wet woods, swamps, thickets and streambanks; more typical of moist hardwood forests (where the region's most common gooseberry).

STATUS MW-FAC | NCNE-FACU | GP-FACU

Ribes glandulosum Grauer
♦ SKUNK-CURRANT

DESCRIPTION Native shrub to 8 dm tall. Stems sprawling, spines and bristles absent;

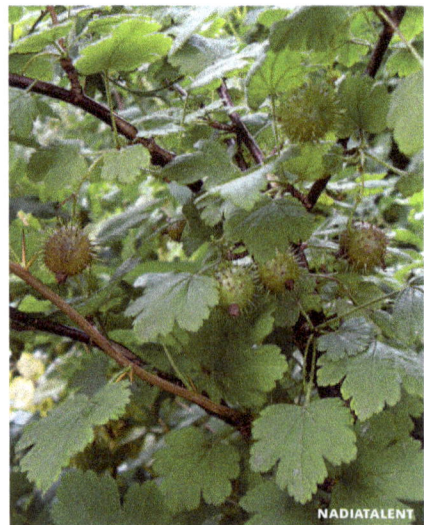

Ribes americanum

Ribes cynosbati

stems and leaves with skunklike odor when crushed; older stems smooth and dark as outer bark peels off; young stems smooth to finely hairy, brown-gray. Leaves 2–8 cm long and 4–8 cm wide, 3–5-lobed, smooth above, paler and finely glandular hairy below (at least along veins); margins toothed or double-toothed; petioles 3–6 cm long, finely hairy. Flowers yellow-green to purple, saucer-shaped, in loose upright clusters 3–6 cm long, on slender stalks; bracts very small, the stalks and bracts with gland-tipped hairs; sepals 2 mm long; petals 1–2 mm long. Fruit a dark red berry with bristles and gland-tipped hairs, 6 mm wide. June.

SYNONYM *Ribes prostratum.*

HABITAT cedar and tamarack swamps, cool wet woods, thickets and streambanks.

STATUS MW-FACW | NCNE-FACW | GP-OBL

Ribes hirtellum Michx.
♦ NORTHERN GOOSEBERRY

DESCRIPTION Native shrub to 9 dm tall. Stems upright, outer bark pale, soon peeling to expose dark inner layer; young stems gray and smooth, or with 1–3 slender spines at nodes and scattered bristles between nodes. Leaves 2.5–5 cm long and 2–5 cm wide, with 3 or 5 pointed lobes, upper surface dark green, smooth to sparsely hairy, lower surface paler, hairy at least along veins, without glands; margins coarsely toothed and fringed with hairs;

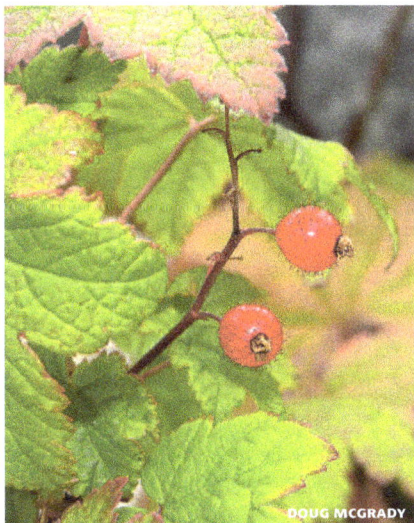

petioles 1–3 cm long, hairy, some of which are gland-tipped. Flowers green-yellow to purple, bell-shaped, 6–9 mm long, in clusters of 2–3 on short, smooth stalks; stamens as long or longer than sepals, the bracts fringed with long hairs. Fruit an edible, smooth, dark blue–black berry, 8–12 mm wide. June.

SYNONYM *Ribes huronense.*

HABITAT cedar and tamarack swamps, thickets, shores, rocky openings.

STATUS MW-FACW | NCNE-FACW | GP-FAC

Ribes hudsonianum Richards.
♦ HUDSON BAY CURRANT

DESCRIPTION Native shrub, 6–9 dm tall. Stems upright, spines and bristles absent; bark gray, with scattered yellow resin dots, peeling to expose inner purple-black bark. Leaves 5–9 cm long and 6–13 cm wide, 3–5-lobed, with unpleasant odor when rubbed, upper surface dark green and mostly hairless, underside paler, smooth to hairy and with yellow resin dots; margins coarsely toothed, the teeth with a hard tip; petioles 2.5–8 cm long, with fine hairs and resin dots. Flowers white, bell-shaped, 4–5 mm long, in small clusters on threadlike

Ribes glandulosum

Ribes hirtellum

stalks. Fruit a smooth, blue-black berry, 8–10 mm wide, barely edible. June.

HABITAT cedar swamps, wet conifer woods and streambanks.

STATUS MW-OBL | NCNE-OBL | GP-OBL

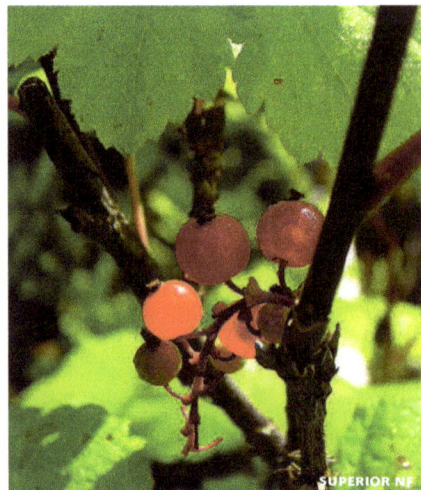

Ribes lacustre (Pers.) Poiret
◆PRICKLY CURRANT

DESCRIPTION Native shrub to 1 m tall. Stems upright or spreading, densely bristly, long-spiny at nodes; older bark gray, peeling to expose dark inner bark. Leaves 4–8 cm long and 4–7 cm wide, with 3–5 deeply parted, pointed lobes, upper surface dark green and mostly smooth, underside paler with scattered gland-tipped hairs; margins cleft into rounded teeth; petioles 2.5–4 cm long, with gland-tipped hairs. Flowers yellow-green to pinkish, saucer-shaped, 4–5 mm wide, on stalks with dark, gland-tipped hairs, in arching or drooping clusters. Fruit a purple-black berry covered with gland-tipped hairs, 9–12 mm wide. Flowering May–June.

HABITAT moist conifer woods, swamps, thickets, and rock outcrops.

STATUS MW-FACW | NCNE-FACW | GP-FACW

Ribes oxyacanthoides L.
NORTHERN or BRISTLY GOOSEBERRY

DESCRIPTION Native shrub to 1 m tall. Stems upright with 1–3 spines to 1 cm long at nodes and smaller spines scattered between nodes; young stems gray-brown and finely hairy. Leaves 2.5–5 cm long and 2–5 cm wide, with 3–5 blunt or rounded lobes, upper surface sparsely hairy, some hairs tipped with glands, underside resin-dotted, hairy, some gland-tipped, es-

pecially along veins; margins coarsely toothed and hairy, some hairs gland-tipped; petioles 0.5–3 cm long, with short hairs and scattered glands. Flowers green-yellow, bell-shaped, 6–9 mm long, in clusters of 2–3 on short stalks; stamens shorter than petals. Fruit a smooth, edible, blue-black berry, 9–12 mm wide. June.

SYNONYM *Ribes setosum.*

HABITAT rocky and sandy shores, rocky openings, stabilized dunes, moist woods.

STATUS MW-FACU | NCNE-FACU | GP-FACU; Wisc (THR).

Ribes triste Pallas
◆SWAMP RED CURRANT

DESCRIPTION Low native shrub, 0.4–1 m tall. Stems spreading or lying on ground and rooting at nodes, spines and bristles absent; older stems smooth, purple–black, young stems short-hairy. Leaves 4–10 cm long and 4–10 cm wide, with 3–5 broad lobes, dark green and mostly smooth above, paler and usually finely hairy below; margins with both rounded and sharp teeth, the teeth with a hard tip; petioles 2.5–6 cm long, with scattered gland-tipped hairs. Flowers green-purple, 4–5 mm wide, on stalks 1–4 mm long, in drooping clusters of 5–12. Fruit a smooth, red berry, 6–9 mm wide. Flowering May–June.

HABITAT wet woods and swamps, alder thickets, seeps.

STATUS MW-OBL | NCNE-OBL | GP-OBL

Ribes lacustre

Ribes triste

Haloragaceae
Water-Milfoil Family

PERENNIAL AQUATIC HERBS. Leaves alternate or whorled, finely dissected. Flowers small, stalkless in axils of leaves or bracts, 3- or 4-parted, regular, perfect (with both staminate and pistillate parts) or imperfect, petals small or absent. Fruit small and nutlike, dividing into 3 or 4 segments.

Haloragaceae WATER-MILFOIL FAMILY

1   Flowers 4-parted; leaves mostly whorled, emersed leaves reduced to small bracts . . . . *Myriophyllum*

1   Flowers 3-parted; leaves alternate, emersed leaves not bractlike . . . . *Proserpinaca*

Myriophyllum
WATER-MILFOIL

Perennial aquatic herbs. Stems submerged, sparsely branched, freely rooting at lower nodes. Leaves mostly whorled (alternate in *Myriophyllum farwellii*), pinnately divided into threadlike segments, upper leaves often reduced to bracts. Flowers small, mostly imperfect, stalkless in axils of upper emersed leaves (the floral bracts) or axils of underwater leaves; staminate flowers above pistillate flowers; perfect flowers (if present) in middle portion of spike; sepals inconspicuous; petals 4 or absent; stamens 4 or 8; pistil 4-chambered. Fruit nutlike, 4-lobed, each lobe (mericarp) with 1 seed, rounded on back or with a ridge or row of small bumps.

ADDITIONAL SPECIES **Parrot's-Feather** (*Myriophyllum aquaticum* (Vell.) Verdc., right), introduced from Brazil and potentially invasive, is known from southern LP of Mich. *Myriophyllum humile* (Raf.) Morong, more common in ne USA, is reported from Chippewa County in Wisc.

DUARTE FRADE

Myriophyllum WATER-MILFOIL

1   Leaves simple, reduced to small, blunt-tipped scales; stems erect from creeping rhizomes . . . . *M. tenellum*

1   Leaves dissected into narrow segments . **2**

2   Leaves alternate, ± opposite, or scattered on stem . . . . *M. farwellii*

2   Foliage leaves all whorled **3**

3   Flowers and bracts below flowers alternate on stem . . . . *M. alterniflorum*

3   Flowers and bracts below flowers whorled **4**

4   Bracts surrounding staminate flowers deeply cleft . . . . *M. verticillatum*

4   Bracts surrounding staminate flowers sharply toothed or entire **5**

5   Bracts sharply toothed and much longer than flowers . . . . *M. heterophyllum*

5   Bracts surrounding staminate flowers entire and not longer than flowers **6**

6   Leaf segments mostly 5–12 on each side of midrib; small bulbs (turions) produced at ends of stems and in upper leaf axils . . . . *M. sibiricum*

6   Leaf segments many, 12–20 on each side of midrib; turions absent . . . . . . *M. spicatum*

Myriophyllum alterniflorum DC.
ALTERNATE-FLOWER WATER-MILFOIL

DESCRIPTION Native perennial aquatic herb. Stems very slender. Leaves in whorls of 3– 5, usually less than 1 cm long and shorter than the stem internodes, pinnately divided. Flower spikes raised above water surface, 2–5 cm long; bracts mostly alternate, linear, shorter than the flowers; staminate flowers with 4 pink petals; stamens 8. Fruit segments 1–2 mm long, rounded on back and base.

HABITAT acidic lakes, Lake Superior shoreline.

STATUS NCNE-OBL

Myriophyllum farwellii Morong
FARWELL'S WATER-MILFOIL

DESCRIPTION Native perennial aquatic herb; plants entirely underwater, turions present at ends of stems. Leaves 1–3 cm long, dissected into threadlike segments, all or most leaves alternate, or ±

opposite, or irregularly scattered on stems. Flowers underwater, single in axils of foliage leaves; pistillate flowers with 4 purple petals; stamens 4, tiny. Fruit 2 mm long, each fruit segment with 2 small, bumpy, longitudinal ridges.

HABITAT ponds and small lakes.

STATUS MW-OBL | NCNE-OBL; rare in Mich UP (THR).

Myriophyllum heterophyllum

Michx. ♦ TWO-LEAF WATER-MILFOIL

DESCRIPTION Native perennial aquatic herb. Stems stout, to 3 mm wide, often red-tinged, to 1 m or more long. Leaves whorled, 1.5–4 cm long, divided into threadlike segments. Flowers in spikes raised above water surface, 5–30 cm long; floral bracts whorled, smaller than foliage leaves, ovate, sharply toothed, spreading or curved downward. Flowers both perfect and imperfect; petals of staminate and perfect flowers 1–3 mm long; stamens 4. Fruit olive, ± round, 2 mm long; fruit segments rounded or with 2 small ridges, beaked by the curved stigma. Flowering June–Aug.

HABITAT lakes, ponds and pools in streams; sometimes where calcium-rich.

STATUS MW-OBL | NCNE-OBL | GP-OBL

Myriophyllum sibiricum Komarov

♦ COMMON WATER-MILFOIL

DESCRIPTION Native perennial aquatic herb; plants often whitish when dried. Stems to 1 m or more long. Leaves in whorls of 3–4, 1–4 cm long, with mostly 5–10 threadlike segments on each side of midrib; internodes between whorls about 1 cm long. Flowers in spikes with whorled flowers and bracts, raised above water surface, red, clearly different than underwater stems, 4–10 cm long; flowers imperfect, the upper staminate, the lower pistillate; floral bracts much smaller than the leaves, oblong to obovate; staminate flowers with pinkish petals (absent in pistillate flowers), 2–3 mm long; stamens 8, the yellow-green anthers conspicuous when flowering. Fruit ± round, 2–3 mm long, the segments rounded on back.

SYNONYM *Myriophyllum exalbescens.*

HABITAT shallow to deep water of lakes, ponds, marshes, ditches and slow-moving streams; sometimes where calcium-rich.

STATUS MW-OBL | NCNE-OBL | GP-OBL

NOTE Eurasian water-milfoil (*Myriophyllum spicatum*), introduced from Eurasia, is similar to *Myriophyllum sibiricum,* but has more finely divided leaves (12–24 threadlike segments on each side of midrib) and larger floral bracts.

Myriophyllum heterophyllum

Myriophyllum sibiricum

Myriophyllum spicatum L.

◆EURASIAN WATER-MILFOIL

DESCRIPTION Introduced, invasive perennial, similar to *M. sibiricum.* Stems widening below head and curved to a horizontal position, usually many-branched near water surface, internodes between leaves mostly 1–3 cm long, turions absent. Leaves with more leaf segments per side (mostly 12–20) than in *Myriophyllum sibiricum;* lower flower bracts often divided into comblike segments and often longer than the flowers. Fruit segments 2–3 mm long. Aug–Sept.

HABITAT lakes, ponds, and streams, where it may form large, dense mats, hindering boating, swimming, and fishing.

STATUS MW-OBL | NCNE-OBL | GP-OBL; spreading across the e USA.

Myriophyllum tenellum Bigelow

SLENDER WATER-MILFOIL

DESCRIPTION Native perennial aquatic herb. Stems slender, 10–30 cm long, mostly upright and unbranched. Leaves absent or reduced to a few spaced scales. Flowers in spikes raised above water surface, 2–5 cm long; flower bracts mostly alternate, oblong to obovate, entire, shorter to slightly longer than the flowers. Fruit segments rounded on back and at base, 1 mm long.

HABITAT acidic lakes; often forming large colonies, especially in deep water.

STATUS MW-OBL | NCNE-OBL | GP-OBL

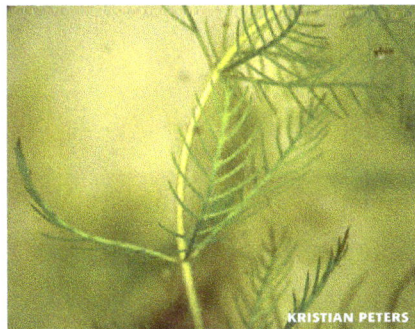

Myriophyllum verticillatum L.

◆WHORLED WATER-MILFOIL

DESCRIPTION Native perennial aquatic herb, similar to *M. sibiricum,* but plants often larger. Stems 5–25 dm long. Leaves in whorls of 4–5, with 9–17 threadlike segments along each side of midrib, 1–5 cm long; lower and middle internodes between whorls mostly less than 1 cm long. Flowers perfect, or the lower pistillate and upper staminate; in spikes 4–12 cm long, the floral bracts much smaller than the leaves, with comblike segments, mostly longer than the flowers; petals blunt-tipped, 2–3 mm long, smaller in pistillate flowers; stamens 8. Fruit ± round, 2–3 mm long, the segments rounded on back. Flowering July–Sept.

HABITAT lakes, ponds and quiet places in rivers.

STATUS MW-OBL | NCNE-OBL | GP-OBL

Proserpinaca MERMAID-WEED

Perennial aquatic herbs. Stems creeping, simple or with few branches. Leaves alternate, pinnately divided, or emersed leaves lance-shaped and sharply toothed. Flowers small, perfect, green or purple-tinged, 1–3 in axils of emersed leaves, stalkless; sepals triangle-shaped, persistent; petals absent, stamens 3, stigmas 3. Fruit nutlike, 3-angled, with 3 seeds.

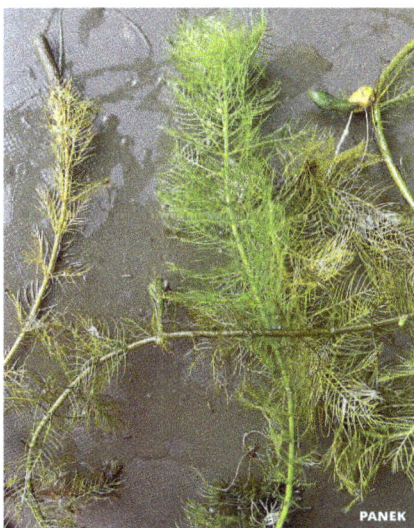

Myriophyllum spicatum

Myriophyllum verticillatum

Proserpinaca **MERMAID-WEED**

1 Upper leaves unlike lower leaves; upper leaves lance-shaped, margins sharply toothed; s Wisc, Mich........*P. palustris*

1 Upper and lower leaves alike, pinnately divided; rare, s Mich only....*P. pectinata*

Proserpinaca palustris L.
◆MARSH MERMAID-WEED

DESCRIPTION Native perennial aquatic herb, often forming large colonies. Stems horizontal at base and often rooting; the flower-bearing branches erect, 1–4 dm tall. Underwater leaves, if present, ovate in outline, 2–4 cm long, deeply divided into linear segments (as in photo below); emersed leaves narrowly lance-shaped, 2–6 cm long, margins with sharp, forward-pointing teeth. Fruit 2–5 mm long and as wide. Flowering June–Aug.

HABITAT Shallow water of ponds, streambanks and ditches, muddy shores, sedge meadows; usually where seasonally flooded.

STATUS MW-OBL | NCNE-OBL | GP-OBL

Proserpinaca pectinata Lam.
COMB-LEAF MERMAID-WEED

DESCRIPTION Native perennial aquatic herb. Underwater leaves similar to those of *P.*

Proserpinaca palustris

palustris; emersed leaves ovate in outline, 1.5–3 cm long, deeply divided into 6–12 pairs of narrow segments. Fruit 2–4 mm long and as wide. Flowering July–Aug.

HABITAT sandy ditches.

STATUS NCNE-OBL | GP-OBL; Atlantic coast species; disjunct in Mich (END).

Hypericaceae
St. John's-Wort Family

SMOOTH ANNUAL OR PERENNIAL HERBS (shrubby in *Hypericum kalmianum*). Stems usually unbranched below, branched in head. Leaves simple, opposite, dotted with dark or translucent glands (visible when held to a light), especially on underside; margins entire; petioles absent. Flowers few to many in clusters at ends of stems or from upper leaf axils, perfect, regular, sepals 5, petals 5, yellow or pink to green or purple; stamens 9–35, separate or joined near base into 3 or more groups; styles 3, ovary superior. Fruit a 3-chambered, many-seeded capsule.

Hypericaceae **ST. JOHN'S-WORT FAMILY**

1 Petals yellow; stamens 15–many ....
............................*Hypericum*

1 Petals pink or purple; stamens 9 ....
............................*Triadenum*

Hypericum ST. JOHN'S-WORT
Shrubs or herbs. Leaves opposite, sometimes dotted with black and/or small transparent glands; margins entire. Flowers in clusters at ends of stems and upper leaf axils, yellow, perfect, regular, sepals 5, petals 5, stamens 5–many, separate or joined into 3 or 5 bundles. Fruit a capsule.

Hypericum **ST. JOHN'S-WORT**

1 Shrub to 1 m tall..........*H. kalmianum*

1 Herbs...................................2

2 Plants 1–2 m tall; flowers 4 cm or more wide; styles 5 ....*H. ascyron*

2 Plants smaller; flowers less than 3 cm wide; styles 3................................3

3 Stamens 20 or more; styles joined, persisting on capsule as a straight beak..4

3 Stamens less than 20; styles separate to base and often spreading, capsules not beaked **5**

4 Leaves 4–6x longer than wide, margins rolled under; rare in Mich LP . . . . *H. adpressum*

4 Leaves 2–3x longer than wide, margins flat; more widespread in region . *H. ellipticum*

5 Sepals broadest above middle; fruit rounded at tip **6**

5 Sepals lance-shaped, broadest below middle; fruit tapered to tip **7**

6 Bracts leafy and oval, uppermost 0.5–2 mm wide; sepals much shorter than fruit . . . . *H. boreale*

6 Bracts narrow and awl-shaped, uppermost to 0.2 mm wide; sepals same length as fruit . . . . *H. mutilum*

7 Leaves 1-nerved (sometimes 3-nerved), tapered to base; sepals 2–4 mm long . . . . *H. canadense*

7 Leaves 5–7-nerved, rounded at base and broadest below middle; sepals 5–6 mm long . . . . *H. majus*

Hypericum adpressum Barton
SHORE ST. JOHN'S-WORT

DESCRIPTION Native perennial herb, spreading by rhizomes. **Stems** 3–8 dm long, sometimes spongy at base. **Leaves** numerous, ascending on stem, narrowly oblong, 3–6 cm long and 5–10 mm wide, tapered to a rounded tip; margins entire, rolled under; petioles absent. **Flowers** in ± flat-topped clusters at end of stems; sepals lance-shaped to ovate, 2–6 mm long; petals yellow, 6–8 mm long, stigmas 3 (sometimes 4). **Fruit** an ovate capsule, 4–6 mm long, gradually tapered to a beak formed by the persistent styles. **Flowering** July–Aug.

HABITAT marsh borders, sandy low prairie, ditches.

STATUS MW-OBL | NCNE-OBL; Mich (THR).

Hypericum ascyron L.
♦GIANT ST. JOHN'S-WORT

DESCRIPTION Native perennial herb. **Stems** upright, branched, 6–20 dm long. **Leaves** lance-shaped to oval, 4–10 cm long and 1–4 cm wide, base often clasping stem; petioles absent. **Flowers** few, 4–6 cm wide, mostly single on stalks from upper leaf axils; stamens numerous, joined at base into 5 bundles; petals bright yellow; styles 5, not persisting. **Fruit** an ovate, 5-chambered capsule, 15–30 mm long. **Flowering** July–Aug.

SYNONYM *Hypericum pyramidatum.*

HABITAT streambanks, ditches, fen and marsh margins.

STATUS MW-FAC | NCNE-FAC

Hypericum boreale (Britton) E. Bickn.
NORTHERN ST. JOHN'S-WORT

DESCRIPTION Native perennial herb. **Stems** round or slightly 4-angled, branched above. **Leaves** oval or oblong, rounded at ends and nearly clasping stem, 3–5-nerved, larger leaves 1–2 cm long and 0.5–1 cm wide; petioles absent. **Flowers** in clusters at ends of stems and from upper leaf axils; sepals blunt-tipped; petals yellow, 3 mm long; stamens 8–15; styles 3 (sometimes 4), less than 1 mm long. **Fruit** a 1-chambered purple capsule, 3–5 mm long. **Flowering** July–Sept.

HABITAT pond and marsh margins, low areas between dunes, open bogs.

STATUS MW-OBL | NCNE-OBL | GP-OBL

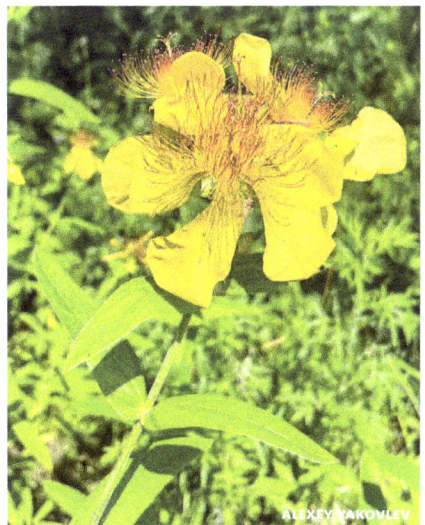

Hypericum ascyron

Hypericum canadense L.
CANADIAN ST. JOHN'S-WORT

DESCRIPTION Native annual or perennial herb, with short leafy stolons from base of plant. Stems upright, branched, 1–6 dm long. Leaves linear, 1–4 cm long and 1–4 mm wide, blunt-tipped, mostly 1-nerved, bracts much smaller; petioles absent. Flowers in open clusters at ends of stems and from upper leaf axils; sepals lance-shaped, 3–5 mm long; petals yellow, 2–3 mm long; stamens 12–22; styles 3 (sometimes 4), less than 1 mm long. Fruit a purple capsule 4–6 mm long. Flowering July–Sept.

HABITAT sandy shores, wetland margins.

STATUS MW-FACW | NCNE-FACW

Hypericum ellipticum Hook.
♦ PALE ST. JOHN'S-WORT

DESCRIPTION Native perennial herb, spreading by rhizomes. Stems 2–5 dm long, branched only in head. Leaves oval, 1–4 cm long and 1–1.5 cm wide, rounded at tip, narrowed at base and sometimes clasping stem; petioles absent. Flowers few to many, in clusters at ends of stems; sepals to 6 mm long; petals pale yellow, 6–7 mm long; stigmas 3 (sometimes 4), small. Fruit a 1-chambered capsule, 5–6 mm long, rounded to a short beak formed by the persistent styles. Flowering July–Aug.

HABITAT streambanks, sandy shores and flats, thickets, bogs.

STATUS MW-OBL | NCNE-OBL

Hypericum kalmianum L.
♦ KALM'S ST. JOHN'S-WORT

DESCRIPTION Branched native shrub to 1 m tall. Branches 4-angled, twigs flattened. Leaves linear, 2–4 cm long and 3–8 mm wide, often waxy on underside; margins sometimes rolled under; petioles absent. Flowers in clusters of 3–7 at ends of stems, yellow, 2–3.5 cm wide; stamens many, not joined; styles 5. Fruit a 5-chambered, ovate capsule, 7–10 mm long, beaked by the persistent style base. Flowering June–Sept.

HABITAT dunes (especially wet areas between dunes) and rocky lakeshores, mostly near Great Lakes, often on limestone or where calcium-rich.

STATUS MW-FACW | NCNE-FACW

Hypericum majus (A. Gray) Britton
♦ LARGE CANADIAN ST. JOHN'S-WORT

DESCRIPTION Native perennial herb, spreading from rhizomes or stolons. Stems upright, unbranched or branched above, 1–6 dm long. Leaves lance-shaped, 2–4 cm long and 3–10 mm wide, dotted with brown sunken glands, 5–7-nerved from base; leaf tip rounded, leaf base rounded or heart-shaped and weakly clasping; petioles absent. Flowers few to many in clusters at ends of stems and from upper leaf axils;

Hypericum ellipticum

Hypericum kalmianum

sepals lance-shaped, 4–6 mm long; petals yellow, equal to sepals but then shriveling to half the length of sepals; stamens 14–21, not joined; styles to 1 mm long. Fruit a red-purple ovate capsule, 5–7 mm long. Flowering July–Sept.

HABITAT streambanks, sandy, mucky or calcareous shores, low areas between dunes, marshes, wetland margins.

STATUS MW-FACW | NCNE-FACW | GP-FACW

Hypericum mutilum L.
SLENDER ST. JOHN'S-WORT

DESCRIPTION Native annual or perennial herb. Stems 1–8 dm long, branched above. Leaves lance-shaped to oval, 1–4 cm long, 3–5 nerved from base, petioles absent. Flowers in branched, leafy clusters at ends of stems and from upper leaf axils, upper leaves bractlike and 1–4 mm long; sepals linear and pointed at tip; petals pale orange-yellow, 2–3 mm long; stamens 5–16; styles 3, less than 1 mm long. Fruit a green capsule, 2–4 mm long. Flowering July–Sept.

SYNONYM *Hypericum parviflorum.*

HABITAT streambanks, wet meadows, marshes, ditches; usually where sandy.

STATUS MW-FACW | NCNE-FACW | GP-FACW

Triadenum
MARSH ST. JOHN'S-WORT

Smooth perennial herbs. Leaves opposite, entire and oval-shaped, ours dotted with small dark and transparent glands. Flowers pink to green-purple, in clusters at ends of stems and from leaf axils; stamens 9, in 3 groups of 3; sepals 5; petals 5; styles 3. Fruit a cylindric capsule.

Triadenum **MARSH ST. JOHN'S-WORT**

1   Sepals 3–4 mm long, oval and rounded at tip; styles mostly less than 1 mm long.....
.... *T. fraseri*

1   Sepals 5–8 mm long, lance-shaped and tapered to a tip; styles 2–3 mm long.......
.... *T. virginicum*

Triadenum fraseri (Spach) Gleason
⧫ MARSH ST. JOHN'S-WORT

DESCRIPTION Native perennial herb, with creeping rhizomes. Stems upright, mostly unbranched, red, smooth, 3–6 dm long. Leaves oval or ovate, 3–6 cm long and 1–3 cm wide, pinnately veined, rounded at tip, rounded or heart-shaped and clasping at the base, with dark dots and transparent glands on underside. Flowers in clusters at ends of stems and from leaf axils, often remaining closed; sepals 3–5 mm long, rounded at tip; petals pink to green-purple, 5–8 mm long; stamens

Hypericum majus

Triadenum fraseri

9, joined at base into 3 bundles, the bundles alternating with orange glands; styles 1–2 mm long. Fruit a purple, cylindric capsule, 7–12 mm long, abruptly narrowed to the 1 mm long persistent style beak. Flowering July–Aug.

SYNONYMS *Hypericum fraseri, Hypericum virginicum* var. *fraseri, Triadenum virginicum* subsp. *fraseri.*

HABITAT marshes, sedge meadows, open bogs, fens, sandy and calcium-rich shores.

STATUS MW-OBL | NCNE-OBL | GP-OBL

Triadenum virginicum (L.) Raf.
MARSH ST. JOHN'S-WORT

DESCRIPTION Native perennial herb spreading by rhizomes, similar to *T. fraseri* but with larger flowers. Stems upright, mostly unbranched, red, smooth, 3–6 dm long. Leaves oblong, oval or ovate, 3–6 cm long and 1–3 cm wide, pinnately veined, rounded at tip, ± heart-shaped and clasping at base, with dark dots and transparent glands on leaf underside. Flowers in clusters at ends of stems and from leaf axils; sepals 5–8 mm long, lance-shaped; petals pink to green-purple, 8–10 mm long; stamens 9, joined at base into 3 bundles; styles 1–2 mm long. Fruit a red-purple cylindric capsule, 8–12 mm long, gradually tapered to the 2–3 mm long persistent style beak. Flowering July–Aug.

SYNONYM *Hypericum virginicum.*

HABITAT sphagnum bogs, wet meadows, shores.

STATUS MW-OBL | NCNE-OBL | GP-OBL; Atlantic coastal species, extending inland to Wisc and Mich.

Juglandaceae
Walnut Family

Carya HICKORY

Carya laciniosa (Michx.) Loudon
◆SHELLBARK-HICKORY

DESCRIPTION Large native tree to 30 m, trunk to 1 m wide; lower branches drooping. Bark light gray, separating into long, vertical plates which curve

away from trunk at tips. Twigs stout, gray or brown, dotted with orange lenticels; buds dark brown, hairy, the outer bud scales tipped with a long, stiff point. Leaves alternate, pinnately divided into 5–9 leaflets; the leaflets lance-shaped to ovate, pointed at tip, rounded and unequal at base; upper surface dark green and smooth, underside paler and with soft hairs; margins finely toothed. Flowers tiny, staminate and pistillate flowers separate but borne on same tree, appearing after leaves begin to unfold in spring; staminate flowers in drooping catkins, pistillate flowers in clusters of 2–5. Fruit round but depressed at top, 3–5 cm wide, with an outer covering (husk) splitting into 4 sections; the nut compressed, ridged, sweet-tasting.

HABITAT floodplain forests.

STATUS MW-FACW | NCNE-FACW | GP-FAC

Lamiaceae
Mint Family

PERENNIAL, OFTEN AROMATIC HERBS. Stems usually 4-angled. Leaves simple, opposite, sharply toothed or deeply lobed. Flowers in leaf axils or in heads or spikes at ends of stems, perfect (with both staminate and pistillate parts), nearly regular to irregular; sepals 5-toothed or sometimes 2-lipped; petals white, pink, blue or purple, often 2-lipped; stamens 2 or 4; ovary 4-lobed, splitting into 4, 1-seeded nutlets when mature.

Carya laciniosa

Lamiaceae MINT FAMILY

1 Corolla regular or nearly so, with 4–5 lobes of equal length 2
1 Corolla irregular, 1- or 2-lipped 3
2 Stamens 2; plants not strongly scented ...
.... *Lycopus*
2 Stamens 4; plants strongly mint-scented .
.... *Mentha arvensis*
3 Upper lip of corolla absent, lower lip large
.... *Teucrium canadense*
3 Upper and lower corolla lips well-developed 4
4 Calyx with a rounded bump on upper side; petals blue .... *Scutellaria*
4 Calyx without bump on upper side; petal colors vary 5
5 Flowers in stalked clusters at branch ends, the clusters forming a flat-topped or rounded inflorescence ....
.... *Pycnanthemum virginianum*
5 Flowers ± stalkless, in spikes or crowded heads 6
6 Leaf margins ± entire 7
6 Leaf margins regularly toothed 8
7 Leaves linear .... *Stachys hyssopifolia*
7 Leaves broader, less than 4x longer than wide.................. *Prunella vulgaris*
8 Flowers paired in slender spikes; leaves and stems smooth ... *Physostegia virginiana*
8 Flowers 3 or more at each node; leaves and stems usually hairy.............. *Stachys*

Lycopus WATER-HOREHOUND

Perennial, ± unscented herbs. Stems erect, 4-angled. Leaves opposite, coarsely toothed or deeply lobed, smaller on upper stems; petioles short or absent. Flowers small, in clusters in middle and upper leaf axils, often appearing whorled; white to pink, the sepals and petals often dotted on outer surface, 4-lobed, stamens 2. Fruit a nutlet.

Lycopus WATER-HOREHOUND

1 Sepal lobes broad, triangular to ovate, to 1 mm long, shorter than or nearly as long as nutlets, midvein not prominent.........2
1 Sepal lobes slender, 1–3 mm long, longer than nutlets, midvein prominent........3
2 Leaves mostly less than 3 cm wide; stamens and styles visible, longer than petals; outer rim of nutlets taller than the inner rim....
.... *L. uniflorus*

2 Larger leaves 3 cm or more wide; stamens and styles hidden by petals; inner and outer rim of nutlets same height, the 4 nutlets appearing flat-topped across tops .....
.... *L. virginicus*
3 Main leaves stalkless............ *L. asper*
3 Leaves stalked 4
4 Upper surface of leaves with appressed hairs .... *L. europaeus*
4 Upper surface of leaves ± smooth 5
5 Plants spreading by rhizomes; lower and middle stem leaves pinnately parted at least near base; leaf undersides smooth, or with appressed hairs on midvein ....
.... *L. americanus*
5 Plants spreading by surface stolons; leaf margins sharp-toothed, not deeply parted; leaf undersides finely hairy ... *L. rubellus*

Lycopus americanus Muhl.
♦AMERICAN WATER-HOREHOUND

DESCRIPTION Native perennial herb, spreading by rhizomes, tubers absent. Stems erect, often branched, 2–8 dm long, upper stems smooth or short-hairy. Leaves opposite, lance-shaped, 3–8 cm long and 1–4 cm wide, with glandular dots, smooth or rough on upper surface, underside veins short-hairy; margins coarsely and irregularly deeply toothed or lobed, the lowest teeth largest; nearly stalkless or on short petioles. Flowers in dense, whorled clusters in leaf

Lycopus americanus

axils; sepal lobes narrow, sharp-tipped, 1–3 mm long, longer than fruits; petals white, sometimes pink to purple-dotted, 4-lobed, the upper lobe wider and notched. Fruit a nutlet, 1–2 mm long. Flowering July–Sept. HABITAT common; marshes, wet meadows, shores, streambanks, ditches, calcareous fens, wetland margins.
STATUS MW-OBL | NCNE-OBL | GP-OBL

Lycopus asper Greene
♦WESTERN WATER-HOREHOUND

DESCRIPTION Perennial emergent herb, spreading by rhizomes (tubers present) and also usually stolons. Stems erect, 2–8 dm long, simple or sometimes branched, hairy, at least on stem angles. Leaves opposite, oval to oblong lance-shaped, 3–10 cm long and 0.5–3 cm wide, smooth or rough; margins coarsely toothed; stalkless. Flowers in dense, whorled clusters in leaf axils; sepal lobes narrow, firm, sharp-tipped, 1–3 mm long, longer than nutlets; petals white, 4-lobed, only slightly longer than sepals. Fruit a nutlet, 1–2 mm long. Flowering July–Sept.
HABITAT shores and ditches, especially where disturbed, often with **American water-horehound** (*Lycopus americanus*); considered introduced in Wisc and Mich; more common in w USA.
STATUS MW-OBL | NCNE-OBL | GP-OBL

ANDREY ZHARKIKH

Lycopus asper

Lycopus europaeus L.
EUROPEAN WATER-HOREHOUND

DESCRIPTION Introduced perennial herb, similar to **American water-horehound** (*L. americanus*) and hybridizing with it, but often with slender stolons as well as rhizomes; plants more hairy, the hairs along veins of lower leaf-surface to 2 mm long. Leaves opposite, mostly wider and more bluntly toothed than in *Lycopus americanus*, upper surface usually with appressed hairs. Flowers white, in dense, whorled clusters in leaf axils; sepals sharp-tipped, 3–5 mm long. Fruit a nutlet, 1–2 mm long. Flowering July–Aug.
HABITAT moist to wet areas, often where disturbed.
STATUS MW-OBL | NCNE-OBL

Lycopus rubellus Moench
STALKED WATER-HOREHOUND

DESCRIPTION Native perennial herb, spreading by stolons, each stolon ending in a slender tuber, producing a single stem the following year; plants smooth to densely hairy. Leaves opposite, lance-shaped to oval, 5–10 cm long and 1–3 cm wide; margins entire to sharply toothed; petioles present or blades below lowest tooth concave and petiolelike. Flowers crowded in whorled clusters in leaf axils; sepal lobes 5, tapered to a tip but not awl-like; petals white, 5-lobed, 2x longer than sepals. Fruit a nutlet to 1.5 mm long and 1 mm wide, flat across the top. Flowering Aug–Sept.
HABITAT floodplain forests, wet woods and wet forest openings.
STATUS MW-OBL | NCNE-OBL | GP-OBL

Lycopus uniflorus Michx.
♦NORTHERN WATER-HOREHOUND

DESCRIPTION Native perennial herb, similar to **western water-horehound** (*L. asper*). Stems smooth or short-hairy, 1–5 dm long. Leaves opposite, lance-shaped to oblong, 3–6 cm long and 1–3 cm wide, margins with a few outward-pointing teeth, petioles short or nearly absent. Flowers in dense, whorled clusters in leaf axils; sepal

lobes broad, triangular to ovate, soft, rounded at tip, to 1 mm long, shorter to as long as nutlets; petals white or pink, 2–3 mm long, 5-lobed, longer than sepals. Fruit a nutlet 1–1.5 mm long. Aug–Sept.

HABITAT swamps, streambanks, thickets, wet meadows, open bogs, calcareous fens, ditches; often with American water-horehound (*Lycopus americanus*).

STATUS MW-OBL | NCNE-OBL | GP-OBL

Lycopus virginicus L.
VIRGINIA WATER-HOREHOUND

DESCRIPTION Native perennial herb, spreading by stolons (tubers usually absent). Stems 2–6 dm long, with dense covering of appressed hairs. Leaves opposite, lance-shaped to oval, 5–10 cm long and 1.5–5 cm wide, long-hairy, lower surface usually also with short, feltlike hairs; margins coarsely toothed, the lowest tooth just below middle of blade, the margin below tooth concave and petiolelike. Flowers in whorled clusters from leaf axils; sepals shorter than nutlets; petals white, 4-lobed, (upper lobe often notched). Fruit a nutlet, 1–2 mm long, the group of 4 nutlets ± flat across tips. Flowering July–Sept.

SYNONYM *Lycopus membranaceus*.

HABITAT floodplain forests.

STATUS MW-OBL | NCNE-OBL | GP-OBL; Mich (THR).

NOTE Probably hybridizes with **northern water-horehound** (*Lycopus uniflorus*) where

their ranges overlap, producing a hybrid swarm called *Lycopus × sherardii*.

Mentha MINT

Mentha arvensis L. ♦ FIELD-MINT

DESCRIPTION Native perennial herb, strongly mint-scented, spreading by rhizomes and often also by stolons. Stems 2–8 dm long, 4-angled, hairy at least on stem angles. Leaves opposite, ovate to lance-shaped or oval–lance-shaped, 2–7 cm long and 0.5–3 cm wide, smooth or hairy; margins with sharp, forward-pointing teeth; petioles short. Flowers small, white or light pink to lavender, hairy, crowded in whorled clusters in middle and upper leaf axils; sepals 2–3 mm long, hairy and glandular; petals ± regular to slightly 2-lipped, 4–6 mm long, glandular on outside, 4- or 5-lobed; stamens and style longer than petals. Fruit a smooth nutlet to 1 mm long, enclosed by the persistent sepals. Flowering July–Sept.

HABITAT wet meadows, marshes, swamps, thickets, streambanks, ditches, springs and other wet places.

STATUS MW-FACW | NCNE-FACW | GP-FACW

NOTE our only native species of *Mentha*.

Physostegia OBEDIENCE

Physostegia virginiana (L.) Benth.
♦ OBEDIENCE, FALSE DRAGON-HEAD

DESCRIPTION Native perennial herb, spreading by rhizomes. Stems erect, 5–15 dm long, often branched near top, 4-angled. Leaves opposite, oval to oblong

Lycopus uniflorus

Mentha arvensis

lance-shaped, 2–15 cm long and 1–4 cm wide, sometimes smaller upward; margins with sharp teeth; stalkless, not clasping. Flowers in several racemes 5–20 cm long, the stalks short-hairy; sepals 4–8 mm long, often with some gland-tipped hairs; petals pink-purple or white with purple spots, 1.5–3 cm long, short-hairy to smooth. Fruit a nutlet, 2–3 mm long. Flowering July–Sept.

SYNONYM *Dracocephalum virginianum.*

HABITAT sedge meadows, low prairie, shores, swamps, floodplain forests, ditches.

STATUS MW-FACW | NCNE-FACW | GP-FACW

Prunella SELF-HEAL

Prunella vulgaris L. ◆SELF-HEAL

DESCRIPTION Native perennial herb. Stems upright or sometimes spreading, 1–5 dm long, 4-angled. Leaves opposite, lance-shaped to oval or ovate, 2–8 cm long and 1–4 cm wide; lower leaves wider than upper; margins entire or with a few small teeth; petioles present. Flowers in dense spikes 2–5 cm long and 1–2 cm wide, with obvious bracts; sepals to 1 cm long, green or purple, with spine-tipped teeth; corolla 2-lipped, the upper lip hoodlike and entire, lower lip shorter and 3-lobed; petals blue-violet (rarely pink or white), 1–2 cm long; stamens 4, about as long as petals. Fruit a smooth nutlet. Flowering June–Oct.

HABITAT common in many types of wetlands (especially where disturbed): swamps, wet forest depressions, wet trails, streambanks, ditches; also in drier forests and fields.

STATUS MW-FAC | NCNE-FAC | GP-FAC

Pycnanthemum
MOUNTAIN-MINT

Pycnanthemum virginianum (L.)
Durand & B. D. Jackson
◆VIRGINIA MOUNTAIN-MINT

DESCRIPTION Native, strongly scented, perennial herb. Stems to 1 m long, branched above, 4-angled, angles short-hairy. Leaves numerous, opposite, narrowly lance-shaped, 3–4 cm long and to 1 cm wide (leaves in heads much smaller), upper surface smooth, with 3–4 pairs of lateral veins, undersides often finely hairy on midvein; margins entire but fringed with short, rough hairs; ± stalkless. Flowers small, 2-lipped, in branched, crowded clusters at ends of stems and branches from upper leaf axils; sepals short woolly-hairy; petals white, purple-spotted. Fruit a 4-parted nutlet. Flowering July–Sept.

HABITAT wet meadows, marshes, tamarack swamps, calcareous fens, low prairie.

STATUS MW-FACW | NCNE-FACW | GP-FAC

ADDITIONAL SPECIES *Pycnanthemum muticum* is known from wet meadows in s LP of Mich (state threatened). Its main leaves

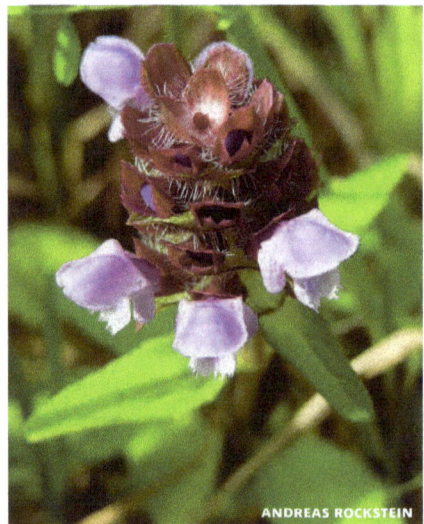

FRITZ REYNOLDS
Physostegia virginiana

ANDREAS ROCKSTEIN
Prunella vulgaris

are ovate. Other species of *Pycnanthemum* in our region typically occur on dry uplands.

Scutellaria SKULLCAP

Perennial herbs, spreading by rhizomes. Stems erect or spreading, 4-angled. Leaves opposite, ovate to lance-shaped, margins toothed, stalked or ± stalkless. Flowers blue or blue with white markings, single on short stalks in axils of middle and upper leaves, or in racemes from leaf axils; calyx 2-lipped, with a rounded bump on upper side; corolla 2-lipped, pubescent on outer surface, upper lip hood-like, lower lip ± flat, 3-lobed; stamens 4, ascending into the upper corolla lip. Fruit a 4-parted nutlet.

Scutellaria SKULLCAP

1 Flowers single in leaf axils; flowers more than 15 mm long .... *S. galericulata*
1 Flowers in racemes from leaf axils; flowers to 8 mm long .... *S. lateriflora*

Scutellaria galericulata L.

♦ MARSH *or* HOODED SKULLCAP

DESCRIPTION Native perennial herb. Stems erect or spreading, 2–8 dm long, unbranched or branched, 4-angled, short-hairy at least on angles of upper stem. Leaves opposite, lance-shaped to narrowly ovate, 2–6 cm long and 0.5–2.5 cm wide, upper surface smooth, underside short-hairy, tapered to a blunt tip; margins with low, rounded, forward-pointing teeth; petioles very short. Flowers 2-lipped, single in leaf axils (and paired at nodes), on stalks 1–3 mm long; sepals 3–6 mm long; petals blue, marked with white, 15–25 mm long. Fruit a nutlet. Flowering June–Sept.

SYNONYM *Scutellaria epilobiifolia.*

HABITAT shores, streambanks, marshes, wet meadows, swamps, thickets, ditches.

STATUS MW-OBL | NCNE-OBL | GP-OBL

Scutellaria lateriflora L.

BLUE SKULLCAP

DESCRIPTION Native perennial herb. Stems 2–6 dm long, usually branched, 4-angled, short-hairy on upper stem angles or smooth. Leaves opposite, ovate to lance-shaped, 3–8 cm long and 1.5–5 cm wide, smooth; margins coarsely toothed; petioles 0.5–2 cm long. Flowers 2-lipped, in elongate racemes from leaf axils; sepals 2–4 mm long; petals blue (rarely pink or white), 5–8 mm long. Fruit a nutlet. Flowering July–Sept.

HABITAT shores, streambanks, wet meadows, marshes, swamps, shaded wet places.

STATUS MW-OBL | NCNE-OBL | GP-FACW

Stachys HEDGE-NETTLE

Erect, perennial herbs, spreading by rhizomes; plants usually hairy. Stems 4-angled. Leaves opposite, margins entire or toothed, stalkless or with short petioles. Flowers in interrupted spikes at ends of stems, appearing whorled in ± evenly spaced clusters; sepals ± regular, with 5 equal teeth; corolla 2-lipped, petals pink, often with purple spots or mottles, upper lip concave, entire, lower lip spreading, 3-lobed; stamens 4, ascending under the upper lip. Fruit a brown, 4-lobed nutlet, loosely enclosed by the persistent sepals.

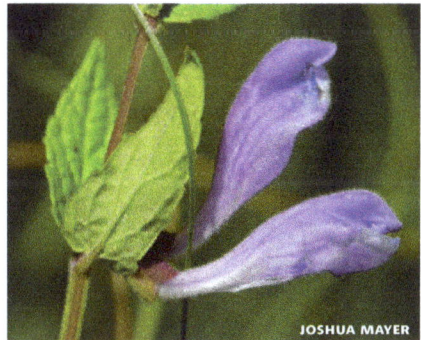

Pycnanthemum virginianum

Scutellaria galericulata

Stachys HEDGE-NETTLE

1 Stems smooth; leaves entire, ± stalkless, linear, less than 1 cm wide *S. hyssopifolia*
1 Stems hairy, at least on stem angles; leaves with sharp, forward-pointing teeth, stalkless or stalked, larger leaves more than 1 cm wide **2**
2 Stems hairy on angles and sides......... .... *S. palustris*
2 Stems hairy only on stem angles .... *S. tenuifolia*

Stachys hyssopifolia Michx.
HYSSOP HEDGE-NETTLE

DESCRIPTION Native perennial herb. Stems 3–5 dm long, often branched from base, 4-angled, smooth, or sometimes with fine hairs at nodes and on stem angles. Leaves opposite, smooth, linear, 2–6 cm long and 3–10 mm wide, uppermost leaves reduced to short bracts; margins entire or with a few low teeth; ± stalkless. Flowers in spaced, several-flowered clusters at ends of stems and on branches from leaf axils; sepals smooth or with a few hairs; petals light purple, smooth. Fruit a nutlet. Flowering July–Sept.

SYNONYM *Stachys atlantica.*

HABITAT sandy shores and wet dune areas, wet meadows, low prairie.

STATUS MW-FACW | NCNE-FACW

Stachys palustris L.
♦ MARSH HEDGE-NETTLE

DESCRIPTION Introduced perennial herb. Stems 3–8 dm long, unbranched or branched, 4-angled, stiffly hairy on angles and with short, gland-tipped hairs on sides. Leaves opposite, lance-shaped to oblong, 4–12 cm long and 2–5 cm wide, softly hairy on both sides; margins with rounded, forward-pointing teeth; stalkless or with short petioles. Flowers in clusters of 6–10 in an interrupted spike at end of stem (sometimes also clustered in upper leaf axils); sepals 5–8 mm long, with long, glandless hairs and shorter gland-tipped hairs; petals purple to pale red with purple spots, 9–13 mm long. Fruit a nutlet. Flowering June–Aug.

HABITAT marshes, wet meadows, ditches, thickets, streambanks, openings in swamps.

STATUS MW-OBL | NCNE-OBL

Stachys tenuifolia Willd.
♦ SMOOTH HEDGE-NETTLE

DESCRIPTION Native perennial herb. Stems 4–10 dm long, 4-angled, smooth, or with downward-pointing bristly hairs on stem angles. Leaves opposite, lance-shaped to ovate, 6–14 cm long and 2–6 cm wide, ± smooth; margins with sharp, forward-pointing teeth; petioles slender, 1–

Stachys palustris

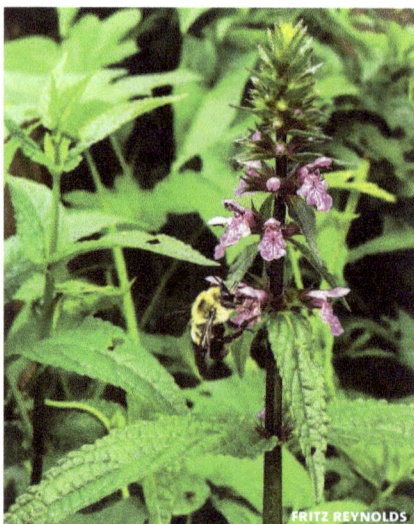

Stachys tenuifolia

2 cm long or absent. Flowers in interrupted spikes at ends of stems or also in upper leaf axils; sepals 5–7 mm long, smooth; petals pale red to purple, 1.5–2.5 cm long. Fruit a nutlet. Flowering July–Sept.

HABITAT floodplain forests, shores, streambanks, thickets, wet meadows.

STATUS MW-OBL | NCNE-FACW | GP-FACW

Teucrium
GERMANDER, WOOD-SAGE

Teucrium canadense L.
◆ AMERICAN GERMANDER

DESCRIPTION Native perennial herb, spreading by rhizomes. Stems 3–10 dm long, mostly unbranched, 4-angled, long-hairy. Leaves opposite, lance-shaped or oblong, 4–12 cm long and 1.5–5 cm wide, upper surface smooth or sparsely hairy, underside with dense, matted hairs, margins irregularly finely toothed, petioles 5–15 mm long. Flowers in a dense spikelike raceme, 5–20 cm long; bracts present and narrowly lance-shaped; flowers on stalks 1–3 mm long; sepals ± regular, purple or green, 4–7 mm long, covered with long silky hairs and very short glandular ones; corolla irregular, 10–16 mm long, with short gland-tipped hairs, upper lip absent, lower lip large; petals pink to purple; stamens 4, arched over the corolla. Fruit a golden nutlet. Flowering July–Sept.

HABITAT marshes, wet meadows, shores, streambanks, thickets, floodplain forests.

STATUS MW-FACW | NCNE-FACW | GP-FACW

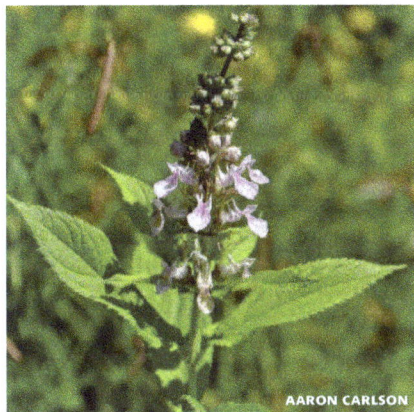

Lauraceae
Laurel Family

Lindera SPICEBUSH

Lindera benzoin (L.) Blume
◆ NORTHERN SPICEBUSH

Much-branched native shrub to 4 m tall; bark gray-brown and rough with age. Leaves alternate, aromatic, oval to obovate, 6–12 cm long and to 6 cm wide, broadest at or above middle, larger leaves abruptly tapered to tip, smaller leaves with rounded tips; margins entire; petioles 7–10 mm long. Flowers yellow, 6 mm wide, in clusters of 4–6 at nodes of previous year stems, appearing before leaves in spring, the staminate and pistillate flowers separate and on different plants. Fruit aromatic, green when young, bright red and shiny when mature, 6–10 mm long, berrylike on a short stalk. Flowering April–May.

HABITAT moist to wet deciduous forests and swamps, streambanks, occasionally in cedar swamps; usually where shaded.

STATUS MW-FACW | NCNE-FACW | GP-FACW

Lentibulariaceae
Bladderwort Family

INSECTIVOROUS HERBS. Leaves in a basal rosette (*Pinguicula*); or floating, or in peat, muck, or wet soil (*Utricularia*). Flowers perfect (with both staminate and pistillate parts), irregular, 2-lipped, sometimes with a spur, 1 to several on an erect stem; stamens 2. Fruit a capsule.

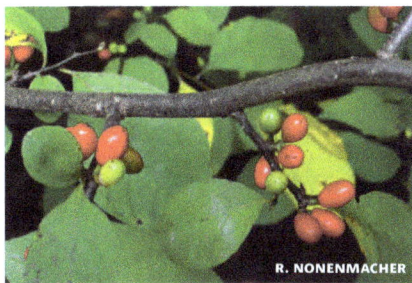

AARON CARLSON
Teucrium canadense

R. NONENMACHER
Lindera benzoin

Pinguicula BUTTERWORT

Pinguicula vulgaris L.

◆COMMON *or* VIOLET BUTTERWORT, BOG VIOLET

DESCRIPTION Native perennial herb. Small insects are trapped by the sticky, slimy surface of the yellow-green leaves. Leaves 3–6 in a basal rosette, ovate or oval, 2–5 cm long, blunt-tipped, narrowed to base, upper surface sticky; margins inrolled. Flowers single atop a leafless stalk (scape) 5–15 cm long; corolla violet-purple, spurred, 2-lipped, the upper lip 3-lobed, the lower lip 2-lobed, 1.5–2 cm long (including spur). Fruit a 2-chambered capsule. Flowering June–July.

HABITAT rock crevices in sandstone cliffs along Lake Superior, wet areas between dunes, marl flats and calcareous fens; usually found with Mistassini primrose (*Primula mistassinica*).

STATUS NCNE-OBL; Wisc (END).

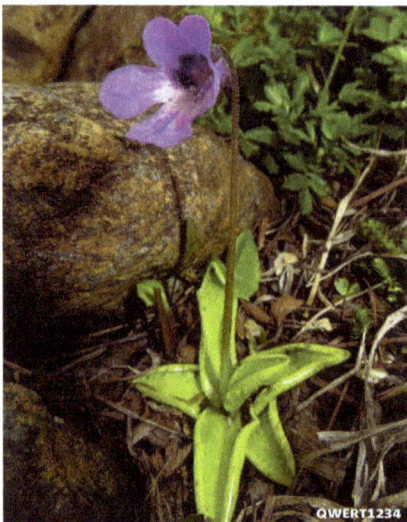

Pinguicula vulgaris

Utricularia BLADDERWORT

Aquatic or wetland, annual or perennial herbs. Leaves underwater, alternate, entire or dissected into many linear segments, some with bladders which trap tiny aquatic invertebrates; or leaves in wet soil and rootlike or absent. Flowers perfect, irregular, 1 to several in a raceme atop stalks raised above water or soil surface, each flower subtended by a small bract; corolla yellow or purple, similar to a snapdragon flower, 2-lipped, the upper lip erect, entire or slightly 2-lobed, lower lip entire or 3-lobed, the corolla tube extended backward into a sac or spur, stamens 2. Fruit a many-seeded capsule.

NOTE Bladderworts have small sacs ("bladders") which capture aquatic invertebrates. When the animal brushes trigger hairs on the bladder, the bladder quickly inflates, drawing in the organism, which is then digested.

ADDITIONAL SPECIES *Utricularia ochroleuca* R. W. Hartm., discovered in 2019 in Keweenaw County in Mich UP, Plants found at the edge of a beaver pond.

5 Scapes with a whorl of emersed leaves near the middle . . . . *U. radiata*

5 Scapes leafless. **6**

6 Leaf divisions flat in section **7**

6 Leaf divisions round or threadlike **8**

7 Bladders borne on leaves; smallest leaf divisions entire (visible with a 10x hand lens); flower with a sac or spur much shorter than lower lip . . . . *U. minor*

7 Bladders on branches separate from leaves; smallest leaf divisions finely toothed, the teeth spine-tipped; flower with a spur as long as lower lip . . . . *U. intermedia*

8 Plants large; leaves floating; scapes 1 mm or more wide; flowers 13 mm or more long, 5 or more per head; larger bladders more than 2 mm wide. . . . . *U. macrorhiza*

8 Plants smaller; leaves floating or creeping on lake bottom or wet shores; scapes threadlike; flowers to 12 mm long, 1–3 per head; larger bladders mostly less than 2 mm wide. **9**

9 Plants forming tangled masses, creeping on bottom in shallow water or on muck or drying pond edges; often with emergent scapes with at least 1 normal flower; cleistogamous flowers absent . . . *U. gibba*

9 Plants forming a delicate mass of floating leaves; emergent scapes with normal flowers rare; cleistogamous flowers common, on stalks 4–8 mm long . . . . *U. geminiscapa*

Utricularia cornuta Michx.
◆LEAFLESS BLADDERWORT

DESCRIPTION Native annual or perennial herb. **Stems** and leaves underground, roots with tiny bladders. **Flowers** yellow, with a downward-pointing spur 6–15 mm long, on stalks 1–2 mm long, 1–6 atop an erect stalk 10 25 cm long; bracts

Utricularia cornuta

ovate, 1–2 mm long. **Fruit** a rounded capsule. **Flowering** June–Sept.

HABITAT acid lakes, shores, peatlands, calcareous pools between dunes, borrow pits.

STATUS MW-OBL | NCNE-OBL | GP-OBL

Utricularia geminiscapa Benj.
MIXED BLADDERWORT

DESCRIPTION Native annual or perennial herb, similar to **common bladderwort** (*U. macrorhiza*) but smaller. **Stems** floating below water surface, sparsely branched. **Leaves** alternate, 1–2 cm long, branched into 4–7 segments and without bladders, or unbranched with bladders. **Flowers** yellow, 2–5 atop a slender stalk, 5–15 cm long, bracts below flowers 2–3 mm long; individual flower stalks 4–8 mm long, these arched when plants fruiting; cleistogamous flowers without petals more commonly produced, these single on leafless stalks 5–15 mm long along stems and often 1 at base of scape. **Flowering** July–Aug.

SYNONYM *Utricularia clandestina*.

HABITAT acid lakes, pools in open bogs.

STATUS MW-OBL | NCNE-OBL

Utricularia gibba L.
◆CREEPING BLADDERWORT

DESCRIPTION Native annual or perennial herb. **Stems** creeping on bottom in shallow water,

Utricularia gibba

mostly less than 10 cm long, radiating from base of flower stalk (scape) and forming mats. Leaves alternate, scattered, to 5 mm long, 1–2-forked into threadlike segments; bladders present. Flowers 1–3, yellow, 5–6 mm long, with a thick, blunt spur shorter than lower lip, atop a single stalk 5–10 cm long. Fruit a rounded capsule. Flowering July–Sept.

HABITAT exposed shores, lakes, ponds, marshes, fens.

STATUS MW-OBL | NCNE-OBL | GP-OBL

Utricularia intermedia Hayne
NORTHERN BLADDERWORT

DESCRIPTION Native annual herb. Stems very slender, creeping along bottom in shallow water. Leaves alternate, 0.5–2 cm long, mostly 3-parted near base, then again divided 1–3x, the segments linear and flat, margins with small, bristly teeth; bladders 2–4 mm wide, borne on branches separate from leaves. Flowers yellow, 2–4 atop an emergent stalk 5–20 cm long; individual flower stalks to 15 mm long, remaining erect in fruit; spur nearly as long as lower lip. Fruit a capsule. Flowering June–Aug.

HABITAT shallow water (usually alkaline), marly pools between dunes, calcareous fens, marshes, ponds and rivers.

STATUS MW-OBL | NCNE-OBL | GP-OBL

Utricularia macrorhiza Le Conte
♦ COMMON BLADDERWORT

DESCRIPTION Native perennial herb. Stems floating below water surface, sparsely branched, often forming large mats. Leaves alternate, 1–5 cm long, 2-forked at base and repeatedly 2-forked into segments of unequal length, the segments ± round in section, becoming smaller with each branching, the final segments threadlike; bladders 1–4 mm wide, borne on leaf segments. Flowers yellow, 6–20 atop a stout stalk 6–25 cm long; lower flower lip 1–2 cm long, sometimes much smaller on late-season flowers, upper lip ± equal to lower lip; spur about 2/3 as long as lower lip; stalks bearing individual flowers curved downward in fruit. Fruit a capsule. Flowering June–Aug.

SYNONYM *Utricularia vulgaris*.

HABITAT shallow water of lakes, ponds, peatlands, marshes and rivers.

STATUS MW-OBL | NCNE-OBL | GP-OBL

Utricularia minor L.
♦ LESSER BLADDERWORT

DESCRIPTION Native perennial herb. Stems with few branches, 10–30 cm long, creeping on bottom in shallow water or on wet soil. Leaves alternate, to 1 cm long, with few divisions, the segments slender, flat, the smallest segments strongly tapered to tip, margins entire; bladders 1–2 mm wide, 1–5 on leaves. Flowers pale yellow, 2–8 atop a threadlike stalk 4–15 cm long; individual flower stalks to 1 cm long, curved downward in fruit; lower lip of flower 4–8 mm long, 2x longer than upper lip; spur small, to half length of lower lip. Fruit a capsule. Flowering June–Aug.

Utricularia macrorhiza

Utricularia minor

HABITAT fens, open bogs, sedge meadows and marshes; often in shallow water and where calcium-rich.

STATUS MW-OBL | NCNE-OBL | GP-OBL

Utricularia purpurea Walter
♦ SPOTTED *or* PURPLE BLADDERWORT

DESCRIPTION Native annual or perennial herb. **Stems** underwater, to 1 m long. **Leaves** in whorls of 5–7, branched into threadlike segments, many segments tipped by a bladder. **Flowers** red-purple, 1–4 atop a stalk 3–15 cm long; corolla 1 cm long, lower lip 3-lobed, with a yellow spot near base; spur short and appressed to lower lip. **Fruit** a capsule. **Flowering** July–Sept.

SYNONYM *Vesiculina purpurea*.

HABITAT acid lakes and ponds in water to 1 m deep, peatlands, marshes.

STATUS MW-OBL | NCNE-OBL

Utricularia radiata Small
FLOATING BLADDERWORT

DESCRIPTION Native annual or perennial herb. **Stems** long, floating below water surface. Lower leaves alternate, 2–3 cm long, divided into threadlike, bladder-bearing segments; uppermost leaves below the flower stalk in whorls of 4–7, with petioles 1–4 cm long and inflated, tipped by finely dissected branches. **Flowers** yellow, mostly 3–5 atop a stalk 3–5 cm long, bracts 2–3 mm long; individual flowers on stalks 1–2 cm long; lower lip of flower shallowly 3-lobed, 8–10 mm long; spur short and appressed to lower lip. **Fruit** a capsule. **Flowering** July–Sept.

SYNONYM *Utricularia inflata* var. *minor*.

HABITAT pools between dunes.

STATUS MW-OBL | NCNE-OBL | GP-OBL; Atlantic coast, rarely inland; Mich (END).

Utricularia resupinata B. D. Greene
♦ LAVENDER BLADDERWORT

DESCRIPTION Native annual or perennial herb. **Stems** delicate, on water surface in shallow water or creeping just below soil surface. **Leaves** alternate, 3-parted from base, the middle segment erect and linear, to 3 cm long; the 2 lateral segments slender, rootlike, with bladders. **Flowers** purple, 1 cm long, single atop an erect stalk 2–10 cm long; bract tubelike, surrounding the stem, its margin notched; flower tipped backward on stalk and facing upward; lower lip 3-lobed; spur ± horizontal. **Fruit** a rounded capsule. **Flowering** July–Aug.

SYNONYM *Lecticula resupinata*.

HABITAT shallow to deep water, wet lake and pond shores where sandy or mucky.

STATUS MW-OBL | NCNE-OBL

Utricularia purpurea

Utricularia resupinata

Utricularia subulata L.
◆ SLENDER BLADDERWORT

DESCRIPTION Native annual or perennial herb, with short, rootlike branches from base of flower stalk (these underground and rarely collected), with small bladders 0.5 mm long. Leaves aboveground leaves short-lived, linear, to 1 cm long. Flowers yellow, 1-10, atop a very slender, erect stalk 3-20 cm long; bracts ovate or oval, 1-2 mm long, attached to flower stalk at or below their middle; individual flowers on stalks 5-15 mm long; lower flower lip 4-7 mm long; spur ± equal to lower lip and appressed to its underside. Fruit a round capsule. Flowering July-Sept.

SYNONYMS *Utricularia cleistogama, Setiscapella cleistogama.*

HABITAT wet places between dunes, wet prairie.

STATUS MW-OBL | NCNE-OBL | GP-OBL; Mich (THR).

Linaceae
Flax Family

Linum FLAX
Perennial herbs (those included here). Leaves simple, alternate or opposite, narrow, margins entire, petioles absent. Flowers regular, perfect, 5-parted; petals yellow (ours) or blue. Fruit a 10-chambered capsule.

Linum FLAX
1 Stems not angled; branches of head stiffly ascending .... *L. medium*
1 Stems sharply angled and winged; branches of head spreading .. *L. striatum*

Linum medium (Planchon) Britton
◆ COMMON YELLOW FLAX

DESCRIPTION Native perennial herb; plants smooth. Stems 2-6 dm long, unbranched below head. Leaves mostly alternate, narrowly lance-shaped, 1-2.5 cm long and 2-5 mm wide, smaller upward; margins entire; petioles absent. Flowers in a branched head, the branches stiffly ascending, the flowers on stalks 2-3 mm long; outer sepals lance-shaped, entire, 2-3 mm long, inner sepals shorter with gland-tipped hairs or teeth on margins; petals yellow, 4-8 mm long. Fruit a persistent, ± round capsule, 2 mm long, often purple-tinged near top. Flowering June-Sept.

HABITAT sandy, calcium-rich shores, moist places between dunes.

STATUS MW-FACU | NCNE-FACU | GP-FAC

Linum striatum Walter
RIDGESTEM YELLOW FLAX

DESCRIPTION Native perennial herb; plants smooth. Stems erect, 3-9 dm long, 1 or several from base of plant, with 3 sharp, narrow wings extending downward from base of each leaf. Leaves mostly opposite, oval to obovate, the larger leaves 1-3 cm long and 4-9 mm wide. Flowers in an open panicle, the panicle branches short and spreading; flower stalks sharply angled, 1-3 mm long; sepal margins mostly entire; petals yellow, 3-5 mm long. Fruit a ± round capsule, soon splitting into 10 segments. Flowering June-July.

HABITAT sandy shores and marshes; also in dry pine woods.

STATUS MW-FACW | NCNE-FACW | GP-FACW

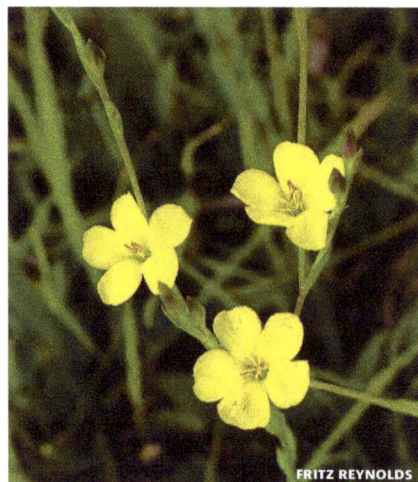

FRITZ REYNOLDS
Linum medium

Linderniaceae
False Pimpernel Family

Lindernia FALSE PIMPERNEL

Lindernia dubia (L.) Pennell
♦ FALSE PIMPERNEL

DESCRIPTION Native annual herb. **Stems** smooth, 1–2 dm long, widely branched. **Leaves** opposite, ovate to obovate, 5–30 mm long and 3–10 mm wide, the upper leaves smaller; margins entire or with small, widely spaced teeth; petioles absent. **Flowers** single, on slender stalks 0.5–2.5 cm long from leaf axils; sepals 5, linear; corolla light blue-purple, 5–10 mm long, 2-lipped, the upper lip 2-lobed, the lower lip 3-lobed and wider than upper lip; fertile stamens 2, staminodes (sterile stamens) 2. **Fruit** an ovate capsule, 4–6 mm long. **Flowering** June–Sept.
SYNONYM *Lindernia anagallidea*.
HABITAT mud flats, sandbars, shores of temporary ponds and marshes, streambanks.
STATUS MW-OBL | NCNE-OBL | GP-OBL

Lythraceae
Loosestrife Family

ANNUAL OR PERENNIAL HERBS, woody at base in *Decodon*. Leaves simple, opposite, or both opposite and alternate, or whorled, margins entire, ± stalkless.

Lindernia dubia

Flowers 1 or several in leaf axils or in spikelike heads at ends of stems; perfect (with both staminate and pistillate parts), regular or irregular; sepal lobes 4 or 6; petals 4 or 6, separate, pink or purple, deciduous; stamens usually 2x number of petals. Fruit a dry, many-seeded capsule.

Lythraceae **LOOSESTRIFE FAMILY**

1   Plants arching, woody near base; leaves with petioles and mostly whorled . . . .
. . . . *Decodon verticillatus*

1   Plants annual or perennial herbs; leaves opposite, or if whorled, leaves without petioles **2**

2   Plants perennial; flowers in spikelike heads at ends of stems; petals and sepals 6 . . . .
. . . . *Lythrum*

2   Plants annual; flowers from leaf axils; petals and sepals 4 or 5 (when present) . . **3**

3   Flowers mostly 2–5 per leaf axil; flowers purple-tinged . . . . *Ammannia*

3   Flowers mostly 1 per axil, not purple **4**

4   Submersed leaves lance-shaped, broadest at base, less than 3 mm wide . . . .
. . . . *Didiplis diandra*

4   Leaves oval, widest near middle, larger leaves 3 mm or more wide . . . .
. . . . *Rotala ramosior*

Ammannia TOOTH-CUP

Ammannia **TOOTH-CUP**

1   Petals deep rose-purple; longest peduncles longer than 3 mm . . . . *A. coccinea*

1   Petals pale lavender; longest peduncles 0–3 mm . . . . *A. robusta*

Ammannia coccinea Rottb.
TOOTH-CUP, SCARLET LOOSESTRIFE

DESCRIPTION Native annual herb; plants smooth. **Stems** erect, 2–8 dm long, often branched from base. **Leaves** opposite, linear, 2–8 cm long and 3–15 mm wide, heart-shaped and clasping at base; margins entire; petioles absent. **Flowers** stalkless, in clusters of 1–3 per leaf axil; petals 4 (rarely 5), 2–3 mm long, rose-purple, sometimes with a purple midvein at base; stamens 4 or 8. **Fruit** a round, 4-parted capsule, 3–5 mm wide, tipped by the persistent style. **Flowering** July–Oct.

HABITAT exposed mud flats and marshes, disturbed open wet areas; sometimes where calcium-rich.

STATUS MW-OBL | NCNE-OBL | GP-OBL

Ammannia robusta Heer & Rege
♦ TOOTH-CUP, SCARLET LOOSESTRIFE

DESCRIPTION Native annual herb; plants smooth and often strongly red-tinged. **Stems** erect, 2–8 dm long, often branched from base. **Leaves** opposite, linear, 2–8 cm long and 3–15 mm wide, heart-shaped and clasping at base; margins entire; petioles absent. **Flowers** stalkless, in clusters of 1–3 per leaf axil; petals 4 (rarely 5), 2–3 mm long, rose-purple, sometimes with a purple midvein at base; stamens 4 or 8. **Fruit** a round, 4-parted capsule, 3–5 mm wide, tipped by the persistent style. **Flowering** July–Oct.

HABITAT exposed mud flats and marshes, disturbed open wetlands with bare soil.

STATUS MW-OBL | NCNE-OBL | GP-OBL

Decodon WATER-WILLOW
Decodon verticillatus (L.) Elliott
♦ SWAMP *or* WHORLED LOOSESTRIFE

DESCRIPTION Native perennial herb, woody near base. **Stems** slender, angled, smooth or slightly hairy, 1–3 m long, arching downward and rooting at tip when in contact with water or mud. **Leaves** in whorls of 3–4 or opposite, lance-shaped, 5–15 cm long and 1–3 cm wide, smooth above, sparsely hairy below; margins entire; petioles short. **Flowers** in dense clusters in upper leaf axils; sepals 5–7, short, triangular; petals pink-purple, tapered to base, 10–15 mm long; stamens 10 (rarely 8), alternately longer and shorter than petals. **Fruit** a ± round capsule, 5 mm wide. **Flowering** July–Sept.

HABITAT shallow water and margins of lakes, ponds, bogs, swamps and marshes; soils mucky.

STATUS MW-OBL | NCNE-OBL | GP-OBL

Didiplis WATER-PURSLANE
Didiplis diandra (Nutt.) A. Wood
WATER-PURSLANE

DESCRIPTION Native annual herb; plants underwater or on exposed shores. **Stems** weak, branched, 1–4 dm long. **Leaves** numerous, opposite; underwater leaves linear, straight across base, 1–2.5 cm long; emersed leaves shorter and wider, tapered at base; petioles absent. **Flowers** few, inconspicuous, green. **Fruit** a small round capsule. **Flowering** July–Aug.

SYNONYM *Peplis diandra.*

HABITAT shallow water and muddy pond margins.

Ammannia robusta

Decodon verticillatus

STATUS MW-OBL | NCNE-OBL | GP-OBL; uncommon in Minn (END).

NOTE Plants somewhat resemble water-star-wort (*Callitriche*), but in water-starwort the underwater leaves have a shallow notch at tip and the capsule is flattened.

Lythrum LOOSESTRIFE

Perennial herbs. Stems erect, sometimes rather woody at base, usually with ascending branches above, upper stems 4-angled. Leaves opposite, entire, alternate, or rarely whorled, lance-shaped, stalkless, reduced to bracts in the head. Flowers in showy, spikelike heads, 1 to several in axils of upper leaves, regular or somewhat irregular, the stamens and styles of 2 or 3 different lengths. Sepals joined into a tube, the calyx tube cylinder-shaped, green-striped with 8–12 nerves; petals 6, purple, not joined; stamens 6 or 12; ovary 2-chambered. Fruit an ovate capsule, enclosed by the calyx tube.

Lythrum **LOOSESTRIFE**

1   Flowers single in upper leaf axils; stamens usually 6 . . . . *L. alatum*
1   Flowers many in spikelike heads at ends of stems; stamens usually 12 (6 long and 6 short) . . . . *L. salicaria*

Lythrum alatum Pursh
⬥ WINGED LOOSESTRIFE

DESCRIPTION Smooth native perennial, spreading by rhizomes. **Stems** 2–8 dm long,

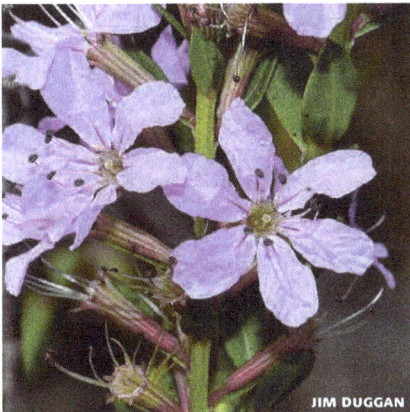

Lythrum alatum

usually branched above, somewhat woody at base. Lower leaves usually opposite, upper leaves alternate; lance-shaped, 1–4 cm long and 3–10 mm wide, rounded at base; margins entire; petioles absent. **Flowers** single in axils of upper, reduced leaves (bracts), short-stalked; calyx tube 4–6 mm long, smooth; petals 6, deep purple, 3–7 mm long; stamens usually 6. **Fruit** a capsule enclosed by the sepals. **Flowering** June–Aug.

SYNONYM *Lythrum dacotanum.*

HABITAT lakeshores, wet meadows, marshes, low prairie, calcareous fens, ditches; especially where sandy.

STATUS MW-OBL | NCNE-OBL | GP-OBL

Lythrum salicaria L.
⬥ PURPLE LOOSESTRIFE

DESCRIPTION Introduced, invasive perennial herb, spreading and forming colonies by thick, fleshy roots which send up new shoots. **Stems** erect, 6–15 dm long, 4-angled, with many ascending branches. **Leaves** opposite or sometimes in whorls of 3, becoming alternate and reduced to bracts in the head; lance-shaped, 3–10 cm long and 0.5–2 cm wide, mostly heart-shaped and clasping at base; margins entire; petioles absent. **Flowers** large and showy, 2 or more in axils of reduced upper leaves (bracts), in spikes 1–4 dm long at ends of branches; sepals joined, the calyx tube 4–6 mm long, hairy; petals 6, purple-magenta, 7–10 mm

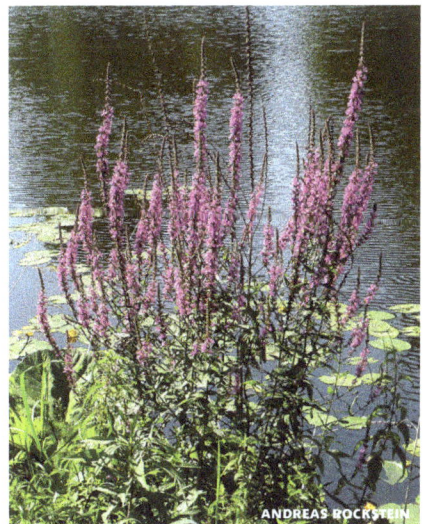

Lythrum salicaria

long; stamens usually 12, the stamens and styles of 3 different lengths. Fruit a capsule enclosed by the sepals. Flowering June–Sept.

HABITAT introduced from Europe and previously planted as an ornamental, escaping to marshes, wet ditches, streambanks, cranberry bogs and shores, where a serious threat to native flora and of little value to wildlife. In addition to spreading vegetatively, a single plant may produce several hundred thousand seeds (or more) each year.

STATUS MW-OBL | NCNE-OBL | GP-OBL; invasive (and naturalized) over much of USA and s Canada.

Rotala TOOTH-CUP

Rotala ramosior (L.) Koehne
♦ TOOTH-CUP, WHEELWORT

DESCRIPTION Small, native annual herb. Stems smooth, 4-angled, to 4 dm long, unbranched or branched from base, the branches spreading to upright. Leaves opposite, linear to oblong, 1–5 cm long and 2–12 mm wide; margins entire; stalkless or tapered to a short petiole. Flowers single and stalkless in leaf axils, calyx tube bell-shaped to cylindric, 2–5 mm long, not strongly nerved, the lobes alternating with appendages of same length; petals small, white to pink, 4, slightly longer than sepals; stamens

DOUG MCGRADY

Rotala ramosior

4. Fruit a round capsule enclosed by the sepals. Flowering July–Oct.

HABITAT muddy or sandy shores, marshes (especially those that dry during growing season), low spots in fields, ditches and other seasonally flooded places.

STATUS MW-OBL | NCNE-OBL | GP-OBL; Minn (THR).

Malvaceae
Mallow Family

ANNUAL OR PERENNIAL HERBS with upright stems. Leaves alternate, entire to lobed or dissected, often round or kidney-shaped, palmately veined. Flowers single or in small, narrow clusters from leaf axils, with 5 united sepals and 5 petals; stamens many and joined near base, forming a tube around the style. Fruit a capsule.

Malvaceae **MALLOW FAMILY**

1 Flowers perfect; pistil 5-parted, the fruit segments (carpels) united, not separating when mature . . . . *Hibiscus*

1 Flowers either staminate or pistillate and borne on separate plants; pistil 10-parted or more, the fruit segments separating when mature . . . . *Napaea dioica*

Hibiscus ROSE-MALLOW

Large, upright perennial herbs. Leaves alternate, smooth or hairy, palmately divided. Flowers large and showy, pink to white; ovary divided into 5 segments (carpels). Fruit an ovate capsule.

Hibiscus **ROSE-MALLOW**

1 Leaves and stems without hairs . *H. laevis*

1 Leaf undersides and upper stems with velvety hairs . . . . *H. moscheutos*

Hibiscus laevis All.
SMOOTH ROSE-MALLOW

DESCRIPTION Native perennial herb; stems and leaves smooth. Stems upright, 1–2 m long. Leaves triangular in outline, heart-shaped at base, sometimes with outward pointing basal lobes; margins with rounded teeth; petioles 3–15 cm long. Flowers large, from leaf axils or clustered

at ends of stems or branches; petals pink with darker center, 5–8 cm long. Fruit an ovate capsule, enclosed by the calyx; seeds silky-hairy. Aug–Sept.

SYNONYM *Hibiscus militaris.*

HABITAT marshes, muddy shores, shallow water.

STATUS MW-OBL | NCNE-OBL | GP-OBL; adventive in Minn, historically known from Mich.

Hibiscus moscheutos L.
♦ SWAMP ROSE-MALLOW

DESCRIPTION Large native perennial herb. Stems upright, numerous from base of plant, 1–2 m long, upper stems gray-hairy. Leaves ovate, upper surface green and smooth, lower surface white velvety hairy; margins toothed; petioles 2–12 cm long. Flowers large and showy, 10–20 cm wide, pink or white, often red or purple in center, at ends of stalks near top of stems. Fruit a hairless, ovate capsule; seeds smooth. Flowering Aug–Sept.

SYNONYM *Hibiscus palustris.*

HABITAT marshes, streambanks and disturbed wet areas.

STATUS MW-OBL | NCNE-OBL | GP-OBL

Napaea GLADE-MALLOW

Napaea dioica L.
GLADE-MALLOW

DESCRIPTION Large native perennial herb.

PEGANUM

Hibiscus moscheutos

Stems erect, 1–2 m long. Leaves round in outline, 1–3 dm wide, deeply 5–9 lobed, the lobes coarsely toothed, on long petioles; upper leaves smaller, with short petioles. Flowers either staminate or pistillate and on separate plants; many in large panicles at ends of stems; petals white, obovate, petals of staminate flowers 5–9 mm long, petals of pistillate flowers smaller. Fruit a 10-parted capsule, the segments (carpels) 5 mm long, ribbed, and irregularly separating when mature. Flowering June–Aug.

HABITAT moist floodplain forests, riverbanks.

STATUS MW-FACW | NCNE-FACW; Minn (THR).

Melastomataceae
Melastome Family

Rhexia
MEADOW-BEAUTY, MEADOW-PITCHERS

Perennial herbs; plants hairy. Stems round or square in cross-section. Leaves opposite, prominently 3–5-veined from base, margins with forward-pointing teeth and fringed with hairs; stalkless or short-petioled. Flowers large, pale to bright purple, rarely white, bell-shaped, in clusters of a few at ends of stems; sepals joined, 4-lobed; petals 4; stamens 8. Fruit a 4-chambered capsule.

Rhexia **MEADOW-BEAUTY, MEADOW-PITCHERS**

1   Stems ± round in section . . . . . . *R. marina*
1   Stems square or angled in section . . . .
        . . . . *R. virginica*

Rhexia mariana L.
DULL MEADOW-BEAUTY

DESCRIPTION Weedy, native perennial herb, with shallow rhizomes; plants densely coarse-hairy. Stems ± round in section, 3–8 dm long, branched. Leaves lance-shaped, 2–6 cm long and 1–2 cm wide, spreading, with bristly hairs on both sides; margins toothed and fringed with hairs; petioles short. Flowers pale purple (sometimes white), stalked, in open clusters at ends of stems; petals 4, rounded at tip, 1–2 cm long. Fruit a capsule, enclosed by the

glandular-hairy sepal tube. Flowering July-Sept.

HABITAT moist openings and shores.

STATUS MW-OBL | NCNE-OBL | GP-FACW; Mich (THR); more common in se USA.

Rhexia virginica L.
◊ VIRGINIA MEADOW-BEAUTY, DEER-GRASS

DESCRIPTION Native perennial herb, roots often with tubers. Stems simple, or branched above, 4-angled and winged, 2–6 dm long, usually with bristly hairs. Leaves opposite, ovate, 2–6 cm long and 1–3 cm wide, smooth or with short, stiff hairs on either side, tapered to a tip; margins finely toothed; stalkless. Flowers stalked, in clusters from ends of stems and upper leaf axils; sepals narrow, 2–4 mm long; petals purple, 10–20 mm long. Fruit a 4-chambered capsule. Flowering July–Sept.

HABITAT open shores, thickets (often of Aronia); soils acidic, sandy or peaty.

STATUS MW-OBL | NCNE-OBL | GP-OBL

Menyanthaceae
Buckbean Family

Menyanthaceae **BUCKBEAN FAMILY**

1 Leaves emergent, not floating, trifoliate; flowers white, in a raceme ....
.... *Menyanthes trifoliata*

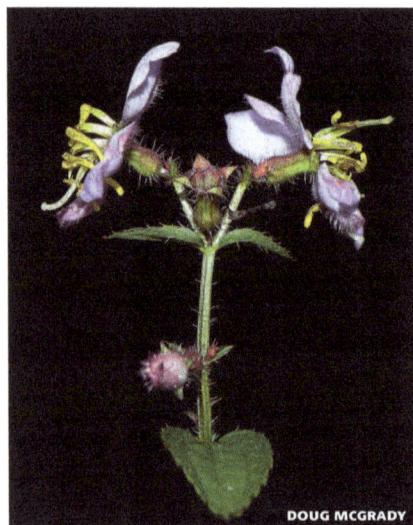

1 Leaves floating, simple; flowers yellow, in an umbel on a long pedicle ....
.... *Nymphoides peltata*

Menyanthes BUCKBEAN

Menyanthes trifoliata L.
◊ BUCKBEAN

DESCRIPTION Native perennial herb, with thick rhizomes covered with old leaf bases; plants smooth. Leaves alternate along rhizomes, palmately divided into 3 leaflets, the leaflets oval to ovate, 3–10 cm long and 1–5 cm wide, entire or sometimes wavy-margined; petioles 5–30 cm long, the base of petiole expanded and sheathing stem. Flowers in racemes on leafless stalks 2–4 dm long and longer than the leaves; bracts mostly 3–5 mm long; individual flowers on stalks 5–20 mm long; flowers perfect, regular, 5-parted, often of 2 types, some with flowers with long stamens and a shorter style, others with a long style and shorter stamens; sepal lobes 2–3 mm long; corolla funnel-shaped, 8–12 mm long, petals white, often purple-tinged, bearded with white hairs on inner surface; stamens 5. Fruit a rounded capsule, 6–10 mm wide; seeds shiny, yellow-brown. Flowering May–July.

HABITAT open bogs and fens (especially in pools and outer moat), cedar swamps, wet thickets.

STATUS MW-OBL | NCNE-OBL | GP-OBL

Rhexia virginica

DOUG MCGRADY

Menyanthes trifoliata

GERTJAN VAN NOORD

Nymphoides
FLOATING-HEART

Nymphoides peltata (Gmel.) Kuntze
◆ YELLOW FLOATING-HEART

DESCRIPTION Introduced perennial herb; leaves floating; potentially invasive. Stems rooted, with long branched stolons (up to 2 meters) just below the water surface; multi-leaved, rooting "plantlets" are produced at each node. Leaves floating, round to heart-shaped (cordate), 3–12 cm in diameter, resembling small waterlilies (*Nuphar, Nymphaea*). Leaves green to yellow-green; margins slightly wavy; leaf underside often purple. Flowers bright yellow, petals 5, 2–4 cm wide, shallowly fringed, borne above the water surface. Fruit a beaked capsule ca. 2.5 cm long, seeds shiny, with stiff hairs along margins. Flowering June–Oct.

HABITAT ponds, lakes.

STATUS MW-OBL | NCNE-OBL | GP-OBL

NOTE A Mediterranean/East Asian species, first reported in the USA (Massachusetts) in 1882, and formerly used in water-gardens (sale now prohibited in our region).

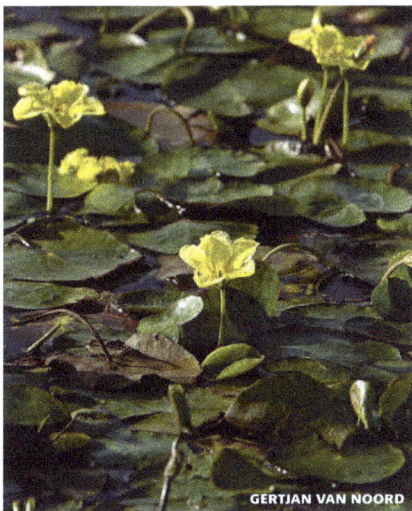

GERTJAN VAN NOORD
Nymphoides peltata

Montiaceae
Candy-Flower Family

Montia MINER'S-LETTUCE

Montia chamissoi (Ledeb.) Greene
MINER'S-LETTUCE

DESCRIPTION Smooth native perennial herb, forming colonies from spreading rhizomes and stolons. Stems upright, 5–20 cm long. Leaves opposite, spatula-shaped to obovate, 2–5 cm long, margins entire. Flowers perfect, 3–10 in drooping racemes from ends of stems or upper leaf axils, flower stalks 1–3 cm long; petals 5, white to pink, 5–8 mm long, stamens 5. Flowering June–July.

HABITAT streambanks, springs and seeps.

STATUS MW-OBL | NCNE-OBL | GP-OBL; se Minn (END) and adjacent ne Iowa; disjunct from Rocky Mts.

Myricaceae
Bayberry Family

Myrica
SWEET GALE, BAYBERRY

Myrica gale L. ◆ SWEET GALE
DESCRIPTION Much-branched native shrub, 6–15 dm tall. Bark dark gray to red-brown with small pale lenticels. Twigs hairy, dotted with glands. Leaves alternate, deciduous, wedge-shaped, tapered to base, broadest above middle, 3–6 cm long and 1–2 cm wide, tip rounded and toothed, dark green on upper surface, paler below, dotted with shiny yellow glands; petioles

HAJOTTHU
Myrica gale

short, 1–3 mm long. Flowers staminate and pistillate flowers separate and on different plants, appearing before or with unfolding leaves; staminate flowers in catkins 10–20 mm long, with dark brown, shiny triangular scales; pistillate flowers in conelike, brown clusters 10–12 mm long. Fruit a flattened, ovate achene, resin-dotted, 2–3 mm long. Flowering April–May.

SYNONYM *Gale palustris.*

HABITAT lakeshores, marshes, swamps and bogs, often at water's edge or in shallow water.

STATUS MW-OBL | NCNE-OBL | GP-OBL

Nelumbonaceae
Lotus-Lily Family

Nelumbo LOTUS-LILY

Nelumbo lutea (Willd.) Pers.
♦ AMERICAN LOTUS-LILY

DESCRIPTION Native perennial aquatic herb, from a large, horizontal rootstock. Leaves large and shield-shaped, 3–7 dm wide, ribbed, floating on water surface or held above water surface, smooth above, somewhat hairy below; petioles thick, attached at center of blade. Flowers pale yellow, single, 15–25 cm wide; petals obovate, blunt-tipped; receptacle flat-topped, to 1 dm wide. Seeds acornlike, 1 cm thick. Flowering July–Aug.

HABITAT lakes, ponds, backwater areas, marshes, where sometimes forming large colonies.

STATUS MW-OBL | NCNE-OBL | GP-OBL; Mich (THR).

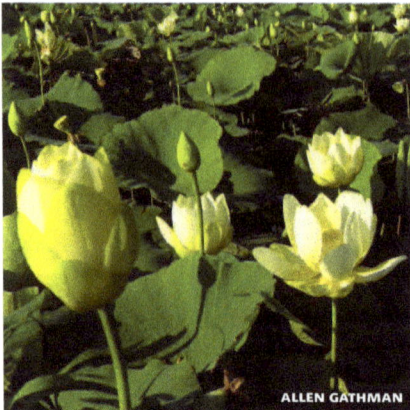

ALLEN GATHMAN

Nelumbo lutea

Nymphaeaceae
Water-Lily Family

AQUATIC PERENNIAL HERBS. Stems long and fleshy, from horizontal rhizomes rooted in bottom mud. Leaves large, leathery, mostly floating or emergent above water surface, heart-shaped to shield-shaped, notched at base, margins entire. Flowers showy, single on long stalks and borne at or above water surface, perfect (with both staminate and pistillate parts), white or yellow, sepals 4–6, green or yellow; petals numerous, small (*Nuphar*) to large and showy (*Nymphaea*). Fruit a many-seeded, berrylike capsule, opening underwater when mature.

Nymphaeaceae WATER-LILY FAMILY

1 Flowers yellow, often red-tinged, sepals petal-like, true petals small; leaf blades oblong to oval or heart-shaped... *Nuphar*

1 Flowers white (rarely pink), sepals green, true petals large and showy; leaf blades nearly round .... *Nymphaea*

Nuphar
YELLOW WATER LILY, SPATTERDOCK

Aquatic herbs. Leaves large and floating or emergent. Sepals 5–6, yellow and petal-like, forming a saucer-shaped flower; petals small and numerous.

Nuphar YELLOW WATER LILY, SPATTERDOCK

1 Leaves small, 5–10 cm wide; anther shorter than its stalk (filament); disk at base of stigma red; uncommon species of n Upper Midwest region.......... *N. microphylla*

1 Leaves larger; anther longer than its stalk; disk at base of stigma green or yellow; widespread species **2**

2 Leaf blades usually raised above water surface, basal lobes of blades close or touching; petioles round or oval in section .... *N. advena*

2 Leaf blades floating, basal lobes widely spreading; petioles flattened and narrowly winged .... *N. variegata*

Nuphar advena (Aiton) Aiton f.
♦YELLOW WATER LILY

DESCRIPTION Native aquatic perennial herb. Leaves mostly emergent and raised above water surface, 2-4 dm wide, with a broadly triangular notch at leaf base, petioles round or oval in section; underwater leaves absent. Flowers 3-6 cm wide; sepals 6, yellow, sometimes purple on inner surface; petals small, numerous; anthers 3-7 mm long and as long or longer than their stalks (filaments); disk at base of stigma green, 10-15 mm wide, with 14-18 rays. Fruit ovate, to 4 cm long. Flowering June-Sept.

SYNONYMS *Nuphar fluviatilis, Nuphar macrophylla.*

HABITAT shallow to deep water of slow-moving streams, lakes and ponds.

STATUS [not rated]

Nuphar microphylla (Pers.) Fern.
YELLOW WATER LILY

DESCRIPTION Native aquatic perennial herb. Leaves both underwater and floating; floating leaves 5-10 cm long and 3-8 cm wide, notch at base usually more than half as long as midvein; petioles flattened on upper side; underwater leaves membranous, somewhat larger. Flowers 1.5-2 cm wide, sepals 5 (sometimes 10), yellow on inner surface; petals small and many; anthers 1-3 mm long, shorter than the filaments; disk at base of stigma red, 3-6 mm wide, with 6-10 rays. Fruit ovate, 15 mm long. Flowering July-Aug.

SYNONYMS *Nuphar luteum, Nuphar pumila.*

HABITAT lakes, ponds and slow-moving streams.

STATUS [not rated]; Mich (END).

Nuphar variegata Durand
♦BULLHEAD LILY

DESCRIPTION Native aquatic perennial herb. Leaves mostly floating, 10-25 cm wide, notch usually less than half as long as midvein, petioles flattened on upper side and narrowly winged; underwater leaves absent or few. Flowers 2.5-5 cm wide; sepals usually 6, yellow, red-tinged on inner surface; petals small and numerous; anthers 4-7 mm long, longer than filaments; disk at base of stigma green, 1 cm wide, with 10-15 rays. Fruit ovate, 2-4 cm long. Flowering June-Aug.

SYNONYM *Nuphar fraterna.*

HABITAT ponds, lakes, quiet streams.

STATUS [not rated]

Nymphaea WATER-LILY

Large aquatic plants, from stout rhizomes, these sometimes with lateral tubers. Leaves floating, round, notched to the petiole, petioles not flattened or winged. Flowers white and showy; sepals 4, green; petals white or pink, showy, numerous and overlapping; stamens many, the outer stamens with broadened, petal-like filaments, anthers yellow, ovary depressed at tip with a rounded projection from center,

Nuphar advena

Nuphar variegata

stigmas 10–25. Fruit round, covered with persistent petal and stamen bases, maturing under water; seeds numerous, each enclosed within a sac (aril).

Nymphaea WATER-LILY

1 Leaves round in outline, narrowly notched; flowers large and showy, usually fragrant; common . . . . *N. odorata*
1 Leaves oval in outline, notch more widely spreading; flowers smaller and scarcely fragrant; rare in n Minn and on Isle Royale, Mich . . . . *N. leibergii*

Nymphaea leibergii Morong
◆ PYGMY WATER-LILY

DESCRIPTION Native aquatic perennial herb, rhizomes ascending. **Stems** arising from tip of rhizomes. **Leaves** 7–12 cm long and to 3/4 as wide, notch fairly wide, upper surface green, green or purple below. **Flowers** white, usually not fragrant, 4–8 cm wide, reported to open in afternoon; sepals 4, green, 2–3 cm long; petals 8–17, about as long as sepals; stamens 20–40. **Fruit** not covered by the erect sepals. **Flowering** Summer.
SYNONYMS *Castalia leibergii, Nymphaea tetragona.*
HABITAT shallow water of ponds and lakes.
STATUS NCNE-OBL | GP-OBL; Minn (THR), Isle Royale, Mich (END).

Nymphaea odorata Aiton
◆ WHITE WATER-LILY

DESCRIPTION Native aquatic perennial herb, rhizomes sometimes with knotty tubers. **Leaves** floating, round, 1–3 dm wide, with a narrow notch, green and shiny on upper surface, usually purple or red below. **Flowers** white (rarely pink), usually fragrant, 7–20 cm wide, often opening in morning and closing in late afternoon (or remaining open on cool, cloudy days); sepals 4, green, 3–10 cm long; petals 17–25, about as long as sepals, oval, tapered to a rounded tip; stamens 40–100. **Fruit** round, mostly covered by the sepals; seeds 2–4 mm long. **Flowering** June–Aug.
SYNONYMS *Castalia tuberosa, Nymphaea tuberosa.*
HABITAT shallow water of ponds and lakes, quiet water of rivers.
STATUS MW-OBL | NCNE-OBL | GP-OBL

Nyssaceae
Tupelo Family

Nyssa SOUR GUM

Nyssa sylvatica Marshall
◆ BLACK TUPELO, BLACK GUM

DESCRIPTION Small to medium native tree, to 15 m tall; trunk 2–4 dm wide. **Bark** dark gray to red-brown, deeply furrowed into blocky plates. **Twigs** red-brown. **Leaves** alternate, thick and firm, oval to obovate, 4–12 cm long and half as wide, usually abruptly tapered to a rounded tip, dark green, smooth and shiny on upper surface, turning bright scarlet in fall, paler and often hairy below; margins entire; petioles short. **Flowers** greenish, on thin, finely hairy stalks; the staminate and pistillate flowers on separate trees or sometimes the flowers perfect; staminate flowers on slender, finely hairy stalks in a many-flowered head, stamens 5–10; pistillate flowers 2–8, at end of a stalk 3–5 cm long, petals 5. **Fruit** a sour, blue-black, oval drupe, 1–1.5 cm long, in clusters of 1–3.

Nymphaea leibergii

Nymphaea odorata

Flowering May–June.
HABITAT swamp margins, shores, wet depressions in forests.
STATUS MW-FAC | NCNE-FAC | GP-FAC

Oleaceae
Olive Family

Fraxinus ASH

Medium to large trees. Leaves deciduous, opposite, pinnately divided into leaflets. Flowers in clusters from axils of previous year's twigs, mostly single-sexed, staminate and pistillate flowers on different trees, rarely perfect, petals absent. Fruit a 1-seeded, winged samara.

Fraxinus **ASH**

1   Lateral leaflets ± stalkless........ *F. nigra*
1   Lateral leaflets short-stalked **2**
2   Wing of samara 4–7 mm broad; leaflet underside smooth or hairy..............
.... *F. pennsylvanica*
2   Wing of samara 7–10 mm broad; leaflet underside hairy............. *F. profunda*

Fraxinus nigra Marshall ◆ BLACK ASH

DESCRIPTION Native tree to 15 m tall; trunk 30–60 cm wide, crown open and narrow. **Bark** gray, thin, flaky. **Twigs** smooth, round in section, dark green, becoming gray. **Leaves** opposite, pinnately divided into 7–11 stalkless (except for terminal) leaflets; leaflets lance-shaped to oblong, 7–13 cm long and 2.5–5 cm wide, long-tapered to a tip; margins with sharp, forward-pointing teeth. **Flowers** appear in spring before leaves, in open clusters on twigs of previous year; some perfect, some single-sexed, staminate and pistillate flowers on different trees. **Fruit** a 1-seeded samara, 2.5–4 cm long and 6–10 mm wide, the wing broad and rounded at tip, deciduous or persisting until following spring. **Flowering** April–May.
HABITAT floodplain forests, cedar swamps, wet depressions in forests.
STATUS MW-FACW | NCNE-FACW | GP-FACW

Fraxinus pennsylvanica Marshall
GREEN ASH, RED ASH

DESCRIPTION Native tree to 15 m tall; trunk 30–60 cm wide. **Bark** dark gray or brown, thick, with shallow furrows and netlike ridges. **Twigs** usually hairy for 1–3 years, becoming light gray or red-brown. **Leaves** opposite, pinnately divided into 7–9 leaflets; leaflets oblong lance-shaped to ovate, 7–13 cm long and 2.5–4 cm wide, upper surface smooth, underside smooth or hairy; margins entire or with few forward-pointing teeth; petioles short, smooth or hairy. **Flowers** appear in spring before or with leaves, in compact, hairy clusters on twigs of previous year; single-sexed, staminate and pistillate flowers on different trees.

Nyssa sylvatica

Fraxinus nigra

Fruit a 1-seeded, slender samara, 2.5–5 cm long, in open clusters persisting until following spring. Flowering April–May. HABITAT floodplain forests, swamps, shores, streambanks. STATUS MW-FACW | NCNE-FACW | GP-FAC NOTE both smooth and hairy twig-forms of *Fraxinus pennsylvanica* occur, with twigs of trees becoming less hairy westward across our region.

Fraxinus profunda (Bush) Bush
PUMPKIN ASH

DESCRIPTION Native tree to 30 m tall; trunk 30–60 cm wide. Bark gray, scaly, with shallow furrows and netlike ridges. Twigs gray or brown, velvety-hairy to smooth. Leaves opposite, pinnately divided into 7–9 leaflets; leaflets lance-shaped to oval, 15–20 cm long and 7–10 cm wide, long-tapered to a tip, upper surface smooth, underside velvety-hairy; margins entire or finely toothed; petioles 8–15 mm long. Flowers appear in spring before leaves, in elongated clusters on twigs of previous year; single-sexed, staminate and pistillate flowers on different trees. Fruit a 1-seeded samara, linear or spatula-shaped, 4–7 cm long and 7–11 mm wide, the wing extending at least to middle of the round seed body. Flowering April–May. SYNONYM *Fraxinus tomentosa*. HABITAT floodplain forests and swamps, especially where seasonally flooded. STATUS MW-OBL | NCNE-OBL; Mich (THR); more common in se USA. NOTE Pumpkin ash is the region's only species of ash with velvety-hairy twigs and mostly smooth leaf margins.

Onagraceae
Evening-Primrose Family

ANNUAL OR PERENNIAL HERBS. Leaves opposite to alternate, simple to pinnately divided, stalkless or short-petioled. Flowers usually large and showy, perfect (with both staminate and pistillate parts), regular, borne in leaf axils or in heads at ends of stems; sepals 8 or 4; petals 4, white, yellow or pink to rose-purple; ovary 4-chambered. Fruit a 4-chambered capsule; seeds many, with or without a tuft of hairs (coma).

Onagraceae EVENING-PRIMROSE FAMILY

1 Petals 2, white; leaves opposite; fruit with bristly hairs . . . . *Circaea alpina*
1 Petals 4 (rarely absent), white, pink, or yellow; leaves opposite or alternate; fruit without bristly hairs 2
2 Petals yellow (or absent); seeds without a tuft of hairs (coma) . . . . *Ludwigia*
2 Petals pink, white or rose-purple; seeds with a tuft of hairs . . . . *Epilobium*

Circaea
ENCHANTER'S NIGHTSHADE

Circaea alpina L.
♦ SMALL or ALPINE ENCHANTER'S NIGHTSHADE

DESCRIPTION Native perennial herb, spreading from rhizomes thickened and tuberlike at ends. Stems weak, 1–3 dm long, mostly smooth. Leaves opposite, ovate, 2–5 cm long and 1–3 cm wide; margins coarsely toothed; petioles flat on upper side, underside thin-winged along center. Flowers white, in short racemes of 10–15 flowers, becoming 1 dm long in fruit; sepals 1–2 mm long; petals to 2 mm long. Fruit a 1-seeded capsule, 2–3 mm long, covered with soft hooked bristles. Flowering June–Aug. HABITAT cedar swamps (where often on rotting logs), low spots in forests. STATUS MW-FACW | NCNE-FACW | GP-FACW

KRZYSZTOF ZIARNEK

Circaea alpina

Epilobium WILLOW-HERB

Perennial herbs, often producing leafy rosettes or bulblike offsets (turions) at base of stem late in growing season. Leaves simple, opposite, alternate, or opposite below and becoming alternate above; stalkless or short-petioled. Flowers white to pink, single in axils of upper reduced leaves, or in spike or racemes at ends of stems; sepals 4; petals 4; stamens 8, the inner 4 stamens shorter than outer 4; ovary 4-chambered, maturing into a linear, 4-parted capsule, splitting from tip to release numerous brown seeds which are tipped with a tuft of fine hairs (coma).

Epilobium WILLOW-HERB

1 Stigma deeply 4-lobed 2
1 Stigma entire or slightly notched, not deeply lobed 3
2 Petals 10-15 mm long, shallowly notched at tip; leaves ± clasping stem . . . *E. hirsutum*
2 Petals 4-9 mm long, deeply notched at tip; leaves not clasping stem . *E. parviflorum*
3 Leaves more than 1 cm wide, margins toothed, not rolled under; stems usually with lines of hairs 4
3 Leaves less than 1 cm wide, margins ± entire, rolled under; stem hairs not in lines 5
4 Tuft of hairs attached to tip of seeds (coma) ± white, seeds with a broad, short beak; margins of stem leaves with mostly 10-30 teeth on a side . . . . *E. ciliatum*
4 Coma brown, seeds beakless; leaf margins with more than 30 teeth on a side . . . . *E. coloratum*
5 Hairs spreading . . . . *E. strictum*
5 Hairs flattened against stems and leaves 6
6 Upper surface of leaves finely hairy . . . . *F. leptophyllum*
6 Upper surface of leaves ± smooth . . . . *E. palustre*

Epilobium ciliatum Raf.

◆ AMERICAN WILLOW-HERB

DESCRIPTION Native perennial herb, with over-wintering leafy rosettes. Stems often branched, 3-10 dm long, smooth below, short-hairy above, especially in the head (where often with gland-tipped hairs). Leaves opposite, usually alternate near top; lance-shaped to ovate, 3-10 cm long and 0.5-3 cm wide; margins with few, small, forward-pointing teeth; stalkless or with short, winged petioles to 6 mm long. Flowers usually nodding when young, on stalks 3-10 mm long, on branches from upper leaf axils; sepals ovate, 2-5 mm long; petals white (or pink), notched at tip, 2-8 mm long. Fruit a linear capsule, 4-8 cm long, with gland-tipped hairs; seeds 1 mm long, the coma white. Flowering July-Sept.
SYNONYM *Epilobium adenocaulon.*
HABITAT shores, streambanks, marshes, wet meadows, seeps, ditches and other wet places.
STATUS MW-FACW | NCNE-FACW | GP-FACW

Epilobium coloratum Biehler

PURPLE-LEAF WILLOW-HERB

DESCRIPTION Native perennial herb, producing basal, leafy rosettes in fall; similar to **American willow-herb** (*Epilobium ciliatum*) but larger. Stems 5-10 dm long, much-branched in the head, smooth below, short-hairy above with hairs often in lines; stems and leaves often purple-tinged. Leaves mostly opposite, becoming alternate and smaller above, lance-shaped, 5-15 cm long and 0.5-3 cm wide, long-tapered to a pointed tip; margins finely toothed, with irregular sharp teeth; short-petioled to stalkless. Flowers many on branches from

Epilobium ciliatum

upper leaf axils; sepals lance-shaped, 2–3 mm long; petals pink or white, 3–5 mm long, notched at tip; individual flowers on stalks to 10 mm long. Fruit a linear capsule, 3–5 cm long; seeds 1.5 mm long, the coma brown when mature. Flowering July–Sept.

HABITAT shores, seeps, swamps and wet woods, wet meadows, fens, ditches.

STATUS MW-OBL | NCNE-OBL | GP-OBL

Epilobium hirsutum L.
HAIRY WILLOW-HERB

DESCRIPTION Introduced perennial herb, spreading by rhizomes. Stems much-branched, 5–15 dm long, upper stems with a dense covering of soft, straight hairs. Leaves opposite (but bracts alternate), lance-shaped or oblong, 5–10 cm long and 1–3 cm wide, hairy on both sides, somewhat clasping at base; margins with sharp, forward-pointing teeth; petioles absent. Flowers upright on stalks from upper leaf axils, petals red-purple, 10–15 mm long, shallowly notched; stigma 4-lobed. Fruit a hairy, linear capsule, 5–8 cm long; coma nearly white. Flowering June–Sept.

HABITAT introduced from Eurasia and sometimes invasive, now established in swamps, thickets, marshes, shores, wet meadows and ditches.

STATUS MW-FACW | NCNE-FACW | GP-FACW

Epilobium leptophyllum Raf.
♦ LINEAR-LEAF WILLOW-HERB

DESCRIPTION Native perennial herb, similar to **marsh willow-herb** (*Epilobium palustre*) but somewhat larger and more hairy. Stems simple or branched, 2–10 dm long, with short, incurved hairs. Leaves opposite or alternate, linear or linear lance-shaped, 2–7 cm long and 1–6 mm wide, upper surface hairy, underside hairy, at least on midvein, lateral veins indistinct; margins entire and rolled under; petioles short or ± absent. Flowers erect in upper leaf axils on short, slender stalks to 1 cm long; petals light pink, 3–5 mm long, entire or slightly notched at tip. Fruit a linear, finely hairy capsule, 4–5 cm long; the coma yellow-white. Flowering July–Sept.

SYNONYM *Epilobium lineare.*

HABITAT swamps, marshes, open bogs, sedge meadows, shores, streambanks, springs.

STATUS MW-OBL | NCNE-OBL | GP-OBL

Epilobium palustre L.
MARSH WILLOW-HERB

DESCRIPTION Native perennial herb, from slender rhizomes or stolons. Stems simple or with a few branches above, 1–6 dm long, upper stem hairy with small incurved hairs. Leaves mostly opposite, lance-shaped, erect or ascending, 2–6 cm long and 3–15 mm wide, tapered to a rounded tip, upper surface smooth or with sparse hairs along midvein, underside smooth or finely hairy along midvein, lateral veins distinct; margins entire and often rolled under; stalkless. Flowers few in upper leaf axils, on short stalks; petals white to pink, 3–5 mm long, notched at tip. Fruit a linear, finely hairy capsule; coma pale. Flowering July–Aug.

SYNONYM *Epilobium oliganthum.*

HABITAT open bogs and swamps.

STATUS MW-OBL | NCNE-OBL | GP-OBL

Epilobium parviflorum Schreber
HAIRY WILLOW-HERB

DESCRIPTION Introduced perennial herb, similar to *E. hirsutum* but plants smaller, and the petals only 4–7 mm

Epilobium leptophyllum

long (vs. mostly 10–15 mm long in *Epilobium hirsutum*). Stems 4–8 dm long. Lower leaves opposite, upper leaves often alternate; 2–8 cm long, not clasping at base; margins with sharp, forward-pointing teeth; ± stalkless. Flowers upright on stalks from upper leaf axils; petals red-purple, 4–7 mm long, deeply notched; stigma 4-lobed. Fruit a hairy, linear capsule; coma ± white. Flowering July–Sept. HABITAT swamps, shores and ditches. STATUS [not rated]

Epilobium strictum Muhl.
♦DOWNY WILLOW-HERB

DESCRIPTION Native perennial herb, spreading by slender rhizomes; plants densely soft white-hairy. Stems erect, simple or branched above, 3–6 dm long. Lower leaves opposite, upper leaves alternate; lance-shaped, ascending, 2–4 cm long and 3–8 mm wide, tapered to a rounded tip; margins mostly entire, rolled under; stalkless. Flowers on slender stalks from upper leaf axils; petals pink, 5–8 mm long, notched at tip. Fruit a linear, densely hairy capsule; coma pale brown. Flowering July–Aug.

HABITAT conifer swamps, sedge meadows, calcareous fens, marshes.
STATUS MW-OBL | NCNE-OBL | GP-OBL

Epilobium strictum

Ludwigia
WATER-PRIMROSE, PRIMROSE-WILLOW

Perennial herbs. Stems floating, creeping, or upright. Leaves simple, opposite or alternate, entire. Flowers single in leaf axils; sepals 4; petals 4 (or absent), yellow or green, large or very small; stamens 4; stigma unlobed. Fruit a 4-chambered, many-seeded capsule; seeds without a tuft of hairs at tip (coma).

ADDITIONAL SPECIES ♦*Ludwigia peploides* (Kunth) P. H. Raven, known from se LP of Mich, is an introduced, potentially invasive plant, found on muddy shores and in shallow water. Plants are mat-forming, with large, 5-petaled flowers (photo below).

Ludwigia
WATER-PRIMROSE, PRIMROSE-WILLOW

1 Leaves opposite, stalked; stems floating or creeping and rooting at nodes *L. palustris*
1 Leaves alternate, ± stalkless; stems mostly erect 2
2 Flower petals showy, yellow; flowers and fruit on stalks 3 mm or more long .... *L. alternifolia*
2 Flower petals very small or absent; flowers and fruit stalkless..................... 3
3 Plants smooth (or leaves sometimes with hairs on margins) .... *L. polycarpa*
3 Plants usually hairy .... *L. sphaerocarpa*

Ludwigia alternifolia L.
SQUARE-POD WATER-PRIMROSE, SEED-BOX

DESCRIPTION Native perennial herb, roots often with tubers; plants smooth or finely hairy.

Ludwigia peploides

Stems erect, 4–10 dm long, branched. Leaves alternate, lance-shaped, 5–10 cm long, margins entire; stalkless or with short petioles. Flowers single from leaf axils, on stalks 3–5 mm long, with 2 lance-shaped bracts near the flower; petals 4, yellow, 3–4 mm long. Fruit a smooth capsule, 4–5 mm long and as wide, square above, sharp-angled and rounded below. Flowering July–Aug.

HABITAT swamps, thickets, marshes, shores, ditches; especially where sandy.

STATUS MW-OBL | NCNE-OBL | GP-OBL

Ludwigia palustris (L.) Elliott
♦ COMMON WATER-PURSLANE,
MARSH PURSLANE

DESCRIPTION Native perennial herb. Stems weak, creeping and rooting at nodes or partly floating, simple to branched, 1–5 dm long, succulent, smooth or with sparse scattered hairs. Leaves opposite, lance-shaped to ovate, 0.5–3 cm long and 0.5–2 cm wide, shiny green or red, margins entire, tapered at base to a winged petiole to 2 cm long. Flowers single in leaf axils, stalkless; sepals broadly triangular, 1–2 mm long; petals usually absent, or small and red. Fruit a capsule, 2–5 mm long and 2–3 mm wide, somewhat 4-angled, with a green stripe on each angle. Flowering July–Sept.

SYNONYM *Isnardia palustris.*

HABITAT shallow water or exposed mud of pond margins, lakeshores, streambanks, ditches, springs.

STATUS MW-OBL | NCNE-OBL | GP-OBL

Ludwigia polycarpa Short & Peter
♦ TOP-POD WATER-PRIMROSE

DESCRIPTION Native perennial herb, producing leafy stolons from base in fall; plants smooth. Stems erect,

1–9 dm long, often branched, usually 4-angled. Leaves alternate, lance-shaped to oblong lance-shaped, 3–12 cm long and 5–15 mm wide; margins entire; ± stalkless. Flowers single in leaf axils, stalkless; sepals triangular, 2–4 mm long, usually persistent; petals green and very small or absent. Fruit a short-cylindric, rounded 4-angled capsule, 4–7 mm long and 3–5 mm wide. Flowering July–Sept.

HABITAT borders of swamps and marshes, muddy shores, wet depressions.

STATUS MW-OBL | NCNE-OBL | GP-OBL

Ludwigia sphaerocarpa Elliott
ROUND-POD WATER-PRIMROSE

DESCRIPTION Native perennial herb, spreading by stolons; plants finely hairy or sometimes smooth. Stems to 1 m long, usually branched above. Leaves alternate, lance-shaped, 5–10 cm long and 1–2 cm wide, upper leaves much smaller; margins with low teeth; petioles absent. Flowers single and ± stalkless in leaf axils, greenish; sepal lobes ovate; petals absent. Fruit a ± round, finely hairy capsule, 3–4 mm long, not angled. Flowering July–Sept.

HABITAT swamp margins, lakeshores, often in shallow water.

STATUS MW-OBL | NCNE-OBL | GP-OBL; LP of Mich (THR); disjunct from the Atlantic coast.

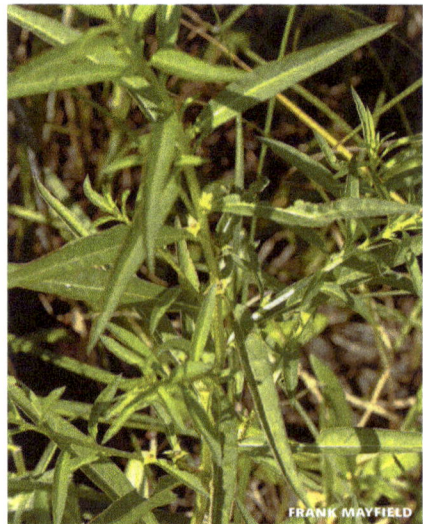

Ludwigia palustris

Ludwigia polycarpa

Orobanchaceae
Broom-Rape Family

MOSTLY PARASITIC PLANTS, now includes some former members of Scrophulariaceae. Three genera of wetland plants in our region.

Orobanchaceae **BROOM-RAPE FAMILY**

1　Most leaves deeply pinnately divided; corolla yellow.... *Pedicularis lanceolata*

1　Stem leaves toothed or entire, not deeply pinnately lobed; corolla various colors... **2**

2　Corolla nearly regular, the 5 lobes similar. .... *Agalinis*

2　Corolla irregular, 2-lipped, the upper lip 2-lobed, the lower lip 3-lobed.............. .... *Melampyrum lineare*

Agalinis AGALINIS

Annual herbs. Stems slender, erect, branched, 4-angled. Leaves opposite, linear, stalkless, smooth, or rough-to-touch on upper surface. Flowers showy, in clusters at ends of branches; sepals joined, the calyx 5-lobed, bell-shaped; petals united, corolla 5-lobed, bell-shaped, and slightly 2-lipped, pink to purple; stamens 4, of 2 different lengths. Fruit a ± round, many-seeded capsule.

Agalinis **AGALINIS**

1　Plants yellow-green; petals pink; uncommon in s Wisc and s Mich .... *A. skinneriana*

1　Plants deep green, often tinged with purple; petals purple; widespread species **2**

2　Flowers on short stalks 2–5 mm long .... *A. purpurea*

2　Flower stalks 1 cm or more long.......... .... *A. tenuifolia*

Agalinis purpurea (L.) Pennell
◆SMOOTH AGALINIS

DESCRIPTION Native annual. Stems slender, 2–8 dm long, 4-angled, smooth to slightly rough, branched and spreading above. Leaves opposite, spreading, linear, 1–5 cm long and 1–3 mm wide; margins entire; petioles absent. Flowers on spreading stalks 2–5 mm long, in racemes on the branches; calyx 4–6 mm long; corolla purple, 2–3 cm long, the lobes spreading, 5–10 mm long. Fruit a round capsule, 4–6 mm wide. Flowering Aug–Sept.

SYNONYM *Gerardia purpurea*.

HABITAT wet meadows, fens, shores of Great Lakes and inland lakes and ponds, moist areas between dunes, ditches; usually where sandy.

STATUS MW-FACW | NCNE-FACW | GP-FACW

Agalinis skinneriana (A. Wood) Britton
PALE FALSE FOXGLOVE

DESCRIPTION Native annual herb; plants rough-to-touch. Stems slender, 1.5–4 dm long, ridged or narrowly 4-winged, usually with a few upright branches. Leaves opposite, upright, narrow and bristlelike, 1–2.5 cm long and 1–2 mm wide, smaller above; margins entire; petioles absent. Flowers on stalks from leaf axils, the stalks longer than the calyx, calyx teeth very small; corolla pink to light purple, to 2 cm long, the lobes to 2 cm wide. Fruit an oblong capsule, longer than the calyx. Flowering Aug–Sept.

HABITAT sandy, calcium-rich, moist prairie; also in drier prairie and pine barrens.

STATUS NCNE-FACU; Wisc (END), Mich (END).

Agalinis tenuifolia (M. Vahl) Raf.
COMMON AGALINIS

DESCRIPTION Native annual. Stems slender, erect, 2–6 dm tall, smooth, usually with

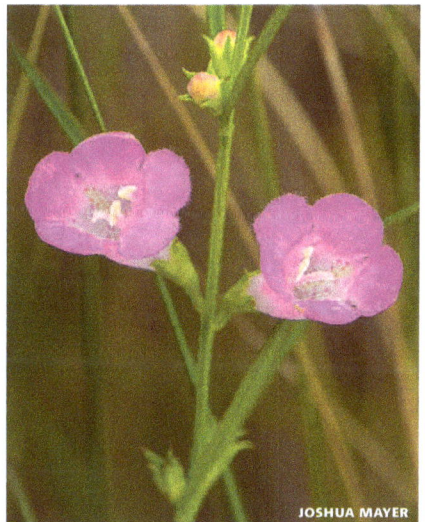

Agalinis purpurea

many branches. Leaves opposite, spreading, linear, 1–5 cm long and 1–3 mm wide, upper surface slightly rough; margins entire; petioles absent. Flowers on slender ascending stalks, 1–2 cm long; calyx 3–5 mm long, with short teeth; corolla purple (rarely white), often spotted, 10–15 mm long, the lobes 3–5 mm long. Fruit a round capsule, 4–6 mm wide. Flowering Aug–Sept.

SYNONYM *Agalinus besseyana, Gerardia tenuifolia.*

HABITAT wet meadows, low prairie, fens, shores, streambanks and ditches, usually where sandy.

STATUS MW-FACW | NCNE-FACW | GP-FAC

Melampyrum COW-WHEAT

Melampyrum lineare Desr.
◆AMERICAN COW-WHEAT

DESCRIPTION Native annual, partially parasitic on other plants, often red-tinged when in open habitats. Stems usually branched, 1–4 dm long. Leaves opposite, lower leaves oblong lance-shaped, upper leaves linear or lance-shaped, often toothed near base; petioles short or absent. Flowers from upper leaf axils; sepals joined, calyx lobes 5 and awl-shaped; corolla about 1 cm long, 2-lipped, the upper lip white, the lower pale yellow. Fruit a capsule to 1 cm long. Flowering Summer.

HABITAT common in a wide variety of habitats, ranging from wet to dry forests and openings; in wetlands occasional in swamps and on hummocks in open fens.

STATUS MW-FAC | NCNE-FACU

Pedicularis LOUSEWORT

Pedicularis lanceolata Michx.
◆SWAMP-LOUSEWORT

DESCRIPTION Native perennial herb; plants at least partially parasitic on other plants. Stems 3–8 dm long, ± smooth, unbranched or few-branched. Leaves opposite, or in part alternate, mostly lance-shaped, 4–9 cm long and 1–2 cm wide, pinnately lobed; margins with small rounded teeth; lower leaves short-petioled, upper leaves stalkless. Flowers ± stalkless, in spikes at ends of stems and from upper leaf axils; the spikes 2–10 cm long; calyx 2-lobed; corolla yellow, about 2 cm long, the upper lip entire and arched, lower lip upright. Fruit an unequally ovate capsule, mostly shorter than the sepals. Flowering July–Sept.

HABITAT wet meadows, calcareous fens, wetland margins, springs, streambanks.

STATUS MW-OBL | NCNE-FACW | GP-OBL

Melampyrum lineare

Pedicularis lanceolata

Oxalidaceae
Wood-Sorrel Family

Oxalis WOOD-SORREL

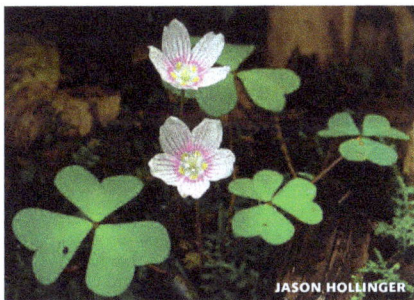

Oxalis montana Raf.
⬧NORTHERN WOOD-SORREL

DESCRIPTION Native perennial herb, from slender, scaly rhizomes. **Leaves** single or 3–6 together, all from base of plant, on stalks 4–15 cm long, these joined at base; palmately divided into 3 leaflets, the leaflets notched at tips, sparsely hairy. **Flowers** perfect, broadly bell-shaped, single atop stalks 6–15 cm long (usually slightly taller than leaves), with a pair of small bracts above middle of stalk; sepals 5, much shorter than petals; petals 5, white or pink, with pink veins, 10–15 mm long. **Fruit** a smooth, nearly round capsule. **Flowering** May–July.

SYNONYM *Oxalis acetosella.*

HABITAT hummocks in swamps, wet depressions in forests, moist wetland margins, usually where shaded.

STATUS MW-FACU | NCNE-FACU

Penthoraceae
Ditch-Stonecrop Family

Penthorum
DITCH-STONECROP

Penthorum sedoides L.
⬧DITCH-STONECROP

DESCRIPTION Native perennial herb, spreading by rhizomes; plants often red-tinged. **Stems** 1–6 dm long, smooth and round in section below, upper stem often angled and with gland-tipped hairs. **Leaves** alternate, lance-shaped to narrowly oval, 2–10 cm long and 0.5–3 cm wide, tapered to tip and base; margins with small, forward-pointing teeth; stalkless or on petioles to 1 cm long. **Flowers** star-shaped, perfect, 3–6 mm wide, on short stalks, in branched racemes at ends of stems; sepals 5, green, triangular, 1–2 mm long; petals usually absent; stamens 10; pistils 5, joined at base and sides to form a ring. **Fruit** a many-seeded capsule, the seeds about 0.5 mm long. **Flowering** July–Sept.

HABITAT streambanks, muddy shores, ditches.

STATUS MW-OBL | NCNE-OBL | GP-OBL

Phrymaceae
Lopseed Family

Mimulus MONKEY-FLOWER

Perennial herbs (or sometimes annual in *Mimulus guttatus*). Leaves opposite, margins shallowly toothed. Flowers often large and showy, single on stalks from leaf axils or in leafy racemes at ends of stems; sepals joined, the calyx tube-shaped; corolla 2-lipped, the upper lip 2-lobed, the lower lip 3-lobed, yellow or blue-violet; stamens 4, of 2 different lengths; stigmas 2. Fruit a cylindric capsule.

Oxalis montana

Penthorum sedoides

Mimulus MONKEY-FLOWER

1 Flowers blue to violet; stems 4-angled; leaves pinnately veined (with lateral veins arising all along the midrib); stems and pedicels glabrous .... *M. ringens*

1 Corolla yellow; stems terete or many-ridged but not square in cross-section; leaves nearly palmately veined (with lateral veins all near base of blade); stems and pedicels glabrous or pubescent **2**

2 Plants ± densely viscid-pubescent (with many hairs longer than 0.5 mm); calyx lobes slightly unequal, half to fully as long as the tube; corolla nearly regular .... *M. moschatus*

2 Plants glabrous or with minute, often gland-tipped hairs to 0.5 mm long; calyx lobes very unequal, only the upper large lobe as much as half as long as the tube; corolla 2-lipped........................ **3**

3 Corolla 3-4.5 cm long, the throat nearly closed by the up-arching lower lip; style 20-25 mm long; plant pubescent with tiny hairs on calyx, pedicels, and stems *M. guttatus*

3 Corolla 1-2.7 cm long, the throat open; style less than 14 mm long; plants glabrous or with minute glandular hairs **4**

4 Style mostly 3-5 mm long; widespread species.................... *M. glabratus*

4 Style mostly 8-11 (-14) mm long; rare in n Mich .... *M. michiganensis*

Mimulus glabratus HBK.
♦ROUND-LEAF MONKEY-FLOWER

DESCRIPTION Native perennial herb, spreading by stolons and often forming large mats. **Stems** succulent, smooth, 0.5-5 dm long, creeping and rooting at nodes, the stem ends angled upward. **Leaves** opposite, nearly round to broadly ovate, 1-2.5 cm wide, palmately veined, rounded at tip, hairy when young, becoming smooth; margins shallowly toothed or entire; petioles short and winged, or the upper leaves stalkless. **Flowers** yellow, on stalks from leaf axils and at ends of stems; calyx 5-9 mm long, barely toothed, irregular, the upper lobe large, the other lobes smaller; corolla 2-lipped, 9-15 mm long, the throat open and bearded on inner surface. **Fruit** an ovate capsule, 5-6 mm long. **Flowering** June-Aug.

SYNONYM *Erythranthe geyeri*.

HABITAT cold springs, seeps, banks of spring-fed streams; usually where calcium-rich.

STATUS MW-OBL | NCNE-OBL | GP-OBL

NOTE Most plants of our region are var. *jamesii* (synonym var. *fremontii*). Plants previously treated as var. *michiganensis* now treated as **Mimulus michiganensis** (next page), a rare plant of calcareous seeps and northern white cedar swamps of e UP (Mackinac County) and n LP of Mich (which is the total range of this species; federally and state endangered). It is distinguished from var. *jamesii* by its longer styles and corollas (see *Michigan Flora*, Vol. 3, by E.G. Voss, 1996).

Mimulus guttatus DC.
COMMON YELLOW MONKEY-FLOWER

DESCRIPTION Native annual or perennial herb, spreading by stolons or rhizomes; plants smooth or short-hairy. **Stems** 0.5-6 dm long, unbranched or branched. **Leaves** opposite, ovate to obovate, 2-8 cm long and 1-4 cm wide, reduced to bracts in the head; margins with irregular, coarse teeth; petioles short on lower leaves, upper leaves stalkless or clasping. **Flowers** yellow, showy, in loose racemes at ends of stems; flowers on stalks 1-2.5 cm long; calyx irregular, 10-15 mm long, the upper lobe largest; corolla yellow, often spotted with red or purple, 2-lipped, 2.5-5 cm long, lower lip bearded at base. **Fruit** an ovate capsule, about as long as calyx tube. **Flowering** July-Sept.

SYNONYM *Erythranthe guttata*.

HABITAT springs and spring-fed streams; wet, seepy woods.

STATUS MW-OBL | NCNE-OBL | GP-OBL; Mich UP, main range w USA.

PATRICK ALEXANDER

Mimulus glabratus

Mimulus michiganensis (Pennell)
Posto & Prather
MICHIGAN MONKEY-FLOWER

DESCRIPTION Native, perennial, mat-forming herb. **Stems** to about 40 cm long, reclining at base, rooting at lower leaf nodes and producing additional shoots via stolons. **Leaves** opposite, broadly ovate, margins nearly entire to coarsely sharp-toothed; petioles shorter than blades. **Flowers** borne on slender pedicels from the upper leaf axils, bright yellow, tubular, two-lipped, 16–27 mm long; lower lip irregularly red-spotted. **Fruit** (seldom produced) an oblong capsule 8–10 mm long, seeds many, with longitudinal striations. **Flowering** July–Aug.

SYNONYM *Erythranthe michiganensis.*

HABITAT marly springs, in streams in cedar swamps, calcareous shores and ditches, usually near shores of Great Lakes.

NOTE Michigan monkey-flower has been on the federal and state list of endangered species since 1990. Its total known range is in the Straits of Mackinac and Grand Traverse regions of Michigan (less than 20 known sites altogether in Mackinac County in the UP, and Benzie, Cheboygan, Emmet, and Leelanau counties (and Beaver Island) in the Lower Peninsula; endemic to Michigan. See note under *Mimulus glabratus.*

Mimulus moschatus Douglas
♦ MUSKY MONKEY-FLOWER

DESCRIPTION Native perennial herb; plants long-hairy, sticky, with a musky odor. **Stems** 2–4 dm long, lower stems creeping, the tips ascending. **Leaves** opposite, ovate, 3–6 cm long and 1–2 cm wide, pinnately veined; margins entire or with spaced, coarse teeth; petioles short.

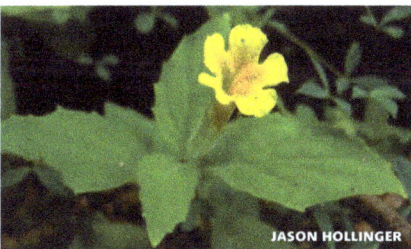

Flowers single from upper leaf axils, on slender stalks 1–2 cm long; calyx about 1 cm long, the lobes about equal; corolla yellow, open in the throat, 1.5–2.5 cm long and 2–3x longer than sepals. **Fruit** a capsule. **Flowering** July–Aug.

SYNONYM *Erythranthe moschata.*

HABITAT shores, swamp margins, streambanks, springs, ditches, wet forest trails.

STATUS NCNE-OBL | GP-OBL

Mimulus ringens L.
♦ ALLEGHENY or SQUARE-STEMMED MONKEY-FLOWER

DESCRIPTION Smooth native perennial herb, from stout rhizomes. **Stems** usually erect, 3–8 dm long, 4-angled and sometimes winged. **Leaves** opposite, oblong to lance-shaped, 4–12 cm long and 1–3.5 cm wide, upper leaves smaller; margins with forward-pointing teeth; petioles absent, the base of leaf clasping stem. **Flowers** single from upper leaf axils, on slender stalks 1–5 cm long and longer than the sepals; calyx regular, angled, 1–2 cm long, the lobes awl-shaped, 3–5 mm long; corolla blue-violet, 2-lipped, 2–3 cm long, the throat nearly closed, the upper lip erect and bent upward, lower lip longer and bent backward. **Fruit** a capsule, about as long as calyx tube. **Flowering** July–Aug.

HABITAT streambanks, oxbow marshes, swamp openings, floodplain forests, muddy shores, ditches; sometimes where disturbed.

STATUS MW-OBL | NCNE-OBL | GP-OBL

JASON HOLLINGER

Mimulus moschatus

MATT LAVIN

Mimulus ringens

Plantaginaceae
Plantain Family

ANNUAL OR PERENNIAL HERBS. Leaves simple, entire, all from base of plant. Flowers perfect in a narrow spike *(Plantago)*, each flower subtended by bracts, or single-sexed, the staminate and pistillate flowers on same plant *(Littorella)*; flower parts mostly in 4s. Fruit a capsule opening at tip.

NOTE Plantaginaceae is the accepted name for the family that includes not only the plantains with their reduced flowers, but also the related larger-flowered genera formerly placed in the Scrophulariaceae, as well as highly reduced aquatic taxa such as *Hippuris* (former Hippuridaceae) and *Callitriche* (former Callitrichaceae).

Plantaginaceae **PLANTAIN FAMILY**

1 Flowers tiny, lacking a corolla, or corolla regular and scarious 2
1 Flowers usually conspicuous, with both calyx and corolla present, the corolla petal-like, usually conspicuously bi-laterally symmetrical 5
2 Leaves in a basal rosette 3
2 Leaves opposite or whorled on an elongate stem 4
3 Leaves terete (to ca. 3 mm thick at middle, then tapering to apex), at most 1-veined, glabrous; flowers unisexual (the staminate long-stalked, the pistillate basal); fruit indehiscent; submersed or on moist shores . . . . *Littorella uniflora*
3 Leaves flat, in most species with at least 3 prominent veins and/or pubescent; flowers bisexual (in heads or spikes); capsule circumscissile; dry or rarely wet habitats . . . . *Plantago*
4 Leaves in whorls of 6–12 (usually 9) . . . . *Hippuris vulgaris*
4 Leaves opposite . . . . *Callitriche*
5 Stem leaves all or mostly alternate on fertile stems (lowermost leaves sometimes opposite and rosette of larger basal leaves sometimes present) . . . . *Veronica*
5 Stem leaves all or mostly opposite (rarely whorled) on fertile stems (may be alternate beneath flowers) 6
6 Leaves (especially middle and lower ones) deeply pinnately toothed or lobed ca. 1/3 or more the distance to the midrib; se LP of Mich . . . . *Leucospora multifida*
6 Leaves of main stem toothed or entire but not so deeply pinnately toothed or lobed (uppermost leaves or bracts may have small basal lobes); mostly widespread species 7
7 Corolla ca. 2.3–3.5 cm long; sepals broadly ovate-orbicular, overlapping; stamens 4 fertile plus a filamentous elongate staminodium . . . . *Chelone*
7 Corolla less than 1.5 cm long; sepals linear-lanceolate to somewhat ovate, not conspicuously overlapping; stamens 2 or 4 (including any staminodia) 8
8 Leaves in whorls of 3–6, sharply toothed; inflorescence of 1–several dense elongate slenderly tapering spikes or spike-like racemes; corolla tube much longer than the lobes . . . . *Veronicastrum virginicum*
8 Leaves opposite, entire or toothed; inflorescence racemose or flowers solitary in axils of alternate or opposite bracts or leaves; corolla tube various 9
9 Corolla 2-lipped, the tube much longer than the lobes; flowers solitary in axils of opposite leaves; sepals 5 . . . . . . *Gratiola*
9 Corolla often nearly regular, the tube shorter than the lobes (usually a flat limb); flowers in axillary racemes or solitary in axils of bracts or leaves; sepals 4 *Veronica*

Bacopa WATER-HYSSOP

Bacopa rotundifolia (Michx.) Wettst.
WATER-HYSSOP

DESCRIPTION Native perennial succulent herb, spreading by stolons. Stems creeping, 0.5–4 dm long, rooting at leaf nodes, smooth when underwater, usually hairy when emersed. Leaves opposite, obovate to nearly round, 1–3.5 cm long and 1–2.5 cm wide, smooth, palmately veined, rounded at tip, clasping at base; margins entire or slightly wavy; petioles absent. Flowers 1–2 from leaf axils, on hairy stalks to 1.5 cm long and shorter than the leaves; sepals 5, unequal, 4–5 mm long; corolla white with a yellow throat, 2-lipped, bell-shaped, 5–10 mm long, the 5 lobes shorter than the tube; stamens

4. Fruit a round capsule, about as long as sepals. Aug–Sept.

SYNONYM *Bacopa simulans, Hydranthelium roundifolium.*

HABITAT mud flats and shallow water of ponds and marshes.

STATUS MW-OBL | NCNE-OBL | GP-OBL

Callitriche
WATER-STARWORT

Small, annual aquatic herbs with weak, slender stems and fibrous roots. Leaves simple, opposite, all underwater or upper leaves floating; underwater leaves linear, 1-nerved, entire except for shallowly notched tip; floating leaves mostly in clusters at ends of stems, obovate to spatula-shaped, 3–5-nerved, rounded at tip. Flowers tiny, staminate and pistillate flowers usually separate on same plant, each flower with 1 stamen or 1 pistil; single and stalkless in middle and upper leaf axils, or 1 staminate and 1 pistillate flower in each axil, subtended by a pair of thin, translucent, deciduous bracts, or the bracts absent; styles 2, ovary flattened, oval to round, 4-chambered, separating when mature into 4 nutlets.

Callitriche **WATER-STARWORT**

1   Leaves all underwater, 1-veined, linear ...
.... *C. hermaphroditica*

1   Leaves both underwater and floating; floating leaves 3-veined, spatula-shaped or obovate **2**

2   Margins of fruit without wings; pits on fruit not in rows............. *C. heterophylla*

2   Margins of fruit with small wings; fruit pitted in rows....................... **3**

3   Fruit obovate, less than 1.5 mm long, slightly longer than wide .... *C. palustris*

3   Fruit nearly round in outline, 1.5–2 mm long and as wide .... *C. stagnalis*

Callitriche hermaphroditica L.
AUTUMNAL WATER-STARWORT

DESCRIPTION Native annual. Stems 10–30 cm long. Leaves all underwater, alike, linear, 1-nerved, 3–12 mm long and to 1.5 mm wide, shallowly notched at tip, clasping at base, the opposite leaf bases not connected; darker green than our other species. Flowers either staminate or pistillate; single in leaf axils, not subtended by translucent bracts. Fruit flattened, rounded, 1–2 mm long, deeply divided into 4 segments. Flowering June–Sept.

SYNONYM *Callitriche autumnalis.*

HABITAT shallow to deep water of lakes, ponds, marshes, ditches and slow-moving streams.

STATUS MW-OBL | NCNE-OBL | GP-OBL

Callitriche heterophylla Pursh
♦LARGER WATER-STARWORT

DESCRIPTION Native annual. Stems 10–20 cm long. Leaves of 2 types; underwater leaves linear, 1–2 cm long and to 1.5 mm wide, 1-nerved, notched at tip, the leaf pairs connected at base by a narrow wing; floating leaves in clusters at ends of stems or opposite along upper stems, 3–5-nerved, obovate to spatula-shaped, rounded at tip, 6–15 mm long and 3–7 mm wide; leaves intermediate between underwater and floating leaves often present. Flowers either staminate or pistillate; usually 1 staminate and 1 pistillate flower together in leaf axils, subtended by a pair of translucent, deciduous bracts. Fruit about 1 mm long and not more than 0.1 mm longer than wide, often broadest above middle, not wing-margined, pits on surface not in rows. Flowering May–Aug.

HABITAT shallow water or mud of springs, stream pools, ponds and wet depressions.

STATUS MW-OBL | NCNE-OBL | GP-OB; Wisc (THR), Mich (THR).

NOTE similar to **spiny water-starwort** (*Callitriche palustris*) but less common, differing mainly in the fruits.

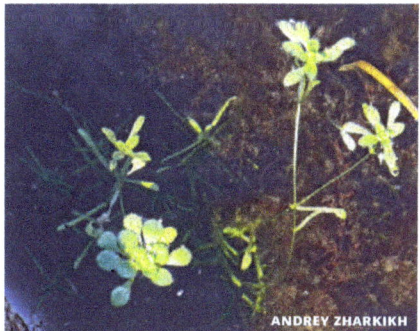

ANDREY ZHARKIKH

Callitriche heterophylla

Callitriche palustris L.
♦ SPINY WATER-STARWORT

DESCRIPTION Native annual. **Stems** 10–20 cm long. **Leaves** of 2 types; underwater leaves mostly linear, 1–2 cm long and to 1 mm wide, shallowly notched at tip, the leaf pairs connected at base by a narrow wing; floating leaves in clusters at ends of stems or opposite along upper stems, 3–5–nerved, obovate to spatula-shaped, rounded at tip, 5–15 mm long and 2–5 mm wide; leaves intermediate between underwater and floating leaves usually present. **Flowers** either staminate or pistillate; usually 1 staminate and 1 pistillate flower together in leaf axils, subtended by a pair of translucent bracts, these soon deciduous. **Fruit** 1–1.5 mm long and about 0.2 mm longer than wide, broadest above middle, narrowly winged near tip, pitted in vertical rows. **Flowering** June–Sept.

SYNONYM Callitriche verna.

HABITAT shallow water of lakes, ponds, streams; exposed mudflats.

STATUS MW-OBL | NCNE-OBL | GP-OBL

Callitriche stagnalis Scop.
POND WATER-STARWORT

DESCRIPTION Introduced annual. **Stems** 1–3 dm long. **Underwater leaves** linear, 4–10 mm long. **Floating leaves** spatulate to obovate, tapering to the base. **Fruit** nearly orbicular, 1.5–2 mm long and wide, the carpels distinctly wing-margined and separated by a V-shaped furrow.

HABITAT in cold ponds and streams.

STATUS MW-OBL | NCNE-OBL; an introduced species more common on the east and west coasts.

MARKUS HOFBAUER

Callitriche palustris

Chelone TURTLEHEAD

Perennial herbs. Leaves opposite, with toothed margins. Flowers large, 5-parted, perfect, white or rose-purple, 2-lipped, the lower lip 3-lobed and woolly on inner surface; stamens 5 (4 fertile and 1 sterile and smaller), anthers woolly; style thread-like. Fruit a many-seeded capsule.

Chelone TURTLEHEAD
1 Petals creamy white; leaves nearly stalkless; widespread species... *C. glabra*
1 Petals rose-purple; leaves stalked; uncommon in se Mich........ *C. obliqua*

Chelone glabra L.
♦ WHITE TURTLEHEAD

DESCRIPTION Native perennial herb. **Stems** erect, 5–10 dm long, rounded 4-angled, unbranched or sometimes branched above. **Leaves** opposite, lance-shaped, to 15 cm long and 1–3 cm wide, tapered to a sharp tip; margins with sharp, forward-pointing teeth; petioles very short or absent. **Flowers** in dense spikes at ends of stems, 3–8 cm long; sepals 5; corolla white or light pink, 2.5–3.5 cm long. **Fruit** an ovate capsule. **Flowering** Aug–Sept.

HABITAT swamp openings, thickets, streambanks, shores, wet meadows, marshes, calcareous fens.

STATUS MW-OBL | NCNE-OBL | GP-OBL

SUPERIOR NF

Chelone glabra

Chelone obliqua L.
PURPLE TURTLEHEAD

DESCRIPTION Native perennial herb. **Stems** 3–6 dm long. **Leaves** opposite, lance-shaped, 5–15 cm long and 2–5 cm wide; margins with sharp, somewhat spreading teeth; petioles 5–15 mm long. **Flowers** in dense spikes at ends of stems, 2–6 cm long; sepals 5, fringed with hairs; corolla 2.5–3.5 cm long, rose-purple. **Fruit** an ovate capsule. **Flowering** Aug–Oct.
HABITAT thickets and riverbanks.
STATUS MW-OBL | NCNE-OBL; Mich (END).

Gratiola HEDGE HYSSOP
Low annual or perennial herbs of shallow water and shores. Leaves opposite. Flowers on stalks from leaf axils; sepals 5; corolla white or yellow, 2-lipped, the upper lip entire or 2-lobed, the lower lip 3-lobed; fertile stamens 2. Fruit a 4-chambered capsule.

Gratiola **HEDGE HYSSOP**

1 Plants perennial, spreading by rhizomes; flowers bright yellow; leaves entire, widest at base .... *G. aurea*
1 Plants annual, rhizomes absent; flowers white; leaves toothed, widest near middle of blade 2
2 Flowers on slender, spreading stalks 1–2 cm long; upper stems with fine, gland-tipped hairs; widespread .... *G. neglecta*
2 Flowers on stout, erect stalks to 5 mm long; stems smooth; uncommon in s portion of region .... *G. virginiana*

Gratiola aurea Pursh
GOLDEN HEDGE HYSSOP

DESCRIPTION Native perennial herb. **Stems** ascending or creeping, 1–3 dm long, somewhat 4-angled, smooth or glandular hairy. **Leaves** opposite, lance-shaped to ovate, 1–2.5 cm long, with dark, glandular dots; margins entire or with a few small teeth; petioles absent. **Flowers** on slender stalks 5–15 mm long from leaf axils; sepals lance-shaped, 4–5 mm long; corolla bright yellow, 10–15 mm long. **Fruit** a round capsule, 2–3 mm long, about as long as sepals. **Flowering** July–Sept.
SYNONYM *Gratiola lutea*.
HABITAT shallow water of lakes, wet sandy or gravelly shores.
STATUS MW-OBL | NCNE-OBL | GP-OBL; Mich (THR).

Gratiola neglecta Torr.
♦ CLAMMY HEDGE HYSSOP

DESCRIPTION Native annual herb. **Stems** erect to horizontal, 5–25 cm long, usually branched, glandular-hairy above. **Leaves** opposite, linear to lance-shaped, 5–25 mm long and 1–10 mm wide, clasping at base; margins entire to wavy-toothed; petioles absent. **Flowers** single in the leaf axils, on slender stalks 1–2 cm long, subtended by a pair of small narrow bracts; sepals 5, unequal, 3–6 mm long, enlarging after flowering; corolla white, tube-shaped, slightly 2-lipped, 6–10 mm long; stamens 2. **Fruit** an ovate capsule, 3–5 mm long. **Flowering** June–Sept.
HABITAT mud flats, shores of ponds and marshes.
STATUS MW-OBL | NCNE-OBL | GP-OBL

Gratiola virginiana L.
ROUND-FRUIT HEDGE HYSSOP

DESCRIPTION Native annual-biennial herb. **Stems** upright, 1–4 dm long, ± smooth, simple or branched. **Leaves**

Gratiola neglecta

opposite, oval or obovate, 2–4 cm long and 1–2 cm wide, 3–5-veined; margins coarsely toothed; petioles absent. Flowers single on stalks from leaf axils, the stalks erect, 1–5 mm long, subtended by a pair of bracts about as long as sepals; sepals narrowly lance-shaped, 5 mm long; corolla white, 8–12 mm long, inner surface with purple lines. Fruit a round capsule, 5–9 mm long. Flowering June–Aug.

SYNONYM *Gratiola sphaerocarpa.*

HABITAT marshes, shores, wet sand flats; sometimes in shallow water.

STATUS MW-OBL | NCNE-OBL | GP-OBL; Mich (THR).

Hippuris MARE'S-TAIL

Hippuris vulgaris L.

♦MARE'S-TAIL

Native perennial herb, from large, spongy rhizomes. Stems 2–6 dm long, unbranched, underwater and lax, or emersed and upright, densely covered by the closely spaced whorls of leaves. Leaves numerous, in whorls of 6–12, linear, 1–2.5 cm long and 1–3 mm wide, stalkless. Flowers very small, perfect, stalkless and single in upper leaf axils, or often absent; sepals and petals lacking; stamen 1, style 1, ovary 1-chambered. Fruit nutlike, oval, 2 mm long. Flowering June–Aug.

HABITAT shallow water or mud of marshes, lakes, streams and ditches.

STATUS MW-OBL | NCNE-OBL | GP-OBL

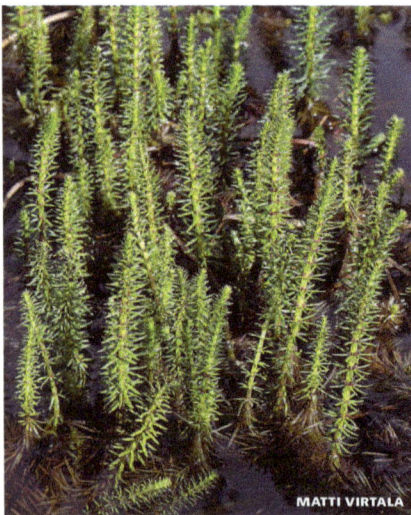

Hippuris vulgaris

Leucospora CONOBEA

Leucospora multifida (Michx.) Nutt.

♦CONOBEA

DESCRIPTION Native annual herb; plants with fine, sticky hairs. Stems leafy, 1–2 dm long, branched. Leaves opposite, 1–3 cm long, pinnately divided into 3–7 segments; margins entire or deeply lobed; petioles present. Flowers mostly single on slender stalks 5–10 mm long from leaf axils; sepals awl-shaped; corolla pale lavender, tube-shaped, 2-lipped, the upper lip 2-lobed, the lower 3-lobed; stamens 4 of 2 different lengths. Fruit a smooth, narrow, many-seeded capsule, about same length as sepals. Flowering July–Sept.

SYNONYM *Conobea multifida.*

HABITAT low prairie, sandy ditches, calcium-rich wetlands.

STATUS MW-FACW | NCNE-FACW | GP-FACW; se Mich.

Limosella MUDWORT

Limosella aquatica L.

♦NORTHERN MUDWORT

DESCRIPTION Small native annual herb (sometimes perennial by stolons); plants smooth, succulent. Stems 3–10 cm long. Leaves from base of plant, linear but wider at tip, the tip emersed or floating, 1–2.5 cm long and 2–8 mm wide, tapered to a long petiole; margins entire. Flowers small, single on stalks from base of plant, the stalks shorter than the leaves; calyx lobes triangular, corolla white or pink, 1–3 mm wide, slightly

Leucospora multifida

longer than sepals; stamens 4. Fruit an ovate capsule, 2–3 mm long. Flowering June–Sept. HABITAT streambanks, shores and mud flats of temporary ponds and marshes.

STATUS MW-OBL | NCNE-OBL | GP-OBL; Minn, main range w USA.

NOTE Sometimes placed in Scrophulariaceae.

Littorella SHOREWEED

Littorella uniflora (L.) Asch.
♦SHOREWEED

DESCRIPTION Low native perennial; plants clumped, often forming mats. Leaves bright green, linear, to 5 cm long and 2–3 mm wide, succulent; margins entire. Flowers only from emersed plants, single-sexed, staminate and pistillate flowers on same plant; staminate flowers 1–2 on stalks to 4 cm long; pistillate flowers stalkless among the leaves; sepals 4 (sometimes 3 in pistillate flowers), lance-shaped, 2–4 mm long, with a dark green midrib and lighter margins; petals joined, 4-lobed; stamens 4, longer than the petals. Fruit a 1-seeded nutlet, 2 mm long and 1 mm wide. Flowering July–Aug.

SYNONYM *Littorella americana*.

HABITAT exposed lakeshores where sandy or mucky, or in water 1 m or more deep.

STATUS NCNE-OBL

Plantago PLANTAIN

Perennial or annual herbs. Leaves all from base of plant, simple. Flowers small, perfect or single-sexed, green, ± stalkless in axils of small bracts, grouped into crowded spikes; sepals 4; petals 4. Fruit a capsule.

Limosella aquatica

Plantago PLANTAIN

1 Plants perennial; stamens 4; e Wisc and s Mich........................*P. cordata*
1 Plants annual; stamens 2; sw Minn.......................................*P. elongata*

Plantago cordata Lam.
♦KING-ROOT, HEART-LEAF PLANTAIN

DESCRIPTION Smooth native perennial, roots long and fleshy (to more than 1 cm wide). Stems stout, hollow, to 6 dm long, longer than leaves, often purple-tinged. Leaves large, broadly ovate, 10–25 cm long and 8–20 cm wide, pinnately veined, rounded or heart-shaped at base; margins entire or shallowly toothed; petioles winged near base of blade. Flowers green, in loose, interrupted spikes to 3 dm long; sepals ovate, about as long as bracts; petals joined, the 4 lobes spreading. Fruit a rounded capsule, 5–10 mm long, opening at or near the middle, usually with 2 seeds. Flowering April–June.

HABITAT floodplain forests, wooded streambanks, sometimes in shallow water; usually where calcium-rich.

STATUS MW-OBL | NCNE-OBL; Wisc (END), Mich (END); uncommon throughout most of its range in e USA.

Plantago elongata Pursh
SLENDER PLANTAIN

DESCRIPTION Small native annual herb from a slender taproot. Stems to 15 cm long, finely hairy. Leaves all from base of plant, narrow and linear, 3–8 cm long, tapered to a blunt tip; margins mostly entire; petioles absent. Flowers green, in slender spikes 3–8 cm long, with mix of staminate, pistillate and some perfect flowers;

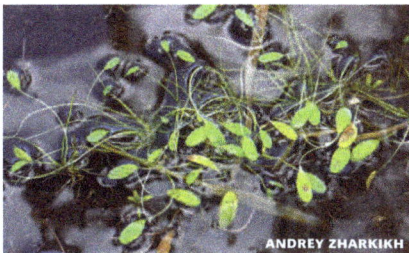

Plantago cordata

sepals to 2 mm long and as long as bracts; petals joined, the lobes to 1 mm long, usually widely spreading. Fruit a capsule 2–3 mm long, opening at or near the middle; seeds 1–2 mm long. Flowering May–Aug.

HABITAT moist, usually alkaline places; entering the region in sw Minn (THR); main range Great Plains to Pacific Coast.

STATUS MW-FACW | GP-FACW

Veronica SPEEDWELL

Annual or perennial herbs. Leaves opposite, or becoming alternate in the head. Flowers single or in racemes from leaf axils or at ends of stems; se-pals deeply 4-parted, enlarging after flowering; corolla blue or white, 4-lobed, somewhat 2-lipped, the tube shorter than the lobes; stamens 2. Fruit a flattened capsule, lobed at tip; styles usually persistent on fruit.

Veronica SPEEDWELL

1 Flowers in racemes at ends of stems; flower stalks 1–2 mm long . . *Veronica peregrina*
1 Flowers in racemes from leaf axils; flower stalks longer than 2 mm 2
2 Leaves with short petioles . *V. americana*
2 Petioles absent 3
3 Leaves with sharp forward-pointing teeth, or entire; upper leaves clasping stem; capsules swollen . . *V. anagallis-aquatica*
3 Leaves with small, gland-tipped, outward-pointing teeth; leaves narrowed to base, not clasping stem; capsules strongly flattened . . . . *V. scutellata*

Veronica americana L.

⧫AMERICAN SPEEDWELL

DESCRIPTION Native perennial herb, spreading by rhizomes; plants smooth and succulent. Stems erect to creeping, 1–6 dm long. Leaves opposite, ovate to lance-shaped (or lower leaves oval), 2–8 cm long and 0.5–3 cm wide, upper leaves tapered to a tip, lower leaves often rounded; margins with forward-pointing teeth; petioles short. Flowers in stalked racemes from leaf axils; the racemes with 10–25 flowers and to 15 cm long; corolla 4-lobed, blue (sometimes white), often with purple stripes. Fruit a ± round, compressed capsule, 3–4 mm long,

slightly notched at tip, the styles persistent, 2–4 mm long. Flowering July–Sept.

SYNONYM *Veronica beccabunga*.

HABITAT streambanks and wet shores, hummocks in swamps, springs.

STATUS MW-OBL | NCNE-OBL | GP-OBL

Veronica anagallis-aquatica L.

WATER SPEEDWELL

DESCRIPTION Introduced, biennial or short-lived perennial herb, spreading by stolons or leafy shoots produced in fall; plants ± smooth. Stems erect to spreading, 1–6 dm long, often rooting at lower nodes. Leaves opposite, lance-shaped to ovate, 2–10 cm long and 0.5–5 cm wide, tapered to a blunt or rounded tip; margins entire or with fine, forward-pointing teeth; petioles absent, the leaves often clasping. Flowers in many-flowered racemes from leaf axils, the racemes 5–12 cm long; corolla 4-lobed, blue or striped with purple, about 5 mm wide. Fruit a round, compressed capsule, 2–4 mm long, notched at tip, the styles persistent, 1–2 mm long. Flowering June–Sept.

SYNONYMS *Veronica catenata, Veronica comosa*.

HABITAT wet, sandy or muddy streambanks and ditches; often in shallow water.

STATUS MW-OBL | NCNE-OBL | GP-OBL; naturalized throughout much of USA.

Veronica americana

Veronica peregrina L.
♦ PURSLANE-SPEEDWELL

DESCRIPTION Small, native annual herb. **Stems** upright, 0.5–3 dm long, unbranched or with spreading branches, usually glandular-hairy. Lower leaves opposite, becoming alternate and smaller in the head, oval to linear, 5–25 mm long and 1–5 mm wide, rounded at tip; margins of lower leaves sparsely toothed, upper leaves entire; petioles short or absent. **Flowers** small, on short stalks from upper leaf axils; corolla 4-lobed, ± white, about 2 mm wide. **Fruit** an oblong heart-shaped capsule, 2–4 mm long, notched at tip, the styles not persistent. **Flowering** May–July.

HABITAT mud flats, shores, ditches, temporary ponds, swales; also weedy in cultivated fields, lawns and moist disturbed areas.

STATUS MW-FACW | NCNE-FAC | GP-FACW

Veronica scutellata L.
NARROW-LEAVED SPEEDWELL

DESCRIPTION Native perennial herb, spreading by rhizomes or leafy shoots produced in fall; plants smooth (or sometimes with sparse hairs). **Stems** slender, erect to reclining, 1–4 dm long, often rooting at lower nodes. **Leaves** opposite, linear to narrowly lance-shaped, 3–8 cm long and 2–10 mm wide, tapered to a sharp tip; margins entire or with small, irregularly spaced teeth; petioles absent. **Flowers** in racemes from leaf axils, the racemes with 5–20 flowers, as long or longer than the leaves; corolla 4-lobed, blue, 6–10 mm wide. **Fruit** a strongly flattened capsule, 3–4 mm long, notched at tip, the styles persistent, 3–5 mm long. **Flowering** June–Sept.

HABITAT marshes, pond margins, hardwood swamps, thickets, springs, streambanks, wet swales and depressions.

STATUS MW-OBL | NCNE-OBL | GP-OBL

Veronicastrum
CULVER'S ROOT

Veronicastrum virginicum (L.) Farw.
♦ CULVER'S ROOT

DESCRIPTION Erect native perennial herb, 1–2 m tall, usually with several upright branches. **Leaves** in whorls of 3–6, lance-shaped; margins with fine, forward-pointing teeth; petioles to 1 cm long. **Flowers** in erect, spikelike racemes to 15 cm long, the flowers crowded and spreading; corolla white, nearly regular, 4–5-parted, the lobes shorter than the tube; stamens 2, long-exserted from the corolla mouth. **Fruit** a capsule, 4–5 mm long. **Flowering** June–Aug.

SYNONYMS *Veronica virginicum, Leptandra virginicum.*

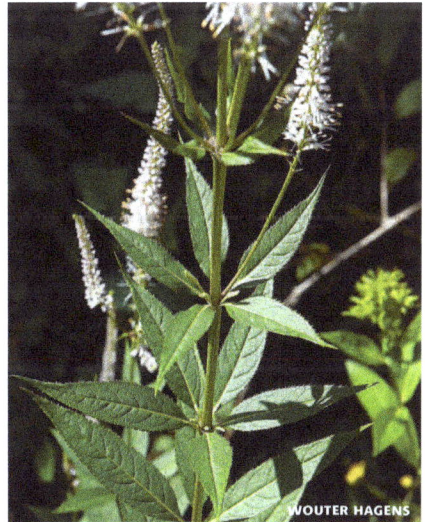

Veronica peregrina

Veronicastrum virginicum

HABITAT moist to wet prairie, fens and streambanks; also in drier deciduous woods and sandy grasslands.
STATUS MW-FAC | NCNE-FAC | GP-FAC

Platanaceae
Planetree Family

Platanus SYCAMORE

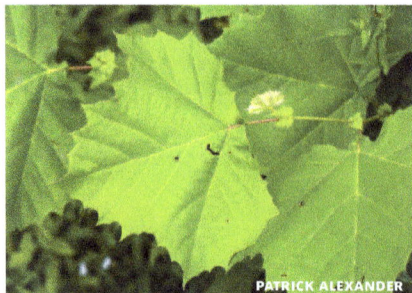

Platanus occidentalis L.
⬥SYCAMORE

DESCRIPTION Large native tree to 30 m tall or more; trunk to 2 m wide, crown broad and spreading. Bark redbrown when young, soon breaking into thin, flat sections which fall away to expose whitegreen inner bark; twigs smooth, light brown; buds light brown and pointed. Leaves alternate, to 20 cm long and about as wide, divided into 3 or 5 shallow, sharp-pointed lobes, bright green and smooth on upper surface, underside paler, smooth except for sparse hairs on veins. Staminate and pistillate flowers tiny, in dense clusters, separate but on same tree. Fruit a round, light brown head, 2–3 cm wide, on a long, drooping stalk. Flowering May.
HABITAT riverbanks, floodplain forests and lakeshores.
STATUS MW-FACW | NCNE-FACW | GP-FAC

Polemoniaceae
Phlox Family

ANNUAL OR PERENNIAL HERBS. Leaves opposite (*Phlox*) or pinnately divided (*Polemonium*). Flowers perfect (with both staminate and pistillate parts), single or in clusters at ends of stems and from leaf axils; sepals and petals 5-parted and joined for part of length. Fruit a 3-chambered capsule, with usually 1 seed per chamber.

Platanus occidentalis

Polemoniaceae **PHLOX FAMILY**
1   Leaves undivided . . . . *Phlox*
1   Leaves pinnately divided into leaflets . . . .
     . . . . *Polemonium occidentale*

Phlox PHLOX
Erect perennial herbs of southern portions of region. Leaves opposite, margins entire. Flowers pink, purple or rarely white, in stalked clusters at ends of stems and from upper leaf axils; sepals joined and tubelike; corolla 5-lobed, tubelike but flared outward at tip; stamens 5. Fruit a 3-chambered capsule.

Phlox **PHLOX**
1   Flower head rounded, the individual flower clusters 1 to few, on stalks of unequal lengths . . . . *P. glaberrima*
1   Flower head ± cylinder-shaped, the clusters several or more, on stalks of about equal length . . . . *P. maculata*

Phlox glaberrima L.
⬥SMOOTH PHLOX

DESCRIPTION Native perennial herb, plants usually smooth. Stems erect, unbranched, 5–10 dm long. Leaves opposite, firm, lance-shaped, long-tapered to a sharp tip, 5–12 cm long and 3–15 mm wide; margins entire; petioles absent (except on lowest leaves). Flowers 1.5–2 cm wide, in stalked clusters (cymes) at end of stems and from upper 1–4 pairs of leaf axils, the lower clusters long-stalked; sepals joined to form a sharp-tipped tube; petals pink, rounded at tip; style elongated. Fruit a 3-chambered capsule. Flowering June-Aug.
HABITAT wet to moist, calcium-rich meadows and low prairies.
STATUS MW-FACW | NCNE-FACW | GP-FAC; Wisc (END).

Phlox maculata L. SPOTTED PHLOX, WILD SWEET-WILLIAM

DESCRIPTION Native perennial herb. Stems erect, 3–8 dm long, simple or branched above, smooth or finely hairy, usually red-spotted. Leaves opposite, smooth, ± firm, lance-shaped, 5–12 cm long and 0.5–1.5 cm wide, long-tapered to a sharp tip; margins entire; petioles absent. Flowers 1–2 cm wide, in stalked clusters (cymes) at ends of stems and from several to many upper leaf axils, these short-stalked, forming a long, narrow head 10–20 cm long, the head finely hairy; sepals smooth, joined to form a sharp-tipped tube 6–8 mm long; petals pink or purple, rounded at tip; style elongated. Fruit a 3-chambered capsule. Flowering July–Sept.

SYNONYM *Phlox pyramidalis.*

HABITAT fens, sedge meadows, springs.

STATUS MW-FACW | NCNE-FACW; Mich (THR).

Polemonium JACOB'S LADDER

Polemonium occidentale Green WESTERN JACOB'S LADDER

DESCRIPTION Native perennial herb. Stems erect, to 10 dm long, single from upturned ends of short, unbranched rhizomes. Leaves alternate, pinnately divided with up to 27 leaflets, the leaflets 1–4 cm long, smaller upward; margins entire; petioles short or absent. Flowers blue, 10–15 mm wide, crowded in a long panicle composed of smaller clusters of flowers; sepals joined to form a tube; petal lobes longer than calyx tube; stamens shorter or equal to corolla; style longer than stamens. Fruit a 3-chambered capsule. Flowering July. Ours are subsp. *lacustre.*

HABITAT cedar swamps and thickets.

STATUS MW-FACW | NCNE-FACW | GP-FACW; Minn (END), Wisc (END); disjunct from main range of Rocky Mountains.

Polygalaceae
Milkwort Family

Polygala MILKWORT

Polygala cruciata L.
♦ DRUM-HEADS, MARSH MILKWORT

DESCRIPTION Native annual herb. Stems erect, 4-angled, 1–4 dm long, usually branched above. Leaves mostly in whorls of 4, linear or oblong lance-shaped, 1–4 cm long and 1–5 mm wide, rounded and often with a short, sharp point at tip; margins entire; petioles short or absent. Flowers ± stalkless in cylindric racemes, 1–5 cm long and 1–1.5 cm wide; flowers pale purple or green purple; sepals 5, the 2 lateral sepals (wings) petal-like, 4–6 mm long and 3–4 mm wide at base; petals 3, joined into a tube; stamens 8 (sometimes 6). Fruit a 2-chambered capsule, with a single, ± hairy seed in each chamber. Flowering July–Sept.

SYNONYM *Polygala aquilonia.*

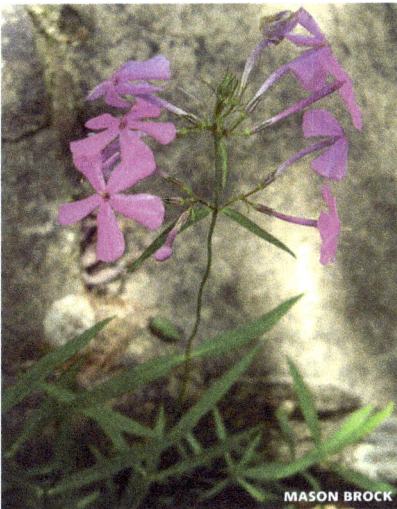

Phlox glaberrima

Polygala cruciata

HABITAT sandy or mucky lakeshores, wet areas between dunes.
STATUS MW-FACW | NCNE-FACW | GP-OBL; Minn (END); main range Atlantic coastal plain from Maine to Texas.

Polygonaceae
Smartweed Family

ANNUAL OR PERENNIAL HERBS, plants sometimes vining. Leaves alternate, simple, sometimes wavy-margined, otherwise entire; the nodes usually enlarged. Stipules joined to form a membranous or papery sheath (ocrea) around stem at each node. Flowers in spike-like racemes or small clusters from leaf axils (*Persicaria, Polygonum*), or in crowded panicles at ends of stems (*Rumex*). Flowers small, perfect (with both staminate and pistillate parts), regular, petals absent. In Rumex the sepals herbaceous, green to brown, in inner and outer groups, each group with 3 sepals, the 3 inner enlarging after flowering, becoming broadly winged, persisting to enclose the achene; stamens 4–8; ovary 1-chambered, styles 2–3; in other genera of family, sepals more or less petal-like, white to pink or yellow, mostly 5 (sometimes 4). Fruit a 3-angled or lens-shaped achene.

NOTE Polygonaceae recognized by presence of a stipular sheath (*ocrea*), which surrounds the stem above the attachment of each leaf. The similar reduced structure in the inflorescence is called an *ocreola*.

Polygonaceae SMARTWEED FAMILY
1 Tepals 6, greenish or reddish, scarcely petaloid, the 3 inner (but not the outer) ones enlarging in fruit and concealing the achene; stigmas a feathery tuft; plants in some species dioecious or polygamous and hence some flowers entirely staminate ... .*Rumex*
1 Tepals 4–5, white to red and ± petaloid at least along the margins, uniform in size or the outer ones larger; stigmas usually not feathery and plants mostly with bisexual flowers.................................2
2 Stem and petioles with retrorse prickles; leaves hastate or sagittate (with acute basal lobes).......................*Persicaria*

2 Stem and petioles without prickles; leaves various..............................3
3 Flowers 1–4 at a node, sessile or pediceled in the axils of foliage leaves or bracts; leaf blades jointed at the base, less than 2 (–2.4) cm broad; summit of ocrea silvery white, becoming lacerate-shredded; annuals.... ....*Polygonum*
3 Flowers numerous in peduncled terminal or axillary spikes, racemes, or panicles, often densely crowded; leaves not jointed at base of blade, in some species over 2.5 cm broad; summit of ocrea tinged with brown, shattering at maturity but not shredding; annuals or perennials ....*Persicaria*

Persicaria
LADY'S-THUMB, SMARTWEED
Annual and perennial herbs. Flowers pink or sometimes white, in terminal spikes. The genus was formerly included in *Polygonum*.

Persicaria LADY'S-THUMB, SMARTWEED
1 Stems with downward-pointing prickles on the stem angles 2
1 Stems smooth to hairy, but not prickly ..3
2 Basal lobes of leaves pointed downward; achenes 3-sided ....*P. sagittata*
2 Basal lobes pointed outward; achenes 2-sided........................*P. arifolia*
3 Perennial herbs from rhizomes or stolons 4
3 Taprooted annual herbs 6
4 Flowers in 1 or 2 terminal racemes ....*P. amphibia*
4 Flowers in several to many terminal and axillary racemes 5
5 Perianth dotted with glands . *P. punctata*
5 Perianth not dotted with glands ....*P. hydropiperoides*
6 Sheathing stipules (ocreae) fringed with bristles at tip 7
6 Ocreae entire or irregularly cut, not fringed with bristles 10
7 Perianth dotted with glands 8
7 Perianth not dotted with glands 9
8 Tepals usually 4; achenes dull ....*P. hydropiper*
8 Tepals 5; achenes shiny .....*P. punctata*
9 Upper stem and peduncles with gland-tipped hairs ....*P. careyi*

9 Upper stem and peduncles not with gland-tipped hairs .... *P. maculosa*

10 Outer sepals strongly 3-nerved, each nerve ending in an anchor shaped fork; racemes nodding to erect .... *P. lapathifolia*

10 Outer sepals with faint, irregularly forked nerves; racemes erect ... *P. pensylvanica*

Persicaria amphibia (L.) A. Gray
◆ WATER SMARTWEED

DESCRIPTION Native perennial herb; plants either aquatic with floating leaves, or emergent and exposed, both types with spreading rhizomes. **Stems** to 1 m or more long, leaves and habit variable. Submerged plants smooth, usually branched, the branches floating, branch tips often upright and raised above water surface. Exposed plants hairy. **Leaves** floating, leathery, oval, 4–20 cm long and 1–4 cm wide, rounded at tip; stipules (ocreae) membranous; petioles 1–8 cm long. Leaves of exposed plants stalkless or with short petioles. **Flowers** pink to red, in 1–2 spikelike racemes from branch tips, the racemes 2–15 cm long and 1–2 cm wide; sepals 5-lobed to below middle, 4–5 mm long; stamens 5. **Fruit** a lens-shaped achene, 2–4 mm long, shiny dark brown. **Flowering** June–Sept.

SYNONYMS *Polygonum amphibium, Polygonum coccineum, Polygonum natans.*

HABITAT ponds, lakes, marshes, bog pools,

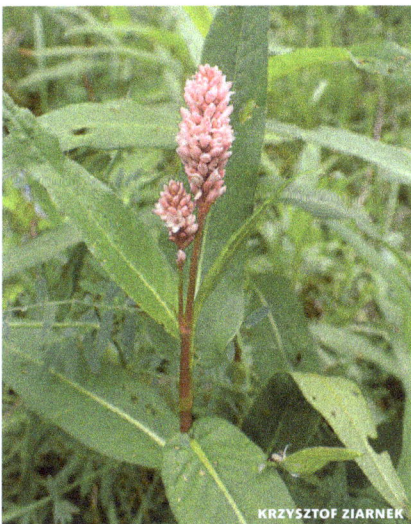

Persicaria amphibia

backwater areas, quiet streams.
STATUS MW-OBL | NCNE-OBL | GP-OBL

Persicaria arifolia (L.) Haralds.
HALBERD-LEAVED TEARTHUMB

DESCRIPTION Native annual herb, similar to **arrow-leaved tearthumb** (*P. sagittata*). **Leaves** to 20 cm long and 15 cm wide, arrowhead-shaped at base but the triangular-shaped basal lobes pointing outward rather than downward as in *P. sagittata.* **Flowers** in rounded heads at ends of stems or from leaf axils, flower stalks with glands; sepals pink, 2–3 mm long. **Fruit** a lens-shaped achene, 4–5 mm long. **Flowering** July–Sept.

SYNONYM *Polygonum arifolium.*

HABITAT swamps, wet woods, streambanks, shores.

STATUS MW-OBL | NCNE-OBL | GP-OBL

Persicaria careyi (Olney) Greene
CAREY'S HEARTS-EASE

DESCRIPTION Native annual herb. **Stems** upright, branched, to 1 m long, with gland-tipped hairs. **Leaves** lance-shaped; stipules (ocreae) fringed with bristles and covered with stiff, spreading hairs. **Flowers** in cylindric, drooping racemes 3–6 cm long; sepals pink or rose, 3 mm long; stamens 5 (sometimes to 8). **Fruit** a black, smooth, shiny achene, 2 mm wide. **Flowering** July–Aug.

SYNONYM *Polygonum careyi.*

HABITAT sandy lakeshores and streambanks, marshes, recently burned wetlands.

STATUS MW-FACW | NCNE-FACW; uncommon in Mich (THR); historical records from Minn.

Persicaria hydropiper (L.) Delarbre
◆ WATER-PEPPER SMARTWEED

DESCRIPTION Introduced annual herb. **Stems** red, erect to sprawling, 2–6 dm long, sometimes rooting at lower nodes, branched or unbranched, peppery-tasting. **Leaves** lance-shaped, 3–8 cm long and to 2 cm wide, hairless except for short hairs on veins and margins, nearly stalkless or with a short petiole; stipules (ocreae) membranous, 5–15 mm long,

swollen and fringed with bristles. Flowers green and usually white-margined, continuous in slender racemes, often nodding at tip; sepals 5, 3–4 mm long, with glandular dots; stamens 4 or 6. Fruit a dull, dark brown achene, 3-angled or lens-shaped, 2–3 mm long. Flowering July–Oct.

SYNONYM *Polygonum hydropiper*.

HABITAT muddy shores, streambanks, floodplains, marshes, ditches and roadsides.

STATUS MW-OBL | NCNE-OBL | GP-OBL

Persicaria hydropiperoides (Michx.)
Small FALSE WATER-PEPPER

DESCRIPTION Native perennial herb, spreading by rhizomes. Stems erect to sprawling with upright tips, to 1 m long, usually branched, nearly smooth or with short hairs. Leaves linear to lance-shaped, 4–12 cm long and to 2.5 cm wide, petioles short; stipules (ocreae) membranous, 5–15 mm long, with stiff hairs and fringed with bristles. Flowers green, white or pink, in 2 to several slender racemes, 1–6 cm long, often interrupted near base; sepals 2–3 mm long, 5-lobed to just below middle, without glandular dots or only the inner sepals slightly glandular; stamens 8. Fruit a black, shiny, 3-angled achene with concave sides, 2–3 mm long. Flowering July–Sept.

SYNONYM *Polygonum hydropiperoides*.

HABITAT shallow water or wet soil; ponds, marshes, swamps, bogs and fens, streambanks, lakeshores and ditches.

STATUS MW-OBL | NCNE-OBL | GP-OBL

Persicaria lapathifolia (L.) S.F. Gray
◆DOCK-LEAVED SMARTWEED

DESCRIPTION Native annual. Stems erect to sprawling, unbranched or few-branched, 2–15 dm long. Leaves lance-shaped, 4–20 cm long and 0.5–5 cm wide, smooth above, often densely short-hairy on leaf undersides; petioles to 2 cm long, smooth to glandular; stipules (ocreae) 5–20 mm long, entire or with irregular, jagged margins. Flowers deep pink, white or green, crowded in erect or nodding racemes 1–5 cm long; sepals 3–4 mm long, 4- or 5-lobed to below middle, the outer 2 sepals strongly 3-nerved; stamens usually 6. Fruit a brown, lens-shaped achene, 2–3 mm long. Flowering July–Sept.

SYNONYM *Polygonum lapathifolium*.

HABITAT marshes, wet meadows, shores, streambanks, ditches, cultivated fields.

STATUS MW-FACW | NCNE-FACW | GP-OBL

Persicaria maculosa S.F. Gray
◆LADY'S-THUMB

DESCRIPTION Introduced annual herb. Stems upright to spreading, 2–8 dm long, often red, unbranched to branched. Leaves lance-shaped, 3–15 cm long and 0.5–3 cm wide, smooth or with few hairs, underside usually dotted with

HUGH KNOTT
Persicaria hydropiper

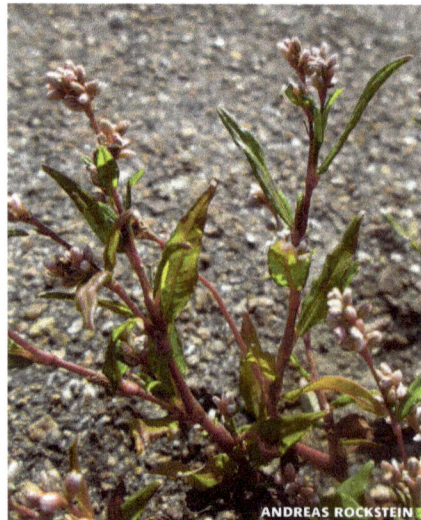
ANDREAS ROCKSTEIN
Persicaria lapathifolia

small glands, leaves stalkless or on petioles to 1 cm long; ocreae 5-15 mm long, fringed with bristles, with short hairs. Flowers pink to rose, crowded in straight, cylindric racemes 1-4 cm long and 0.5-1 cm wide; sepals 2-4 mm long, 5-lobed to near middle; stamens 6. Fruit a black, shiny achene, lens-shaped or sometimes 3-angled, 2-3 mm long. Flowering July–Sept.

SYNONYM *Polygonum persicaria.*

HABITAT muddy shores, streambanks, ditches and cultivated fields, often weedy; introduced from Europe and now throughout much of North America.

STATUS MW-FACW | NCNE-FAC | GP-FACW

Persicaria pensylvanica (L.) M. Gómez
PENNSYLVANIA SMARTWEED

DESCRIPTION Native annual herb. Stems erect, 3-20 dm long, unbranched to widely branching. Leaves lance-shaped, 3-15 cm long and 1-4 cm wide, smooth except for short hairs on margins; petioles to 2.5 cm long; stipules (ocreae) 0.5-1.5 cm long, entire or with an irregular, jagged margin, hairless, not fringed with bristles. Flowers pink to white, in dense racemes 2-3 cm long, the flower stalks with gland-tipped hairs; sepals 3-5 mm long, 5-parted to below middle, the outer sepals faintly nerved; stamens 8 or less. Fruit a dark brown to black, shiny achene, lens-shaped, to 3 mm long. Flowering June–Sept.

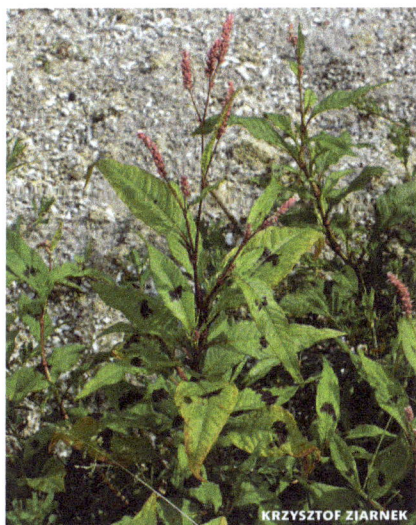

SYNONYM *Polygonum pensylvanicum.*

HABITAT streambanks, shores, marshes, fens, ditches, cultivated fields.

STATUS MW-FACW | NCNE-FACW | GP-FACW

Persicaria punctata (Ell.) Small
DOTTED SMARTWEED

DESCRIPTION Native annual or perennial herb. Stems erect to spreading, 4-10 dm long, unbranched to branched. Leaves narrowly lance-shaped or oval, 4-15 cm long and 1-2 cm wide, smooth except for small short hairs on margins, underside usually dotted with small glands; petioles short; stipules (ocreae) 5-15 mm long, smooth or with stiff hairs and fringed with bristles. Flowers green-white; in numerous slender, loosely flowered racemes, interrupted in lower portion, to 10 cm long; sepals 3-4 mm long, with glandular dots, 5-parted to about middle; stamens 6-8. Fruit a dark, shiny achene, lens-shaped or 3-angled, 2-3 mm long. Flowering Aug–Sept.

SYNONYM *Polygonum punctatum.*

HABITAT floodplain forests, marshes, shores, streambanks and cultivated fields.

STATUS MW-OBL | NCNE-OBL | GP-OBL

Persicaria sagittata (L.) Gross
◆ARROW-LEAVED TEARTHUMB

DESCRIPTION Slender native annual herb.

Persicaria maculosa

Persicaria sagittata

Stems 4-angled, weak, usually supported by other plants, 1–2 m long, with downward pointing prickles on stem angles, petioles, leaf midribs and flower stalks. **Leaves** lance-shaped to oval, arrowhead-shaped at base, 3–10 cm long and to 2.5 cm wide, the basal lobes pointing downward; petioles long on lower leaves, shorter above; stipules (ocreae) 5–10 mm long, with a few hairs on margins. **Flowers** white or pink; in round racemes to 1 cm long, on long slender stalks at ends of stems or from leaf axils; sepals 3 mm long, 5-parted to below middle. **Fruit** a brown to black, shiny achene, 3-angled, 2–3 mm long. **Flowering** July–Sept.

SYNONYM *Polygonum sagittatum.*

HABITAT swamps, marshes, wet meadows and burned wetlands.

STATUS MW-OBL | NCNE-OBL | GP-OBL

Rumex DOCK

Perennial, sometimes weedy, herbs (annual in *Rumex fueginus*). Leaves large and clustered at base of plants, or leafy-stemmed; mostly oblong to lance-shaped, flat to wavy-crisped along margins, usually with petioles. Membranous sheath around stem present at each node (ocrea). Flowers in crowded whorls in panicles at ends of stems; flowers small and numerous, green but turning brown; sepals in 2 series of 3, the inner 3 sepals (valves) enlarging, becoming winged and loosely enclosing the achene, giving the appearance of a 3-winged fruit, the midvein of the valve often swollen to produce a grainlike tubercle on the back; stamens 6; styles 3. Fruit a brown, 3-angled achene, tipped with a short slender beak.

Rumex DOCK

1   Margins of mature valves with coarse or spine-tipped teeth **2**

1   Margins of mature valves entire or shallowly lobed, not toothed **4**

2   Plants annual, fibrous-rooted or with slender taproots; margins of valves dissected into spine-tipped teeth. . . . .*R. fueginus*

2   Plants perennial from stout taproots; valve margins toothed or spine-tipped **3**

3   Grains 3, margins of valves coarsely toothed; w Minn. . . . .*R. stenophyllus*

3   Grains 1; margins of valves with spine-tipped teeth; e Minn and eastward . . . .*R. obtusifolius*

4   Flower stalks without a large swollen joint; grains 3, base of grain distinctly above base of valve . . . .*R. britannica*

4   Flower stalks with a large swollen joint below the middle or near base; grains 1–3, base of grain even with base of valve **5**

5   Fruit with 3 grains, the grains projecting below the valves; flower stalks 2–5x longer than fruit. . . . .*R. verticillatus*

5   Fruit with 1–3 grains, the grains not projecting below the valves; flower stalks 1–2x longer than fruit **6**

6   Leaves crisp-margined (crinkled); grains 2/3 as wide as long . . . .*R. crispus*

6   Leaf margins flat; grains narrower, up to half as wide as long **7**

7   Grains usually 1 (sometimes 2–3); leaves mostly less than 4x longer than wide . . . .*R. altissimus*

7   Grains usually 3; leaves mostly more than 4x longer than wide . . .*R. triangulivalvis*

Rumex altissimus A. Wood
PALE DOCK

DESCRIPTION Native perennial herb, similar to **willow dock** (*R. salicifolius*). **Stems** 3–10 dm long, usually branched from base and with short branches above. **Leaves** all from stem, ovate to lance-shaped, 6–20 cm long and 2–6 cm wide, margins flat or slightly wavy. **Flowers** in panicles 1–3 dm long, the panicle branches short and ± upright; flower stalks short, 3–5 mm long, swollen and jointed near base; valves rounded, 4–6 mm long and as wide, flattened across base, margins smooth or irregularly toothed; grains usually well-developed on only 1 of the 3 valves, although sometimes present on 2–3 valves; the largest grain lance-shaped. **Fruit** a brown achene, 2–3 mm long. **Flowering** May–Aug.

HABITAT marshes, shores, streambanks, ditches, disturbed areas.

STATUS MW-FACW | NCNE-FACW | GP-FAC

Rumex britannica L.
GREAT WATER-DOCK

DESCRIPTION Native perennial. **Stems** stout, unbranched, 2–2.5 m long. **Leaves** lance-

shaped or oblong lance-shaped, lower leaves 30–60 cm long, upper leaves 5–15 cm long; margins flat. Flowers in panicles to 5 dm long; valves rounded, flat at base, 5–8 mm long and as wide, smooth or with small teeth; grains 3, narrowly lance-shaped, the base distinctly above base of valve. Flowering June–Aug.

SYNONYM *Rumex orbiculatus.*

HABITAT marshes, fens, streambanks and ditches, often in shallow water.

STATUS MW-OBL | NCNE-OBL | GP-OBL

Rumex crispus L. ♦CURLED DOCK

DESCRIPTION Introduced perennial, from a thick taproot. Stems stout, upright, usually single, 5–15 dm long. Basal leaves large, 10–30 cm long and 1–5 cm wide, on long petioles, often drying early in season; stem leaves smaller and with shorter petioles, oval to lance-shaped, margins strongly wavy-crisped (crinkled). Flowers in large branched panicles, the panicle branches ± upright; flower stalks drooping at tips, 5–10 mm long, swollen-jointed near base; valves heart-shaped to broadly ovate, 4–5 mm long and as wide, margins ± smooth; grains 3, swollen, often of unequal size, rounded at ends. Fruit a brown achene, 2–3 mm long. Flowering July–Sept.

HABITAT wet meadows, shores, ditches, old fields, and other wet and disturbed areas; weedy; introduced from Eurasia, naturalized throughout USA and much of world; considered a noxious weed in Mich.

STATUS MW-FAC | NCNE-FAC | GP-FAC

Rumex fueginus Phil. ♦ GOLDEN DOCK

DESCRIPTION Native annual. Stems hollow, to 8 dm long, much-branched. Leaves mostly on stems, smaller upward, lance-shaped to linear, 5–20 cm long and 0.5–4 cm wide, wedge-shaped or heart-shaped at base, margins flat to wavy-crisped. Flowers in large open panicles, the panicle branches ± upright, leafy, the flower stalks jointed near base; valves triangular-ovate, 2–3 mm long, the margins lobed into 2–3 spine-tipped teeth on each side; grains 3. Fruit a light brown achene, 1–2 mm long. Flowering July–Aug.

SYNONYM *Rumex maritimus.*

HABITAT marshes, shores, streambanks and ditches, sometimes where brackish.

STATUS MW-FACW | NCNE-FACW | GP-FACW; Mich (THR).

Rumex obtusifolius L.
BITTER DOCK

DESCRIPTION Introduced perennial herb. Stems stout, to 12 dm long, usually unbranched. Lower leaves oblong or ovate, to 30 cm long and 15 cm wide, heart-shaped or rounded at base; upper

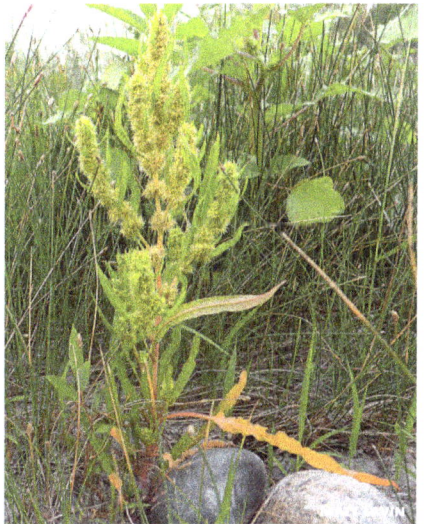

Rumex crispus

Rumex fueginus

leaves smaller. Flowers in much-branched panicles, flower stalks longer than fruit, jointed near base; valves triangular-ovate, 4–5 mm long, with 2–4 spine-tipped teeth on each side; grains large and with tiny wrinkles. Fruit a shiny, red-brown achene. Flowering June–Aug.

HABITAT floodplain forests and openings, cultivated fields and disturbed areas; introduced, now across most of USA.

STATUS MW-FACW | NCNE-FAC | GP-FACW

Rumex stenophyllus Ledeb.
NARROW-LEAF DOCK

DESCRIPTION Introduced perennial herb, similar to **curled dock** (*R. crispus*) but leaves less wavy-crisped and the valves with many small teeth on margins. Valves triangular to rounded, 4–6 mm long and as wide; grains 3. Fruit a brown achene, 2–3 mm long. Flowering July–Sept.

HABITAT wet meadows, shores, streambanks, ditches, and disturbed places, usually where brackish; more common westward in Great Plains.

STATUS MW-FACW | NCNE-FACW | GP-FACW

Rumex triangulivalvis (Danser) Rech. f.
WILLOW DOCK

DESCRIPTION Native perennial taprooted herb. Stems smooth, 3–10 dm long, usually branched from base and with short branches on stem. Leaves mostly on stems, not much smaller upward, narrowly lance-shaped, tapered at both ends, pale waxy green, 5–16 cm long and 1–3 cm wide, margins mostly flat. Flowers in panicles 1–3 dm long, panicle branches few and ± upright, with small linear leaves at base; flower stalks 2–4 mm long, swollen and jointed near base; valves thick, triangular, 3–6 mm long and wide, margins smooth or shallowly toothed; grains usually 3. Fruit a brown achene, 2 mm long. Flowering June–Aug.

SYNONYMS *Rumex mexicanus, Rumex salicifolius.*

HABITAT wet meadows, marshes, shores, streambanks, ditches and other low areas, sometimes where brackish.

STATUS MW-FACW | NCNE-FAC | GP-FACW

Rumex verticillatus L.
WATER-DOCK, SWAMP-DOCK

DESCRIPTION Native perennial taprooted herb. Stems stout, 1–1.5 m long, with many short branches from leaf axils. Leaves narrowly lance-shaped, tapered to base, margins flat. Flowers in leafless panicles 2–4 dm long, the panicle branches few and ± upright; flower stalks 10–15 mm long, jointed near base; valves triangular-ovate, 4–6 mm long and wide, thickened at center; grains 3, lance-shaped, the base blunt and projecting 0.5 mm below base of valve. Flowering June–Sept.

SYNONYM *Rumex floridanus.*

HABITAT marshes, swamps, wet forests, backwater areas and muddy shores, often in shallow water.

STATUS MW-OBL | NCNE-OBL | GP-FACW

Primulaceae
Primrose Family

PERENNIAL HERBS (ours). Leaves simple, opposite (sometimes whorled in *Lysimachia*), or leaves all basal. Flowers perfect (with both staminate and pistillate parts), regular, single from leaf axils, or in clusters at ends of stems; sepals 4–5, petals mostly 5 (varying from 4–9), joined, tube-shaped below and flared above, deeply cleft to shallowly lobed at tip; ovary superior, style 1; stamens 5. Fruit a 5-chambered capsule.

Primulaceae **PRIMROSE FAMILY**

1 Leaves all from base of plant, leaf underside strongly whitened . . . . *Primula mistassinica*

1 Leaves from stem, green on both sides . . **2**

2 Flowers white to pink, single and stalkless in leaf axils; petals absent; sepals petal-like; nw Minn . . . . *Lysimachia maritima*

2 Flowers white or yellow, single and stalked from leaf axils or in racemes from axils; sepals and petals present; widespread. . . **3**

3 Flowers white; leaves basal and alternate on stem . . . . *Samolus valerandi*

3 Flowers yellow; leaves opposite . . . . *Lysimachia*

Lysimachia LOOSESTRIFE

Perennial herbs, spreading by rhizomes. Stems erect. Leaves opposite (sometimes appearing whorled), ovate or lance-shaped. Flowers 5-parted, yellow, single on stalks from leaf axils or in racemes or panicles; sepals green; petals bright to pale yellow (white to pink in *L. maritima)*. Fruit a capsule.

ADDITIONAL SPECIES Garden loosestrife (*Lysimachia vulgaris*), with densely hairy stems and leaves and mostly whorled leaves, is occasional on mudflats along rivers and in wet meadows, mostly in s portion of our region. This introduced species sometimes escapes from cultivation in ne and c USA. Sea-Milkwort (*Lysimachia maritima*), a plant of wet meadows and seeps, is reported from Kittson County in nw Minn, becoming more common in w USA.

Lysimachia LOOSESTRIFE
1 Plants creeping; leaves round, on short, smooth petioles . . . . *L. nummularia*
1 Plants upright, leaves longer than wide; petioles hairy or absent 2
2 Flowers in dense clusters at ends of stems or from leaf axils. 3
2 Flowers single from leaf axils 4
3 Flowers in clusters at ends of stems . . . . *L. terrestris*
3 Flowers in clusters from leaf axils . . . . *L. thyrsiflora*
4 Leaves rounded or heart-shaped at base; petioles 1–3 cm long, fringed with hairs . . . . *L. ciliata*
4 Leaves tapered at both ends; petioles absent or short, smooth or fringed with hairs 5
5 Leaves narrowly linear, to 5 mm wide; margins smooth and rolled under . . . . *L. quadriflora*
5 Leaves lance-shaped to oval, usually greater than 9 mm wide; margins finely hairy or rough-to-touch 6
6 Leaves and flowers with red or black dots or streaks; petals entire at tip . . . . *L. quadriflora*
6 Leaves and flowers without red or black dots or streaks; petals ragged-fringed at tip . . . . *L. hybrida*

Lysimachia ciliata L.
♦FRINGED LOOSESTRIFE
DESCRIPTION Native perennial herb, spreading by rhizomes. Stems upright, 3–12 dm long, unbranched or with few branches above. Leaves ovate to lance-shaped, 4–15 cm long and 2–6 cm wide, rounded to heart-shaped at base, green above, slightly paler below; margins fringed with short hairs; petioles 0.5–5 cm long, fringed with hairs. Flowers yellow, single from upper leaf axils, on stalks 2–7 cm long; sepal lobes lance-shaped, 4–8 mm long, often with 3–5 parallel red-brown veins; petal lobes rounded and finely ragged at tip, 4–10 mm long and 3–9 mm wide, with a short slender tip. Fruit a capsule, 4–7 mm wide. Flowering June–Aug.
SYNONYM *Steironema ciliata*.
HABITAT usually shaded wet areas, such as shores, streambanks, wet meadows, ditches, floodplains, wet woods and thickets.
STATUS MW-FACW | NCNE-FACW | GP-FACW

Lysimachia hybrida Michx.
LOWLAND LOOSESTRIFE
DESCRIPTION Native perennial herb, spreading by rhizomes. Stems usually erect, 2–8 dm long, unbranched or sometimes branched from base, usually branched above. Leaves narrowly lance-

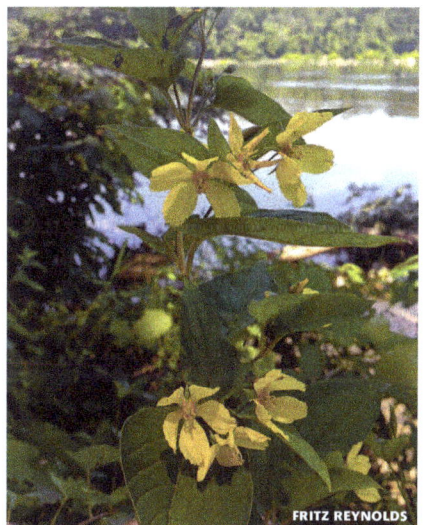

Lysimachia ciliata

shaped to ovate, 3–10 cm long and 1–2 cm wide, tapered to base, upper surface green, underside green or slightly paler; lower leaves opposite, stalked, withering, petioles fringed with hairs at least near base; upper leaves ± whorled and stalkless, persistent. Flowers yellow, single from leaf axils but often appearing crowded, on stalks 1–4 cm long; sepal lobes lance-shaped, 3–6 mm long; petal lobes rounded and finely fringed at tip, 5–10 mm long and 4–10 mm wide, with a short slender tip. Fruit a capsule, 4–6 mm wide. Flowering July–Aug.

SYNONYM *Lysimachia verticillata*.

HABITAT wet meadows, marshes, streambanks, ditches and shores, sometimes in shallow water.

STATUS MW-OBL | NCNE-OBL | GP-OBL; historical records from Mich.

Lysimachia nummularia L.
◆MONEYWORT, CREEPING JENNIE

DESCRIPTION Introduced perennial herb, often forming mats. Stems creeping, to 5–6 dm long. Leaves opposite, dotted with black glands, round or broadly oval, 1–2.5 cm long; petioles short. Flowers single in leaf axils, on stalks to 2.5 cm long; sepals leaflike, triangular, 6–8 mm long; petals yellow, dotted with dark red, 10–15 mm long. Fruit a capsule, shorter than sepals. Flowering June–Aug.

HABITAT swamps, floodplain forests, stream-

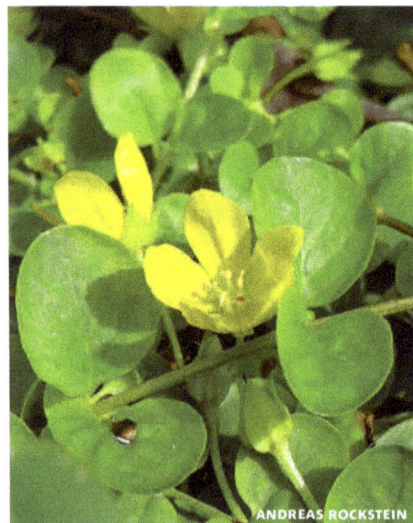

banks, shores, meadows, ditches; native of Europe and potentially invasive.

STATUS MW-FACW | NCNE-FACW | GP-FACW

Lysimachia quadriflora Sims
◆SMOOTH LOOSESTRIFE

DESCRIPTION Native perennial herb, spreading by rhizomeswhich form clusters of basal rosettes. Stems upright, 3–10 dm long. Leaves opposite, sometimes appearing whorled; stem leaves stalkless, often ascending, linear, 3–8 cm long and 2–7 mm wide, margins smooth or rolled under, sometimes fringed with a few hairs near base. Flowers yellow, single in clusters at ends of stems and branches, on stalks 1–4 cm long; sepal lobes lance-shaped, 4–6 mm long; petal lobes oval, 7–12 mm long and 5–9 mm wide, entire or finely ragged at tip. Fruit a capsule, 3–5 mm wide. Flowering July–Aug.

SYNONYM *Lysimachia longifolia*.

HABITAT wet meadows, pond and marsh margins, low prairie, calcareous fens; often where sandy and calcium-rich.

STATUS MW-OBL | NCNE-OBL | GP-FACW

Lysimachia terrestris (L.) BSP.
◆SWAMP CANDLES

DESCRIPTION Native perennial herb, spreading by shallow rhizomes. Stems smooth,

Lysimachia nummularia

Lysimachia quadriflora

4–8 dm long, usually branched. Leaves opposite, dotted with glands, narrowly lance-shaped, 5–10 cm long and 2–4 cm wide; small bulblike structures produced in leaf axils late in season; bracts awl-like, 3–8 mm long. Flowers yellow, in a single, crowded, upright raceme, 1–3 dm long; sepals lance-shaped; petal lobes oval, 5–7 mm long, with dark lines, on stalks 8–15 mm long. Fruit a capsule, 2–3 mm wide. Flowering June–Aug. HABITAT marshes, fens, thickets, muddy shores, and ditches.
STATUS MW-OBL | NCNE-OBL | GP-OBL

Lysimachia thyrsiflora L.
♦SWAMP LOOSESTRIFE

DESCRIPTION Native upright herb, spreading by rhizomes; plants conspicuously dotted with dark glands. Stems smooth or with patches of brown hairs, 3–7 dm long, unbranched or branched on lower stem. Leaves opposite, linear to lance-shaped, 4–12 cm long and 0.5–4 cm wide, smooth above, smooth or sparsely hairy below; petioles absent. Flowers yellow, crowded in dense racemes from leaf axils, on spreading stalks 2–5 cm long; mostly 6-parted; sepal lobes awl-shaped, 1–3 mm long; petal lobes linear, 3 mm long; stamens 2x longer than petals. Fruit a capsule, 2–4 mm wide. Flowering June–Aug.
HABITAT many types of wetlands: thickets, shores, fens and bogs, marshes, low places

in conifer and deciduous swamps, often in shallow water.
STATUS MW-OBL | NCNE-OBL | GP-OBL

Primula PRIMROSE

Primula mistassinica Michx.
♦MISTASSINI PRIMROSE

DESCRIPTION Native perennial herb. Stems to 25 cm long. Leaves all at base of plant, oblong lance-shaped, 2–7 cm long, long tapered to base, smooth on upper surface, smooth or often white-yellow powdery below; margins with outward pointing teeth; bracts below flowers awl-shaped, 3–6 mm long. Flowers 1–2 cm wide, 2–10 in a cluster atop a leafless stalk; sepals joined, shorter than petals; petals joined, tubelike and flared at ends, pink and sometimes with a yellow center. Fruit an oblong, upright capsule to 1 cm long. Flowering May–June.
SYNONYM *Primula intercedens*.
HABITAT shorelines of Lakes Michigan and Superior; in Wisc, also on moist ledges in St. Croix, Wisc and Kickapoo River valleys; moist, calcium-rich meadows and interdunal wetlands, also inland in calcareous fens and on sandstone cliffs; often with **violet butterwort** (*Pinguicula* vulgaris).
STATUS MW-FACW | NCNE-FACW | GP-FACW

Lysimachia terrestris

Lysimachia thyrsiflora

Samolus WATER PIMPERNEL

Samolus valerandi L.

◆ WATER PIMPERNEL

DESCRIPTION Native perennial herb. **Stems** 1–3 dm long, branched above. **Leaves** in a basal cluster and alternate on stem, obovate, blunt or rounded at tip, 2–10 cm long; margins entire; lower leaves with petioles, upper leaves often stalkless. **Flowers** white, 2–3 mm wide, on slender spreading stalks 5–15 mm long, in loosely flowered racemes 3–15 cm long. **Fruit** a round capsule, 2–3 mm wide. **Flowering** June–Sept.

SYNONYMS *Samolus floribundus, Samolus parviflorus.*

HABITAT muddy and sandy streambanks, where often in shade; ditches and salt marshes.

STATUS MW-OBL | NCNE-OBL | GP-OBL

Ranunculaceae
Buttercup Family

ANNUAL OR PERENNIAL, aquatic or terrestrial herbs (or vines in *Clematis*). Leaves simple to compound, usually alternate, sometimes opposite or whorled, or all at base of plant. Flowers mostly white or yellow, usually with 5 (occasionally more) separate petals and sepals, or petals absent and then with petal-like sepals; sepals leafy and green or petal-like and colored; flowers perfect (with both staminate and pistillate parts), stamens usually numerous; pistils several to many, ripening into beaked achenes or dry capsules (follicles).

Ranunculaceae **BUTTERCUP FAMILY**

1 Vines; leaves opposite; fruit with a long, feathery style . . . . *Clematis*

1 Herbs; leaves alternate or from base of plant; fruit not with a long, feathery style **2**

2 Leaves linear, 1–2 mm wide, all from base of plant; sepals spurred at base; achenes in a spikelike cluster to 6 cm long . . . .
. . . . *Myosurus minimus*

2 Leaves not linear; sepals not spurred; achenes in round to short-cylindric heads
. . . . **3**

3 Flowers irregular, purple to blue; uncommon plant of sw Wisc . . . .
. . . . *Aconitum columbianum*

3 Flowers regular, yellow to white; mostly widespread species **4**

4 Leaves from base of plant except for 2–3 whorled, leafy bracts below flowers . . . . . .
. . . . *Anemone canadensis*

4 Stems leafy, the leaves alternate, or leaves all from base of plant **5**

5 Flowers yellow, *or* leaves simple and unlobed, *or* plants aquatic **6**

5 Flowers not yellow; leaves divided into leaflets; plants not aquatic **8**

6 Leaves all alike, unlobed; sepals yellow (rarely pink or white), large and petal-like; petals absent . . . . *Caltha*

Primula mistassinica

Samolus valerandi

6 Leaves usually of 2 types (stem leaves different from basal leaves), *or* leaves deeply lobed or divided; sepals green; petals yellow or white **7**

7 Flowers solitary on long slender peduncles from the nodes; sepals 3. . . . . *Coptidium lapponicum*

7 Flowers usually at least 2 in a branched inflorescence; sepals 5 . . . . **8**

8 All leaves simple and shallowly lobed with rounded teeth . . *Halerpestes cymbalaria*

8 All, or at least stem leaves, deeply lobed, divided, or compound . . . . . . *Ranunculus*

9 Leaves all from base of plant, flowers single at ends of leafless stalks. . . *Coptis trifolia*

9 Stems leafy; flowers many in panicles at ends of stems . . . . *Thalictrum*

Aconitum MONKSHOOD

Aconitum columbianum Nutt.
⬦NEW YORK *or* NORTHERN MONKSHOOD

DESCRIPTION Native perennial herb, from a thickened rootstock; rare. **Stems** upright, to 1.5 m long. **Leaves** divided into 5–7 coarsely toothed segments. **Flowers** few in a raceme, showy, irregular, purple to blue, the upper sepal rounded, domelike, covering the petals; stamens many, pistils 3–5. **Fruit** a follicle. **Flowering** July–Aug.
SYNONYM *Aconitum noveboracense*.

Aconitum columbianum

HABITAT ledges and bases of shaded cliffs, streambanks.
STATUS MW-FACW | NCNE-FAC; driftless area of sw Wisc (THR) and ne Iowa; also known from ne Ohio and New York; federally listed as threatened.

Anemone ANEMONE

Anemone canadensis L.
⬦CANADA ANEMONE

DESCRIPTION Native perennial herb, from slender rhizomes, often forming large patches. **Stems** erect, 1–6 dm long, unbranched below the head. **Leaves** all from base of plant and with long petioles except for 2–3 stalkless leafy bracts below the head; 4–15 cm wide, deeply 3–5-lobed, round to kidney-shaped in outline, underside with long silky hairs, margins sharp-toothed. **Flowers** mostly single at ends of stalks, white and showy, 2–5 cm wide; sepals 5, petal-like, 1–2 cm long; petals absent; stamens and pistils many. Achenes clustered in a round, short-hairy head; achene body flat, 3–5 mm long and wide, beak 2–4 mm long. **Flowering** May–Aug.
SYNONYMS *Anemonidium canadense*, *Anemonastrum canadense*.
HABITAT wet openings, streambanks, thickets, low prairie, ditches, roadsides.
STATUS MW-FACW | NCNE-FACW | GP-FACW

Caltha
MARSH-MARIGOLD, COWSLIP
Succulent perennial herbs. Leaves simple, mostly from base of plant, becoming smaller upward, heart-shaped; margins entire or rounded-toothed. Flowers mostly bright yellow (*Caltha palustris*), to

Anemone canadensis

pink or white (*Caltha natans*), single at ends of stalks; sepals large and petal-like; petals absent; stamens many. Fruit a follicle.

Caltha **MARSH-MARIGOLD, COWSLIP**

1 Flowers bright yellow; common ....
.... *C. palustris*
1 Flowers pink or white; uncommon in ne Minn, nw Wisc, Mich UP ...... *C. natans*

Caltha natans Pallas
♦ FLOATING MARSH-MARIGOLD

DESCRIPTION Native perennial herb. **Stems** floating or creeping, branched, rooting at nodes. **Leaves** heart- or kidney-shaped, 2–5 cm wide, notched at base, upper leaves smaller. **Flowers** pink or white, 1 cm wide; sepals oval; petals absent; stamens 12–25. **Fruit** a follicle, 4–5 mm long, in dense heads of 20–40. **Flowering** July–Aug.
HABITAT in shallow water and shores of ponds and slow-moving streams.
STATUS NCNE-OBL; Minn (END), Wisc (END); reported from Baraga County in Michigan UP (THR).

Caltha palustris L.
♦ COMMON MARSH-MARIGOLD

DESCRIPTION Loosely clumped native perennial herb. **Stems** smooth, 2–6 dm long, hollow. **Leaves** heart-shaped to kidney-shaped, 4–10 cm wide, usually with 2 lobes at base; margins smooth or shallowly toothed; lower leaves with long petioles, stem leaves with shorter petioles. **Flowers** bright yellow, showy at ends of stems or in leaf axils, 2–4 cm wide; sepals 4–9, petal-like, 12–20 mm long; petals absent;

stamens many; pistils 4–15, with short styles. **Fruit** a follicle, 10–15 mm long. March–June.
HABITAT shallow water, swamps, wet woods, thickets, streambanks, calcareous fens, marshes, springs.
STATUS MW-OBL | NCNE-OBL | GP-OBL

Clematis CLEMATIS

Herbaceous or woody plants, erect, or climbing by the prehensile leaf-rachis. Leaves opposite, simple or compound. Flowers solitary or panicled, usually dioecious. Sepals petal-like, commonly 4. Petals none. Stamens numerous. Pistils numerous; style elongate. Fruit a flattened achene, terminated by the elongate persistent style.

Clematis **CLEMATIS**

1 Sepals whitish, less than 1 cm long, in a branched inflorescence ... *C. virginiana*
1 Sepals purple, 4–5 cm long, solitary ......
.... *C. occidentalis*

Clematis occidentalis (Hornem.) DC.
PURPLE CLEMATIS

DESCRIPTION Native, prennial, woody vine. **Stems** trailing or climbing, to 2 m long. **Leaves** opposite, divided into 3 leaflets, long-stalked, ovate in outline; entire, crenately toothed, or lobed. **Flowers** chiefly axillary, solitary, on peduncles about equaling the subtending petiole; sepals 4, blue, ovate lance-shaped, 3–5 cm long, softly villous. **Fruit** a villous achene, in a dense globular head; styles long-villous, 3–4 cm

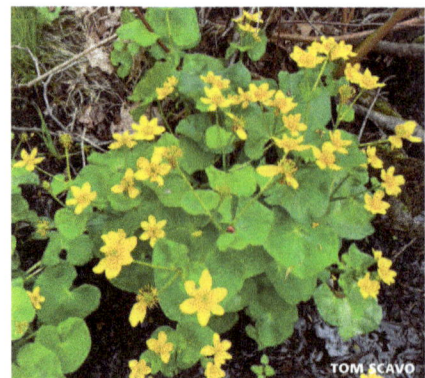

Caltha natans

Caltha palustris

long. Flowering May.

SYNONYM *Clematis verticillaris*.

HABITAT Rocky woods and streambanks.

NOTE the 1–3 large, purple flowers open early in spring with the unfolding leaves.

Clematis virginiana L.
♦ VIRGIN'S BOWER

DESCRIPTION Native perennial woody vine. Stems slender, to 5 m long or more, trailing on ground or over shrubs, smooth, brown to red-purple. Leaves opposite, divided into 3 leaflets, the leaflets ovate, 4–8 cm long and 2.5–5 cm wide; margins sharp-toothed or lobed; petioles 5–9 cm long. Flowers Staminate and pistillate flowers separate and on separate plants, in many-flowered, open clusters from leaf axils, on stalks 1–8 cm long, usually shorter than leaf petioles; sepals 4, creamy-white, 6–10 mm long; petals absent. Fruit a rounded head of hairy brown achenes tipped with feathery, persistent styles 2.5–4 cm long. Flowering July–Sept.

HABITAT Thickets, streambanks, moist to wet woods, rocky slopes.

STATUS MW-FAC | NCNE-FAC | GP-FAC

Coptidium
ARCTIC-BUTTERCUP

Coptidium lapponicum (L.) Gand. ex Rydb. ♦ LAPPLAND BUTTERCUP

DESCRIPTION Native perennial herb, spreading by rhizomes. Stems prostrate, 1–2 dm long, sending up 1 shoot from each node, the shoots with 1–2 basal leaves, sometimes with a single smaller leaf above. Leaves kidney-shaped, deeply 3-cleft, margins with rounded teeth or shallowly lobed. Flowers single at ends of shoots; petals yellow with orange veins, 8–12 mm wide; sepals 3, curved downward. Achenes in a round head; achene body 2–3 mm long, swollen near base, flattened above, beak slender, sharply hooked. Flowering June–July.

SYNONYM *Ranunculus lapponicus*.

HABITAT Cedar swamps and bogs.

STATUS NCNE-OBL | GP-OBL; Wisc (END), UP of Mich (THR); more common north of Upper Midwest region.

NOTE Resembles **Alaska gold-thread** (*Coptis trifolia*) but the leaves are lobed, not compound, lighter green and deciduous.

Coptis GOLDTHREAD

Coptis trifolia (L.) Salisb.
♦ ALASKA GOLDTHREAD

DESCRIPTION Native perennial herb, with slender, bright yellow rhizomes. Leaves from base of plant on long petioles, evergreen, divided into 3-leaflets, the leaflets shallowly lobed, with rounded teeth tipped by an abrupt point. Flowers single, white, 10–15 mm wide, on a stalk 5–15 cm long from base of plant; sepals 4–7,

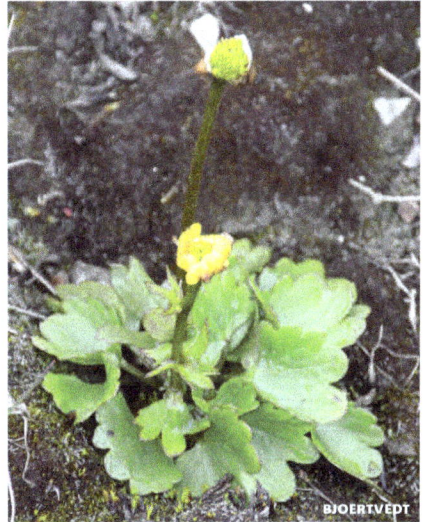

Clematis virginiana

Coptidium lapponicum

petal-like; petals absent; pistils 3–7, narrowed to a short, slender style. Fruit a beaked follicle 8–13 mm long. Flowering May–June.

SYNONYM *Coptis groenlandica.*

HABITAT wet conifer woods and swamps, often on mossy hummocks.

STATUS MW-FACW | NCNE-FACW | GP-FACW

Halerpestes
ALKALI BUTTERCUP

Halerpestes cymbalaria (Pursh) Greene
SEASIDE CROWFOOT

DESCRIPTION Native perennial herb, spreading by stolons and forming dense mats. Stems 3–20 cm long, smooth. Leaves all from base of plant, ovate to kidney-shaped, 5–25 mm long and 4–30 mm wide, heart-shaped at base; margins with rounded teeth, often with 3 prominent lobes at tip; petioles sparsely hairy. Flower stalks longer than leaves, unbranched or with a few branches, with 1 to several flowers; sepals 5, green-yellow, 3–5 mm long, deciduous; petals usually 5, yellow, turning white with age, 3–5 mm long; stamens 10–30. Achenes numerous in a cylindric head to 10 mm long; achene body 1.5–2 mm long, longitudinally nerved, beak short and straight. Flowering June–Sept.

SYNONYMS *Cyrtorhyncha cymbalaria, Ranunculus cymbalaria.*

HABITAT wet meadows, streambanks, shores, ditches and seeps, often in wet mud or sand; Lake Michigan shores; often where brackish.

STATUS MW-OBL | NCNE-OBL | GP-OBL; Wisc (THR), Mich (THR).

Myosurus MOUSE-TAIL

Myosurus minimus L. ♦ MOUSE-TAIL

DESCRIPTION Small, inconspicuous, native annual herb. Stems 4–15 cm long. Leaves in a basal tuft, hairless, linear, mostly less than 1 mm wide. Flowers small, few to many, in a spike above leaves on a slender stalk to 6 cm long when mature; sepals 5, green, upright, with a spur at base; petals 5 or sometimes absent, small, white or pink; stamens 5–10; pistils many, on an elongate receptacle. Fruit an achene, 2–3 mm long and 1 mm wide, with a small beak. Flowering April–June.

SYNONYM *Ranunculus minimus.*

HABITAT wet to moist places such as streambanks and floodplains, sometimes temporarily in shallow water; also in disturbed drier areas.

STATUS MW-FACW | NCNE-FAC | GP-FACW

Ranunculus
BUTTERCUP, CROWFOOT, SPEARWORT

Aquatic, semi-aquatic, or terrestrial annual and perennial herbs. Stems erect to sprawling, sometimes floating in water. Leaves simple, or compound and finely dissected, often variable on same plant; alternate on stem or all from base of plant; petioles short to long. Flowers borne above water surface in aquatic species; sepals usually 5, green; petals usually 5, yellow or white, often fading to white, usually with a small nectary pit covered by a scale near base of petal; stamens and pistils numerous. Achenes many in a

Coptis trifolia

Myosurus minimus

round or cylindric head; achene body thick or flattened, tipped with a beak.

Ranunculus
BUTTERCUP, CROWFOOT, SPEARWORT

1 Flowers white; leaves divided into linear or threadlike segments; plants typically aquatic . . . . *R. aquatilis*

1 Flowers yellow; leaves simple to deeply lobed or divided into narrow segments; plants aquatic, emergent, or terrestrial . . **2**

2 All leaves linear, simple and entire . . . . *R. flammula*

2 All, or at least stem leaves, deeply lobed, divided or compound **3**

3 Basal and stem leaves distinctly different in shape, the basal leaves mostly entire or with rounded teeth, the stem leaves deeply divided . . . . *R. abortivus*

3 Basal and stem leaves similar, all deeply lobed, divided or compound **4**

4 Achenes swollen, without a sharp margin **5**

4 Achenes flattened, with a sharp or winglike margin **7**

5 Petals 2-4 mm long; achenes to 1.2 mm long, nearly beakless; plants terrestrial or in water only part of season . *R. sceleratus*

5 Petals 4-14 mm long; achenes 1.2-2.5 mm long, beaked; plants underwater or exposed later in season **6**

6 Petals more than 7 mm long; achene body more than 1.6 mm long, achene margin thickened and white-corky below middle . . . . *R. flabellaris*

6 Petals less than 7 mm long; achene body less than 1.6 mm long, achene margin rounded but not thickened . . . *R. gmelinii*

7 Petals 7-15 mm long; stems often recurved and rooting at nodes **8**

7 Petals 2-5 mm long; stems not rooting at nodes **9**

8 Leaves deeply divided but outer segment not on a petiole; style short, curved; plants weedy . . . . *R. acris*

8 Leaves compound, the outermost lobe on a petiole; style long and straight; plants not weedy . . . . *R. hispidus*

9 Beak of achene strongly hooked . . . . *R. recurvatus*

9 Beak of achene straight or only slightly curved **10**

10 Petals shorter than sepals; achenes in cylindric heads longer than wide; widespread . . . . *R. pensylvanicus*

10 Petals equal or longer than sepals; heads ovate or round; uncommon in n Minn and on Isle Royale . . . . *R. macounii*

Ranunculus abortivus L.
◆ SMALL-FLOWERED CROWFOOT

DESCRIPTION Native biennial or perennial herb. **Stems** upright, 2-5 dm long, branched above, smooth or with fine hairs. **Leaves** at base of plant round to kidney-shaped, margins with rounded teeth, some leaves lobed; petioles long; **stem leaves** 3-5-divided into linear segments, margins entire or broadly toothed, petioles absent. **Flowers** yellow, petals 2-3 mm long, shorter than sepals. **Achenes** in a short, round head; achene body swollen, 1-2 mm long, with a very short, curved beak. **Flowering** April-June.

HABITAT wet to moist woods, floodplains, wet meadows, thickets, ditches; especially where soils disturbed or compacted.

STATUS MW-FACW | NCNE-FAC | GP-FAC

Ranunculus acris L.
◆ COMMON *or* TALL BUTTERCUP

DESCRIPTION Introduced perennial herb, with fibrous roots. **Stems** hairy, to 1 m long, with few branches, most leaves on lower part of stem. **Leaves** kidney-shaped, deeply 3-7-divided, the segments again lobed or dissected; branch leaves much smaller, 3-parted. **Flowers** numerous; sepals 5, half length of petals; petals 5, bright yellow, 6-15 mm long, obovate, often with a rounded notch at tip. **Achenes** in a round head; achene body flat, 2-3 mm long, beak 0.5 mm long. **Flowering** June-Aug.

Ranunculus abortivus

HABITAT common weed of fields, thickets, ditches and shores.

STATUS MW-FAC | NCNE-FAC | GP-FACW; introduced from Europe, weedy and naturalized.

Ranunculus aquatilis L.
♦ WHITE WATER-CROWFOOT

DESCRIPTION Native perennial aquatic herb; plants mostly smooth. Stems underwater or floating (sometimes stranded on muddy shores in late summer), 2–8 dm long, unbranched or with a few branches, rooting from lower nodes. Leaves round to kidney-shaped in outline, 2–3x divided into narrow threadlike segments 1–2 cm long; stiff and not collapsing when removed from water; petioles absent or to 4 mm long. Flowers at or below water surface, single from upper leaf axils, 1–1.5 cm wide; sepals 5, purple-green, spreading, 2–4 mm long; petals 5, white, yellow at base, 4–9 mm long. Achenes many in a round head; achene body obovate, ridged, the beak thin and straight, 1–1.5 mm long. Flowering May–Aug.

HABITAT ponds, lakes, streams, rivers, ditches.

STATUS MW-OBL | NCNE-OBL | GP-OBL

NOTE Our plants sometimes treated as two species as follows:

1 Styles (at least the longest) and achene beaks 0.6–1.1 mm long, more than 1/3 the length of the achene body . *R. longirostris*

1 Styles and achene beaks very short, less than 0.6 mm long, less than about 1/3 the length of the body . . . . *R. trichophyllus*

Ranunculus flabellaris Raf.
♦ YELLOW WATER-CROWFOOT

DESCRIPTION Native perennial herb; plants smooth or sometimes hairy when growing out-of-water. Stems floating, or upright from a sprawling base when exposed, branched, rooting at lower nodes, 3–7 dm long. underwater Leaves 3-parted into linear segments 1–2 mm wide, exposed leaves (when present) round to kidney-shaped in outline, 2–10 cm long and 2–12 cm wide, divided into 3 segments, the segments again 3-divided. Flowers 1 to several at ends of stems; sepals 5, green-yellow, 4–8 mm long; petals 5–8, bright yellow, 6–15 mm long. Achenes 50–75 in a round to ovate head; achene body obovate, to 2 mm long, the margin thickened and corky below middle, beak broad, flat, 1–1.5 mm long. Flowering May–July.

HABITAT shallow water or muddy shores of ponds, quiet streams, swamps, woodland pools, marshes and ditches.

STATUS MW-OBL | NCNE-OBL | GP-OBL

Ranunculus flammula L.
CREEPING SPEARWORT

DESCRIPTION Native perennial herb, spreading by stolons; plants often covered with appressed hairs. Stems sprawling, rooting at nodes, unbranched or few-branched, with upright shoots 4–15 cm long. Leaves in small clusters at nodes, simple, linear or threadlike, 1–5 cm long and 1.5 mm wide, margins ± entire; upper leaves smaller and with shorter petioles than lower.

Ranunculus acris

Ranunculus aquatilis

Flowers single at ends of stems; sepals 5, yellow-green, 2–4 mm long, with stiff hairs; petals 5, yellow, obovate, 3–5 mm long. Achenes 10–25 in a round head; achene body swollen, obovate, 1–1.5 mm long, smooth, the beak short, to 0.5 mm long. Flowering June–Aug.

HABITAT sandy, gravelly, or muddy shores; shallow to deep water, water usually acidic.

STATUS MW-FACW | NCNE-FACW | GP-FACW

NOTE sterile underwater plants may be recognized by the arching green stolons and threadlike, blunt-tipped leaves.

Ranunculus gmelinii DC.
SMALL YELLOW WATER-CROWFOOT

DESCRIPTION Native perennial herb, similar to **yellow water-crowfoot** (*R. flabellaris*) but plants aquatic or at least partly under-water; smooth or sometimes with coarse hairs. Stems usually sprawling and rooting at nodes, 1–5 dm long, sparsely branched. Leaves all on stem or with a few basal leaves on long petioles, deeply 3-lobed or dissected, the segments again forked 2–3 times; underwater leaf segments 2–4 mm wide; exposed leaves to 2 cm long and 1.5–2.5 cm wide. Flowers usually 1 to several at ends of stems; sepals 5, green-yellow, 3–6 mm long; petals 5–8, yellow, 4–8 mm long. Achenes 50–70 in a round to ovate head; achene body obovate, 1–1.5 mm long, the margin rounded, not corky-thickened, the beak broad and thin, 0.4–0.7 mm long, somewhat curved. Flowering May–Aug.

HABITAT streambanks and lakeshores, springs, pools in swamps and bogs.

STATUS MW-FACW | NCNE-FACW | GP-FACW; Wisc (END).

Ranunculus hispidus Michx.
♦ NORTHERN SWAMP BUTTERCUP

DESCRIPTION Native perennial herb; stems and leaves variable. Stems upright, 2–9 dm long, smooth or strongly coarse-hairy. Leaves from base of plant and on stems, the basal leaves larger and with longer petioles than stem leaves; 3-lobed, heart-shaped in outline, 3–14 cm long and 4–20 cm wide, with appressed hairs on veins, upper leaves usually strongly toothed. Flowers 1 to several; sepals 5, yellow-green, 5–11 mm long, hairy; petals 5–8, yellow, fading to white, 7–15 mm long and 3–10 mm wide. Achenes 15–30 or more in a round head; achene body obovate, 2–4 mm long, smooth, winged on margin, the beak straight, 2–3 mm long. Flowering May–July.

SYNONYM *Ranunculus septentrionalis.*

HABITAT wet woods, floodplains and swamps, thickets, lakeshores, wet meadows and fens.

STATUS MW-FAC| NCNE-FAC | GP-FACW

Ranunculus macounii Britton
MARSH CROWFOOT

DESCRIPTION Native annual or short-lived perennial herb, similar to **bristly crowfoot** (*R. pensylvanicus*) but uncommon; plants smooth to densely hairy. Stems erect or reclining, hollow, 2–7 dm long, branched, the branches again branched. Leaves from base of plant and on stems, the basal leaves larger and with longer petioles than stem leaves; triangular in outline, 4–14 cm long and 6–16 cm wide, 3-lobed or divided into 3 segments, the segments themselves 3-lobed and coarsely toothed. Flowers several at ends of branches; sepals 5, yellow, 3–5 mm

Ranunculus flabellaris

Ranunculus hispidus

long; petals 5, yellow, 3–5 mm long, equal or longer than sepals; stamens 15–35. Achenes 30–50 in an ovate to round head; achene body flat, 3 mm long, smooth or shallowly pitted, with a stout, slightly curved or straight beak 1–2 mm long. Flowering June–Aug.
HABITAT wet meadows, marshes, shores, streambanks and ditches.
STATUS MW-OBL | NCNE-OBL | GP-OBL; Mich (THR; Isle Royale).

Ranunculus pensylvanicus L. f.
BRISTLY CROWFOOT
DESCRIPTION Native annual or short-lived perennial herb. Stems erect, hollow, 3–8 dm long, branched or unbranched. Leaves at base of plant withering early, larger and with longer petioles than the few stem leaves; 4–12 cm long and 4–15 cm wide, with appressed hairs, 3-lobed and coarsely toothed, the terminal leaflet stalked. Flowers few, on short stalks; sepals 5, yellow, 4–5 mm long; petals 5, pale yellow, fading to white, shorter than the sepals, 2–4 mm long; stamens 15–20. Achenes many, in a rounded cylindric head 10–15 mm long; achene body flattened, 2–3 mm long, smooth, the beak stout, 0.5–1.5 mm long. Flowering July–Aug.
HABITAT marshes, wet meadows, ditches and streambanks, often in muck.
STATUS MW-OBL | NCNE-OBL | GP-FACW
range Minn, Wisc, Mich.

Ranunculus recurvatus Poiret
◆HOOKED CROWFOOT
DESCRIPTION Native perennial herb. Stems 2–7 dm long, usually hairy, branches few.

Leaves broadly kidney-shaped or round in outline, 3-parted to below middle, covered with long, soft hairs; petioles present on all but uppermost leaves. Flowers on stalks at ends of stems; sepals curved downward, to 6 mm long; petals pale yellow, 4–6 mm long; styles strongly hooked. Achenes in a short-cylindric head; achene body flat, round, sharp-margined, to 2 mm long; beak 1 mm long, hooked or coiled. Flowering May–June.
HABITAT moist deciduous forests (especially in openings), swamps; also in drier woods; southward in our region also in partial shade in calcareous fens.
STATUS MW-FACW | NCNE-FACW | GP-FACW

Ranunculus sceleratus L.
◆CURSED CROWFOOT
DESCRIPTION Weedy native annual herb; plants smooth, sometimes partly submersed in shallow water. Stems upright, hollow, 1–6 dm long, branched above and with many flowers. Leaves from base of plant less deeply parted and with longer petioles than stem leaves; upper stem leaves small; leaves deeply 3-parted, the main lobes again lobed, heart-shaped at base, rounded at tip, 1–6 cm long and 3–8 cm wide. Flowers numerous at ends of stalks from upper leaf axils and branches; sepals 5, 2–3 mm long, yellow-green, tips curved downward; petals 5, light yellow, fading to white, 3–5 mm long. Achenes numerous in a short-cylindric head 4–11 mm long; achene body obovate, 1 mm long, slightly corky-thickened on margins; beak tiny, blunt. Flowering May–Sept.
HABITAT muddy shores, streambanks, wet meadows, marshes and other wet places.
STATUS MW-OBL | NCNE-OBL | GP-OBL

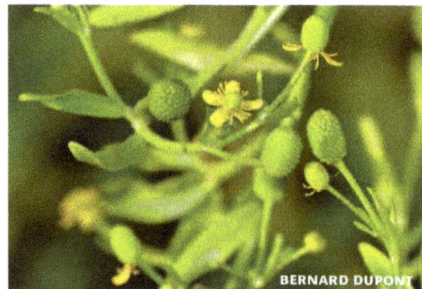

Ranunculus recurvatus

Ranunculus sceleratus

Thalictrum MEADOW-RUE

Perennial herbs. Leaves alternate, compound. Staminate and pistillate flowers separate, in panicles on separate plants; sepals 4-5, green or petal-like but soon deciduous; petals absent; stamens numerous, the stalks (filaments) long and slender; pistils several to many. Fruit a ribbed or nerved achene.

Thalictrum MEADOW-RUE

1   Leaflets mostly with 4 or more teeth or lobes, often appearing 3-lobed, each lobe tipped with 1-3 teeth . . . .
. . . . *Thalictrum venulosum*

1   Leaflets with 2 or 3 lobes, the lobes usually not toothed. **2**

2   Underside of leaflets with very short hairs (rarely smooth); not glandular; leaves odorless; common . . . .
. . . . *Thalictrum dasycarpum*

2   Underside of leaflets with small beads and hairs tipped with gray or amber exudate; leaves with strong odor when crushed; uncommon (more frequent south of Upper Midwest region) . *Thalictrum revolutum*

Thalictrum dasycarpum Fischer & Ave-Lall. ⬥ PURPLE MEADOW-RUE

DESCRIPTION Native perennial herb, from a short rootstock. Stems purple-tinged, 1-2 m long, branched above.

Leaves divided into 3-4 groups of leaflets; leaflets 15 mm or more long, mostly tipped with 3 pointed lobes, dark green above, underside sparsely short-hairy, not waxy and without gland-tipped hairs; margins usually slightly turned under; stem leaves mostly without petioles. Flowers in panicles at ends of stems; staminate and pistillate flowers separate and on different plants (sometimes with some perfect flowers); sepals 3-5 mm long, lance-shaped; anthers linear and sharp-tipped, 2-3 mm long, filaments white; stigmas straight, 2-4 mm long. Achenes 4-6 mm long, ribbed, in a round cluster. Flowering June-July.

HABITAT wet to moist meadows, low prairie, swamps, thickets, streambanks.

STATUS MW-FACW | NCNE-FACW | GP-FAC

Thalictrum revolutum DC.
⬥ SKUNK MEADOW-RUE

DESCRIPTION Native perennial herb, from short rootstocks, with strong odor when crushed. Stems ± smooth, often purple-tinged, 0.5-1.5 m long. Lowest leaves on petioles, middle and upper leaves stalkless; leaves divided into 3-4 groups of leaflets; leaflets variable in shape and size, usually 3-lobed, some 1-2 lobed, upper surface smooth, underside leathery and conspicuously net-veined, finely hairy with gland-tipped hairs, margins turned under. Flowers in panicles at ends of stems; staminate and pistillate flowers separate and on different plants (sometimes with some perfect flowers); anthers linear, 2-3 mm long, filaments threadlike, 2-5 mm long; pistils 6-12, stigmas 2-3 mm long. Achenes oval or lance-shaped, 4-5 mm long, ridged,

Thalictrum dasycarpum

Thalictrum revolutum

with tiny gland-tipped hairs. Flowering June–July.

SYNONYM *Thalictrum amphibolum.*

HABITAT streambanks, thickets, moist meadows and prairies.

STATUS MW-FAC | NCNE-FAC | GP-FACW

Thalictrum venulosum Trelease
NORTHERN MEADOW-RUE

DESCRIPTION Native perennial herb, spreading by rhizomes; plants pale green, waxy. Stems erect, 3–10 dm long. Leaves divided into 3–4 groups of leaflets; leaflets firm, nearly circular or obovate in outline, tipped by 3–5 lobes, underside veiny, appearing wrinkled, usually sparsely covered with gland-tipped hairs; lower leaves on petioles, upper leaves stalkless. Flowers in narrow panicles at ends of stems, the panicle branches nearly erect; staminate and pistillate flowers separate and on different plants; stamens 8–20, anthers linear and pointed at tip, filaments slender. Achenes ovate, 4–6 mm long, tapered to a short beak. Flowering June–July.

SYNONYM *Thalictrum confine.*

HABITAT streambanks, thickets and wet, calcium-rich shores of Lakes Huron and Michigan.

STATUS MW-FAC | NCNE-FACW | GP-FAC

NOTE distinguished from **skunk meadow-rue** (*Thalictrum revolutum*) by its less glandular leaflets and its elongate horizontal rhizomes (*Thalictrum revolutum* has an erect rootstock). Hybrids with *Thalictrum dasycarpum* have been reported.

Rhamnaceae
Buckthorn Family

SHRUBS OR SMALL TREES, with simple, opposite or alternate leaves and small flowers. Flowers perfect or unisexual, regular, 4–5-merous. Petals present or lacking, small, separate. Stamens as many as and alternate with the sepals, opposite and often enfolded by the petals. Ovary 1, sessile on the disk or immersed in it; styles 2–5, united for all or part of their length. Fruit a capsule or drupe.

Rhamnaceae BUCKTHORN FAMILY

1 Leaf margins entire or nearly so . . . .
. . . .*Frangula alnus*
1 Leaf margins toothed . . . .*Rhamnus*

Frangula FALSE BUCKTHORN

Frangula alnus P. Mill.
◆GLOSSY FALSE BUCKTHORN

DESCRIPTION Introduced shrub or tree to 5 m tall. Stems with pale lenticels. Leaves mostly alternate but some leaves often nearly opposite, oval or obovate, 5–8 cm long and 3–5 cm wide, tapered to a blunt or sharp tip; margins entire or slightly wavy; petioles stout, 1–2 cm long. Flowers appearing after leaves in spring, single or in clusters of 2–8 in leaf axils, perfect, green-yellow, 5-parted, to 5 mm wide; petals 1–2 mm long. Fruit a purple-black, berrylike drupe, 7 mm wide, with 2–3 nutlike stones. Flowering May–Aug.

SYNONYM *Rhamnus frangula.*

HABITAT conifer swamps, thickets, calcareous fens, lakeshores, especially where disturbed or cleared; also invading drier woods.

STATUS MW-FACW | NCNE-FAC | GP-FAC

NOTE introduced from Eurasia; escaping and often invasive in disturbed wetlands and upland woods in ne and c USA.

Rhamnus BUCKTHORN

Shrubs or small trees. Leaves simple, alternate (sometimes opposite in *Rhamnus cathartica*), pinnately veined, usually with stipules. Flowers perfect, or staminate or pistillate, regular, single or few from leaf axils; sepals joined, 4- or 5-parted; petals

OLD MUZZLE

Frangula alnus

4 or 5. Fruit a purple-black, berrylike drupe with 2–4, 1-seeded stones.

Rhamnus BUCKTHORN

1 Leaves with 2–4 obvious pairs of lateral veins . . . . *R. cathartica*
1 Leaves with mostly 5 or more pairs of lateral veins **2**
2 Leaves less than 8 cm long; petals 4 . . . . *R. lanceolata*
2 Leaves more than 8 cm long; petals absent . . . . *R. alnifolia*

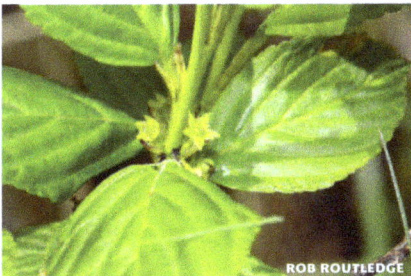

Rhamnus alnifolia L'Her.
◆ALDER-LEAF BUCKTHORN

DESCRIPTION Native shrub to 1 m tall, forming low thickets. Leaves alternate, oval to ovate, 6–10 cm long and 3–5 cm wide, green above, paler green below; margins with low, rounded teeth; petioles grooved, 5–12 mm long; stipules linear, to 1 cm long, deciduous before fruits mature. Flowers appearing with leaves in spring, in clusters of 1–3 flowers from leaf axils; yellow-green, usually 5-parted, 3 mm wide, on short stalks, with both stamens and pistils but one or other is nonfunctional, sepals 1–2 mm long, petals absent. Fruit a purple-black, berrylike drupe, 6–8 mm wide, with 1–3 nutletlike stones. Flowering May–June.

SYNONYM *Endotropis alnifolia*.

HABITAT conifer swamps, thickets, sedge meadows, wet depressions in deciduous forests; usually where calcium-rich.

STATUS MW-OBL | NCNE-OBL | GP-FACW

Rhamnus cathartica L.
◆EUROPEAN BUCKTHORN

DESCRIPTION Introduced, invasive shrub or small tree to 5 m tall. Stems with pale lenticels, often tipped with thorns. Leaves mostly alternate but some leaves often nearly opposite, oval or obovate, 5–8 cm long and 3–5 cm wide; margins entire or slightly wavy; petioles stout, 1–2 cm long. Flowers appearing after leaves in spring, perfect, single or in clusters of 2–8 in leaf axils, green-yellow, 5-parted, to 5 mm wide; petals 1–2 mm long. Fruit a purple-black, berrylike drupe, 7 mm wide, with 2–3 nutlike stones. Flowering May–Aug.

HABITAT conifer swamps, thickets, calcareous fens, lakeshores, moist to dry woods, especially where disturbed, heavily grazed, or cleared. Introduced from Eurasia; escaping from cultivation in ne and c North America.

STATUS MW-FAC | NCNE-FAC | GP-FACU

Rhamnus lanceolata Pursh
LANCE-LEAVED BUCKTHORN

DESCRIPTION Native shrub to 1–2 m tall. Leaves alternate, lance-shaped to oval, 3–7 cm long and 1–3 cm wide, tapered to a short tip; margins with small, incurved, forward-pointing teeth; petioles to 1 cm long. Flowers appearing with leaves in spring, staminate and pistillate flowers separate, staminate flowers usually 2–3 in leaf axils, pistillate flowers usually single; sepals green-yellow, 4-lobed, petals to 1 mm long. Fruit a black, berrylike drupe, with 2 nutlike stones.

SYNONYM *Endotropis lanceolata*.

HABITAT calcareous fens.

STATUS MW-FACW | NCNE-FACW | GP-FAC

Rhamnus alnifolia

Rhamnus cathartica

Rosaceae
Rose Family

SHRUBS; or perennial, biennial, or annual herbs. Leaves evergreen or deciduous, mostly alternate and simple or compound. Flowers perfect (with both staminate and pistillate parts), regular, with 5 sepals and petals; stamens numerous. Fruit an achene, capsule, or fleshy fruit with numerous embedded seeds (a *drupe,* as in a strawberry), or a fleshy fruit with seeds within (a *pome,* as in apples and pears).

Rosaceae **ROSE FAMILY**

1 Leaves simple **2**
1 Leaves divided into leaflets. **5**
2 Fruit dry; ovary superior **3**
2 Fruit fleshy; ovary inferior. **4**
3 Leaves 3-5 lobed *Physocarpus opulifolius*
3 Leaves simple. . . . . *Spiraea*
4 Leaf margins with gland-tipped teeth; flowers less than 1 cm wide . . . .
. . . . *Aronia prunifolia*
4 Leaves without gland-tipped teeth; flowers 1 cm or more wide . . . .
. . . . *Amelanchier bartramiana*
5 Plants shrubs or brambles **6**
5 Plants herbs **8**
6 Plants without thorns or bristles; flowers yellow . . . . *Dasiphora fruticosa*
6 Plants with thorns or bristles; flowers white or pink **7**
7 Leaves pinnately divided into 7 leaflets; petals pink . . . . *Rosa palustris*
7 Leaves palmately divided into 3-5 leaflets; petals white . . . . *Rubus*
8 Leaves divided into 3 equal leaflets. . . . . . .
. . . . *Rubus pubescens*
8 Leaves not divided into 3 equal leaflets. . **9**
9 Flowers white or pink, in large panicles . . .
. . . . *Filipendula rubra*
9 Flowers yellow, white, or red-purple, single or in few-flowered clusters, or in narrow, spikelike racemes **10**
10 Flowers in long, narrow, spikelike racemes . . . . *Agrimonia parviflora*
10 Flowers not in long, narrow, spikelike racemes **11**
11 Flowers white, in many-flowered cylindric spikes; sepals 4, petals absent. . . . .
. . . . *Sanguisorba canadensis*

11 Flowers yellow or red-purple, single or in few-flowered clusters; sepals 5, petals present **12**
12 Leaves deeply parted or divided; styles elongating and becoming longer than achene, persistent as a beak atop the achene . . . . *Geum*
12 Leaves various; styles short, deciduous **13**
13 Petals deep maroon to purple; sepals red tinged; stem usually decumbent, the lower portion in water or wet ground, rooting at nodes. . . . . *Comarum palustre*
13 Petals yellow; sepals green; stem usually erect, or with slender stolons; drier places . . . . *Potentilla*

Agrimonia AGRIMONY

Agrimonia parviflora Aiton
SWAMP AGRIMONY

DESCRIPTION Native perennial herb. **Stems** stout, 1-1.5 m long, densely covered with coarse, brown hairs; the stem in the inflorescence is also glandular. **Leaves** pinnately divided into mostly 11-23 leaflets, the leaflets lance-shaped, with much smaller leaflets intermixed with larger leaflets, smooth above, hairy and glandular below; margins with sharp, forward-pointing teeth; stipules present and deeply toothed. **Flowers** on short, erect stalks, in a long, narrow, spikelike raceme at end of stem; petals 5, small, yellow. **Fruit** a dry achene, at ends of spreading stalks. **Flowering** July-Aug.

HABITAT streambanks, wet meadows, wet woods, ditches.

STATUS MW-FACW | NCNE-FAC | GP-FACW

Amelanchier
SERVICEBERRY, SHADBUSH

Amelanchier bartramiana (Tausch) Roemer MOUNTAIN JUNEBERRY

DESCRIPTION Native shrub to 2 m tall, often forming clumps. **Twigs** purplish, ± smooth. **Leaves** alternate, ovate

to oval, 2–5 cm long and 1–2.5 cm wide, tapered to a blunt or sharp tip and often tipped with a small spine, green above, paler below, often purple-tinged when unfolding; margins with small, sharp, forward-pointing teeth; petioles to 1 cm long. Flowers 1 cm or more wide, single or in groups of 2–4 at ends of branches or on stalks 1–2 cm long from leaf axils; sepals lance-shaped; petals white, oval to oblong, 6–10 mm long. Fruit a dark purple, edible, berrylike pome, 1 cm long. Flowering May–Aug.

HABITAT conifer swamps, bogs, thickets.

STATUS NCNE-FAC

NOTE hybrids are frequent between various species of *Amelanchier,* making positive identification difficult. Most other species of serviceberry occur in drier habitats, especially where rocky or sandy.

Aronia CHOKEBERRY

Aronia prunifolia (Marsh.) Rehd.
♦ PURPLE CHOKEBERRY

DESCRIPTION Native shrub, 1–2.5 m tall. Twigs gray to purple, smooth or hairy. Leaves alternate, oval or obovate, 3–8 cm long and 1–4 cm wide, upper surface dark green and smooth (except for dark, hairlike glands along midveins), underside paler, smooth or hairy; margins with small, rounded, forward-pointing teeth, the teeth gland-tipped; petioles to 1 cm long. Flowers 5–10 mm wide, in clusters of 5–15 at ends of stems and short, leafy branches; sepals usually glandular; petals white, 4–6 mm long. Fruit a dark purple, berrylike pome, 5–10 mm wide. Flowering May–June.

SYNONYMS *Aronia arbutifolia, Aronia prunifolia, Photinia melanocarpa.*

Aronia prunifolia

HABITAT tamarack swamps, open bogs, thickets, marshes and shores.

STATUS MW-FACW | NCNE-FACW | GP-OBL

NOTE Completely glabrous plants are sometimes recognized as *Aronia melanocarpa* (Michx.) Elliott.

ADDITIONAL SPECIES *Aronia arbutifolia* (L.) Pers., more common eastward, is reported from several locations in the Mich LP. It differs from *A. prunifolia* in the fruit turning bright red when fully ripe.

Comarum MARSHLOCKS

Comarum palustre L.
♦ MARSH-CINQUEFOIL

DESCRIPTION Native perennial herb, from long, stout rhizomes. Stems 3–8 dm long, ascending to sprawling or floating in shallow water, often rooting at nodes, ± woody at base; lower stems smooth, upper stems sparsely hairy. Leaves all from stem, pinnately divided or nearly palmate, with 3–7 leaflets; leaflets oblong to oval, 3–10 cm long and 1–3 cm wide, mostly rounded at tip, underside waxy; margins with sharp, forward-pointing teeth; lower leaves long-petioled, upper leaves nearly stalkless; stipules forming wings around petioles of lower leaves, becoming shorter upward. Flowers single or paired from leaf axils, or in open clusters; sepals dark red or purple (at least on inner surface), ovate to lance-shaped, 6–

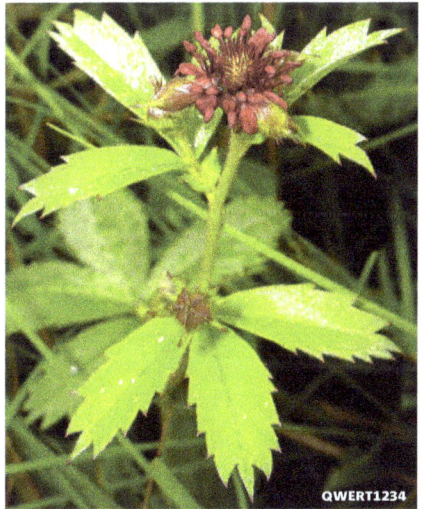

Comarum palustre

20 mm long; petals 5 (sometimes 10), very dark red, 3–5 mm long, with a short slender tip; stamens about 25, dark red. Achenes red to brown, smooth, 1 mm long. Flowering June–Aug.

SYNONYM *Potentilla palustris.*

HABITAT open bogs (especially in pools and wet margins), conifer swamps, wet shores.

STATUS MW-OBL | NCNE-OBL | GP-OBL

Dasiphora
SHRUBBY CINQUEFOIL

Dasiphora fruticosa (L.) Rydb.
♦ SHRUBBY CINQUEFOIL

DESCRIPTION Much-branched native shrub, 0.5–1 m tall. Twigs brown to red, covered with long, silky- white hairs; bark of older branches shredding. Leaves alternate, pinnately divided; leaflets 3–7 (mostly 5), the terminal 3 leaflets often joined at base, oval to oblong, 1–2 cm long and 3–7 mm wide, tapered at each end, upper surface dark green, underside paler, with silky hairs on both sides or at least on underside; margins entire, often rolled under; short-stalked. Flowers 5-parted, bright yellow, 1–2.5 cm wide, 1 to few in clusters at ends of branches; bracts lance-shaped, much narrower than the ovate sepals; stamens 15–20. Fruit a small head of hairy achenes surrounded by the 10-parted calyx. Flowering

Dasiphora fruticosa

June–Sept.

SYNONYMS *Pentaphylloides floribunda, Potentilla fruticosa.*

HABITAT calcareous fens, lakeshores, open bogs, conifer swamps, wet meadows.

STATUS MW-FACW | NCNE-FACW | GP-FACW

Filipendula
QUEEN-OF-THE-PRAIRIE

Filipendula rubra (Hill) B. L. Robinson
♦ QUEEN-OF-THE-PRAIRIE

DESCRIPTION Large native perennial herb. Stems smooth, 1–2 m long. Leaves large, lower leaves to 8 dm long and to 2 dm wide, pinnately parted or divided into 5–9 segments, the segments opposite, stalkless, with 3–5 deep or shallow lobes; margins sharply toothed. Flowers pink-purple, fragrant, 7–10 mm wide, in a panicle 1–2 dm wide at ends of stems; petals 5, 2–4 mm long; stamens many. Fruit an erect, smooth capsule, 6–8 mm long. Flowering June–July.

HABITAT wet meadows and shores, calcareous fens; soils usually calcium-rich.

STATUS MW-OBL | NCNE-FACW; ne Minn, Wisc; uncommon in Mich (THR); some populations may represent escapes from gardens.

Geum AVENS
Perennial herbs. Lower leaves pinnately lobed or divided, upper leaves smaller, less divided or entire. Flowers yellow, white or purple; 1 to many in clusters at ends of stems; petals 5; stamens 10 to many. Fruit an achene.

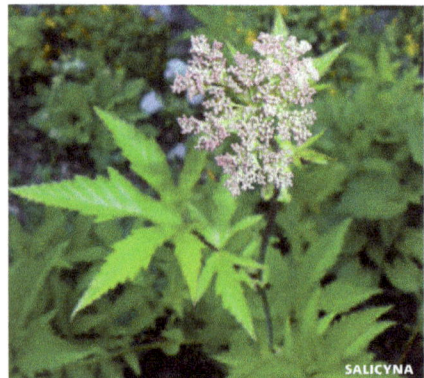

Filipendula rubra

Geum AVENS

1 Flowers nodding; sepals red-purple, upright or ascending at flowering time ... .*G. rivale*
1 Flowers erect; sepals green, curved downward at tip **2**
2 Petals white ....*G. laciniatum*
2 Petals yellow **3**
3 Terminal and lateral segment of basal leaves similar in size and shape ....*G. allepicum*
3 Terminal segment much larger than lateral segments............*G. macrophyllum*

Geum allepicum Jacq.
♦YELLOW AVENS

DESCRIPTION Native perennial herb. Stems erect or ascending, to 1 m long, branched above, covered with coarse hairs. Leaves variable, basal leaves pinnately divided into 5–7 oblong leaflets, wedge-shaped at base, petioles long-hairy; stem leaves divided into 3–5 segments, stalkless or short-petioled; margins coarsely toothed. Flowers 1 to several, short-stalked, on branches at ends of stems; sepals lance-shaped; petals 5, yellow; style jointed. Fruit an achene, usually with long hairs. Flowering June–July.

HABITAT swamps, wet forests, wet meadows, marshes, calcareous fens, ditches, roadsides.

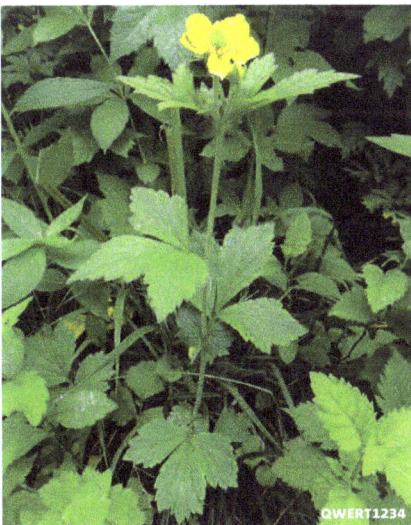

Geum allepicum

STATUS MW-FACW | NCNE-FAC | GP-FACU

Geum laciniatum Murray
♦ROUGH AVENS

Native perennial herb. Stems 4–10 dm long, covered with long, mostly downward-pointing hairs. Leaves lower leaves pinnately divided, the segments pinnately lobed; upper leaves divided into 3 leaflets or lobes; margins coarsely toothed; petioles hairy. Flowers mostly single at ends of densely hairy stalks from ends of stems; sepals triangular, 4–10 mm long; petals 5, white, 3–5 mm long. Fruit an achene, 3–5 mm long (excluding style), grouped into round heads 1–2 cm long. Flowering May–June.

HABITAT wet woods, floodplain forests, ditches.

STATUS MW-FACW | NCNE-FACW | GP-FACW

Geum macrophyllum Willd.
LARGE-LEAF AVENS

DESCRIPTION Native perennial herb. Stems to 1 m long, un-branched, or branched above, bristly-hairy. Leaves pinnately divided, basal leaves stalked, the terminal segment large, 3–7-lobed, with much smaller segments intermixed; stem leaves smaller, deeply 3-lobed or divided into 3 leaflets, short-stalked or stalkless; margins sharply toothed. Flowers 1 to several on branches at ends of stems; sepals triangular, bent backward; petals yellow, obovate, 4–7 mm long; style jointed. Fruit a finely hairy achene. Flowering May–July.

HABITAT moist to wet forest openings, streambanks, wet meadows, ditches.

STATUS MW-FACW | NCNE-FACW | GP-FACW

Geum laciniatum

Geum rivale L. ♦ PURPLE AVENS

DESCRIPTION Native perennial herb. **Stems** erect, 3–8 dm long, mostly unbranched, hairy. **Leaves** basal leaves large, 1–4 dm long, pinnately divided, the terminal 1–3 leaflets much larger than other segments, stalked; stem leaves smaller, 2–5 on stem, pinnately divided or 3-lobed, short-stalked or stalkless; margins shallowly lobed and coarsely toothed. **Flowers** mostly nodding, few on branches at ends of stems, the branches with short, gland-tipped hairs and long, coarse hairs; sepals 5, purple, triangular, 6–10 mm long, ascending; petals 5, yellow to pink with purple veins, tapered to a clawlike base; stamens many; styles jointed above middle, the portion above joint deciduous, lower portion persistent and curved in fruit. **Fruit** a long-beaked, hairy achene, 3–4 mm long, grouped into round heads. **Flowering** May–July.

HABITAT conifer swamps, wet forests, bogs, fens, wet meadows; often where calcium-rich.

STATUS MW-OBL | NCNE-OBL | GP-FACW

Physocarpus NINEBARK

Physocarpus opulifolius (L.) Maxim.
♦ EASTERN NINEBARK

DESCRIPTION Much-branched native shrub, 2–3 m long. **Twigs** greenish, slightly an-

gled or ridged, smooth or finely hairy; bark of older stems shredding in long thin strips. **Leaves** alternate, ovate in outline, mostly 3-lobed, dark green above, paler and often sparsely hairy below; margins irregularly toothed; petioles 1–2 cm long, with a pair of small, deciduous stipules at base. **Flowers** 5-parted, white, 5–10 mm wide, many in stalked, rounded clusters at ends of branches. **Fruit** a red-brown pod, 5–10 mm long, in round clusters; seeds 1–2 mm long, shiny, 3–4 in each pod. **Flowering** June–July.

HABITAT streambanks, lakeshores, swamps, rocky shores of w Lake Superior.

STATUS MW-FACW | NCNE-FACW | GP-FACU

Potentilla CINQUEFOIL

Annual or perennial herbs. Leaves pinnately or palmately divided, alternate or mostly from base of plant. Flowers perfect, regular; sepals 5, alternating with small bracts, the sepals and bractlets joined at base to form a saucer-shaped hypanthium; petals 5, yellow; stamens many; pistils numerous. Fruit a group of many small achenes, surrounded by the persistent hypanthium.

Potentilla CINQUEFOIL

1 Plants spreading by stolons; leaves densely white-hairy on underside .... *P. anserina*

1 Stolons absent; leaves green, smooth to hairy below .... *P. rivalis*

Potentilla anserina L.
♦ SILVER-WEED

DESCRIPTION Native, perennial, clumped herb, with a stout rootstock and spreading by

Geum rivale

Physocarpus opulifolius

stolons to 1 m long. Leaves all at base of plant except for a few clustered leaves on stolons, pinnately divided into 7–25 leaflets; leaflets oblong or obovate, 1.5–5 cm long and 0.5–2 cm wide, lower leaflets much smaller; upper surface green and smooth to gray-green and silky-hairy, underside densely white-hairy; margins with deep, sharp, forward-pointing teeth; stipules brown, membranous, at base of petiole. Flowers single from leafy axils of stolons, on stalks 5–15 cm long; sepals ovate, white silky-hairy; petals yellow, oval to obovate, 5–10 mm long; stamens 20–25. Fruit a light brown achene. Flowering May–Sept.

SYNONYM *Argentina anserina*.

HABITAT wet meadows, shallow marshes, sandy and gravelly shores and streambanks, ditches; soils often calcium-rich; most common along Great Lakes shorelines.

STATUS MW-FACW | NCNE-FACW | GP-FACW

Potentilla rivalis Nutt.
DIFFUSE CINQUEFOIL

DESCRIPTION Native annual or biennial herb, taprooted. Stems upright to spreading, 2–9 dm long, with soft, long hairs, unbranched or branched from base, upper stems branched. Leaves mostly on stems, palmately divided, with 3–7 leaflets, or lower leaves pinnately divided; leaflets obovate to oval, 1.5–5 cm long and 0.5–2.5 cm wide, sparsely hairy; margins with coarse, forward-pointing teeth; stipules ovate, to 1.5 cm long; petioles present, but uppermost leaves nearly stalkless. Flowers in leafy, branched clusters at ends of stems; sepals triangular, 3–6 mm long; petals yellow, obovate, 1–2 mm long, half as long as sepals (or shorter); stamens 10–15. Achenes yellow, less than 1 mm long, smooth. Flowering

June–Aug.

SYNONYMS *Potentilla millegrana, Potentilla pentandra*.

HABITAT wet meadows, streambanks, shores and ditches.

STATUS MW-FACW | NCNE-FACW | GP-FACW

Rosa ROSE

Rosa palustris Marshall ⚬SWAMP ROSE

DESCRIPTION Much-branched, native prickly shrub to 2 m tall. Twigs red-brown, smooth, with a pair of broad-based, downward-curved prickles at nodes; bristles between nodes absent. Leaves alternate, pinnately divided into usually 7 leaflets; leaflets oval or obovate, 2–6 cm long and 1–2 cm wide; underside midrib often soft-hairy; margins finely toothed; stipules narrow; petioles present. Flowers single at ends of leafy branches, or in small clusters of 2–5; petals pink, 2–3 cm long; flower stalks, sepals and hypanthium with stalked glands. Fruit ± round, red-orange, 6–10 mm wide, with gland-tipped hairs. Flowering July–Aug.

HABITAT open bogs, conifer swamps, thickets, shores and streambanks; increasing in disturbed wetlands.

STATUS MW-OBL | NCNE-OBL | GP-OBL

Rubus
RASPBERRY, DEWBERRY, BLACKBERRY

Perennials, woody at least at base, usually with bristly stems. Stems biennial in some species, the first year's canes called primocanes, the second year's growth

Potentilla anserina

Rosa palustris

212 Rosaceae • Rose Family

termed floricanes. Leaves alternate, palmately lobed or divided. Flowers 5-parted, usually perfect, white to pink or rose-purple; stamens many. Fruit a group of small, 1-seeded drupes forming a berry.

Rubus RASPBERRY, DEWBERRY, BLACKBERRY

1 Stems without bristles or prickles 2
1 Stems with bristles or prickles 3
2 Flowering stems 1 or several from a short base; petals light to deep pink, 1–2 cm long; northern portions of region... *R. arcticus*
2 Flowering stems single from a creeping stem; petals green-white, 0.5–1 cm long; widespread species .... *R. pubescens*
3 Leaves gray-hairy on underside; fruit separating from receptacle when ripe (raspberries) .... *R. idaeus*
3 Leaves green on both sides, underside veins hairy; fruit detaching with receptacle when ripe 4
4 Plants tall (to 1.5 m); stems erect or arching; flowers mostly more than 10 in a cluster; fruit black (blackberries) ...... *R. setosus*
4 Plants low and trailing (less than 0.5 m tall); flowers 1 to several in a cluster; fruit red to red-purple (dewberries)...... *R. hispidus*

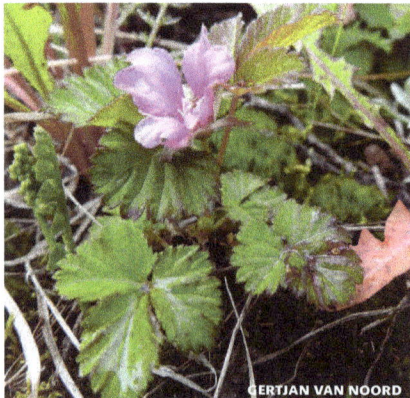

Rubus arcticus L.
◆NORTHERN DWARF RASPBERRY, ARCTIC RASPBERRY

DESCRIPTION Native perennial, woody at base. Stems herbaceous, 5–10 cm long, bristles or prickles absent. Leaves alternate, divided into 3 leaflets, 1–4 cm long and 0.5–3 cm wide, terminal leaflet stalked, lateral pair of leaflets nearly stalkless, lateral leaflets often with a shallow lobe, upper surface smooth, underside finely hairy; margins with blunt, forward-pointing teeth; petioles long, finely hairy; stipules small, ovate. Flowers single at ends of erect stems; sepals lance-shaped, to 1 cm long; petals 5, light to dark pink, 1–2 cm long. Fruit red, nearly round, 1 cm wide, edible. Flowering June–Aug.

SYNONYM *Rubus acaulis*.

HABITAT conifer swamps, open bogs.

STATUS MW-FACW | NCNE-FACW | GP-FACW; Mich UP (END).

Rubus hispidus L.
◆BRISTLY BLACKBERRY, SWAMP DEWBERRY

DESCRIPTION Native shrub. Stems trailing or low-arching, often rooting at tip, with slender bristles or spines 2–5 mm long, these sometimes gland-tipped, not much widened at base. Leaves alternate, divided into 3 leaflets (rarely 5);, the leaflets ovate to obovate, 2–5 cm long and 1–3 cm wide, upper surface dark green and slightly glossy, slightly paler and ± smooth below, some leaves persisting through winter; margins with rounded teeth; petioles finely hairy and bristly; stipules linear, persistent. Flowers single in upper leaf axils or in open clusters of 2–8 at ends of short branches; sepals joined, the lobes ovate, tipped with a small

GERTJAN VAN NOORD
Rubus arcticus

JOSHUA MAYER
Rubus hispidus

dark gland; petals 5, white, 5–10 mm long. Fruit red-purple, less than 1 cm wide, sour, not easily separated from receptacle. Flowering June–Aug.

HABITAT conifer swamps, wet hardwood forests, thickets, wetland margins, sandy interdunal swales.

STATUS MW-FACW | NCNE-FACW

Rubus idaeus L.
WILD RED RASPBERRY

DESCRIPTION Native shrub. Stems erect or spreading, to 1.5 m long, biennial; young stems bristly with slender, often gland-tipped hairs; older stems brown, smooth. Leaves alternate, pinnately divided; primocane leaves divided into 3 or 5 leaflets, floricane leaflets usually 3; leaflets ovate to lance-shaped, upper surface dark green and smooth or sparsely hairy, underside gray-hairy; margins with sharp, forward-pointing teeth; petioles with bristly hairs; stipules slender, soon deciduous. Flowers in clusters of 2–5 at ends of stems and 1–2 from upper leaf axils; sepals with gland-tipped hairs; petals 5, white, shorter than the sepals. Fruit red, about 1 cm wide, edible, separating from receptacle when ripe. Flowering May–Aug.

SYNONYM *Rubus strigosus.*

HABITAT thickets, moist to wet openings, streambanks; often where disturbed.

STATUS MW-FACU | NCNE-FACU | GP-FACU

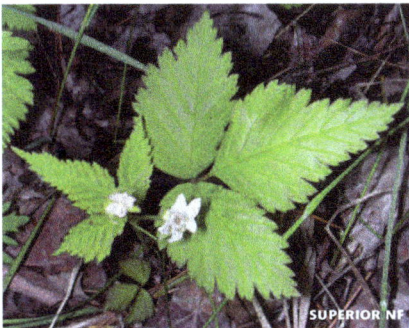

Rubus pubescens Raf.
♦ DWARF RASPBERRY

DESCRIPTION Low native perennial. Stems long-creeping at or near soil surface, with upright, hairy branches 1–3 dm long; the branches herbaceous but woody at base, bristles absent; sterile branches arching to trailing, often rooting at nodes; flowering branches erect, with few leaves. Leaves alternate, divided into 3 leaflets; leaflets oval, 2–6 cm long and 1–4 cm wide, tapered to a sharp point; margins with coarse, forward-pointing teeth, often entire near base; petioles hairy; stipules ovate. Flowers on glandular-hairy stalks, 1–3 in loose clusters at ends of erect branches, sometimes with 1–2 flowers from leaf axils; petals 5, white or pale pink, to 1 cm long. Fruit bright red, round, 5–15 mm wide, the drupelets large, juicy, edible, not separating easily from receptacle. Flowering May–July.

SYNONYM *Rubus triflorus.*

HABITAT conifer swamps, wet deciduous woods, rocky shores.

STATUS MW-FACW | NCNE-FACW | GP-FACW

Rubus setosus Bigel.
BRISTLY BLACKBERRY

DESCRIPTION Native shrub. Stems erect to spreading or arching, to 1.5 m long; branches covered with spreading bristles 1–4 mm long; older canes red-brown, ridged, not rooting at tip. Leaves alternate; primocane leaves divided into 3–5 leaflets; floricane leaves 3-divided; leaflets ovate to obovate, upper and lower surface ± smooth but often hairy on underside veins; margins with sharp, forward-pointing teeth; petioles bristly; stipules linear, 1–2 cm long. Flowers few to many in elongate clusters at ends of stems, with small, leafy bracts throughout the head; petals 5, white, to 1 cm long. Fruit red, ripening to black, round, to 1 cm wide, dry, poor eating quality. Flowering June–Aug.

SYNONYM *Rubus wheeleri.*

HABITAT wetland margins, shores, occasional in open bogs; also in drier sandy prairie.

STATUS MW-FACW | NCNE-FACW | GP-FACW

NOTE *Rubus setosus* sometimes considered a variety of *Rubus hispidus.*

Rubus pubescens

Sanguisorba BURNET

Sanguisorba canadensis L.
AMERICAN BURNET

DESCRIPTION Native perennial herb, from a thick rhizome. **Stems** erect, to 1.5 m long, usually simple below, branched above, ± smooth. **Leaves** alternate, to 3–4 dm long, pinnately divided into 7–15 leaflets; leaflets ovate to oval, 3–7 cm long; margins with sharp, forward-pointing teeth; petioles long on lower leaves, becoming shorter upward; stipules large. **Flowers** perfect, numerous, in 1 to several spikes 4–15 cm long, the spikes erect on long stalks; calyx 4-lobed, white, 2-3 mm long; petals absent; stamens 4, the filaments white, to 1 cm long. **Fruit** an achene, 2 mm long, enclosed by the persistent sepals. **Flowering** Aug–Sept.

HABITAT wet meadows, low prairie; soils often calcium-rich.

STATUS MW-FACW | NCNE-FACW; s LP of Mich (END).

Spiraea SPIRAEA

Shrubs with alternate, undivided leaves. Flowers 5-parted, white to pink, perfect, numerous in clusters at ends of stems. Fruit a cluster of dry, 1-chambered follicles containing small seeds.

Spiraea SPIRAEA

1　Leaves smooth on both sides; flowers white
　................................... *S. alba*

1　Leaf underside densely covered with light brown woolly hairs; flowers rose-pink ....
　........................... *S. tomentosa*

Spiraea alba Duroi ◆ MEADOWSWEET
DESCRIPTION Much-branched native shrub, often forming colonies. **Stems** somewhat angled or ridged, 0.5–1.5 m long, smooth or short-hairy when young, becoming red-brown and smooth. **Leaves** alternate, often crowded on stems, oval to oblong lance-shaped, 3–7 cm long and 1–2 cm wide, smooth on both sides; margins with sharp, forward-pointing teeth; petioles 2–8 mm long; stipules absent. **Flowers** small, 6–8 mm wide, many in a narrow, pyramid-shaped panicle 5–25 cm long at ends of branches; sepals 5, triangular; petals 5, white. **Fruit** a group of 5–8 small follicles, each with several seeds; the fruiting branches often persistent over winter. **Flowering** June–Aug.

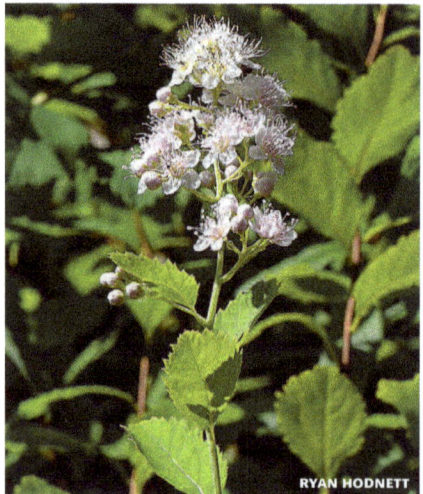

HABITAT wet meadows, streambanks, lakeshores, conifer swamps; soils often sandy.

STATUS MW-FACW | NCNE-FACW | GP-FACW

Spiraea tomentosa L.
◆ HARDHACK, STEEPLE-BUSH

DESCRIPTION Sparsely branched native shrub to 1 m tall. **Stems** young stems covered with brown woolly hairs, becoming smooth and red-brown. **Leaves** alternate, lance-shaped to ovate, 2–5 cm long and 0.5–2 cm wide; ± smooth above, underside gray-green to tan, densely covered with feltlike hairs, the veins prominent; tapered to a pointed or blunt tip; margins with coarse, forward-pointing teeth; petioles 1–4 mm long or absent. **Flowers** small, 3–4 mm wide, in narrow panicles 5–15 cm long at ends of stems, the panicle branches covered with reddish woolly hairs; petals 5, pink or rose (rarely white). **Fruit** a cluster of small, hairy follicles, often persisting over winter. **Flowering** July–Sept.

HABITAT open bogs, conifer swamps, thickets, lakeshores, wet meadows; soils usually sandy.

STATUS MW-FACW | NCNE-FACW | GP-FACW

RYAN HODNETT

Spiraea alba

Rubiaceae
Madder Family

SHRUBS (*Cephalanthus*), or herbs (*Galium*). Leaves simple, opposite or whorled. Flowers small, perfect (with both staminate and pistillate parts), white to green, single or in loose or round clusters; petals joined, 3-4-lobed; stamens 3-4; ovary 2-chambered. Fruit a round head of cone-shaped nutlets (*Cephalanthus*), or a bristly to smooth capsule (*Galium*).

Rubiaceae MADDER FAMILY

1 Shrub .... *Cephalanthus occidentalis*
1 Herbs .... *Galium*

Cephalanthus BUTTONBUSH
Cephalanthus occidentalis L.
⬧BUTTONBUSH

DESCRIPTION Native shrub or small tree, 1-4 m tall. Stems young stems green-brown, with lighter lenticels; older stems gray-brown. Leaves opposite or in whorls of 3, oval to ovate, 8-20 cm long and to 7 cm wide, upper surface bright green and shiny, paler or finely hairy below; margins entire or slightly wavy; petioles grooved, to 2 cm long. Flowers small, perfect, in round, many-flowered heads 2-4 cm wide, on long stalks at ends of stems or from upper leaf axils; petals 4, creamy white, 5-8 mm long; styles longer than petals and swollen at tip. Fruit a round head of brown, cone-shaped nutlets, tipped by 4 teeth of persistent sepals. Flowering June-Aug.

HABITAT hardwood swamps, floodplain forests, thickets, streambanks, marshes, open bogs; often in standing water or muck. **STATUS** MW-OBL | NCNE-OBL | GP-OBL

Galium BEDSTRAW

Perennial herbs (our species), from slender rhizomes. Stems 4-angled, ascending to reclining, smooth or bristly. Leaves entire, in whorls of 4-6. Flowers small, perfect, regular, 1 to several from leaf axils or in clusters at ends of stems; sepals absent; petals joined, 3-4-lobed, white; stamens 3-4; styles 2, ovary 2-chambered and 2-lobed, maturing as 2 dry, round fruit segments which separate when mature.

Galium BEDSTRAW

1 Fruit with bristly hairs 2
1 Fruit smooth or nearly so.............. 4
2 Main leaves in whorls of 5 or more ....
.... *G. triflorum*
2 Leaves in whorls of 4 or less 3
3 Leaves linear to linear lance-shaped, usually less than 5 mm wide; flowers white in a large panicle; common .... *G. boreale*
3 Leaves broader, lance-shaped to ovate, often more than 5 mm wide; flowers

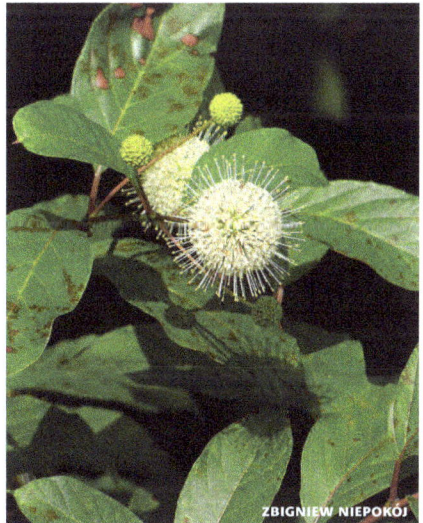

Spiraea tomentosa

Cephalanthus occidentalis

yellowish white; rare in Chippewa County, Mich UP . . . . *G. kamtschaticum*

4 Leaves tipped with a short spine or at least sharp-pointed . . . . *G. asprellum*

4 Leaves rounded or blunt at tip **5**

5 Lobes of corolla 3, mostly wider than long. **6**

5 Lobes of corolla 4, mostly longer than wide **8**

6 Leaves in whorls of 4; flowers and fruit on long, curved, rough-hairy pedicels . . . . *G. trifidum*

6 Leaves usually in whorls of 5 or more; flowers and fruit on straight glabrous pedicels **7**

7 Pedicels 0.5–4 mm long and often curved at maturity, solitary or in pairs in leaf axils or at ends of branches but not on a common peduncle; corolla less than 1 mm wide; mature fruit to 1 mm long; leaves mostly 2.5–7 mm long . . . . *G. brevipes*

7 Pedicels (at least the longest) 3–8 mm long and nearly always straight at maturity, often on a peduncle; corolla 1–1.8 mm wide; mature fruit 1–2 mm long; leaves mostly 5.5–14 (–22) mm long . . . . . . *G. tinctorium*

8 Leaves linear, bent downward, less than 2 mm wide . . . . *G. labradoricum*

8 Leaves linear to oblong, spreading but not angled downward, mostly more than 2 mm wide . . . . *G. obtusum*

Galium asprellum Michx.
ROUGH BEDSTRAW

DESCRIPTION Native perennial herb. **Stems** spreading or reclining on other plants, much-branched, to 2 m long, 4-angled, with rough, downward-pointing hairs on stem angles (which cling tightly to clothing). **Leaves** 6 in a whorl or 5-whorled on branches, narrowly oval, usually widest above middle, 1–2 cm long and 4–6 mm wide, tapered to a sharp tip; underside midvein and margins with rough hairs; petioles absent. **Flowers** in loose, few-flowered clusters at ends of stems and from upper leaf axils; corolla 4-lobed, white, 3 mm wide. **Fruit** smooth. **Flowering** July–Sept.

HABITAT swamps, streambanks, thickets, marshes, wet meadows, calcareous fens.

STATUS MW-OBL | NCNE-OBL | GP-OBL

Galium boreale L.
♦ NORTHERN BEDSTRAW

DESCRIPTION Native perennial herb. **Stems** erect, 2–8 dm long, 4-angled, smooth or with short hairs at leaf nodes, sometimes slightly rough-to-touch. **Leaves** in whorls of 4, linear to lance-shaped, 1.5–4 cm long and 3–8 mm wide, 3-nerved, tapered to a small rounded tip; margins sometimes fringed with hairs; petioles absent. **Flowers** many, 3–6 mm wide, in branched clusters at ends of stems; corolla lobes 4, white. **Fruit** with short, bristly hairs, or smooth when mature. **Flowering** June–Aug.

HABITAT streambanks, shores, thickets, swamps, moist meadows; also in drier woods and fields.

STATUS MW-FAC | NCNE-FAC | GP-FACU

Galium brevipes Fern. & Weig.
LIMESTONE SWAMP BEDSTRAW

DESCRIPTION Native perennial herb. **Stems** scabrous, forming sprawling, tangled mats. **Leaves** whorled, 4 at each node. **Flowers** 1 per peduncle, the peduncles very short, to only 4 mm long. **Fruit** smooth, lacking bristles. **Flowering** July–Aug.

SYNONYM *Galium trifidum* subsp. *brevipes*.

HABITAT marshes, thickets; exposed calcareous shores, interdunal hollows, ditches.

Galium boreale

NOTE The very small pedicels (usually ± re-curved), fruits, corollas, and leaves, if all are present, are distinctive.

Galium kamtschaticum Steller ex J.A. & J.H. Schultes BOREAL BEDSTRAW

DESCRIPTION Low na-tive perennial herb. Flowering stems with only 3-5 whorls of leaves. Leaves in whorls of 4, ovate with 3 prominent veins, the upper leaves larger than the lower leaves. Flowers white, on short pedicels. Fruit with hooked bristles.

HABITAT low places in deciduous forests; a boreal species, rare in Mich (END) where known only from several locations in Chippe-wa County.

NOTE more common north of our region, Boreal bedstraw is distinguished from our other species of *Galium* by its low trailing habit, and the broadly ovate leaves grouped in whorls of 3-5.

Galium labradoricum (Wieg.) Wieg. LABRADOR-BEDSTRAW

DESCRIPTION Native perennial herb. Stems simple or branched, 1-3 dm long, 4-angled, hairy at leaf nodes, smooth on stem angles. Leaves in whorls of 4, soon curveddownward, oblong lance-shaped, 1-1.5 cm long and 1-2 mm wide, blunt-tipped; underside midvein and margins with short, bristly hairs; petioles absent. Flowers single or in small groups on stalks from leaf axils; corolla lobes 4, white. Fruit smooth, dark. Flowering June-July.

HABITAT conifer swamps, sphagnum bogs, fens, sedge meadows.

STATUS MW-OBL | NCNE-OBL | GP-OBL

Galium obtusum Bigel. BLUNTLEAF-BEDSTRAW

DESCRIPTION Native perennial herb. Stems branched, 2-6 dm long, 4-angled, hairy at leaf nodes, otherwise smooth. Leaves mostly in whorls of 4 (some-times 5 or 6), ascending to spreading, linear to lance-shaped or oval, 1-3 cm long and 3-5 mm wide, blunt-tipped; margins with short,

bristly hairs and often somewhat rolled un-der; petioles absent. Flowers in clusters at ends of stems; corolla lobes 4, white. Fruit smooth, dark, often with only 1 segment maturing. Flowering May-July.

HABITAT wet deciduous forests, wet mead-ows, streambanks, floodplains, low prairie.

STATUS MW-FACW | NCNE-FACW | GP-FACW

Galium tinctorium L. SOUTHERN THREE-LOBED BEDSTRAW

DESCRIPTION Native perennial herb. Stems slender, weak, 4-an-gled, with rough hairs on angles. Leaves in whorls of 4 or sometimes 5-6, linear to ob-long lance-shaped, 1-2.5 cm long, tapered to a narrow base, dark green and dull; un-derside midvein and margins with rough hairs; petioles absent. Flowers in clusters of 2-3, on slender, smooth, straight stalks at ends of stems; corolla lobes 3, white. Fruit smooth. Flowering July-Sept.

SYNONYM *Galium claytonii.*

HABITAT conifer swamps, open bogs, fens, thickets, wet shores and marshes.

STATUS MW-OBL | NCNE-OBL | GP-OBL

Galium trifidum L. ♦NORTHERN THREE-LOBED BEDSTRAW

DESCRIPTION Native perennial herb. Stems slender, weak, 2-6 dm long, much-branched, sharply 4-angled, with rough, downward-pointing hairs on stem angles. Leaves in whorls of 4, linear to oblong lance-shaped, 5-20 mm long and 1-3 mm wide, blunt-tipped, dark green and dull on both sides; underside midvein and margins often rough-hairy; petioles absent. Flowers small, on 2-3 slender stalks from leaf axils or at ends of stems, the stalks much longer

Galium trifidum

than the leaves; corolla lobes 3, white. Fruit dark, smooth. Flowering June–Sept.
HABITAT lakeshores, streambanks, swamps, marshes, bogs, springs.
STATUS MW-FACW | NCNE-FACW | GP-OBL
NOTE includes plants sometimes considered a separate species: *Galium brevipes*.

Galium triflorum Michx.
♦ SWEET-SCENTED BEDSTRAW

DESCRIPTION Native perennial herb. Stems prostrate or scrambling, 2–8 dm long, 4-angled, smooth or with rough, downward-pointing hairs on stem angles. Leaves shiny, in whorls of 6 (or 4 on smaller branches), narrowly oval to oblong lance-shaped, 2–5 cm long and to 1 cm wide, l-nerved, tipped with a short, sharp point, slightly vanilla-scented, underside midvein with rough hairs, margins with rough, forward-pointing hairs; petioles absent. Flowers 2–3 mm wide, on slender stalks from leaf axils and at ends of stems, the stalks with 3 flowers or branched into 3 short stalks, each with 1–3 flowers; corolla lobes 4, green-white. Fruit 2-lobed, covered with hooked bristles. Flowering June–Aug.
HABITAT moist to wet woods, hummocks in cedar swamps, wetland margins and shores, clearings.
STATUS MW-FACU | NCNE-FACU| GP-FACU

Salicaceae
Willow Family

DECIDUOUS TREES OR SHRUBS. Leaves alternate, margins entire or toothed; stipules often present at base of leaf petiole, these usually soon falling. Flowers borne in catkins near ends of branches, imper-

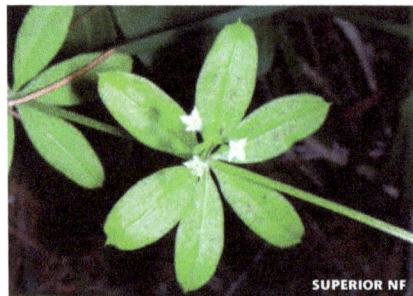

SUPERIOR NF

Galium triflorum

fect (the staminate and pistillate flowers on separate plants), usually appearing before leaves open, or in a few species after leaves open; flowers without petals or sepals, each flower with either 1 or 2 enlarged basal glands (*Salix*) or a cup-shaped disk (*Populus*). Fruit a dry, many-seeded capsule; seeds small, covered with long, silky hairs.

Salicaceae WILLOW FAMILY
1 Large trees; leaves heart-shaped to ovate, mostly less than 2x as long as wide; buds often sticky and covered by 2 or more overlapping scales; catkins drooping, flowers subtended at base by a cup-shaped disk; stamens many, 12–80 ...... *Populus*
1 Shrubs and trees; leaves ovate, lance-shaped or linear, 2x or more longer than wide; buds covered by 1 scale; catkins upright or drooping, flowers subtended by 1 or 2 enlarged glands; stamens 2–8 . *Salix*

Populus
POPLAR, COTTONWOOD

Trees with deciduous, ovate to triangular leaves. Flowers in drooping catkins that develop and mature before and with leaves in spring; staminate and pistillate flowers on separate trees; base of flower with a cup-shaped disk; stamens 10–80. Fruit a 2–4 chambered capsule with many small seeds, these covered with long, white hairs which aid in dispersal by the wind.

NOTE Quaking aspen (*Populus tremuloides*) sometimes occurs as scattered trees in wetlands of the region. It is more common in uplands where it may form large groves.

Populus POPLAR, COTTONWOOD
1 Leaf petioles strongly flattened ....
........................*P. deltoides*
1 Leaf petioles round in section 2
2 Leaves rounded at tip, underside veins hairy, underside not brown-stained ......
........................*P. heterophyll*
2 Leaves tapered to a sharp tip, smooth, underside often stained brown from resin
........................*P. balsamifera*

Populus balsamifera L.
◆ BALSAM-POPLAR

DESCRIPTION Medium to large native tree, to 20 m or more tall, trunk 30–60 cm wide, crown open, somewhat narrow. Bark smooth when young, becoming dark gray and furrowed. Twigs red-brown when young, becoming gray. Leaf buds fragrant, very resinous and sticky. Leaves resinous, ovate to broadly lance-shaped, 8–13 cm long and 4–7 cm wide, tapered to a long tip, rounded or somewhat heart-shaped at base, dark green and somewhat shiny above, white-green or silvery and often stained with rusty brown resin below; margins with small, rounded teeth; petioles round in section, 3–4 cm long. Catkins densely flowered, drooping, appearing before leaves; scales fringed with long hairs, early deciduous; pistillate catkins 10–13 cm long; pistillate flowers with 2 spreading stigmas; stamens 20–30. Capsules ovate, 6–8 mm long, crowded on short stalks. Flowering April–May.

HABITAT swamps, floodplain forests, shores, streambanks, forest depressions, moist dunes.

STATUS MW-FACW | NCNE-FACW | GP-FACW

Populus deltoides Marsh.
◆ EASTERN COTTONWOOD

DESCRIPTION Large native tree to 30 m or more tall, with a large trunk (often 1 m or more wide) and a broad, rounded crown. Bark gray to nearly black, deeply furrowed. Twigs olive-brown to yellow, turning gray with age. Leaf buds very resinous and sticky, shiny, covered by several tan bud scales. Leaves smooth, broadly triangular, 8–14 cm long and 6–12 cm wide, short-tapered to tip, heart-shaped or squared-off at base; margins with forward-pointing, incurved teeth, 2–5 large glands usually present at base of blade near petiole; petioles strongly flattened, 3–10 cm long; stipules tiny, early deciduous. Catkins loosely flowered, drooping, appearing before leaves; scales fringed, soon falling; flowers subtended by a cup-shaped disk 2–4 mm wide; pistillate catkins green, 7–12 cm long in flower, to 20 cm long in fruit; pistillate flowers with 3–4 spreading stigmas; staminate catkins dark red, soon deciduous; stamens 30–80. Capsules ovate, 6–12 mm long, on stalks 3–10 mm long. Flowering April–May.

HABITAT floodplains, streambanks and bars, shores, wet meadows, ditches.

STATUS MW-FAC | NCNE-FAC | GP-FAC

Populus heterophylla L.
SWAMP-COTTONWOOD, DOWNY POPLAR

DESCRIPTION Large native tree, to 20 m or more tall. Bark dull brown and shaggy. Leaf buds hairy, not resinous. Leaves broadly ovate, 12–20 cm long, blunt or rounded at tip, heart-shaped at base; upper surface densely hairy when young, becoming smooth except near base of blade, underside smooth except hairy on

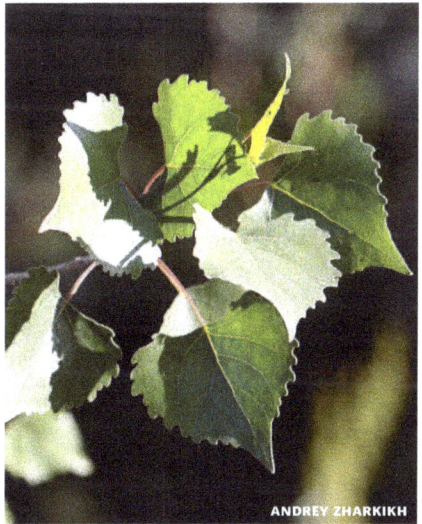

Populus balsamifera

Populus deltoides

main veins; margins with incurved teeth; petioles round in section. Catkins loosely flowered, scales fringed with long hairs; pistillate flowers with 2-3 spreading stigmas; stamens 12-20. Capsules ovate, 7-12 mm long, on stalks 10-15 mm long. May.
HABITAT floodplain forests.
STATUS MW-OBL | NCNE-OBL; LP of Mich (END); main range Atlantic Coast and in the Ohio and southern Miss River valleys.

Salix WILLOW

Shrubs and trees. Leaves variable in shape, petioles glandular in some species; stipules early deciduous or persistent, sometimes absent. Catkins stalkless or on leafy branchlets, usually shed early in season. Staminate and pistillate flowers on separate plants; staminate flowers with mostly 2-3 stamens (to 8 in some species). Fruit a 2-chambered, stalked or stalkless capsule.

Salix **WILLOW**

1 Leaves opposite; young branches often dark purple . . . . *S. purpurea*
1 Leaves alternate; branches various colors. **2**
2 Leaf petioles with glands at or near base of leaf blade **3**
2 Petioles without glands **8**
3 Trees, usually with a single trunk; leaves narrow. **4**
3 Small trees or shrubs, usually with several to many stems; leaves broader **5**
4 Leaves often curved sideways (scythe-shaped), tapered to a long, slender tip; stipules large on vigorous shoots . *S. nigra*
4 Leaves not curved sideways, tapered to a short tip; stipules small and early deciduous . . . . *S. alba*
5 Leaves not waxy on underside . . *S. lucida*
5 Leaves waxy-coated on underside **6**
6 Leaf tips rounded or with a short point; leaf base heart-shaped or rounded; young leaves thin and translucent; buds and leaves with a balsam-like scent . . . . *S. pyrifolia*
6 Leaves tapered to tip; leaf base blunt or rounded; young leaves not translucent; buds and leaves not balsam-scented **7**
7 Young leaves sparsely hairy; margins with small forward-pointing teeth; flowering in early summer . . . . *S. amygdaloides*
7 Young leaves without hairs; margins with small, gland-tipped, forward-pointing teeth; flowering summer or fall . . . . *S. serissima*
8 Mature leaves hairy, at least on underside **9**
8 Mature leaves without hairs (sometimes hairy on petiole and midvein) **16**
9 Leaves linear or narrowly lance-shaped **10**
9 Leaves broadly lance-shaped, oblong or oval **13**
10 Underside of leaves with feltlike covering of white tangled hairs; young twigs white-hairy; plant of peatlands, often where calcium-rich . . . . *S. candida*
10 Leaves not with feltlike hairs; twigs smooth or sparsely hairy **11**
11 Leaf margins smooth and turned under; leaf underside with shiny white hairs; uncommon in northern portions of region . . . . *S. pellita*
11 Leaf margins with gland-tipped teeth; leaf underside with few long hairs **12**
12 Leaf margins with widely spaced sharp teeth; petioles 1-5 mm long; colony-forming shrub of sandy banks *S. interior*
12 Leaf margins with small teeth at least above middle of blade; petioles 3-10 mm long; stems clustered but not forming large colonies. *S. petiolaris*
13 Leaves rounded or heart-shaped at base; margins with gland-tipped, forward-pointing teeth; stipules present and persistent . . . . *S. eriocephala*
13 Leaves tapered to base; margins smooth or toothed; stipules usually deciduous early in season **14**
14 Leaves linear to lance-shaped, more than 5x longer than wide, underside velvety with shiny white hairs; uncommon in n portions of region . . . . *S. pellita*
14 Leaves oval or obovate, less than 5x longer than wide, underside hairs not shiny . . . **15**
15 Small branches widely spreading; young leaves with white hairs; catkins appearing with leaves in spring; catkin bracts yellow or straw-colored . . . . *S. bebbiana*
15 Small branches not widely spreading; young leaves with some red or copper-colored hairs; catkins appearing before leaves in spring; catkin bracts dark brown to black. . . . . *S. discolor*
16 Leaves green on both sides or slightly paler on underside, not waxy or white below. **17**

16 Leaves waxy or white-hairy on underside
.... **19**

17 Single-stemmed tree; stipules large ......
.... *S. nigra*

17 Many-stemmed shrub; stipules small or absent **18**

18 Many-stemmed, colony-forming shrub; upper leaf surface not conspicuously shiny; common .... *S. interior*

18 Clumped shrub, not forming colonies; upper leaf surface very shiny; rare in nw Minnesota .... *S. maccalliana*

19 Leaf margins entire to shallowly lobed or with irregular teeth, sometimes rolled under **20**

19 Leaf margins distinctly and regularly toothed **24**

20 Leaf margins entire and somewhat rolled under **21**

20 Leaf margins irregularly toothed **22**

21 Stems to 1 m tall, or creeping and rooting in moss; upper surface of leaves with raised, netlike veins; catkins appearing with leaves; capsules not hairy . *S. pedicellaris*

21 Stems 1–4 m tall; leaf veins not netlike; catkins appearing before leaves; capsules hairy; northern portions of region........
.... *S. planifolia*

22 Leaves dull green above, wrinkled below; catkins appearing with leaves; bracts of pistillate catkins green-yellow to straw-colored.................... *S. bebbiana*

22 Leaves dark green and shiny above; catkins appearing before leaves; bracts of pistillate catkins dark brown to black **23**

23 Stipules large on vigorous shoots; capsules on distinct stalks 2 mm or more long; widespread species........... *S. discolor*

23 Stipules lance-shaped and soon deciduous, or absent or very small; capsules stalkless or on short stalks less than 2 mm long; n Minn and Wisc............. *S. planifolia*

24 Leaves lance-shaped **25**

24 Leaves broadly oval, ovate, or oblong lance-shaped........................ **28**

25 Leaves ± equally tapered at tip and base **26**

25 Leaves unequally tapered, the tip tapered to a point, the base usually rounded or heart-shaped........................ **27**

26 Young leaves with few to many coppery hairs mixed with white hairs; mature leaves smooth on both sides or hairy above; lateral veins not prominent; capsules lance-shaped .... *S. petiolaris*

26 Young leaves densely silky hairy, without copper-colored hairs; underside of mature leaves silky-hairy with conspicuous, riblike, lateral veins; capsules ovate and rounded at tip; s portions of region .... *S. sericea*

27 Young twigs hairless; stipules small or absent; bracts of pistillate catkins pale yellow and soon deciduous ....
.... *S. amygdaloides*

27 Young twigs gray-hairy; stipules large; bracts of pistillate catkins dark brown to black, persistent.......... *S. eriocephala*

28 Leaves balsam-scented (especially when dried), underside net-veined; stipules tiny or absent; catkins appearing with leaves, on leafy or leafless branches .... *S. pyrifolia*

28 Leaves not balsam-scented, not net-veined; stipules large on vigorous shoots; catkins appearing before or with leaves, stalkless or on short, leafy branches......
.... *S. myricoides*

Salix alba L. WHITE WILLOW

DESCRIPTION Introduced tree, to 20 m tall. **Twigs** golden-yellow, often with long, silky hairs. **Leaves** lance-shaped, 4–10 cm long and 1–2.5 cm wide, dark green and shiny above, waxy white below, smooth to sparsely hairy on both sides, margins with small gland-tipped teeth; petioles 2–8 mm long, with silky hairs; stipules lance-shaped, 2–4 mm long, early deciduous. **Catkins** appearing with leaves in spring; pistillate catkins 3–6 cm long, on leafy branches 1–4 cm long; staminate catkins 3–5 cm long, stamens 2; catkin bracts pale yellow, hairy near base, early deciduous. **Capsules** ovate, 3–5 mm long, without hairs, stalkless or on stalks to 1 mm long. **Flowering** May–June.

HABITAT introduced from Europe, sometimes escaping to streambanks and other wet areas.

STATUS MW-FACW | NCNE-FACW | GP-FACW

Salix amygdaloides Andersson
PEACH-LEAF WILLOW

DESCRIPTION Native shrub or tree, to 15 m tall, often with several trunks. **Twigs** gray-brown to light yellow, shiny and flexible. **Leaves** smooth, lance-

shaped, long-tapered to tip, 5–12 cm long and 1–3 cm wide, yellow-green above, waxy-white below, margins finely toothed; petioles 5–20 mm long and often twisted; stipules small and early deciduous. Catkins appearing with leaves, linear and loosely flowered; pistillate catkins 3–12 cm long, on leafy branches 1–4 cm long; catkin bracts deciduous, pale yellow, long hairy especially on inner surface; stamens 3–7 (usually 5). Capsules smooth, ovate, 3–7 mm long, on stalks 1–3 mm long. Flowering May–June. HABITAT floodplains, streambanks, lake and pond borders. STATUS MW-FACW | NCNE-FACW | GP-FACW

Salix bebbiana Sarg.
◊ BEBB'S *or* BEAKED WILLOW

DESCRIPTION Native shrub or small tree, to 8 m tall; stems 1 to several. Twigs yellow-brown to dark brown, usually with short hairs. Leaves oval to ovate or obovate, tapered to tip, 4–8 cm long and 1–3 cm wide, dull gray-green, hairy or sometimes smooth on upper surface, waxy-gray, hairy and wrinkled below, the veins distinctly raised on lower surface; margins entire to shallowly toothed; petioles 5–15 mm long; stipules deciduous or persistent on vigorous shoots. Catkins appearing before leaves in spring; pistillate catkins loose, 2–6 cm long, on short leafy branches to 2 cm long; catkin bracts persistent, red-tipped when young,

turning brown, long hairy; stamens 2. Capsules ovate, 5–8 mm long, finely hairy, on stalks 2–6 mm long. Flowering May–June. HABITAT swamps, thickets, wet meadows, streambanks, marsh borders. STATUS MW-FACW | NCNE-FACW | GP-FACW

Salix candida Fluegge
◊ HOARY *or* SAGE-LEAVED WILLOW

DESCRIPTION Low native shrub, to 1.5 m tall. Twigs much-branched, covered with dense, matted white hairs. Leaves linear-oblong, tapered at tip, 4–10 cm long and 0.5–2 cm wide, dull, dark green and sparsely hairy above, veins sunken, densely white-hairy below; margins entire and rolled under; petioles 3–10 mm long; stipules persistent, 2–10 mm long, white-hairy. Catkins appearing with leaves in spring; pistillate catkins 1–5 cm long, on leafy branches 0.5–2 cm long; catkin bracts persistent, brown, hairy; stamens 2. Capsules ovate, 4–8 mm long, white-hairy, on stalks to 1 mm long. Flowering May–June. HABITAT fens, bogs, open swamps, streambanks, usually where calcium-rich. STATUS MW-OBL | NCNE-OBL | GP-OBL

Salix discolor Muhl. PUSSY-WILLOW

DESCRIPTION Native shrub or small tree, to 5 m tall. Twigs yellow-brown to red-brown, dull, smooth with age or with patches of fine hairs. Leaves oval and short-tapered to tip, 3–10 cm long and 1–4 cm wide, dark green and smooth above, underside red-hairy when young, becoming white-waxy, smooth and not wrinkled; margins entire or with few rounded teeth; petioles without glands; stipules deciduous, or often

Salix bebbiana

Salix candida

persistent on vigorous shoots. Catkins appearing and maturing before leaves in spring; pistillate catkins 4–8 cm long, stalkless, sometimes with 2 or 3 small, brown, bractlike leaves at the base; stamens 2. Capsules ovate with a long neck, 6–10 mm long, densely gray-hairy, on stalks 2–3 mm long. Flowering April–May.

HABITAT swamps, fens, streambanks, floodplains, marsh borders.

STATUS MW-FACW | NCNE-FACW | GP-FACW

Salix eriocephala Michx.

♦ DIAMOND WILLOW

DESCRIPTION Native shrub or small tree, to 6 m tall. Twigs redbrown to dark brown, hairy when young.

Leaves lance-shaped or oblong lance-shaped, 5–12 cm long and 1–3 cm wide, red-purple and hairy when young, upper surface becoming smooth and dark green, underside becoming pale-waxy; margins finely toothed; petioles without glands, 3–15 mm long; stipules persistent (especially on vigorous shoots), ovate or kidney-shaped, to 12 mm long, hairless, toothed. Catkins appearing with or slightly before leaves in spring; pistillate catkins 2–6 cm long, on short leafy branches to 1 cm long; catkin bracts persistent, brown to black, hairy; stamens 2. Capsules ovate with a long neck, 4–6 mm long, without hairs, on stalks 1–2 mm long. Flowering April–May.

Salix eriocephala

SYNONYM *Salix lutea, Salix rigida*.

HABITAT shores, streambanks, floodplains, ditches and wet meadows, especially along major rivers.

STATUS MW-FACW | NCNE-FACW | GP-FACW

Salix interior Rowlee

♦ SANDBAR-WILLOW

DESCRIPTION Native shrub to 4 m tall, spreading by rhzomes and often forming dense thickets. Twigs yellow-orange to brown, smooth. Leaves linear to lance-shaped, tapered at tip and base, 5–14 cm long and 5–15 mm wide, green on both sides but paler below, at first hairy but soon usually smooth; margins with widely spaced, large teeth; petioles without glands, 1–5 mm long; stipules tiny or absent. Catkins appearing with leaves in spring on short leafy branches (and plants sometimes again flowering in summer); pistillate catkins loosely flowered, 2–8 cm long; catkin bracts deciduous, yellow; stamens 2. Capsules narrowly ovate, 5–8 mm long, hairy when young, smooth when mature, on stalks to 2 mm long. Flowering May–June.

SYNONYM *Salix exigua* ssp. *interior*.

HABITAT shores, streambanks, sand and mud bars, ditches and other wet places; often colonizing exposed banks.

STATUS MW-FACW | NCNE-FACW | GP-FACW

Salix lucida Muhl. ♦ SHINING WILLOW

DESCRIPTION Native shrub or small tree, to 5 m tall. Twigs yellowbrown or dark brown, smooth and shiny.

Leaves lance-shaped to ovate, long-tapered and asymmetric at tip, 4–12 cm long and 1–

Salix interior

4 cm wide, shiny green above, pale below, red-hairy when young, but soon smooth; margins with small, gland-tipped teeth; petioles with glands near base of leaf; stipules often persistent, strongly glandular. Catkins appearing with leaves in spring; pistillate catkins 2–5 cm long, on leafy branches 1–3 cm long; catkin bracts deciduous, yellow, sparsely hairy; stamens 3–6. capsules ovate with a long neck, 4–7 mm long, not hairy, on short stalks to 1 mm long. Flowering May.

HABITAT swamps, shores, wet meadows, moist sandy areas.

STATUS MW-FACW | NCNE-FACW | GP-FACW

Salix maccalliana Rowlee
MCCALLA'S WILLOW

DESCRIPTION Native shrub, 2–4 m tall. Twigs upright, red- to yellow-brown, sparsely hairy or smooth, glossy. Leaves strap-shaped to narrowly oblong, to 8 cm long and 2.5 cm wide, dark green, upper surface glossy; leaf underside not waxy-coated; margins usually finely toothed; young leaves smooth or with white or reddish hairs; petioles sparsely hairy, without glandular dots; stipules small or leaflike. Catkins appearing with leaves in spring; pistillate catkins 2–6 cm long, on leafy branches 1–2.5 cm long. Capsules densely hairy with age.

HABITAT uncommon in shrubby wetlands,

Salix lucida

fens, sedge meadows, soils usually sedge-derived peat and not strongly acid; typically found with other willows, dogwoods (*Cornus* spp.), and bog birch (*Betula pumila*).

STATUS MW-OBL | NCNE-OBL | GP-OBL

Salix myricoides Muhl.
BAYBERRY WILLOW

DESCRIPTION Native shrub to 4 m tall. Twigs yellow to dark brown, hairy when young. Leaves thickened, lance-shaped to ovate or oval, 4–12 cm long and 1.5–5 cm wide, dark green above, strongly waxy-white below; margins with gland-tipped teeth; petioles 5–12 mm long; stipules 5–10 mm long. Catkins appearing shortly before or with leaves; pistillate catkins 2–8 cm long, on leafy branches 5–15 mm long; catkin bracts deciduous, 1–2 mm long, brown-black and long hairy; stamens 2. Capsules lance-shaped, 5–8 mm long, not hairy, on stalks 1–3 mm long. Flowering May.

SYNONYM *Salix glaucophylloides*.

HABITAT dune hollows and sandy shorelines, fens, mostly near Great Lakes; inland on wet, calcium-rich sites.

STATUS MW-FACW | NCNE-FACW

Salix nigra Marshall BLACK WILLOW

DESCRIPTION Medium-sized native tree, to 15 m tall, trunks 1 or several, crown rounded and open. Bark dark brown, furrowed, becoming shaggy. Twigs bright red-brown, often hairy when young. Leaves commonly drooping, linear lance-shaped, 6–15 cm long and 0.5–2 cm wide, long tapered to an often curved tip, green on both sides but satiny above and paler below, lateral veins upturned at tip to form a ± continuous vein near leaf margin; margins finely toothed; petioles 3–8 mm long, hairy, usually glandular near base of blade; stipules to 12 mm long, heart-shaped, usually deciduous. Catkins appearing with leaves in spring; pistillate catkins 3–8 cm long, on leafy branches 1–3 cm long; stamens usually 6 (varying from 3–7); catkin bracts yellow, hairy, deciduous. Capsules ovate, 3–5 mm long, without hairs, on a short stalk to 2 mm long. Flowering May.

HABITAT streambanks, lakeshores and wet

depressions; not tolerant of shade.
STATUS MW-OBL | NCNE-OBL | GP-FACW

Salix pedicellaris Pursh
⧫ BOG-WILLOW

DESCRIPTION Short, sparsely branched native shrub, to 15 dm tall. **Twigs** dark brown and smooth. **Leaves** oblong-lance-shaped to obovate, tapered to tip or blunt and often with a short point, 3–6 cm long and 0.5–2 cm wide, silky hairy when young, becoming hairless and thick and leathery with age, green on upper surface, white-waxy below, veins pale-colored and slightly raised on both sides; margins entire, often slightly rolled under; petioles without glands, 2–8 mm long; stipules absent. **Catkins** appearing with leaves in spring; pistillate catkins 2–4 cm long, on leafy branches 1–3 cm long; catkin bracts persistent, yellow-brown, hairy on inner surface near tip; stamens 2. **Capsules** lance-shaped, 4–7 mm long, without hairs, on stalks 2–3 mm long. **Flowering** May–June.

HABITAT bogs, fens, sedge meadows, interdunal wetlands.

STATUS MW-OBL | NCNE-OBL | GP-OBL

Salix pellita Andersson
SATINY WILLOW

DESCRIPTION Native shrub, 3–5 m tall. **Twigs** easily broken, yellow to olive-brown or red-brown, smooth or sparsely hairy when young, becoming waxy. **Leaves** lance-shaped, 4–12 cm long and 1–2 cm wide, short-tapered to a tip, upper surface without hairs, veins sunken, underside waxy and satiny hairy but becoming smooth with age, with numerous, parallel lateral veins; margins rolled under, entire or with rounded teeth; petioles to 1 cm long; stipules absent. **Catkins** appearing and maturing before leaves in spring; pistillate catkins 2–5 cm long, stalkless or on short branches to 1 cm long; catkin bracts black, long hairy; staminate catkins uncommon. **Capsules** lance-shaped, 4–6 mm long, silky hairy, ± stalkless. **Flowering** May.

HABITAT streambanks, sandy shores and rocky shorelines.

STATUS NCNE-FACW | GP-FACW; uncommon in n Minn, n Wisc (END), and UP and n LP of Mich.

Salix petiolaris J. E. Smith
⧫ MEADOW-WILLOW

DESCRIPTION Native shrub to 5 m tall. **Twigs** red-brown to dark brown, sometimes with short, matted hairs when young, smooth with age. **Leaves** narrowly lance-shaped, 4–10 cm long and 1–2.5 cm wide, hairy when young, becoming smooth, dark green above, white-waxy below,; margins entire or with small, gland-tipped teeth; petioles without glands, 3–10 mm long; stipules absent. **Catkins** appearing with leaves in spring; pistillate catkins 1–4 cm long, stalkless or on short branches to 2 cm long; catkin bracts persistent, brown, with a few long, soft hairs; stamens 2. **Capsules** narrowly lance-shaped, 4–8 mm long, finely hairy, on stalks 2–4 mm long. **Flowering** May.

ROB ROUTLEDGE

Salix pedicellaris

QUINTEN WIEGERSMA

Salix petiolaris

SYNONYM *Salix gracilis*.

HABITAT wet meadows, fens and bogs, floating sedge mats, streambanks, shores, ditches.

STATUS MW-OBL | NCNE-FACW | GP-OBL

Salix planifolia Pursh

◆ TEA-LEAF WILLOW

DESCRIPTION Native shrub, to 3 m tall. Twigs dark red-brown, short hairy when young, soon smooth and shiny. Leaves oval to oblong-lance-shaped, 3–6 cm long and 1–3 cm wide, short-hairy when young, becoming smooth, green above, paler or waxy below; margins entire or with a few small, rounded teeth; petioles 3–6 mm long, without glands; stipules small and deciduous. Catkins appearing before leaves in spring; pistillate catkins 2–5 cm long, stalkless; catkin bracts 2–3 mm long, persistent, black, with long, soft hairs; stamens 2. Capsules lance-shaped, 4–8 mm long, finely hairy, stalkless or on short stalks to 0.5 mm long. Flowering May.

SYNONYM *Salix phylicifolia* subsp. *planifolia*.

HABITAT rocky lakeshores, cedar swamps, black spruce bogs, streambanks, and margins of sedge meadows.

STATUS MW-OBL | NCNE-OBL | GP-OBL; Minn; rare in n Wisc (THR); historically known from UP of Mich (THR).

Salix purpurea L.

◆ BASKET WILLOW, PURPLE-OSIER

DESCRIPTION Introduced shrub, to 2.5 m tall. Twigs smooth, green-yellow to purple. Leaves ± opposite

(unique among our willows), smooth, linear to oblong lance-shaped, 4–9 cm long and 7–16 mm wide, purple-tinged, somewhat waxy below, veins raised and netlike on both sides; margins entire near base, irregularly toothed near tip; petioles short; stipules absent. Catkins appearing with and maturing before leaves in spring; pistillate catkins 2–3.5 cm long, stalkless; catkin bracts black; stamens 2 but often joined. Capsules ovate, 3–4 mm long, short-hairy, stalkless. Flowering May–June.

HABITAT introduced from Europe, occasionally escaping to lakeshores and streambanks.

STATUS MW-FACW | NCNE-FACW | GP-OBL

Salix pyrifolia Andersson

BALSAM-WILLOW

DESCRIPTION Native shrub or small tree, to 5 m tall. Twigs smooth, yellow when young, becoming shiny red. Leaves smooth, ovate to lance-shaped, often rounded at tip, rounded to heart-shaped at base, 4–12 cm long and 2–4 cm wide, red-tinged and translucent when unfolding; green on upper surface, waxy and finely net-veined below; with balsam fragrance (especially when dried); margins with small gland-tipped teeth; petioles 1–2 cm long; stipules absent or small and 1–2 mm long. Catkins appearing with or after leaves in spring; pistillate catkins loosely flowered, 2–6 cm long, on leafy branches 1–3 cm long;

Salix planifolia

Salix purpurea

catkin bracts red-brown, white-hairy, 2 mm long; stamens 2. Capsules lance-shaped, beaked at tip, 6–8 mm long, smooth, on stalks 2–4 mm long. Flowering May–June. SYNONYM *Salix balsamifera*. HABITAT conifer swamps, bogs, rocky shores, wet depressions in boreal forests. STATUS MW-FACW | NCNE-FACW | GP-OBL

Salix sericea Marshall SILKY WILLOW

DESCRIPTION Native shrub, to 4 m tall. Twigs brown, brittle, densely gray or brown hairy when young, becoming smooth except at leaf nodes. Leaves lance-shaped, short tapered to tip, 6–12 cm long and 1–2.5 cm wide, upper surface dark green and smooth or finely hairy, waxy and with short silky hairs below; margins with small, gland-tipped teeth; petioles finely hairy, 5–10 mm long; stipules broadly lance-shaped, to 1 cm long, mostly deciduous. Catkins appearing before leaves in spring; pistillate catkins 1–4 cm long, stalkless or on leafy branches to 1 cm long; catkin bracts black, long-hairy, 1 mm long; stamens 2. Capsules obovate, blunt-tipped, 3–5 mm long, short-hairy, on stalks to 1 mm long. Flowering May. HABITAT streambanks, lakeshores, bogs, ditches; often in water. STATUS MW-OBL | NCNE-OBL; historical records from Wisc.

Salix serissima

Salix serissima (L. H. Bailey) Fernald
♦AUTUMN-WILLOW

DESCRIPTION Native shrub, to 4 m tall. Twigs gray, yellow or dark brown, shiny and smooth. Leaves smooth, oval to lance-shaped, 4–10 cm long and 1–3 cm wide, red and hairless when young; green and shiny above, usually white-waxy below; margins with small gland-tipped teeth; petioles with glands near base of leaf; stipules usually absent. Catkins appearing with or after leaves in spring; pistillate catkins 2–4 cm long, on leafy branches 1–4 cm long; catkin bracts deciduous, light yellow, long hairy; stamens 3–7. Capsules narrowly cone-shaped, 7–10 mm long, smooth, on stalks to 2 mm long. Flowering Late May–July; our latest blooming willow. HABITAT fens, cedar and tamarack swamps, marshes, floating sedge mats, streambanks and shores, often where calcium-rich. STATUS MW-OBL | NCNE-OBL | GP-OBL

Santalaceae
Sandalwood Family

Geocaulon TOADFLAX

Geocaulon lividum (Richardson) Fern.
♦NORTHERN RED-FRUIT TOADFLAX

DESCRIPTION Native perennial herb, from a slender rhzome; at least partially parasitic on other plants. Stems smooth, 1–3 dm long. Leaves alternate, oval or ovate, 1–3 cm long and 1–1.5 cm wide, rounded at tip; margins entire; petioles short.

Geocaulon lividum

Flowers usually 3 on slender stalks from leaf axils, the lateral 2 flowers typically staminate, the middle flower perfect; sepals 4–5, triangular, 1–2 mm long; petals absent; style very short. Fruit a round, orange or red drupe, about 6 mm wide. Flowering June–Aug.

SYNONYM *Comandra livida*.

HABITAT cedar swamps, open bogs; more commonly in sandy conifer woods and forested dune edges.

STATUS NCNE-FAC; Wisc (END; Door County).

Sapindaceae
Soapberry Family

Acer MAPLE

Trees or shrubs. Leaves opposite, simple or compound. Staminate and pistillate flowers borne on same or separate plants; flowers with 5 sepals and 5 petals (sometimes absent), clustered into a raceme or umbel). Fruit a samara with 2 winged achenes joined at base.

ADDITIONAL SPECIES **Mountain maple** (*Acer spicatum* Lam.), a large shrub, sometimes occurs in conifer swamps of the region; it has large 3-lobed leaves.

Acer MAPLE

1 Leaves compound . . . . *A. negundo*
1 Leaves simple 2
2 Leaves shallowly lobed, the terminal lobe broadest at its base; flowers with petals . . . . *A. rubrum*
2 Leaves deeply lobed to middle of blade or below, the terminal lobe narrowed at its base; flowers without petals . . . . *A. saccharinum*

Acer negundo L.
BOXELDER, ASH-LEAVED MAPLE

Native tree, to 20 m tall, the trunk soon dividing into widely spreading branches. Bark brown, ridged when young, becoming deeply furrowed. Twigs smooth, green and often waxy-coated. Leaves opposite, compound, leaflets 3–7, oval to ovate, coarsely toothed or shallowly lobed, upper surface light green and smooth, underside pale green and smooth or hairy. Flowers either staminate or pistillate and on separate trees, appearing with leaves in spring; petals absent; staminate flowers in drooping, umbel-like clusters, pistillate flowers in drooping racemes. Fruit a paired samara 3–4.5 cm long.

HABITAT floodplain forests, streambanks, shores; also fencerows, drier woods and disturbed places.

STATUS MW-FAC | NCNE-FAC | GP-FAC

NOTE Boxelder distinguished from the **ashes** (*Fraxinus*) by its paired fruits (vs. single in ash) and its waxy-green twigs.

Acer rubrum L. ♦RED MAPLE

DESCRIPTION Native tree, to 25 m tall. Bark gray and smooth when young, becoming darker and scaly. Twigs smooth, reddish with pale lenticels. Leaves opposite, 3–5-lobed (but not lobed to middle of blade), coarsely doubly toothed or with a few small lobes, upper surface green and smooth, underside pale green to white, smooth or hairy. Flowers either staminate or pistillate, usually on different trees but sometimes on same tree, in dense clusters, opening before leaves in spring; sepals oblong, 1 mm long, petals narrower and slightly longer. Fruit a paired samara, 1–2.5 cm long.

HABITAT floodplain forests, swamps; also common in drier forests.

STATUS MW-FAC | NCNE-FAC | GP-FAC

NOTE distinguished from **silver maple** (*Acer saccharinum*) by its shallowly lobed leaves vs. deeply lobed in silver maple.

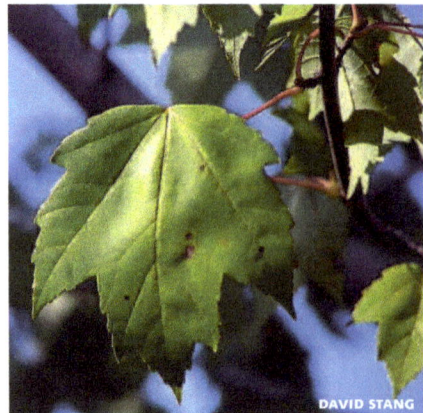

DAVID STANG

Acer rubrum

Acer saccharinum L.

◆ SILVER-MAPLE, SOFT MAPLE

DESCRIPTION Native tree, to 30 m tall. Bark gray or silvery when young, becoming scaly. Twigs red-brown, smooth. Leaves opposite, deeply 5-lobed to below middle of blade, sharply toothed, upper surface pale green and smooth, underside silvery white; petioles usually red-tinged. Flowers either staminate or pistillate, usually on different trees but sometimes on same tree, in dense clusters, opening before leaves in spring. Fruit a paired samara, each fruit 3–5 cm long, falling in early to mid-summer.

HABITAT floodplain forests, swamps, streambanks, shores, low areas in moist forests.

STATUS MW-FACW | NCNE-FACW | GP-FAC

Sarraceniaceae
Pitcherplant Family

Sarracenia PITCHER-PLANT

Sarracenia purpurea L.

◆ PITCHER-PLANT

DESCRIPTION Perennial insectivorous native herb. Flower stalks leafless, 3–6 dm long. Leaves clumped, hollow and vaselike, curved and upright from base of plant, 1–2 dm long and 1–5 cm wide, green or veined with red-purple, winged, smooth on outside, upper portion of inside with downward-pointing hairs, tapered to a short petiole at base. Flowers large and nodding, 5–6 cm wide, single at ends of stalks, perfect; sepals 5; petals 5, obovate, dark red-purple, curved inward over yellow style; ovary large and round. Fruit a 5-chambered capsule; seeds small and numerous.

Flowering May–July.

HABITAT sphagnum bogs, floating bog mats, occasional in calcium-rich wetlands.

STATUS MW-OBL | NCNE-OBL | GP-OBL

NOTE plants lacking the maroon pigment and with yellow flowers have been reported from n LP and Mackinac County of Michigan. These plants have been called *Sarracenia purpurea* form *heterophylla*, and are listed as state threatened.

Saururaceae
Lizard's-Tail Family

Saururus LIZARD'S TAIL

Saururus cernuus L. ◆ LIZARD'S TAIL

DESCRIPTION Perennial, succulent native herb. Stems branched, jointed, 5–12 dm tall. Leaves alternate, heart-shaped to ovate, 6–15 cm long, base of main veins loosely hairy with jointed hairs; margins entire; base of petiole surrounds stem. Flowers in 1–2 fuzzy spikes at ends of stems, 6–15 cm long, often nodding at tip when young; petals and sepals absent; stalks of stamens (filaments) white, much longer than pistils.

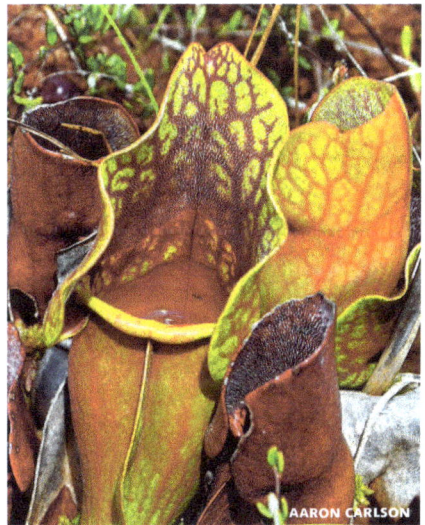

Acer saccharinum

Sarracenia purpurea

Fruit a rough nutlet, 2–3 mm wide. Flowering June–Aug.

HABITAT wet or swampy deciduous woods, cedar swamps, shallow pools, mudflats, ditches, marshes.

STATUS MW-OBL | NCNE-OBL | GP-OBL

Saxifragaceae
Saxifrage Family

PERENNIAL HERBS with alternate, opposite or basal leaves. Flowers perfect (with both staminate and pistillate parts), regular, single on stalks or in narrow heads; sepals 5; petals 5 or absent (*Penthorum*); stamens 5 or 10, stigmas 2 or 4. Fruit mostly a 2-parted capsule.

Saxifragaceae SAXIFRAGE FAMILY
1 Leaves all on stem 2
1 Leaves all (or nearly all) from base of plant
.... 3
2 Leaves mostly opposite; plants low and trailing .. *Chrysosplenium americanum*
2 Leaves alternate; plants upright..........
.... *Penthorum sedoides*
3 Leaves with rounded teeth . *Mitella nuda*
3 Leaves entire.. *Micranthes pensylvanica*

Chrysosplenium
GOLDEN-SAXIFRAGE

Chrysosplenium americanum
Schwein. ◆ AMERICAN GOLDEN SAXIFRAGE

DESCRIPTION Small native perennial herb, often forming large mats. Stems creeping, branched, 5–20 cm long. Leaves lower leaves opposite, the upper often alternate, broadly ovate, 5–15 mm long and as wide, margins entire or with rounded teeth or lobes; petioles short. Flowers single and stalkless from leaf axils, 4–5 mm wide; sepals 4, green-yellow or purple-tinged; petals absent; stamens usually 8 from a red or green disk, anthers red. Fruit a 2-lobed capsule. Flowering April–June.

HABITAT springs, shallow streams, shady wet depressions; soils mucky.

STATUS MW-OBL | NCNE-OBL

ADDITIONAL SPECIES *Chrysosplenium iowense,* more common in arctic regions, occurs on wet, mossy, rocky slopes in the driftless area of ne Iowa (state threatened) and se Minn (state endangered). It differs from *Chrysosplenium americanum* by having leaf margins with 5–7 large rounded teeth.

DOUG MCGRADY

Chrysosplenium americanum

ERIC HUNT

Saururus cernuus

AGNIESZKA KWIECIEŃ

Micranthes pensylvanica

Micranthes SAXIFRAGE

Micranthes pensylvanica (L.) Haw.
♦ SWAMP-SAXIFRAGE

DESCRIPTION Native perennial herb. Stems stout, erect, 3–10 dm long, with sticky hairs. Leaves all from base of plant, ovate to oblong ovate, 1–2 dm long and 4–8 cm wide, smooth or hairy; margins entire to slightly wavy or with irregular rounded teeth; petioles wide. Flowers small, in clusters atop stem, the head elongating with age; sepals bent backward, 1–2 mm long; petals green-white or purple-tinged, lance-shaped, 2–3 mm long; stamens 10, the filaments threadlike. Fruit a follicle. Flowering May–June.

SYNONYM *Saxifraga pensylvanica*

HABITAT swamps, wet deciduous forests, marshes, moist meadows and low prairie; often where calcium-rich.

STATUS MW-OBL | NCNE-OBL | GP-FACW

Mitella
MITREWORT, BISHOP'S-CAP

Mitella nuda L.
♦ SMALL BISHOP'S-CAP

DESCRIPTION Small native perennial herb, spreading by rhizomes or stolons. Leaves all from base of plant, or with 1 small leaf on flower stalk, rounded heart-shaped, 1–3.5 cm wide, both sides with sparse coarse hairs; margins with rounded teeth; petioles 2–8 cm long. Flowers small, green, on short stalks, in racemes of 3–12

flowers, on a glandular-hairy stalk 10–25 cm tall; calyx lobes 5, 1–2 mm long; petals green, pinnately divided into usually 4 pairs of threadlike segments, the segments 2–4 mm long; stamens 10. Fruit a capsule, splitting open to reveal the black, shiny, 1 mm long seeds. Flowering June–July.

HABITAT hummocks in swamps and alder thickets, ravines, seeps, moist mixed conifer and deciduous forests.

STATUS MW-FACW | NCNE-FACW | GP-OBL

Ulmaceae
Elm Family

Ulmus ELM

Ulmus americana L.
♦ AMERICAN ELM

DESCRIPTION Tree to 25 m tall, trunk to 1 m wide, crown broadly rounded or flat-topped, smaller branches usually drooping. Bark gray, furrowed, breaking into thin plates with age. Twigs brown, smooth or with sparse hairs, often zigzagged; buds red-brown. Leaves alternate, simple, to 15 cm long and 7–8 cm wide, oval, pointed at tip, base strongly asymmetrical, upper surface dark green and smooth, lower surface pale and smooth or soft-hairy; margins coarsely double-toothed; petioles short, usually yellow. Flowers small, green-red, hairy, in drooping clusters of 3–4; appearing before leaves unfold in spring. Fruit 1-seeded, oval, 1 cm wide, with a winged, hairy margin, notched at tip.

HABITAT floodplain forests, streambanks and moist, rich woods; less common now than formerly due to losses from Dutch elm disease.

STATUS MW-FACW | NCNE-FACW | GP-FAC

Mitella nuda

Ulmus americana

Urticaceae
Nettle Family

ANNUAL OR PERENNIAL HERBS with watery juice, sometimes with stinging hairs. Leaves alternate or opposite, simple, with petioles. Flowers small, green, in simple or branched clusters from leaf axils, staminate and pistillate flowers usually separate, on same or separate plants; sepals joined, 3–5-lobed; petals absent; ovary superior, 1-chambered. Fruit an achene, often enclosed by the sepals which enlarge after flowering.

CAUTION Stems and leaves of **wood-nettle** (*Laportea*) and **stinging nettle** (*Urtica*) have very irritating hairs – avoid contact!

Urticaceae **NETTLE FAMILY**

1 Plants with stiff stinging hairs 2
1 Plants without stinging hairs; smooth or with sparse small hairs 3
2 Leaves alternate .. *Laportea canadensis*
2 Leaves opposite .... *Urtica dioica*
3 Flowers in cylindric spikes from leaf axils; achene shorter than and hidden by sepals .... *Boehmeria cylindrica*
3 Flowers in dense short clusters from leaf axils; achene equal or longer than sepals.. .... *Pilea*

Boehmeria FALSE NETTLE

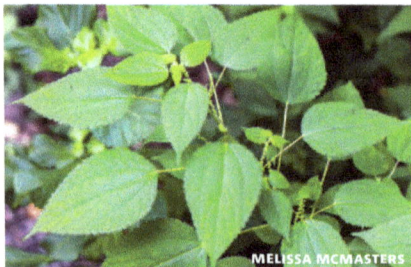

Boehmeria cylindrica (L.) Swartz
♦ FALSE NETTLE

DESCRIPTION Native perennial nettle-like herb; stinging hairs absent. Stems upright, 4–10 dm long, usually unbranched. Leaves opposite, rough-textured, ovate to broadly lance-shaped, narrowed to a pointed tip, with 3 main veins; margins coarsely toothed; petioles shorter than blades. Flowers tiny, green, staminate and pistillate flowers usually on separate plants, in small clusters along unbranched stalks from upper leaf axils, forming cylindric, interrupted spikes of staminate flowers or continuous spikes of pistillate flowers. Fruit an achene, enclosed by the enlarged bristly sepals and petals, ovate and narrowly winged. Flowering July–Aug.
HABITAT floodplain forests, swamps, marshes and bogs.
STATUS MW-OBL | NCNE-OBL | GP-FACW

Laportea WOOD-NETTLE

Laportea canadensis (L.) Wedd.
♦ WOOD-NETTLE

DESCRIPTION Native perennial herb, spreading by rhizomes. Stems somewhat zigzagged, 5–10 dm long. Leaves alternate, 8–15 cm long, ovate and narrowed to a tip, with small stinging hairs, margins coarsely toothed. Flowers small, green, staminate and pistillate flowers separate but borne on same plant; staminate flowers in branched clusters from lower leaf axils, shorter than leaf petioles; pistillate flowers in open, spreading clusters from upper axils, usually much longer than petioles. Fruit a flattened achene, longer than the 2 persistent sepals. Flowering July–Sept.
HABITAT floodplain forests, rich moist woods, low places in hardwood forests, streambanks.
STATUS MW-FACW | NCNE-FACW | GP-FAC
NOTE differs from **stinging nettle** (*Urtica dioica*) by its broader, alternate leaves, and the longer spikelike heads from the upper leaf axils.

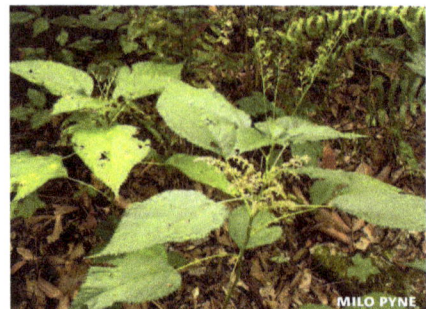

Boehmeria cylindrica

Laportea canadensis

Pilea CLEARWEED

Annual herb, sometimes forming colonies from seeds of previous year. Stems erect to sprawling, smooth, translucent and watery. Leaves opposite, stinging hairs absent, thin and translucent, ovate, with 3 major veins from base of leaf, margins toothed. Flowers green, staminate and pistillate flowers separate, borne on same or different plants, in clusters from leaf axils; staminate flowers with 4 sepals and 4 stamens; pistillate flowers with 3 sepals, ovary superior. Fruit a flattened, ovate achene.

Pilea CLEARWEED

1 Achenes olive-green to dark purple with a narrow pale margin, about as long as wide, covered with low bumps; leaf petioles 1/5–1/2 length of blade............ *P. fontana*

1 Achenes to 1 mm wide, green to yellow, longer than wide, often marked with purple spots, smooth; petioles 1/3 to as long as blade........................ *P. pumila*

Pilea fontana (Lunell) Rydb.
BOG CLEARWEED

DESCRIPTION Native annual herb. Stems 1–4 dm long, often sprawling. Leaves opposite, 2–6 cm long and 1–4 cm wide; petioles 0.5–5 cm long. Flowers in clusters, staminate flowers usually innermost when mixed with pistillate flowers. Fruit a dark olive-green to purple achene, 1–1.5 mm wide, with a narrow pale margin; sepals persistent, shorter to slightly longer than achene. Aug–Sept.
HABITAT lakeshores, riverbanks, swamps, marshes and springs.
STATUS MW-FACW | NCNE-FACW | GP-OBL

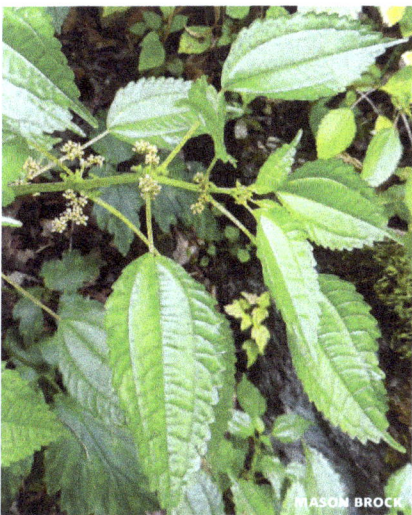

Pilea pumila (L.) A. Gray ◆CLEARWEED

DESCRIPTION Native annual herb; plants similar to *P. fontana*, but sometimes taller (to 5 dm) and with green achenes, often marked with purple and to only 1 mm wide. Leaves opposite, usually larger (to 12 cm long and 8 cm wide), thinner and more translucent than in *Pilea fontana;* petioles to 8 cm long. Flowering July–Sept.
HABITAT swampy woods (often on logs), wooded streambanks, floodplain forests, wet depressions, rocky hollows; usually in partial shade.
STATUS MW-FACW | NCNE-FACW | GP-FAC

Urtica NETTLE

Urtica dioica L. ◆STINGING NETTLE

DESCRIPTION Stout native perennial herb, often forming dense patches from spreading rhizomes. Stems 8–20 dm tall, usually unbranched, with stinging hairs on stems and leaves, the hairs irritating

Pilea pumila

Urtica dioica

to skin. **Leaves** opposite, ovate to lance-shaped, 5–15 cm long and 2–8 cm wide; margins coarsely toothed; petioles 1–6 cm long; stipules lance-shaped, 5–15 mm long. **Flowers** small, green, staminate and pistillate flowers separate but mostly on same plants; flower clusters branched and spreading from leaf axils, the clusters usually longer than petioles, all of one sex or a mix of staminate and pistillate flowers, the pistillate clusters usually above the staminate clusters when both are present. **Fruit** an ovate achene, 1–2 mm long, enclosed by the inner pair of sepals. **Flowering** July–Sept.

SYNONYM *Urtica procera*.

HABITAT moist woods, thickets, ditches, streambanks, disturbed areas.

STATUS MW-FACW | NCNE-FAC | GP-FAC

CAUTION irritating hairs on stems and leaves.

Verbenaceae
Verbena Family

PERENNIAL HERBS with 4-angled, erect or prostrate stems. Leaves opposite, toothed. Flowers small, numerous, perfect (with both staminate and pistillate parts), in branched or unbranched spikes or heads at ends of stems or from upper leaf axils, the spikes elongating as flowers open upward from the base. Calyx 5-toothed (*Verbena*) or 2-parted (*Phyla*); corolla 5-lobed (*Verbena*) or 4-lobed (*Phyla*), somewhat 2-lipped; stamens 4, of 2 lengths. Fruit dry, enclosed by the sepals, splitting lengthwise into 2 or 4 nutlets when mature.

Verbenaceae **VERVAIN FAMILY**

1 Flowers in round heads or short-cylindric spikes on leafless stalks from leaf axils ... .*Phyla lanceolata*
1 Flowers in long spikes at ends of stems and from upper leaf axils ... *Verbena hastata*

Phyla FOGFRUIT

Phyla lanceolata (Michx.) Greene
⧫FOGFRUIT

DESCRIPTION Native perennial herb, sometimes forming mats; plants smooth or with sparse, short, forked

hairs. **Stems** slender, weak, 4-angled, creeping to ascending, often rooting at nodes, the stem tips and lateral branches upright. **Leaves** opposite, ovate to oblong lance-shaped, 2–7 cm long and 0.5–3 cm wide, bright green, tapered to a sharp tip; margins with coarse, forward-pointing teeth to below middle of blade; tapered to a short petiole. **Flowers** small, crowded in spikes from leaf axils, the spikes single, at first round, becoming short-cylindric, 0.5–2 cm long and 5–7 mm wide, on slender stalks 2–9 cm long; calyx 2-parted and flattened, about as long as corolla tube; corolla pale blue or white, 3–4 mm long, 4-lobed and 2-lipped, the lower lip larger than upper lip; withering but persistent in fruit. **Fruit** round, enclosed by the sepals, separating into 2 nutlets. **Flowering** June–Sept.

SYNONYM *Lippia lanceolata*.

HABITAT margins of lakes, ponds, streams, ditches, mud flats; often where seasonally flooded.

STATUS MW-OBL | NCNE-OBL | GP-FACW

Verbena VERVAIN

Verbena hastata L.
⧫COMMON VERVAIN, WILD HYSSOP

DESCRIPTION Native perennial herb; plants with short, rough hairs. **Stems** stout, erect, 4–12 dm tall, 4-angled, sometimes branched above. **Leaves** opposite, lance-shaped to oblong lance-shaped, 4–12 cm long and 1–5 cm wide; margins with

Phyla lanceolata

FRITZ REYNOLDS

coarse, forward-pointing teeth and some-times lobed near base; petioles short. Flowers small, numerous, slightly irregular, in long, narrow spikes 5–15 cm long at ends of stems, the spikes elongating as flowers open upward from base; calyx unequally 5-toothed, 1–3 mm long; corolla dark blue to purple, 5-lobed, trumpet-shaped, slightly 2-lipped, 2–4 mm wide. Fruit 4-angled, splitting into 4 nutlets. Flowering July–Sept.

HABITAT marshes, wet meadows, shores, streambanks, swamp openings, ditches.

STATUS MW-FACW | NCNE-FACW | GP-FACW

Violaceae
Violet Family

Viola VIOLET
Perennial herbs, with or without leafy stems. Leaves all at base of plant or alternate on stems; petioles with membranous stipules. Flowers perfect, nodding and single at ends of stems, with 5 unequal sepals, 2 upper petals, 2 lateral, bearded petals, and 1 lower petal prolonged into a nectar-holding spur at the petal base. Fruit an ovate capsule which splits to eject the seeds.

Viola VIOLET

1 Plants with stems; leaves and flowers borne on the upright stems 2
1 Plants without stems; leaves and flowers directly from rootstock 3

JOSHUA MAYER

Verbena hastata

2 Flowers light blue or lavender . . . . *V. labradorica*
2 Flowers creamy-white . . . . *V. striata*
3 Flowers white 4
3 Flowers purple 8
4 Leaves more than 1.5x longer than wide . 5
4 Leaves often wider than long, less than 1.5x longer than wide 6
5 Leaves lance-shaped, tapered to a narrow base . . . . *V. lanceolata*
5 Leaves broader, ovate, narrowly heart-shaped at base . . . . *V. primulifolia*
6 Leaves dull, not shiny, upper and lower surface without hairs, lower surface not paler than upper; margins ± entire or with rounded teeth; petioles often with long soft hairs . . . . *V. macloskeyi*
6 Leaves shiny and smooth on upper surface, or dull and hairy on either upper or lower surface; underside paler than upper surface; margins with sharp, forward-pointing teeth 7
7 Plants with stolons and horizontal rhizomes; upper and lower surface of leaves sparsely to densely hairy with short hairs less than 1 mm long . . . . . . *V. blanda*
7 Plants without stolons, rhizomes upright; leaves often shiny and smooth on upper surface, or densely hairy on upperside with hairs about 1–2 mm long and smooth below . . . . *V. renifolia*
8 Leaves longer than wide 9
8 Leaves as wide or wider than long 11
9 Leaves glabrous or nearly so; sepal margins not fringed with hairs . . *V. novae-angliae*
9 Leaves sparsely to densely hairy; sepal margins usually fringed with hairs 10
10 Lateral petals with long, threadlike hairs on inner surface; spurred petal densely hairy within . . . . *V. affinis*
10 Lateral petals with short, knob-tipped hairs on inner surface; spurred petal without hairs . . . . *V. cucullata*
11 Sepals long-tapered to a sharp tip; lateral petals with short, knob-tipped hairs on inner surface; spurred petal without hairs . . . . *V. cucullata*
11 Sepals oblong to broadly lance-shaped, rounded at tip; lateral petals with long, threadlike hairs on inner surface 12
12 Flowers held above leaves; leaves and stems without hairs, leaves rounded at tip, margins with rounded teeth; spurred petal densely hairy within; plant of open wetlands and peatlands . *V. nephrophylla*

12 Flowers overtopped by leaves; leaves and stems usually hairy, leaves tapered to a pointed tip, margins with sharp, forward-pointing teeth; spurred petal smooth to slightly hairy within; plant of moist forests .... *V. sororia*

Viola affinis Leconte
LECONTE'S VIOLET

DESCRIPTION Native perennial herb, spreading by rhizomes. **Leaves** all from base of plant, hairless, narrowly heart-shaped; margins with rounded teeth. **Flowers** violet, bearded within with long, threadlike hairs, atop stalks slightly longer than leaves. **Fruit** a purple-flecked capsule on horizontal or arching stalks, seeds dark. Flowering April–May.

HABITAT swamps, floodplain forests, streambanks and lakeshores, low prairie.

STATUS MW-FACW | NCNE-FACW

Viola blanda Willd.
♦SWEET WHITE VIOLET

DESCRIPTION Native perennial herb, spreading by short rhizomes (and stolons later in season). **Stems** smooth. **Leaves** all from base of plant, heart-shaped, dark green and satiny, 2–5 cm wide, upper surface near base of blade usually with short, stiff white hairs; petioles usually red. **Flowers** white, fragrant, on stalks shorter than longer than leaves; lower 3 petals with purple veins near base, all ± beardless; upper 2 petals narrow, twisted backward, 2 side petals forward-pointing. **Fruit** a purple capsule 4–6 mm long, seeds dark brown. Flowering April–May.

SYNONYM *Viola incognita*.

HABITAT hummocks in swamps and bogs, low wet areas in deciduous and conifer forests.

STATUS MW-FACW | NCNE-FACW | GP-FACW

Viola cucullata Aiton
♦BLUE MARSH VIOLET

DESCRIPTION Native perennial herb, spreading by short, branched rhizomes; plants smooth. **Leaves** all from base of plant, ovate to kidney-shaped, to 10 cm wide, heart-shaped at base; margins coarsely toothed; blade angled from the upright petioles. **Flowers** light purple or white, dark at center, on slender stalks longer than leaves; the 2 side petals densely bearded with short hairs, the hairs mostly knobbed or club-tipped. **Fruit** a cylinder-shaped capsule, seeds dark. Flowering April–June.

SYNONYM *Viola obliqua*.

HABITAT swamps, sedge meadows, shady seeps; occasionally in bogs and low areas in forests.

STATUS MW-OBL | NCNE-OBL

Viola labradorica Schrank
LABRADOR VIOLET

DESCRIPTION Native perennial herb; plants smooth. **Leaves** in clumps from rhizomes, at first all from base of plants, later with leafy, horizontal stems to 15 cm long; light green, ovate to kidney-shaped, 1–2.5 cm wide; margins with rounded teeth; petioles 2–6 cm long. **Flowers** pale blue; side petals bearded on inner surface. **Fruit** 4–5 mm long, seeds dark brown. Flowering April–June.

JOSHUA MAYER

Viola blanda

BO GORDY-STITH

Viola cucullata

SYNONYMS *Viola adunca* var. minor, *Viola consperma*.

HABITAT swamps, streambanks, moist hardwood forests.

STATUS MW-FACW | NCNE-FACW | GP-FAC

Viola lanceolata L.
◊ LANCE-LEAVED VIOLET

DESCRIPTION Native perennial herb, spreading by rhizomes and stolons. Leaves from base of plant, narrowly lance-shaped (distinctive in our *Viola* species), more than 2x longer than wide, tapered to base; margins toothed. Flowers white, all beardless; lower 3 petals purple-veined near base. Fruit a green capsule 5–8 mm long, seeds brown. Flowering April–June.

HABITAT open bogs, sedge meadows; soils sandy or mucky.

STATUS MW-OBL | NCNE-OBL | GP-OBL; Minn (THR).

Viola macloskeyi F. Lloyd
WILD WHITE VIOLET

DESCRIPTION Small perennial herb (our smallest violet), spreading by rhizomes and stolons. Leaves all from base of plant, heart-shaped to kidney-shaped, 1–3 cm wide at flowering, later to 8 cm wide, underside orange-tinged; margins with rounded teeth. Flowers white, on upright stalks equal or longer than leaves, 3 lower petals purple-veined near base, 2 side petals beardless or with sparse hairs. Fruit a green capsule 4–6 mm long, seeds olive-black. Flowering April–July.

SYNONYM *Viola pallens*.

HABITAT marshes, sedge meadows, open bogs and swamps, alder thickets; sometimes in shallow water.

STATUS MW-OBL | NCNE-OBL | GP-FACW

Viola nephrophylla Greene
◊ NORTHERN BOG VIOLET

DESCRIPTION Low native perennial herb, spreading by short rhizomes. Leaves all from base of plant, smooth, heart-shaped to kidney-shaped, 1–4 cm long and 2–6 cm wide, rounded at tip; margins with rounded teeth; petioles slender, 2–16 cm long. Flowers single, nodding on slender stalks, the stalks longer than leaves. Flowers violet, bearded near base on inside, or upper pair of petals not bearded. Fruit a capsule 5–10 mm long. Flowering May, sometimes again flowering in Aug or Sept.

HABITAT wet meadows, fens, calcium-rich wetlands, low areas between dunes, streambanks, rocky shores.

STATUS MW-FACW | NCNE-FACW | GP-FACW

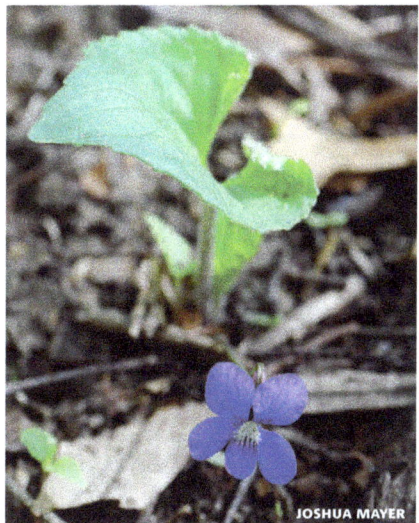

ROB ROUTLEDGE

Viola lanceolata

JOSHUA MAYER

Viola nephrophylla

Viola novae-angliae House
NEW ENGLAND BLUE VIOLET

DESCRIPTION Native perennial. Leaves ovate, longer than wide, cordate, crenate-serrate near the base, distantly so toward the acuminate apex; petioles and lower surface of the leaves villous or pubescent. Flowers violet-purple, the three lower petals villous at the base; sepals obtuse, glabrous; cleistogamous flowers on long ascending peduncles. capsules nearly globose, mottled with purple; seeds light brown to buff. Flowering April–June.

SYNONYM *Viola sororia* var. *novae-angliae*.

HABITAT gravelly and sandy shores and in rock crevices along streams.

STATUS MW OBL | NCNE OBL; Mich (THR).

Viola primulifolia L.
PRIMROSE-LEAVED VIOLET

DESCRIPTION Native perennial herb, spreading by rhizomes and stolons. Leaves all from base of plant, oblong to ovate, rounded at tip, longer than wide; margins with small rounded teeth. Flowers white, on stalks shorter or equal to leaves, 3 lower petals purple-veined at base, 2 side petals beardless or with few hairs. Fruit a capsule 7–10 mm long; seeds red-brown to black. May.

SYNONYMS *Viola lanceolata x pallens, Viola x primulifolia*.

HABITAT wet meadows and bogs, often in sphagnum moss; sandy streambanks; more common along Atlantic Coast from Maine to Tex.

STATUS MW-OBL | NCNE-OBL | GP-OBL

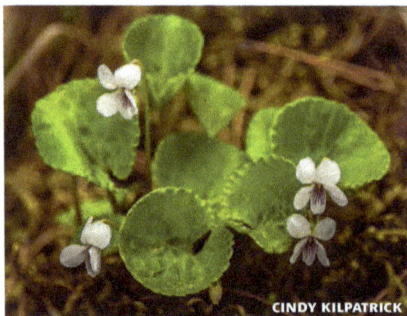

Viola renifolia A. Gray
◆ **KIDNEY-LEAVED VIOLET**

DESCRIPTION Native perennial herb, spreading by long rhizomes. Leaves all from base of plant, mostly kidney-shaped, rounded at tip, varying from smooth and shiny above to hairy on lower surface only; margins with few rounded teeth. Flowers white, all bearded or beardless, 3 lower petals purple-veined at base. Fruit a capsule 4–5 mm long, seeds brown and dark-flecked. Flowering May–July.

HABITAT cedar swamps, sphagnum hummocks in peatlands.

STATUS MW-FACW | NCNE-FACW | GP-FACW

Viola sororia Willd.
◆ **COMMON BLUE VIOLET**

DESCRIPTION Native perennial herb, spreading by short rhizomes. Leaves all from base of plant, ovate to heart-shaped, sometimes expanding to 10 cm wide in summer, with long hairs; margins with rounded teeth; blades angled from the upright petioles. Flowers blue-violet, on stalks about as high as leaves, the 2 side petals densely bearded with hairs 1 mm long and not club-tipped. Fruit a purple-flecked capsule, seeds dark brown. Flowering April–June.

HABITAT moist hardwood forests; occasionally in swamps, floodplain forests and along rocky streambanks.

STATUS MW-FAC | NCNE-FAC | GP-FAC

CINDY KILPATRICK

Viola renifolia

JOSHUA MAYER

Viola sororia

Viola striata Aiton CREAMY VIOLET

DESCRIPTION Native perennial herb, spreading by rhizomes. **Leaves** from base of plant and on leafy stems, smooth, leaves at base of plant and on lower stems rounded at tip, upper leaves heart-shaped and tapered to tip; margins with small teeth. **Flowers** many, creamy white with purple veins at center, on stalks raised well above leaves; side petals bearded, style tip bent. **Fruit** a rounded capsule 4–5 mm long, seeds light brown. **Flowering** April–June.

HABITAT floodplain forests, moist deciduous woods, streambanks, thickets; sometimes weedy.

STATUS MW-FACW | NCNE-FACW | GP-FACW

Vitaceae
Grape Family

Vitis GRAPE

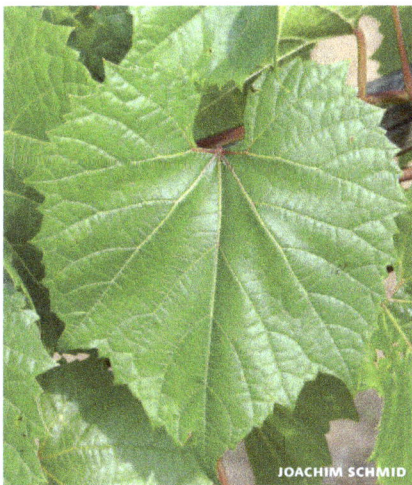

Vitis riparia Michx.
◆RIVERBANK GRAPE

DESCRIPTION Native perennial, vine; woody, twining to 5 m or more long. **Young branches** green or red, hairy, becoming smooth. **Leaves** alternate, heart-shaped in outline, 1–2 dm long and as wide, with a triangular tip and 2 smaller lateral lobes, leaf base with a U-shaped indentation, upper surface smooth, bright green, underside paler and sparsely hairy along veins; margins with coarse, forward-pointing teeth; petioles shorter than blades. **Flowers** small, sweet-scented, green-white to creamy, in stalked clusters 5–10 cm long. **Fruit** a dark blue to black berry, 6–12 mm wide, with a waxy bloom, sour when young, becoming sweeter when ripe in fall. **Flowering** May–July.

SYNONYM *Vitis vulpina* subsp. *riparia*.

HABITAT floodplain forests, moist sandy woods, streambanks, thickets; also on sand dunes.

STATUS MW-FAC | NCNE-FAC | GP-FAC

Vitis riparia

MONOCOTS
Monocotyledoneae

THE MONOCOTYLEDONEAE (or Liliopsida, the Monocots) include families with showy flowers such as members of the Liliaceae (now subdivided into a number of families) and Orchidaceae as well as families lacking petals and sepals such as the Cyperaceae (Sedge Family), Poaceae (Grass Family), and Juncaceae (Rush Family). Identification of species within these three families is frequently dependent on characters of small floral parts, and a hand lens or dissecting microscope is often useful.

Acoraceae
Calamus Family

Acorus SWEETFLAG

Perennial herbs of wetlands; rhizomes and leaves pleasantly scented. Rhizomes branched, creeping at or near surface. Leaves sword-shaped, equitant, bright green, with 1–6 prominent veins parallel along length of leaf. Inflorescence a solitary spadix, borne from near midway of leaf, nearly cylindric, tapering, apex obtuse; true spathe absent. Flowers bisexual; tepals 6, light brown; stamens 6; ovaries 1, usually 3-locular. Fruit light brown to reddish berries with darker streaks. Seeds 1–6(–14), embedded in a mucilagenous jelly.

Acorus SWEETFLAG

1 Leaves lacking a single prominent raised midvein, but with 2–several clearly separate veins of ± equal strength along with numerous fainter veins, the leaf thus not obviously ridged to the naked eye; fruit maturing . . . . *A. americanus*

1 Leaves with a single prominent raised midvein (best observed at about mid-portion of the leaf) that is much more conspicuous than any other vein, appearing as a ridge to the naked eye; fruit not maturing. . . . . *A. calamus*

Acorus americanus (Raf.) Raf.
AMERICAN SWEETFLAG

DESCRIPTION Similar to the less common, introduced *Acorus calamus*, and long considered a variety of it. Best distinguished using leaf venation differences as described in the key.

SYNONYM *Acorus calamus* var. *americanus*.

HABITAT marshes, wet meadows, edges of rivers.

STATUS MW OBL | NCNE OBL | GP OBL

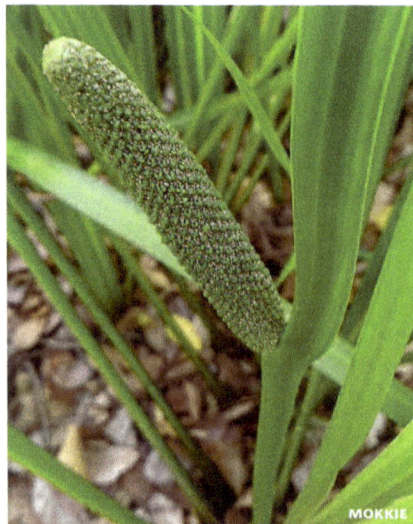

Acorus calamus L. ♦SWEETFLAG

DESCRIPTION Introduced perennial herb, from stout, aromatic rhizomes. **Leaves** linear, long and swordlike, bright-green, leathery, 2-ranked, 5–15 dm long and 1–2 cm wide, sweet-scented when crushed; margins entire, sharp-edged, translucent near base. **Flowers** small, in a cylindric, yellow-green spadix, appearing lateral from a leaflike, tapered stalk; the spadix upright, 5–10 cm long and 1–2 cm wide; flowers perfect, yellow or brown, composed of 6 papery tepals and 6 stamens. **Fruit** a 1–3-seeded berry, dry outside and jellylike on inside. **Flowering** June–July.

HABITAT marshes (often with cat-tails, *Typha* spp.), shores, moist depressions, stream-banks; introduced and naturalized in e and c USA.

STATUS MW-OBL | NCNE-OBL | GP-OBL

NOTE Native plants are distinguished as *Acorus americanus* and may be separated from *Acorus calamus* by having a raised leaf midvein plus 1–5 other veins raised above the leaf surface. *In Acorus calamus* only the midvein is prominently raised. However, the distinctions are not always clear, and populations of both species may be present in the same location.

Alismataceae
Water-Plantain Family

PERENNIAL, AQUATIC OR EMERGENT herbs; plants swollen and tuberlike at base. Leaves all from base of plant and clasping an erect stem; underwater leaves often ribbonlike; emergent leaves broader. Flowers perfect (with both staminate and pistillate parts) or imperfect, in racemes or panicles at ends of stems, with 3 sepals and 3 petals; stamens 6 or more. Fruit a compressed achene, usually tipped by the persistent style.

Alismataceae WATER-PLANTAIN FAMILY

1 Leaves never arrowhead-shaped; flowers perfect, in a panicle-like head; pistils or achenes in a single whorl on a small, flat receptacle . . . . *Alisma*

1 Leaves often arrowhead-shaped; flowers mostly imperfect, whorled on stem; staminate flowers above pistillate in the head; pistils or achenes in several series around a large, round receptacle, and forming a dense, round head . . *Sagittaria*

MOKKIE

Acorus calamus

Alisma WATER-PLANTAIN

Perennial herbs, from cormlike rootstocks. Leaves emersed or floating, ovate to lance-shaped, never arrowhead-shaped; underwater leaves sometimes ribbonlike (in *Alisma gramineum*). Flowers perfect, in whorled panicles, sepals 3, green; petals 3, white or light pink; stamens 6. Fruit a flattened achene in a single whorl on a flat receptacle, style beak small or absent.

NOTE plants formerly classified as *Alisma plantago-aquatica* L. are now termed either *A. subcordatum* or *A. triviale,* with *A. plantago-aquatica* considered a European species.

Alisma **WATER-PLANTAIN**

1 Leaves lance-shaped to oval, or if underwater, leaves long and ribbonlike; flower stalks rarely longer than leaves; petals usually pink .... *A. gramineum*

1 Leaves ovate; flower stalks much longer than leaves; petals white **2**

2 Achenes 1.5–2 mm long; larger fruiting heads 3–4 mm in diameter (excluding sepals)................ *A. subcordatum*

2 Achenes 2.2–3 mm long; larger fruiting heads 4–7 mm in diameter (excluding sepals)...................... *A. triviale*

Alisma gramineum Lej.
NARROW-LEAVED WATER-PLANTAIN

DESCRIPTION Native perennial herb. Leaves emersed leaves lance-shaped to oval, 2–10 cm long and 1–3 cm wide, petioles mostly longer than blades; underwater leaves reduced to linear, ribbonlike petioles, to 6 dm long and 1 cm wide. Flowers many on spreading stalks, the stalks shorter to somewhat longer than the leaves; sepals 3; petals pink, 1–3 mm long. Fruit an achene, 2–3 mm long, with a central ridge and 2 lateral ridges. Flowering July–Sept.

SYNONYM *Alisma geyeri.*

HABITAT shallow, often brackish water, muddy shores, streambanks.

STATUS MW-OBL | NCNE-OBL | GP-OBL

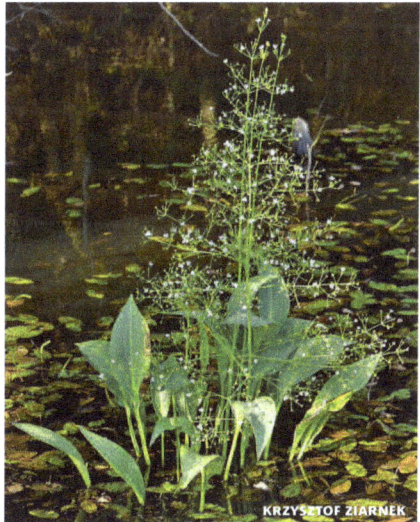

Alisma subcordatum Raf.
SOUTHERN WATER-PLANTAIN

DESCRIPTION Native perennial herb. Leaves ovate to oval, 3–15 cm long and 2–12 cm wide, rounded to nearly heart-shaped at base; petioles long. Flowers clustered on slender stalks 1–10 dm long, in whorls of 3–10; sepals 3; petals white, 3–5 mm long. Fruit an achene, 2–3 mm long, with a central groove. Flowering July–Sept.

SYNONYM *Alisma plantago-aquatica* var. *parviflorum.*

HABITAT shallow water marshes, shores, ditches.

STATUS MW-OBL | NCNE-OBL | GP-OBL

Alisma triviale Pursh
◆NORTHERN WATER-PLANTAIN

DESCRIPTION Native perennial herb. Leaves usually long-petioled, the blade elliptic to broadly ovate, rounded to subcordate at base, 3–18 cm long. Flowers on a scape 1–10 dm long; flower pedicels in whorls of 3–10; sepals obtuse, 2–3 mm long; petals white, about 4 mm long. Fruit an achene 2–3 mm long, usually with a median dorsal groove. Flowering June–Sept.

SYNONYMS *Alisma brevipes, Alisma plantago-aquatica* var. *americanum.*

HABITAT marshes, ponds, and streams.

STATUS MW-OBL | NCNE-OBL | GP-OBL

Alisma triviale

Sagittaria ARROWHEAD

Perennial or annual herbs, with fleshy rootstocks. Leaves sheathing, all from base of plant, variable in shape and size. Emersed and floating leaves usually arrowhead-shaped with large lobes at base, or sometimes ovate to oval and without lobes; underwater leaves often linear in a basal rosette, normally absent by flowering time. Flowers in a raceme of mostly 3-flowered whorls; upper flowers usually staminate, lower flowers usually pistillate or sometimes perfect; sepals 3, green, persistent; petals 3, white, deciduous; stamens 7 to many; pistils crowded on a rounded receptacle. Fruit a crowded cluster of achenes in ± round heads, the achenes flattened and winged, beaked with a persistent style.

Sagittaria ARROWHEAD

1　Emersed leaves not arrowhead-shaped, basal lobes absent 2
1　Emersed leaves all or mostly arrowhead-shaped, with large basal lobes 4
2　Female flowers and fruiting heads ± stalkless . . . . *S. rigida*
2　Female flowers and fruiting heads obviously stalked 3
3　Achenes with beak 0.4–0.6 mm long; anthers clearly shorter than the filaments. . . . . *S. cristata*
3　Achenes with beak minute, scarcely discernable, ca. 0.2 mm long; anthers as long as or longer than filaments. . . *S. graminea*
4　Plants annual, rhizomes absent; sepals appressed to fruiting heads; stalks of fruiting heads stout . . . . *S. calycina*
4　Plants perennial, with rhizomes; sepals reflexed on fruiting heads; stalks of fruiting heads slender 5
5　Bracts below flowers mostly less than 1 cm long; achene beak projecting horizontally from tip of achene . . . . *S. latifolia*
5　Bracts below flowers usually more than 1 cm long; achene beak erect or ascending. 6
6　Achene beak short, erect, to 0.4 mm long; basal lobes of leaves mostly shorter than terminal lobe . . . . *S. cuneata*
6　Achene beak larger, curved and ascending, 0.5 mm long or more; basal lobes of leaves usually equal or longer than terminal lobe . . . . *S. brevirostra*

Sagittaria brevirostra Mackenzie & Bush
MIDWESTERN ARROWHEAD

DESCRIPTION Native perennial herb. **Leaves** arrowhead-shaped, mostly 10–30 cm long and to 20 cm wide; basal lobes lance-shaped and usually equal or longer than terminal lobe; petioles long. **Flowers** in heads 1–2 cm wide, on a mostly unbranched stalk usually longer than leaves, the staminate flowers above the pistillate flowers; bracts below flowers 1–5 cm long; pistillate flowers on ascending stalks 0.5–2 cm long; sepals bent backward in fruit; petals white, 1–2 cm long. **Fruit** a winged achene, 2–3 mm long, separated from style beak by a saddlelike depression, the beak usually curved-ascending. **Flowering** July–Sept.
SYNONYM *Sagittaria engelmanniana* var. *brevirostra*.
HABITAT shallow water and muddy shores, marshes.
STATUS MW-OBL | NCNE-OBL | GP-OBL

Sagittaria calycina Engelm.
MISSISSIPPI *or* LONG-LOBED ARROWHEAD

DESCRIPTION Native annual herb, rhizomes and tubers absent. **Leaves** erect to spreading; emersed blades arrowhead-shaped (the lobes sometimes outward spreading) or oval to ovate, 3–40 cm long and 2–25 cm wide, the basal lobes usually longer than terminal lobe; petioles and flower stalks spongy, round in cross-section. **Flowers** lower flowers usually perfect; upper flowers usually staminate, in heads 1–2 cm wide, from a stalk 1–10 dm tall, heads leaning when fruiting; bracts membranous, short and rounded at lower nodes, upper bracts longer (to 1 cm long) and tapered to a tip; sepals blunt-tipped, bent backward in flower but appressed to the head in fruit; petals white with a yellow base. **Fruit** an achene, 2–3 mm long, beak ± horizontal from top of achene. **Flowering** July–Sept.
SYNONYMS *Lophotocarpus calycinus*, *Sagittaria montevidensis* subsp. *calycinus*.
HABITAT muddy shores.
STATUS MW-OBL | NCNE-OBL | GP-OBL; Mich (THR); historically known from Wisc along Miss River.

Sagittaria cristata Engelm.
CRESTED ARROWHEAD

DESCRIPTION Native, erect or floating, perennial herb. Leaves flat, usually long and narrow and not arrow-shaped. Flowers white to pink, 3-parted; inflorescence of 2–12 whorls of flowers, the staminate flowers above the pistillate flowers. Fruit a winged achene; beak 0.4–0.7 mm long. Flowering July–Sept

SYNONYM *Sagittaria graminea* var. *cristata*. **HABITAT** in shallow water or muddy soil.
STATUS MW-OBL | NCNE-OBL | GP-OBL

Sagittaria cuneata Sheldon
◆ NORTHERN ARROWHEAD, WAPATO

DESCRIPTION Native perennial herb, with rhizomes and large, edible tubers. Leaves submerged leaves (if present) often awl-shaped or reduced to bladeless, expanded petioles (phyllodes); emersed leaves long-stalked, usually arrowhead-shaped, 5–20 cm long and 2–15 cm wide, the basal lobes much shorter than terminal lobe; floating leaves often heart-shaped (unlike our other species of *Sagittaria*). Flowers imperfect, the staminate flowers above the pistillate, in ± round heads 5–12 mm wide, with 2–10 whorls of heads on a stalk 1–6 dm tall, the stalks often branched at lowest node; bracts tapered to tip, 1–4 cm long; sepals ovate, bent backward in flower and fruit; petals white, 7–15 mm long. Fruit an achene, 2–3 mm long; beak erect, small, 0.1–0.4 mm long. Flowering June–Sept.
HABITAT shallow water, shores, streambanks.
STATUS MW-OBL | NCNE-OBL | GP-OBL

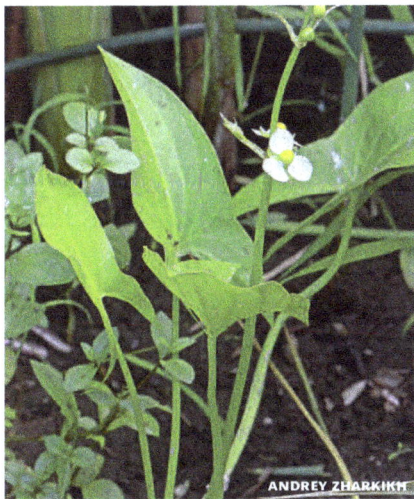

Sagittaria graminea Michx.
◆ GRASS-LEAVED ARROWHEAD

DESCRIPTION Native perennial herb, with rhizomes. Leaves underwater plants sometimes only a rosette of bladeless, ribbonlike petioles (phyllodes) to 1 cm wide; emergent leaves lance-shaped to oval, never arrowhead-shaped, 3–20 cm long and 0.5–3 cm wide, tapered to a blunt tip. Flowers imperfect, the staminate flowers usually above the pistillate, clustered in ± round heads, 5–12 mm wide, the heads on spreading stalks 1–4 cm long; with 2–10 whorls of flowers along an unbranched stalk mostly shorter than leaves; bracts broadly ovate, joined in their lower portion, 2–8 mm long; sepals ovate, bent backward in fruit; petals white, equal or longer than sepals. Fruit a winged achene, 1–2 mm long, beak small or absent. Flowering June–Sept.
HABITAT shallow water and shores.
STATUS MW-OBL | NCNE-OBL | GP-OBL

Sagittaria latifolia Willd.
◆ COMMON ARROWHEAD, WAPATO

DESCRIPTION Native perennial herb, with rhizomes and edible tubers in fall. Leaves variable; emersed leaves arrowhead-shaped, mostly 8–40 cm long and 1–15 cm wide, lobes typically narrow on plants in deep water to broad on emersed plants; plants sometimes with bladeless, ex-

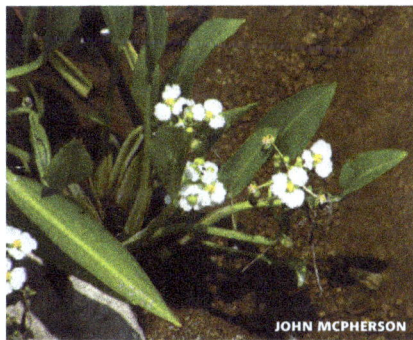

Sagittaria cuneata

Sagittaria graminea

panded petioles (phyllodes). Flowers staminate above and pistillate below, clustered in ± round heads 1–2.5 cm wide, at ends of slender, spreading stalks 0.5–3 cm long, in whorls of 2–15 along a stalk 2–10 dm tall; bracts tapered to a tip or blunt, 0.5–1 cm long; sepals ovate, bent backward by fruiting time; petals white, 7–20 mm long. Fruit a winged achene, 2–4 mm long, the beak projecting horizontally, 1–2 mm long. Flowering July–Sept.

HABITAT shallow water, shores, marshes, and pools in bogs.

STATUS MW-OBL | NCNE-OBL | GP-OBL

Sagittaria rigida Pursh
SESSILE-FRUITED ARROWHEAD

DESCRIPTION Native perennial herb, rhizomes present. Stems erect or lax. Leaves emersed leaves lance-shaped to ovate, rarely with short, narrow basal lobes, (but not arrowhead-shaped), 4–15 cm long and to 7 cm wide; petioles sometimes bent near junction with blades; deep water plants often with only linear, bladeless, expanded petioles (phyllodes). Flowers in ± round heads to 1.5 cm wide, the heads stalkless and bristly when mature due to achene beaks; in 2–8 whorls on a stalk 1–8 dm tall, the stalk often bent near lowest node; flowers imperfect, staminate flowers above the pistillate, staminate flowers on threadlike stalks 1–3 cm long, pistillate flowers ± stalkless; bracts ovate, 5 mm long, joined at base; sepals ovate, 4–7 mm long, bent backward when in fruit; petals white, 1–3 cm long. Fruit a narrowly winged achene, 2–4 mm long; beak ascending, 1–1.5 mm long. Flowering June–Sept.

HABITAT shallow water, shores, streambanks.

STATUS MW-OBL | NCNE-OBL | GP-OBL

Sagittaria latifolia

Araceae
Arum Family

PERENNIAL HERBS with alternate, simple or compound leaves. In our traditional genera of Araceae (*Arisaema, Calla, Peltandra, Pistia, Symplocarpus*), flowers are small and numerous, mostly single-sexed, staminate flowers usually above pistillate, crowded in a cylindric or rounded spadix subtended by a leaflike spathe; sepals 4–6 or absent; petals absent; stamens mostly 2–6; pistils 1- to 3-chambered. Fruit a usually fleshy berry, containing 1 to few seeds, or the entire spadix ripening as a fruit.

Now included in the Araceae are the tiny **duckweeds** (*Lemna, Spirodela,* and *Wolffia*), aquatic genera formerly treated within Lemnaceae. These are small perennial herbs, floating at or near water surface, single or forming colonies. Plants thallus-like (not differentiated into stems and leaves), the thallus (or frond) flat or thickened; the roots, if present, unbranched, 1 or several from near center of leaf underside; reproducing vegetatively by buds from 1–2 pouches on the sides, the parent and budded plants often joined in small groups. Flowers rare, either staminate or pistillate, in tiny reproductive pouches on margins (*Lemna, Spirodela*) or upper surface (*Wolffia*) of the leaves. Fruit a utricle with 1 to several seeds.

Araceae ARUM FAMILY

1 Plants tiny floating or submerged aquatic species less than 1 mm long, without differentiation into leaves or stems **2**

1 Plants large, with clearly differentiated normal leaves and with rhizomes or tubers **4**

2 Roots absent; leaves thickened, less than 1.5 mm long.................... *Wolffia*

2 One to several roots present on leaf underside; leaves flat, mostly more than 1.5 mm long **3**

3 Each leaf with 1 root; leaf underside green or purple-tinged .... *Lemna*

3 Each leaf with 3 or more roots; leaf underside solid purple ....
.... *Spirodela polyrhiza*

4 Leaves compound .... *Arisaema*

4 Leaves simple, not divided **5**

5 Plants forming free floating rosettes, rooting in water; leaves densely hairy, sessile.................. *Pistia stratiotes*

5 Plants not free floating, leaves not arranged in a rosette, rooted in soil near shores or terrestrially; leaves glabrous, petiolate .. **6**

6 Leaves arrowhead-shaped, lobes equaling 1/3 or more length of blade ....
.... *Peltandra virginica*

6 Leaves heart-shaped or rounded at base **7**

7 Leaves broadly heart-shaped, abruptly tapered to a tip; spathe white, long-stalked; flowering in spring ...... *Calla palustris*

7 Leaves rounded ovate, tapered to a rounded tip; spathe green-yellow to purple-brown, short-stalked or stalkless; flowering late winter to early spring................
.... *Symplocarpus foetidus*

Arisaema
JACK-IN-THE-PULPIT

Perennial herbs. Leaves compound. Flowers either staminate or pistillate, on same or different plants; staminate flowers with 2-5, ± stalkless stamens, above the pistillate flowers on a fleshy spadix, the spadix subtended by a green or purple-brown spathe; sepals and petals absent. Fruit a cluster of round, red berries, each berry with 1-3 seeds.

Arisaema JACK-IN-THE-PULPIT

1 Leaflets 7-13; spadix longer than spathe ..
.... *A. dracontium*

1 Leaflets usually 3; spathe arching over spadix.................... *A. triphyllum*

Arisaema dracontium (L.) Schott
♦ GREEN DRAGON

DESCRIPTION Native perennial herb, from corms. Leaf usually single, palmately branched into 7-15 leaflets, the leaflets oval to oblong lance-shaped, tapered to a point and narrowed at base, the central leaflets 1-2 dm long and to 8 cm wide, the outer leaflets progressively smaller; petioles 2-10 dm long. Flowers staminate or pistillate and on different plants, or plants with both staminate and pistillate flowers, the stami-

nate above pistillate, on a long, slender spadix exserted 5-10 cm beyond spathe; the spathe green, slender, rolled inward, 3-6 cm long; the flower stalk shorter than the leaf petiole. Fruit a cluster of red-orange berries. Flowering May-July.

HABITAT wet woods and floodplain forests.

STATUS MW-FACW | NCNE-FACW | GP-FACW

Arisaema triphyllum (L.) Schott
♦ JACK-IN-THE-PULPIT

DESCRIPTION Native perennial herb, from bitter-tasting corms. Stems 3-12 dm long. Leaves usually longer than the flower stalk, mostly 2, divided into 3 leaflets, the terminal leaflet oval to ovate, the lateral leaflets often asymmetrical at base. Flowers staminate or pistillate and usually on separate plants, borne near base of a cylindric, blunt-tipped spadix, subtended by a green, purple-striped spathe, rolled inward below, expanded and arched over the spadix above, abruptly tapered to a tip. Fruit a cluster of shiny red berries. Flowering April-July.

SYNONYM *Arisaema atrorubens.*

HABITAT moist forests, cedar swamps.

STATUS MW-FACW | NCNE-FAC | GP-FAC

Calla WATER-ARUM

Calla palustris L.
♦ WATER-ARUM

DESCRIPTION Native perennial herb, from

Arisaema dracontium

rhizomes, the rhizomes creeping in mud or floating. Leaves broadly heart-shaped, abruptly tapered to a tip, 5–15 cm long and about as wide; petioles stout, 1–2 dm long (or longer when underwater). Flowers perfect or the uppermost staminate, on a short-cylindric spadix, 1.5–3 cm long, shorter than the spathe; the spathe white, ovate, tipped with a short, sharp point to 1 cm long; sepals and petals absent; stamens 6. Fruit a fleshy, few-seeded berry, turning red when ripe. Flowering May–July.
HABITAT bog pools, swamps, shores, ditches.
STATUS MW-OBL | NCNE-OBL | GP-OBL

Lemna DUCKWEED

Small perennial floating herbs, with 1 root per frond (or roots sometimes absent on oldest and youngest leaves). Fronds single or 2 to several and joined in small colonies, floating on water surface or underwater (star-duckweed, *Lemna trisulca*), varying from round, ovate, to obovate or oblong, tapered to a long point (stipe) in star-duckweed; green or often red-tinged; upper surface flat to slightly convex, underside flat or convex. Reproductive pouches 2, on margins of frond. Flowers uncommon, consisting of 2 stamens (staminate flowers) and a single pistil (pistillate flower) in each pouch. Fruit an utricle with 1 to several seeds. Reproduction mostly by budding of new leaves from the reproductive pouches.

Lemna DUCKWEED

1 Fronds denticulate toward the tip, tapered to a slender stipitate base, the stipe often as long as the main body and commonly attached to the parent frond; colonies star-shaped, usually submersed ... *L. trisulca*

1 Fronds entire on the margin, nearly rounded and not obviously stipitate at the base, solitary or in tight colonies, these not star-shaped, floating on the water surface or stranded on mud 2

2 Root sheath winged at the base; root tip sharply pointed; roots not longer than 3 cm; fronds completely green 3

2 Root sheath not winged at the base; root tip mostly rounded; roots often longer than 3 cm; fronds often red-tinged beneath or with red spots on either surface......... 4

3 Fronds very often with 2-3 papillae in a row on the upper surface above the node (the level at which daughter fronds attach); seeds whitish, with 35-60 faint ribs, not escaping the fruit wall when ripening ....
.... *L. perpusilla*

3 Fronds with only 1 prominent papilla above the node; seeds brownish, with 8-22 prominent ribs, falling out of the fruit wall when ripening .... *L. aequinoctialis*

4 Fronds with several about equal sized, small papillae on the upper surface from the midline to the tip (often obscure), very often red-tinged on the lower surface, forming small, obovate to orbicular, rootless, dark green to brown turions under unfavorable conditions, these sinking to the bottom of the water. ... *L. turionifera*

Arisaema triphyllum

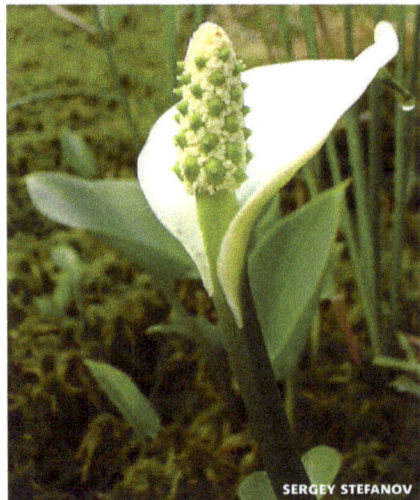
Calla palustris

4 Fronds lacking papillae or with one prominent papilla at the tip and another just above the node and with smaller papillae between them, rarely forming turions; if formed, the turionlike fronds have short roots and are slowly forming daughter fronds **5**

5 Papilla at tip of the frond very prominent; fronds often red beneath. . . . . . *L. obscura*

5 Papilla at tip of the frond not very prominent; fronds never red beneath. . . . .
. . . . *L. minor*

Lemna aequinoctialis Welw.
LESSER DUCKWEED

DESCRIPTION Small native perennial floating herb, very similar to *L. perpusilla* except with only 1 prominent papilla above the node on the upper surface. Seeds brownish, with 8–22 prominent ribs, falling from the fruit at maturity.

STATUS MW OBL | NCNE OBL | GP OBL; sw Wisc, main range s USA.

Lemna minor L.
◆ COMMON DUCKWEED

DESCRIPTION Small native perennial floating herb, with one root from middle of underside of frond. Fronds ± round to oval, 2–4 mm long, often in groups of 2–5, underside often red-tinged; both sides flat or slightly convex. Fruit a 1-seeded utricle. Flowering July–Sept.

HABITAT quiet or stagnant water of ponds, oxbows, shores, slow-moving rivers, ditches.

STATUS MW-OBL | NCNE-OBL | GP-OBL

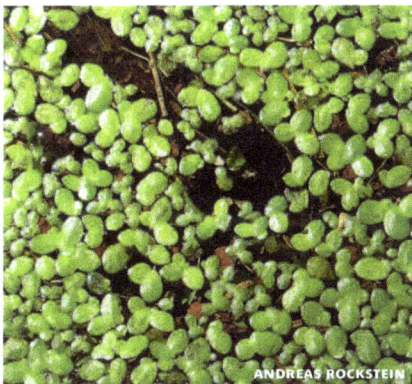

Lemna minor

NOTE the frond underside of the similar **turion duckweed** *(Lemna turionifera)*, found throughout the region, is usually strongly red-tinged, and plants form small, rootless turions (fleshy, overwintering structures that detach, and then start growth in the spring). However, if turions and reddish color absent, plants of turion duckweed may not be separable from *L. minor*.

Lemna obscura (Austin) Daubs
LITTLE DUCKWEED

DESCRIPTION Small native perennial floating herb; quite similar to *L. minor* and differing mainly as follows: fronds obovate to oblong-orbicular, 1.5–3 mm long, slightly asymmetric, often reddish beneath, obscurely 3-nerved; upper surface flat to slightly convex, with a prominent papilla at the apex, the lower surface convex.

STATUS MW- OBL | NCNE-OBL | GP OBL

Lemna perpusilla Torr.
MINUTE DUCKWEED

DESCRIPTION Small native perennial floating herb; roots single from underside of frond or absent. Fronds single or several together, obovate to elliptic, asymmetrical at base, 2–3 mm long, upper surface convex or keeled, with small bumps (papilla) near center and tip of frond. Fruit a 1-seeded utricle.

HABITAT quiet water of ponds, ditches.

STATUS MW-OBL | NCNE-OBL | GP-OBL

Lemna trisulca L. ◆ STAR-DUCKWEED

DESCRIPTION Small native perennial floating herb, forming tangled

Lemna trisulca

colonies just below water surface, floating at surface only when flowering; roots single from underside of frond or absent. Fronds several to many, joined to form star-shaped colonies; fronds oblong lance-shaped, 5–20 mm long, tapered to a slender base (stipe), flat on both sides. Fruit a 1-seeded utricle. HABITAT ponds, streams, ditches. STATUS MW-OBL | NCNE-OBL | GP-OBL

Lemna turionifera Landolt
TURION DUCKWEED

DESCRIPTION Small native perennial floating herb. Fronds single or in groups of several, obovate, usually flat and not humped, 1–4 mm long and 1–1.5 times longer than wide, veins 3, small white dots (papillae) present on midline of upper surface (visible with naked eye but clearer with 10x hand lens); underside of frond usually red or purple and redder than upper side point of root, upper surface (especially near tip) sometimes red-spotted. Turions sometimes present, dark-green to brown, 1-1.6 mm wide, without roots, sinking to bottom and forming new plants. HABITAT quiet water of ponds and lakes. STATUS MW-OBL | NCNE-OBL | GP-OBL

Peltandra ARROW-ARUM

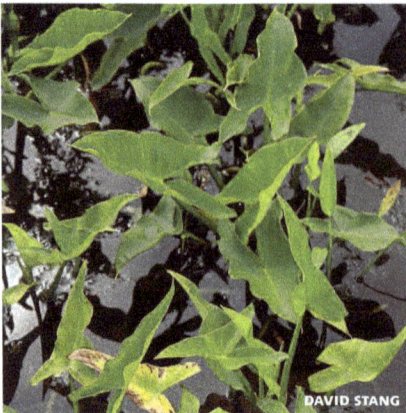

Peltandra virginica (L.) Schott & Endl.
♦ARROW-ARUM, TUCKAHOE

DESCRIPTION Native perennial herb, with thick fibrous roots. Leaves all from base of plant on long petioles, bright green, oblong to triangular in outline, 1–3 dm long and 8–15 cm wide at flowering, to 8 dm long later; leaf base with pair of lobes. Flowers in white to orange spadix about as long as the spathe, atop a curved stalk 2–4 dm long; flowers either staminate or pistillate, the staminate flowers covering upper 3/4 of the spadix, the pistillate flowers on lower portion; spathe green with a pale margin, 1–2 dm long, the lower portion covering the fruit. Fruit a head of green-brown berries, the berries with 1–3 seeds surrounded by a jellylike material. Flowering June–July.
SYNONYM *Peltandra luteospadix*.
HABITAT shallow water, shores, bog pools; often where shaded.
STATUS MW-OBL | NCNE-OBL | GP-OBL

Pistia WATER-LETTUCE

Pistia stratiotes L. ♦WATER-LETTUCE

DESCRIPTION Introduced, free-floating aquatic perennial herb. Leaves light-green, densely hairy, in rosettes to 15 cm wide. Flowers in a short inflorescence to 1.5 cm long; unisexual, and enclosed in a leaflike spathe.
HABITAT lakes, ponds, rivers.
STATUS MW-OBL | NCNE-OBL | GP-OBL
NOTE likely native of South America, and while not typically overwintering in our region, used in water-gardens and perhaps dumped into native habitats; capable of rapidly spreading.

Peltandra virginica

Pistia stratiotes

Spirodela
GREATER DUCKWEED

Spirodela polyrhiza (L.) Schleiden
⬧ GREATER DUCKWEED

DESCRIPTION Native perennial herb, floating on water surface; roots 5–12 per frond. Fronds usually in clusters of 2–5, flat, round to obovate, 3–6 mm long, upper surface green, underside red-purple. Flowers uncommon, comprised of 2–3 stamens (staminate flowers) and 1 pistil (pistillate flower) in each pouch. Fruit a 1–2-seeded utricle. Reproduction mainly by budding of new leaves from reproductive pouches (1 pouch on each margin of frond).

HABITAT stagnant or slow-moving water of lakes, ponds, marshes and ditches, often with common duckweed (*Lemna minor*).

STATUS MW-OBL | NCNE-OBL | GP-OBL

Symplocarpus
SKUNK-CABBAGE

Symplocarpus foetidus (L.) Nutt.
⬧ SKUNK-CABBAGE

DESCRIPTION Native perennial foul-smelling herb, from thick rootstocks. Leaves all from base of plant, ovate to heart-shaped, 3–8 dm long and to 3 dm wide, strongly nerved; petioles short, channeled. Flowers appearing before leaves in late winter or early spring, perfect; the spathe ovate, curved over spadix, 8–15 cm long, green-purple and often mottled; sepals 4. Fruit round, 8–12 cm wide; seeds 1 cm thick. Feb–May.

HABITAT floodplain forests, swamps, streambanks, calcareous fens, moist wooded slopes.

STATUS MW-OBL | NCNE-OBL | GP-OBL

NOTE Skunk-cabbage is the region's earliest flowering plant, the flowers often appearing while still partially snow-covered.

CAUTION leaves will burn the mouth if tasted.

Wolffia WATERMEAL

Tiny perennial herbs, without roots, floating at or just below water surface, sometimes abundant and forming a granular scum across surface, usually mixed with other duckweeds. Fronds single or often paired, globe-shaped or ovate, flat or rounded on upper surface. Flowers uncommon, consisting of 1 stamen (staminate flower) and 1 pistil (pistillate flower) in the pouch. Fruit a round, 1-seeded utricle. Reproduction mainly by budding from the single pouch near base of frond.

NOTE Water-meals are the world's smallest flowering plants. The fronds feel granular or mealy and tend to stick to skin. The species in the Upper Midwest region often occur together, usually in stagnant and/or polluted water.

Wolffia WATERMEAL

1 Leaves rounded on upper surface, not brown-dotted . . . . *W. columbiana*

1 Leaves flattened on upper surface, brown-dotted (under 10x) **2**

Spirodela polyrhiza

CHRISTIAN FISCHER

Symplocarpus foetidus

CEPHAS

2 Leaves rounded at tip, with a wartlike bump in center of upper surface .... *W. brasiliensis*

2 Leaves with an upturned point at tip, wartlike bump absent........ *W. borealis*

Wolffia borealis Griseb.
NORTHERN WATERMEAL

DESCRIPTION Small, native perennial floating herb. Fronds oval to oblong when viewed from above, 0.1–1 mm long and to 0.5 mm wide, with a raised pointed tip, usually brown-dotted; upper surface floating just above water surface, bright green, underside paler.
SYNONYM *Wolffia punctata.*
HABITAT quiet water of ponds, marshes and ditches, often with other species of *Wolffia* and *Lemna.*
STATUS MW-OBL | NCNE-OBL | GP-OBL

Wolffia brasiliensis Weddell
BRAZILIAN WATERMEAL

DESCRIPTION Small, native perennial floating herb. Fronds broadly ovate, 0.5–1.5 mm long and 0.5–1 mm wide, rounded at tip, brown-dotted; upper surface floating just above water surface, with a wartlike bump near center.
SYNONYM *Wolffia papulifera.*

HABITAT quiet water of ponds, often occurring with other species of Wolffia.
STATUS MW-OBL | NCNE-OBL | GP-OBL; Mich (THR).

Wolffia columbiana Karsten
◊COLUMBIAN WATERMEAL

DESCRIPTION Small, native perennial floating herb. Fronds float low in water, only small upper surface exposed, round to broadly ovate and 1–1.5 mm long when viewed from above; nearly round when viewed from side, not raised and pointed at tip; green, not brown-dotted.
HABITAT stagnant water of ponds and marshes.
STATUS MW-OBL | NCNE-OBL | GP-OBL

Butomaceae
Flowering-Rush Family

Butomus FLOWERING RUSH
Butomus umbellatus L.
◊FLOWERING-RUSH

DESCRIPTION Introduced perennial herb, from creeping rhizomes and forming colonies. Leaves all from base of plant, erect when emersed, or floating when in deep water, linear, to 1 m long and 5–10 mm wide, parallel-veined; petioles absent.

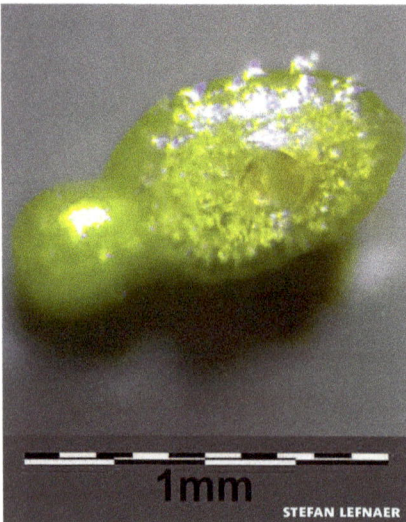
Wolffia columbiana

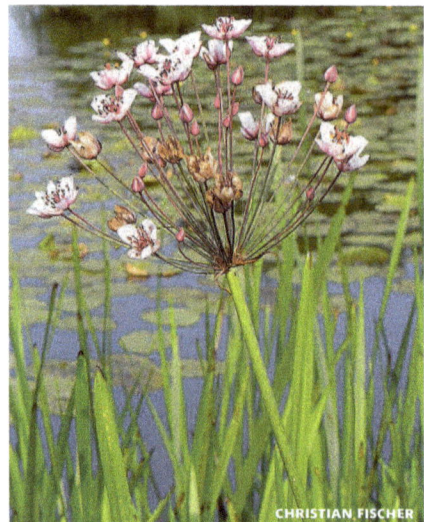
Butomus umbellatus

Flowers pink, perfect, 2–3 cm wide, stalked, in a many-flowered umbel, borne on a round stalk 1–1.2 m tall; with 4 lance-shaped bracts subtending the umbel; sepals 3, petal-like; petals 3; stamens 9; pistils 6. Fruit a dry, many-seeded capsule, splitting open on inner side. Flowering June–Aug.

HABITAT marshes and lakeshores.

STATUS MW-OBL | NCNE-OBL | GP-OBL, introduced from Eurasia, and now established (and often invasive) across much of northern USA and s Canada.

Cyperaceae
Sedge Family

MOSTLY PERENNIAL, grasslike, rushlike or reedlike plants. Stems 3-angled or ± round in section, solid or pithy. Leaves 3-ranked or reduced to sheaths at base of stem; leaf blades, when present, grasslike, parallel-veined, often keeled; sheaths mostly closed around the stem. Flowers small, perfect (with both staminate and pistillate parts), or single-sexed, each flower subtended by a bract (scale); perianth of 1 to many (often 6) small bristles, or a single perianth scale, or absent; stamens usually 3; ovary 2-3-chambered, contained in a saclike covering (perigynium) in *Carex*, maturing into an achene, stigmas 3 or 2. Flowers arranged in spikelets (termed spikes in *Carex*), the spikelets single as a terminal or lateral spike, or several to many in various types of heads, the head often subtended by 1 to several bracts.

Cyperaceae SEDGE FAMILY

1 Achenes enclosed in a closed sac (perigynium) subtended by a scale, the style protruding through the apex; flowers strictly unisexual (sedges with exclusively staminate flowers should be keyed here) .

. . . . *Carex*

1 Achenes not enclosed in a closed sac, naked beside the subtending scale; at least some flowers bisexual (except in *Scleria*) **2**

2 Achenes white, hard (bone-like), ± spherical; flowers all unisexual ... *Scleria*

2 Achenes yellow, brown, or black, rarely whitish, not spherical; at least some flowers bisexual. **3**

3 Scales of spikelets 2-ranked; spikelets ± flattened in cross-section and always more than one per inflorescence **4**

3 Scales of spikelets spirally arranged (or if 2-ranked, the spikelet solitary); spikelets round or several-angled in cross-section, solitary or several to many per inflorescence **5**

4 Stems usually ± angled, solid; inflorescences terminal; achenes without subtending bristles . . . . *Cyperus*

4 Stems round, hollow; inflorescences in the axils of stem leaves; achenes with subtending bristles. . . . .
. . . . *Dulichium arundinaceum*

5 Perianth bristles 6, 3 slender and 3 with an expanded, ± spongy, spoon-like portion at the tip. . . . . *Fuirena*

5 Perianth bristles absent or 1 to many, all slender. **6**

6 Spikelet or cluster of spikelets borne on one side of the stem at the base of a single ± erect to somewhat angled or curved involucral bract that appears to be a continuation of the stem **7**

6 Spikelet or spikelets terminating the stem or borne both terminally and laterally; if more than one spikelet, the inflorescence with 1 to several spreading to reflexed, leaflike involucral bracts **9**

7 Stems less than 0.5 mm thick; plants tiny, less than 10 cm tall . . . . *Lipocarpha*

7 Stems thicker than 0.5 mm; plants usually much taller than 10 cm. **8**

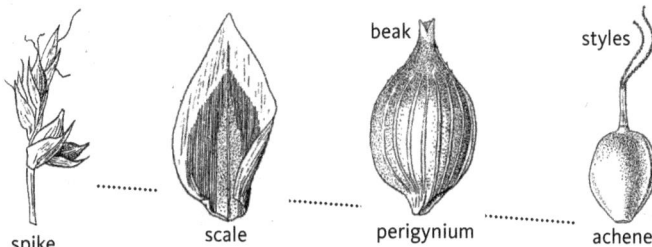

CAREX FLORAL PARTS (*Carex chordorrhiza* shown)

spike   scale   beak · perigynium   styles · achene

8 Plants colonial from elongate rhizomes, perennial; usually more than 0.8 m tall (except in the submerged aquatic *Schoenoplectus subterminalis*); anthers 1–3.5 mm long............. *Schoenoplectus*

8 Plants tufted, annual, mostly less than 0.8 m tall; anthers 0.3–0.9 mm long .... *Schoenoplectiella*

9 Spikelet solitary and terminal on the stem (very rarely a few smaller accessory spikelets occur at the base of the terminal spikelet in the bladeless genus *Eleocharis*) **10**

9 Spikelets several to many on the stem, terminal or lateral **13**

10 Sheaths totally bladeless or at most with an apical tooth up to 1 mm long; achenes usually with an apical tubercle formed by the expanded and persistent base of the style........................ *Eleocharis*

10 Upper sheaths with short green blades 0.3–12 cm long; achenes blunt at apex, tubercle absent **11**

11 Achenes subtended by 1–8 bristles less than twice as long as the achenes, or bristles absent .... *Trichophorum*

11 Achenes subtended by conspicuous silky, white or tawny, hair-like bristles many times as long as the achenes.......... **12**

12 Bristles numerous, (12–) 15–50 or more; rhizomes erect, very short .. *Eriophorum*

12 Bristles 6; rhizomes horizontal and short-creeping .... *Trichophorum alpinum*

13 Achenes subtended by (12–) 15–50 conspicuous, silky, white or tawny, hair-like bristles many times as long as the achenes .... *Eriophorum*

13 Achenes subtended by 1–8 bristles, or bristles absent....................... **14**

14 Leaves flat or folded; with a definite, ± keeled midrib **15**

14 Leaves inrolled and wiry; rounded on the back and without a definite midrib **18**

15 Achenes with a conspicuous tubercle formed by the expanded, persistent style base.................... *Rhynchospora*

15 Achenes blunt at apex, without a tubercle; style base, if expanded, not persistent to maturity **16**

16 Widest leaves 0.5–3 mm wide; achenes lacking bristles .... *Fimbristylis*

16 Widest leaves 4–15 mm wide; achenes subtended by 1–8 bristles.............. **17**

17 Spikelets (10–) 15–36 mm long; achenes 3–5 mm long, including apiculus; anthers 4–5 mm long; stems sharply 3-angled nearly or quite to the base; colonial from rhizomes with large corm-like thickenings .... *Bolboschoenus*

17 Spikelets 2–10 (–12) mm long, achenes 0.9–1.2 mm long; anthers 0.5–1.3 mm long; stems terete, obtusely 3-angled, or sharply 3-angled only toward summit; tufted or with rhizomes lacking corm-like enlargements .... *Scirpus*

18 Styles 2-cleft; achenes subtended by slender bristles .... *Rhynchospora*

18 Styles 3-cleft; achenes lacking bristles.. **19**

19 Achenes 2.2–3.5 mm long; perennials 4–11 dm tall; colonial from elongated rhizomes .... *Cladium mariscoides*

19 Achenes less than 1 mm long; tufted annuals 0.2–4 dm tall .... *Bulbostylis*

Bolboschoenus CLUB-RUSH

Former members of genus *Scirpus*, and sometimes included in *Schoenoplectus*.

Bolboschoenus **CLUB-RUSH**

1 Styles 3-parted; achenes 3-angled; perianth bristles strong, distinctly retrorse-barbed; leaf sheaths convex at tip ... *B. fluviatilis*

1 Styles 2-parted; achenes lens-shaped; perianth bristles weak, scarcely half as long as achene, obscurely barbed; leaf sheaths straight or concave at tip. .. *B. maritimus*

Bolboschoenus fluviatilis (Torr.) Soják
♦RIVER CLUB-RUSH

DESCRIPTION Native perennial, spreading by rhizomes and often forming large colonies. **Stems** stout, erect, 6–15 dm long, sharply 3-angled, the sides ± flat. **Leaves** several on stem, smooth, 6–15 mm wide, upper leaves often longer than the head; bracts 3–5, leaflike, erect to spreading, to 3–4 dm long. **Spikelets** 1–3 cm long and 6–12 mm wide, clustered in an umbel with 10–20 spikelets at end of stem, several of the spikelets nearly stalkless in 1–2 clusters, others single or in groups of 2–5 at ends of spreading or drooping stalks to 8 cm long; scales gold-brown, short-hairy on back, 6–10 mm long, the midvein extended into a curved awn 1–3 mm long; bristles 6, unequal, white to copper-brown, downwardly barbed, persistent, about as long as body of achene;

style yellow, 3-parted. Achenes 3-angled, dull, tan to gray-green, 3-5 mm long, with a beak to 0.5 mm long. Flowering June–Aug.

SYNONYMS *Schoenoplectus fluviatilis, Scirpus fluviatilis.*

HABITAT usually in shallow water of streams, ditches, marshes, lakes and ponds; sometimes where brackish.

STATUS MW-OBL | NCNE-OBL | GP-OBL

NOTE the sharply triangular leafy stems are distinctive.

Bolboshoenus maritimus (L.) Palla
PRAIRIE BULRUSH

DESCRIPTION Native perennial, from tuber-bearing rhizomes. Stems single, sharply 3-angled, 5-15 dm long.

Leaves several on stem, smooth, to 1 cm wide; bracts 3-5, the longest bract sometimes erect, to 30 cm long. Spikelets cylindric, 10-25 mm long and 6-9 mm wide, in clusters of 3-20 in a dense, stalkless head, or some spikelets single or in groups of 2-4 on stalks to 4 cm long; scales ovate, notched at tip, pale brown, 5-7 mm long, with an awn to 2 mm long; bristles 2-6, coppery, about half as long as achene; style 2-parted. Achenes lens-shaped, brown to black, 3-4 mm long, the beak small, to 0.3 mm long. Flowering June–Aug.

SYNONYMS *Schoenoplectus maritimus, Schoenoplectus pungens, Scirpus maritimus, Scirpus paludosus.*

HABITAT marshes, shores, ditches; especially where brackish; considered adventive from western USA.

STATUS MW-OBL | NCNE-OBL | GP-OBL.

NOTE plants more slender than in *B. fluviatilis*, the inflorescence more congested and the scales paler brown.

Bulbostylis HAIR SEDGE
Bulbostylis capillaris (L.) Kunth
⧫HAIR SEDGE

DESCRIPTION Small tufted native annual. Stems threadlike, to 3 dm tall. Leaves short, filiform, involute, mostly at or near the base. Flowers in capitate or umbel-like cymes; longest bract usually exceeding the spikelets; peduncles none or to 1 cm long.

Spikelets few-flowered, ovoid, 3-6 mm long; scales boat-shaped, ovate, purple-brown with strong green midvein, minutely pubescent and ciliate.

Achenes trigonous, to 1 mm long, straw-colored, obscurely rugose, capped with a tiny tubercle.

HABITAT wet, sandy or muddy soil.

STATUS MW FACU | NCNE FACU | GP FACU

Carex SEDGE
Perennial grasslike plants, and the largest genus of wetland plants in the Upper Midwest region. Stems mostly 3-angled. Leaves 3-ranked, margins often finely toothed. Flowers either staminate or pistillate, with both sexes in same spike, or in separate spikes on same plant, or the staminate and pistillate flowers on different plants. Staminate flowers with 3 or rarely 2 stamens; pistillate flowers with style divided into 2 or 3 stigmas. Achenes lens-

Bolboschoenus fluviatilis

Bulbostylis capillaris

shaped or flat on 1 side and convex on other (in species with 2 stigmas), or achenes 3-angled or nearly round (in species with 3 stigmas), enclosed in a sac called the perigynium (singular) or perigynia (plural).

NOTE As evident in the following keys, identification of *Carex* species is often based on characteristics of the *perigynium*—the sac around the achene, and the pistillate scale subtending each perigynium; *perigynia* (plural) are illustrated for nearly all species. For small plants, a hand lens or dissecting scope may be needed to see these diagnostic features. The Sections and species descriptions are arranged alphabetically.

Carex **Sections**

1 Spike solitary, terminal (entirely staminate, entirely pistillate, or mixed) 2

1 Spikes 2 or more, sometimes crowded but distinguishable by the lobed appearance of inflorescence or protruding bracts or visible short segments of rachis between spikes. 8

2 Styles 2-cleft; achenes 2-sided (lenticular); basal sheaths brown 3

2 Styles 3-cleft; achenes 3-sided (or nearly terete); basal sheaths brown or purple-red 4

3 Plants with slender rhizomes; perigynia obscurely or not at all serrate, plump (usually at least as convex on upper face as on the lower), the lowermost tending to be remote (as much as 1 mm apart at points of attachment); spikes without empty basal scales; anthers to 2.5 (-3) mm long . . . . *Carex* sect. *Physoglochin*

3 Plants densely tufted, not rhizomatous; perigynia minutely but strongly and regularly serrate on upper portion and beak, ± flattened, crowded; spikes usually with 1-2 empty basal scales; anthers 2-3.5 mm long *Carex* sect. *Stellulatae* (*C. exilis*)

4 Spikes unisexual (either staminate or pistillate); perigynia pubescent . . . . *Carex* sect. *Acrocystis*

4 Spikes containing both staminate and pistillate flowers; perigynia usually glabrous 5

5 Perigynia minutely pubescent . . . . *Carex* sect. *Acrocystis*

5 Perigynia glabrous. 6

6 Spikes staminate at base, pistillate toward tip, densely flowered, mostly 1 cm or more thick; perigynia inflated (much larger than the included achene), abruptly contracted to a long, very slender beak . . . . *Carex* sect. *Squarrosae*

6 Spikes pistillate at base, staminate above, more slender and sparsely flowered (fewer than 10 perigynia); perigynia various (but neither inflated nor with a long slender beak) 7

7 Perigynia slender, linear-lanceolate (more than 5 times as long as thick), strongly reflexed at maturity . . . . *Carex* sect. *Leucoglochin*

7 Perigynia broad, less than 5 times as long as thick, appressed-ascending . . . . *Carex* sect. *Leptocephalae*

8 All spikes staminate 9

8 At least some spikes bisexual or pistillate 11

9 Plants with long-creeping rhizomes . . . . *Carex* sect. *Divisae* (*C. praegracilis*)

9 Plants densely tufted 10

10 Leaves flat, lax and spreading; usually in swamps and marshes . . . . *Carex* sect. *Deweyanae* (*C. bromoides*)

10 Leaves channeled, stiff and erect; fens and other calcareous open wetlands. . . . . *Carex* sect. *Stellulatae*

11 Styles 2-cleft; achenes 2-sided. 12

11 Styles 3-cleft; achenes 3-sided (or nearly terete) 30

12 Lateral spikes peduncled, or if sessile, then elongate; terminal spike often entirely staminate 13

12 Lateral spikes sessile, short, often crowded; terminal spike at least partly pistillate (rarely staminate) 15

13 Plants coarse, the stems over (3-) 5 dm tall and usually over 1 mm thick, at least toward base; staminate spikes often 2 or more, mostly 2.5-7 cm long; lowermost bract essentially sheathless (rarely with very short sheath); perigynia neither white-pulverulent nor golden-yellow . . . . *Carex* sect. *Phacocystis*

13 Plants slender, the stems to 3 dm tall and less than 1 mm thick (excluding leaf bases) even near the base; terminal (staminate or sometimes mixed) spike solitary, ca. 1 cm long; lowermost bract usually with a short sheath 2-7 mm long; perigynia white-pulverulent or golden-yellow at maturity 14

14 Lowermost pistillate spike sessile or nearly so (except rarely one arising from near base of plant); terminal spike staminate; perigynia green or slightly glaucous, crowded........*Carex* sect. *Phacocystis*

14 Lowermost pistillate spike nearly always peduncled; terminal spike often pistillate near apex, or the pistillate spikes ± loosely flowered; fresh perigynia white-pulverulent or golden-yellow............ ....*Carex* sect. *Bicolores*

15 Stems arising mostly singly from rhizome or stolon 16

15 Stems tufted, the tufts with or without connecting rhizomes 20

16 Perigynia plumply plano-convex to nearly terete in cross-section, not winged or sharply margined; plants of sphagnum bogs, cedar swamps, etc..............17

16 Perigynia strongly flattened, with distinctly winged or sharply edged margins; plants mostly of wet or dry open habitats 18

17 Scales pale-hyaline with green midrib; perigynia apiculate or with very small beak; at least the lower few-flowered spikes ± separated; plants clumped from short, slender rhizomes .. *Carex* sect. *Glareosae*

17 Scales rich brown; perigynia with distinct beak ca. 0.5 mm long; spikes crowded as if in a single head; stems arising from axils of old decumbent stems (stolons) .... *Carex* sect. *Chordorrhizae*

18 Perigynia mostly over 2 mm wide; staminate flowers only at the base of some or all spikes...........*Carex* sect. *Ovales*

18 Perigynia mostly not over 2 mm wide; staminate flowers not restricted to base of spikes.............................19

19 Sheaths of upper leaves green-nerved ventrally, usually not covering the inconspicuous nodes.................... .... *Carex* sect. *Holarrhenae*

19 Sheaths of upper leaves with broad white-hyaline stripe on ventral side covering the included nodes....... *Carex* sect. *Divisae*

20 Staminate flowers at the base of some or all spikes, not at the apex (note especially the terminal spike) 21

20 Staminate flowers at the apex of some or all spikes (even when anthers have fallen, protruding filaments are usually visible).. 26

21 Perigynia with thin-winged margins, at least narrowly so along apical part of body and basal part of beak, strongly flattened and scale-like (in some species elongate), ± appressed and overlapping (or in some species spreading at the tips) 22

21 Perigynia at most with a ridge along the margin, not winged, the achene plumply filling at least the apical part of the body all the way to the margins 23

22 Bracts not resembling the leaves, narrower than 2 mm most or all their length and not over twice as long as the inflorescence; perigynia various...... *Carex* sect. *Ovales*

22 Bracts leaflike, the broadest 2–4 mm wide, many times exceeding the spikes (which are crowded in a dense head); perigynia very narrowly lanceolate, not over 1 mm wide .... *Carex* sect. *Cyperoideae*

23 Body of perigynium elliptic or nearly so (except in *C. arcta*) with at most a very short beak, and with rounded or slightly margined edges, nearly or entirely filled by the achene .... *Carex* sect. *Glareosae*

23 Body of perigynium ovate or lanceolate or prominently beaked, sharp-edged, only 1/2 to 2/3 filled by achene (very spongy around and below base of achene) 24

24 Mature perigynia loosely to strongly appressed-ascending, 4–5.7 mm long; anthers 1.3–2.6 mm long................. .... *Carex* sect. *Deweyanae*

24 Mature perigynia strongly spreading to reflexed, 2–3.6 mm long; anthers 0.8–2 mm long 25

25 Spikes 7–15, usually crowded, except sometimes the lowest, the inflorescence axis mostly concealed; beaks not bidentate .... *Carex* sect. *Glareosae* (*C. arcta*)

25 Spikes 3–8, not usually crowded, inflorescence axis clearly visible; beaks clearly bidentate with teeth 0.1–0.4 mm long .... *Carex* sect. *Stellulatae*

26 Stems stout (often 1.5 mm thick at ca. 3 cm below inflorescence) and very sharply angled (or even narrowly winged), ± soft and easily compressed (flattened in pressing); wider leaves 5–10 mm broad, with rather loose sheaths; perigynia spongy-thickened basally, on short slender stalks; anthers 1.3–2.6 mm long ....*Carex* sect. *Vulpinae*

26 Stems slender (not over 1.5 mm thick at ca. 3 cm below inflorescence, or rarely so in some species), firm, not wing-angled nor easily compressed (hence, not flattened in pressing); leaves, perigynia, and anthers various 27

27 Spikes 10 or fewer, usually greenish at maturity, crowded or remote in a simple inflorescence (one spike, no branches, at each node of it) **28**

27 Spikes numerous (10–many), yellowish or brownish at maturity; inflorescence tending to be compound, at least its lower nodes with 2 or more spikes crowded on a lateral branch **29**

28 Perigynia elliptic, essentially beakless, very plump (nearly terete) and filled by the achene; at least the lower spikes well separated, containing 1–5 perigynia . . . . *Carex* sect. *Dispermae*

28 Perigynia ± ovate, beaked, plano-convex or lenticula . . . . *Carex* sect. *Stellulatae*

29 Pistillate scales terminating in a distinct rough awn; bracts, at least lower ones, very slender and exceeding spikes or branches; ventral surface of leaf sheaths usually transversely wrinkled or puckered (very rarely smooth) . . . *Carex* sect. *Multiflorae*

29 Pistillate scales acute or minutely cuspidate; bracts mostly short, inconspicuous, or absent; leaf sheaths smooth ventrally. . . . . *Carex* sect. *Heleoglochin*

30 Perigynia at least sparsely puberulent, pubescent, hispidulous, or scabrous . . . **31**

30 Perigynia glabrous (in some species, papillose or granular, but not even sparsely puberulent or scabrous). **36**

31 Perigynia 12–18 mm long, in 1–2 short-oblong to spherical spikes 2–3.5 cm wide. . . . . *Carex* sect. *Lupulinae*

31 Perigynia 2–11 mm long, in 2–5 ± elongate, cylindrical spikes less than 2 cm wide . . **32**

32 Leaves hairy. . . . . *Carex* sect. *Carex*

32 Leaves glabrous (often rough or scabrous, but not hairy) **33**

33 Pistillate spikes not over 10 mm long; achenes mostly with very convex or rounded sides (the angles thus obscured), at least apically, very tightly enveloped by the perigynium, especially on the apical half; anthers 1.5–3.7 mm long . . . . *Carex* sect. *Acrocystis* (*C. deflexa*)

33 Pistillate spikes mostly over 10 mm long; achenes with flattish to slightly concave sides (the angles thus ± evident), the summit (especially around base of style) ± loosely enveloped by the perigynium; anthers 2.5–4.7 mm long; plants of dry to wet habitats. **34**

34 Perigynium beak usually more than half as long as the body, the apex not or weakly and obscurely toothed; perigynia scabrous or with short stiff ascending hairs . . . . *Carex* sect. *Anomalae*

34 Perigynium beak less than half as long as the body, with two firm apical teeth; perigynia ± densely short-hairy. **35**

35 Perigynia 6–11 mm long, beak teeth 1.2–2.3 mm long, inner band of upper sheaths strongly purple-red tinged and thickened at apex, the thickened reddish portion opaque, smooth . . . . *Carex* sect. *Carex*

35 Perigynia 2.5–6.5 mm long, beak teeth 0.2–0.8 mm long, inner band of upper sheaths whitish to brown, brown- or purple-dotted, but not uniformly colored, not strongly opaque-thickened at apex, often scabrous . . . . *Carex* sect. *Paludosae*

36 Leaf sheaths (at least at apex) finely pubescent; blades often also pubescent or at least strongly hispidulous, especially toward base of plant **37**

36 Leaf sheaths and blades completely glabrous (though sometimes scabrous) **38**

37 Beak of perigynium with firm teeth ca. 1.5–3 mm long; perigynia ca. 8–10 mm long, in spikes 4–12 cm long . . . . *Carex* sect. *Carex* (*C. atherodes*)

37 Beak of perigynium with teeth scarcely 0.5 mm long or absent; perigynia less than 6 mm long, in spikes less than 3 cm long . . . . *Carex* sect. *Hymenochlaenae*

38 Perigynia ± rounded to broadly tapered at summit, beakless or essentially so (the tiny beak or apiculus less than 0.5 mm long if distinct, or up to 0.8 mm long if vaguely defined, often strongly bent or curved); beak or apiculus (if present) never toothed (or teeth scarcely 0.1 mm long) **39**

38 Perigynium abruptly contracted or more gradually tapering to a definite slender beak 0.5 mm or more long, or to an indistinct tapering beak 1 mm or more long; beak in some species with short apical teeth. **47**

39 Bract of lowest pistillate spike sheathless (at most with a thin scarious sheath 1–3 mm long). **40**

39 Bract of lowest pistillate spike with a sheath 4 mm or more long **43**

40 Terminal spike partly pistillate; pistillate spikes nearly or quite sessile and erect or ascending; roots glabrous or nearly so. *Carex* sect. *Racemosae*

40 Terminal spike normally entirely

staminate; spikes and roots various 41

41 Pistillate spikes mostly drooping at maturity on slender peduncles; species of wet peat lands with roots with dense felt-like pubescence . . . . . *Carex* sect. *Limosae*

41 Pistillate spikes erect or ascending, sessile or peduncled; roots glabrous 42

42 Perigynia 2.5–3.4 mm wide; leaves involute, 0.5–2.5 mm wide . . . . *Carex* sect. *Vesicariae* (*C. oligosperma*)

42 Perigynia 1–2.5 mm wide; leaves flat or folded, 1–35 mm wide . . . . go to couplet 56

43 Terminal spike bearing some perigynia (very rarely a few individuals with one entirely staminate); plants very strongly reddish tinged at base . . . . *Carex* sect. *Hymenochlaenae*

43 Terminal spike entirely staminate; plants reddish or not at base 44

44 Perigynia concave- or at least cuneate-tapering toward the base, ± 3-angled and often somewhat broadly spindle-shaped . . . . *Carex* sect. *Paniceae*

44 Perigynia convex-rounded toward the base, nearly or quite circular in cross-section (or very obscurely triangular), ellipsoid-cylindric to nearly spherical 45

45 Larger perigynia 4–5 mm long, the nerves not raised above the surface at maturity . . . . *Carex* sect. *Griseae*

45 Larger perigynia 2–3.5 mm long; nerves various 46

46 Perigynia with the nerves not raised above the surface, usually ± impressed; staminate spike usually long-peduncled; plants not strongly rhizomatous nor with any pistillate spikes on basal peduncles . . . . *Carex* sect. *Griseae* (*C. conoidea*)

46 Perigynia with the nerves slightly raised above the surface; staminate spike nearly or quite sessile or, if long-peduncled, the plants strongly rhizomatous and with basal pistillate spikes . . . *Carex* sect. *Granulares*

47 Body of perigynium obovoid or obconic, ± truncately contracted into a distinct long slender beak; terminal spike often mostly pistillate (staminate at base only) . . . . *Carex* sect. *Squarrosae*

47 Body of perigynium ovoid to lanceolate or ellipsoid, tapered or contracted into the beak; terminal spike usually staminate, at least apically 48

48 Perigynia in densely crowded spherical to very short-cylindric spikes, spreading and with the lowermost usually reflexed,

usually strongly few-ribbed; at least the uppermost pistillate spikes ± sessile and often crowded; the terminal spike (staminate or partly pistillate) often sessile or short-peduncled 49

48 Perigynia in elongate or long-peduncled spikes or both, all ascending, 2-ribbed or variously many-nerved; inflorescences various, but the upper spikes often not crowded and the terminal spike often long peduncled 50

49 Perigynia 2–6.2 mm long; basal sheaths brown . . . . *Carex* sect. *Ceratocystis*

49 Perigynia 11–18 mm long; basal sheaths reddish purple tinged . . . . *Carex* sect. *Lupulinae*

50 Bract of lowest pistillate spike sheathless; check several stems; rarely, a pistillate spike will be borne abnormally low on the stem and this spike may then have a sheath, which should be disregarded in keying . 51

50 Bract of lowest pistillate spike consistently with sheath 4 mm or more long 56

51 Pistillate scales subtending at least some of the perigynia terminated by a distinct slender scabrous awn; perigynia 3–9 mm long 52

51 Pistillate scales smooth-margined and awnless or very short-awned, or at most with a scabrous margin toward an acuminate (sometimes inrolled) apex (occasionally a long rough awn in species with perigynia more than 9 mm long); perigynia (4–) 4.5–18 mm long 53

52 Scales toward apex of pistillate spikes merely acuminate or with awns shorter than their bodies (the latter easily visible, about half as long as perigynia or longer); staminate spikes 2 or more; body of perigynium rather gradually tapered to a beak 1.5 mm long, including the short (not over 0.8 mm) teeth *Carex* sect. *Paludosae*

52 Scales toward apex of pistillate spikes ordinarily with awns (as on the other pistillate scales) nearly or fully as long as their bodies (the latter small and mostly hidden among the bases of the densely crowded perigynia); staminate spike solitary (or very rarely a second smaller one present); body of perigynium tapered or strongly contracted into a beak 1.2–3.5 mm long, including teeth up to 2.2 mm long . . . . *Carex* sect. *Vesicariae*

53 Basal sheaths pale brown; perigynia very narrowly lanceolate, 4–6.5 times as long as wide and not over 3 mm wide, many-

nerved, tapering to apex (not strongly contracted into a beak); staminate spike solitary (pistillate spikes may be staminate at apex) . . . . *Carex* sect. *Rostrales*

53 Basal sheaths reddish purple tinged, at least on the youngest shoots; perigynia lanceolate or broader, less than 4 times as long as wide, or more than 3 mm wide, or strongly contracted into a conspicuous beak (or all of these); staminate spikes solitary or 2 or more **54**

54 Perigynia not inflated, ± tightly enclosing achene, 1–1.6 mm wide . . . . *Carex* sect. *Hymenochlaenae* (*C. prasina*)

54 Perigynia strongly inflated, not tight around the achene, 2–8 mm wide **55**

55 Perigynia 4–12 mm long, 6–12 (–15)-nerved . . . . *Carex* sect. *Vesicariae*

55 Perigynia 12–17 (–18) mm long, 15–20-nerved . . . . *Carex* sect. *Lupulinae*

56 Perigynia (6–) 9–17 (–18) mm long; beak teeth usually conspicuous and stiff . . . . go to couplet **51**

56 Perigynia 2–6.5 (–9) mm long; beak teeth absent or weak and inconspicuous **57**

57 Perigynia with several to many conspicuous fine nerves on each side . . **58**

57 Perigynia with 2 (–3) main ribs, the sides otherwise nerveless or with much less prominent nerves **59**

58 Nerves of perigynia very numerous (20–65) and impressed, giving a longitudinally corrugated appearance; awns of pistillate scales rough or even ciliate . . . . *Carex* sect. *Griseae*

58 Nerves of perigynia several to many (5–40) and slightly raised; awns of pistillate scales absent, smooth, or rough . . . . *Carex* sect. *Hymenochlaenae*

59 Lowermost pistillate spikes erect or ascending at maturity . . . . *Carex* sect. *Paniceae* (*C. vaginata*)

59 Lowermost pistillate spikes drooping on long slender peduncles at maturity **60**

60 Pistillate spikes not over 15 mm long . . . . . *Carex* sect. *Chlorostachyae* (*C. capillaris*)

60 Pistillate spikes mostly 20 mm or more long . . . . *Carex* sect. *Hymenochlaenae*

CAREX SECTION **Acrocystis**

One wetland species in Upper Midwest region.

Carex deflexa Hornem.
NORTHERN SEDGE

DESCRIPTION Loosely clumped, native grass-like perennial. Stems 1–2 dm long, purple-tinged at base, shorter than the leaves. Leaves soft, 1–3 mm wide. Spikes either staminate or pistillate; staminate spike short, to 5 mm long; pistillate spikes on long, slender stalks near base of plant and also 2–4 spikes on stem near staminate spike; bract leaflike, to 2 cm long; pistillate scales ovate, shorter than perigynia. Perigynia green, oblong-ovate, 2–3 mm long, covered with short hairs, abruptly tapered to a small beak about 0.5 mm long. Achenes 3-angled; stigmas 3. Flowering June–Aug.

HABITAT moist woods and swamps.

STATUS MW-FACW | NCNE-FACW | GP-FACW

CAREX SECTION **Anomalae**

One wetland species in Upper Midwest region.

Carex scabrata Schwein.
♦ROUGH SEDGE

DESCRIPTION Colony-forming, native grass-like perennial; stems rough-to-touch. Stems loosely clustered, 4–9 dm long. Leaves 4–14 mm wide, lowest leaves not reduced to scales. Spikes either staminate or pistillate; staminate spike single, 2–4 cm long, short-stalked; pistillate spikes 3–6, cylindric, 2–4 cm long, upright, the lower on long stalks, the upper stalkless or short-stalked; bracts leaflike; pistillate scales lance-shaped, about as long as the perigynia, tapered to a tip. Perigynia obovate, 3-angled, 2-ribbed, 3–5 mm long, finely coarse-hairy, few-nerved, abruptly tapered to a slightly curved, notched beak. Achenes 3-angled; stigmas 3. Flowering May–Aug.

HABITAT low shaded areas in forests, streambanks, seeps.

STATUS MW-OBL | NCNE-OBL

CAREX SECTION **Bicolores**

Plants short, colonial, loosely tufted, shoots arising singly or few in a clump; rhizomes elongate; bases brown. Terminal spike staminate or gynecandrous, hidden by the crowded lateral spikes. Perigynia plump, golden to whitish, weakly veined; margins and apex rounded, beakless to short-beaked. Stigmas 2. Usually on calcium-rich sites where somewhat disturbed.

CAREX SECTION *Bicolores*

1 Mature perigynia golden-orange when fresh (drying dark brown or, especially if immature, ± white); terminal spikes mostly all staminate (occasionally with a very few perigynia); pistillate scales ± loosely spreading, distinctly shorter than the mature perigynia (usually averaging 3/4 or less as long), most of them acute to cuspidate . . . . *C. aurea*

1 Mature perigynia white-pulverulent when fresh; terminal spikes usually staminate at base only, with several to numerous perigynia apically; pistillate scales ± appressed, nearly (averaging about 3/4) to quite as long as the perigynia, most of them blunt to acute . . . . *C. garberi*

Carex aurea Nutt. *Bicolores*
◆ GOLDEN SEDGE

DESCRIPTION Small, loosely clumped native grasslike perennial. **Stems** upright, 3-angled, 5–30 cm long. **Leaves** 1–4 mm wide. **Spikes** 2–5 per stem, the lower spikes stalked; spikes at ends of stems staminate, 3–18 mm long; lateral spikes pistillate, 8–20 mm long; bract of lowest spike longer than the head; pistillate scales white-tinged to yellow-brown, with a green midvein, tipped with a short, sharp point, shorter than the perigynia. **Perigynia** with short white hairs when young, becoming a distinctive gold-orange when mature (drying paler), round to obovate, beakless or with a very short beak, several-ribbed, 2–3 mm long. **Achenes** dark brown to black, lens-shaped, 1–1.5 mm long; stigmas 2. **Flowering** May–July.

HABITAT moist to wet meadows, low prairie, swales, wet woods and along sandy or gravelly shores; often where calcium-rich.

STATUS MW-FACW | NCNE-FACW | GP-OBL

Carex garberi Fern. *Bicolores*
ELK SEDGE

DESCRIPTION Native grasslike perennial. Similar to *Carex aurea*; one distinction between the 2 species is terminal spike of *C. garberi* is tipped with pistillate flowers (with staminate flowers below); in *Carex aurea*, terminal spike is of staminate flowers only. Also, in *C. garberi*, the perigynia are more granular, more crowded, and more overlapping than in *C. aurea*.

HABITAT Wet sandy, gravelly, or marly shores, limestone pavements, interdunal flats, and edges of cedar thickets.

STATUS MW-FACW | NCNE-FACW | GP-FACW; rare in Minn (THR) and Wisc (THR).

Carex scabrata

Carex aurea

CAREX SECTION **Carex**

Plants typically colonial; rhizomes elongate. Vegetative stems prominent. Perigynia long-beaked with prominent beak teeth.

CAREX SECTION *Carex*

1 Perigynia covered with hairs............. ....*C. trichocarpa*
1 Perigynia smooth and hairless..........**2**
2 Inner band of the uppermost leaf sheaths red to purple and thickened at the summit, glabrous.................*C. trichocarpa*
2 Inner band of leaf sheaths pale or brown, not thickened at the summit, glabrous or pubescent............................**3**
3 Vegetative stems hollow, easily flattened; inner band of the leaf sheaths pubescent, rarely glabrous, not obviously veined; basal leaf sheaths ladder fibrillose *C. atherodes*
3 Vegetative stems solid; inner band of the leaf sheaths strongly veined, glabrous or the veins scabrous; upper and lower leaf sheaths ladder fibrillose .. *C. laeviconica*

Carex atherodes Sprengel *Carex*
♦ SLOUGH SEDGE

DESCRIPTION Loosely clumped, native grass-like perennial from long scale-covered rhizomes. **Stems** 3-angled, 5–12 dm long. **Leaves** 3–12 mm wide; sheaths hairy on back, brown to purple-tinged at the mouth, the lower sheaths shredding into narrow strands. **Spikes** either staminate or pistillate; staminate spikes 2–6 at ends of stems; pistillate spikes 2–4, widely spaced, cylindrical, 2–11 cm long; bracts leaflike, longer than the stems; pistillate scales thin, translucent or pale brown, shorter than the perigynia, tipped with a slender awn. **Perigynia** ovate, 6–11 mm long, long-tapered to a smooth beak, with many distinct nerves, the beak with spreading teeth 1.5–3 mm long. **Achenes** 3-angled, 2–2.5 mm long; stigmas 3. **Flowering** June–Aug.
HABITAT marshes, wet meadows, prairie swales, stream and pond margins, usually in shallow water where may form dense colonies.
STATUS MW-OBL | NCNE-OBL | GP-OBL

Carex laeviconica Dewey *Carex*
♦ LONG-TOOTHED LAKE SEDGE

DESCRIPTION Loosely clumped, native grasslike perennial, from scaly rhizomes. **Stems** stout, 3-angled, 3–12 dm long. **Leaves** shorter to longer than the stem, 2–8 mm wide; sheaths smooth, often purple-tinged below and splitting into fibers. **Spikes** either all staminate or pistillate, the upper 2–6 staminate, 1–4 cm long; the lower 2–4 spikes pistillate, erect, separate, stalkless or short-stalked, cylindric, 3–10 cm long and 6–10 mm wide; bracts leaflike, equal or longer than the head; pistillate

Carex atherodes

Carex laeviconica

scales acute or awn-tipped, the scale body shorter than the perigynium, translucent or brown on the sides. Perigynia green-yellow, broadly ovate, inflated, round in section, 4–9 mm long, strongly many-nerved, tapered to a slender beak 1.5–2 mm long.

HABITAT wet meadows, marshes, lakeshores and streambanks.

STATUS MW-OBL | NCNE-OBL | GP-OBL

Carex trichocarpa Muhl. *Carex*
HAIRY-FRUIT LAKE SEDGE

DESCRIPTION Loosely clumped, native grasslike perennial, with short rhizomes. Stems stout, 6–12 dm long, smooth below, rough-to-touch above. Leaves 2–6 mm wide, rough-to-touch on margins, upper leaves and bracts often longer than stems. Spikes either all staminate or pistillate, the upper 2–6 spikes staminate, long-stalked; pistillate spikes 2–4, cylindric, 4–10 cm long, the upper spikes ± stalkless, the lower spikes on slender stalks; pistillate scales ovate, with white translucent margins, about half as long as perigynia. Perigynia ovate, usually covered with short white hairs, prominently ribbed, gradually tapered to a 2-toothed beak. Achenes 3-angled; stigmas 3. Flowering May–Aug.

HABITAT riverbanks and old river channels, marshes, wet meadows, low prairie.

STATUS MW-OBL | NCNE-OBL | GP-OBL

NOTE similar to **slough sedge** (*Carex atherodes*) but sheaths strongly purple-tinged at tip, the leaf blades not hairy on underside, and the perigynia with short white hairs (vs. smooth in *Carex atherodes*).

CAREX SECTION **Ceratocystis**

Plants tufted; rhizomes short; bases brown. Terminal spike staminate, occasionally androgynous. Lateral spikes pistillate, densely flowered, globose to oblong. Perigynia strongly veined, abruptly beaked; beak toothed, generally reflexed. Stigmas 3. Usually where wet and calcareous.

CAREX SECTION *Ceratocystis*

1 Larger perigynia ca. 2–3 mm long, horizontally spreading, the beak about 1/4 to nearly 1/2 as long as the body . .*C. viridula*

1 Larger perigynia (3–6.2 mm long, at least the beaks becoming conspicuously reflexed on lower half of spike, the beak nearly or fully half as long as the body. . . . **2**

2 Pistillate scales at maturity strongly flushed with shiny brown or reddish color, hence conspicuous in the spike; widest leaves 3–5 mm wide . . . . *C. flava*

2 Pistillate scales greenish or yellowish, the same color as the perigynia and essentially invisible in the spikes; widest leaves 1.5–4 mm wide . . . . *C. cryptolepis*

Carex cryptolepis Mackenzie
Ceratocystis
♦ SMALL YELLOW SEDGE

DESCRIPTION Clumped, native grasslike perennial. Stems 2–6 dm long and longer than leaves. Leaves 2–4 mm wide. Spikes staminate or pistillate; staminate spikes short-stalked or stalkless, the stalk shorter than the pistillate spikes; pistillate spikes 3–4, the upper 2 spikes grouped, the third separate, the fourth spike lower on stem, short-cylindric, 1–2 cm long, stalkless; bracts leaflike and spreading; pistillate scales narrowly ovate, same color as perigynia and as long as perigynia body. Perigynia yellow-brown when mature, lower ones curved outward and downward, body obovate, 3–5 mm long, 2-ribbed and several nerved, contracted into a smooth beak 1–1.5 mm long. Achenes 3-angled; stigmas 3. Flowering June–Aug.

REUVEN MARTIN

Carex cryptolepis

HABITAT wet meadows and marshy areas, peatlands, swamp margins; often where calcium-rich.

STATUS MW-OBL | NCNE-OBL | GP-OBL

Carex flava L. *Ceratocystis*
♦LARGE YELLOW SEDGE

DESCRIPTION Densely clumped, native grasslike perennial from short rootstocks. Stems stiff, 1–7 dm high, usually longer than the leaves. Leaves 4–8 to a stem, mostly near base, 3–5 mm wide. Terminal spike staminate (or rarely partly pistillate), stalkless or short-stalked, linear, 0.5–2 cm long and 2–3 mm wide; pistillate spikes 2–5, sometimes with staminate flowers at tip, the uppermost spikes nearly stalkless, the lower stalked, short-oblong to nearly round, 6–18 mm long and 7–12 mm wide, perigynia 15–35, crowded in several to many rows, their beaks turned downward; bracts conspicuous, leaflike, spreading outward, much longer than the head; pistillate scales ovate, narrower and much shorter than the perigynia, red-tinged except for the pale, three-nerved middle and the narrow translucent margins. Perigynia 4–6 mm long and 1–2 mm wide, obovate, yellow-green becoming yellow with age, conspicuously ribbed, tapered to a slender, finely toothed beak about as long as the body, the tip notched. Achenes 1.5 mm long, obovate, 3-angled, yellow-brown; stigmas 3. Flowering May–Aug.

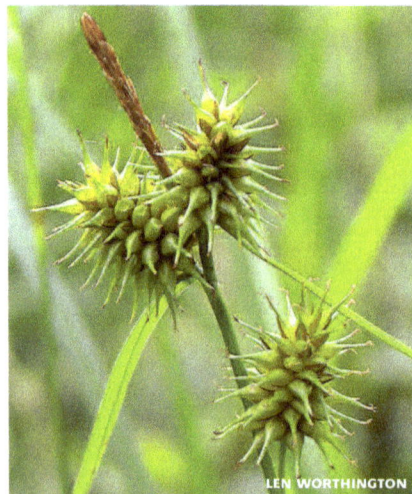

LEN WORTHINGTON

Carex flava

HABITAT wet, peaty meadows, often where calcium-rich.

STATUS MW-OBL | NCNE-OBL | GP-OBL

Carex viridula Michx. *Ceratocystis*
GREEN YELLOW SEDGE

DESCRIPTION Clumped, native grasslike perennial. Stems stiff, slightly 3-angled, 0.5–4 dm long, longer than leaves. Leaves 1–3 mm wide; sheaths white-translucent. Spikes either staminate or pistillate (or sometimes mixed), the terminal spike staminate or with a few pistillate flowers at tip or middle, 3–15 mm long, short-stalked or stalkless, longer than the pistillate spikes or clustered with them; lateral spikes pistillate, 2–6, ovate to short-cylindric, 5–10 mm long, clustered and stalkless above, the lower spikes often separate and on short stalks; bracts leaflike, usually upright, much longer than the heads; pistillate scales brown on sides, rounded or with a short, sharp point, about equal to perigynia. Perigynia yellow-green to brown, rounded 3-angled, obovate, 2–4 mm long, 2-ribbed, tapered to a slightly notched beak 0.5–1 mm long. Achenes 3-angled, to 1.5 mm long; stigmas 3. Flowering May–Aug.

HABITAT wet meadows, sandy lake margins, fens and seeps; often where calcium-rich.

STATUS MW-OBL | NCNE-OBL | GP-OBL

CAREX SECTION
Chlorostachyae

One wetland species in Upper Midwest region.

Carex capillaris L.
♦HAIRLIKE SEDGE

DESCRIPTION Small, densely clumped, native grasslike perennial. Stems slender, 3-angled, 1.5–4 dm long. Leaves mostly at base of plant and much shorter than stems, 1–3 mm wide; sheaths tight. Spikes either staminate or pistillate; terminal spike staminate, 4–8 mm long; lateral spikes 1–4, separated on stem, loosely flowered, short-cylindric, 5–15 mm long, on threadlike, spreading to drooping stalks 5–15 mm long; pistillate scales white, translucent on outer edges, green or light brown in

middle, blunt or acute at tip, shorter but usually wider than perigynia, deciduous. Perigynia shiny brown to olive-green, ovate, round in section, 2–4 mm long, 2-ribbed, otherwise without nerves, tapered to a translucent-tipped beak 0.5 mm or more long. Achenes 3-angled with concave sides, 1–1.5 mm long; stigmas 3. June–July.

HABITAT alder thickets, wetland margins, usually in shade.

STATUS MW-FACW | NCNE-FACW | GP-FACW

CAREX SECTION
Chordorrhizae
One wetland species in Upper Midwest region.

Carex chordorrhiza L. f.
◆ CORDROOT SEDGE

DESCRIPTION Native grasslike perennial, from long, creeping stems. Flowering stems upright, rounded 3-angled in section, 1–3 dm tall, single or several together, arising from axils of dried leaves on older, reclining sterile stems. Leaves several on stem, the lower ones often bladeless, 1–2 mm wide; sheaths translucent. Spikes 3–8, with both staminate and pistillate flowers, staminate flowers borne above pistillate, crowded in an ovate head 5–15 mm long; bracts absent; pistillate scales dark brown, ovate, about equaling the perigynia. Perigynia brown, compressed, ovate, 2–3.5 mm long, leathery, with many nerves on both sides; beak short. Achenes lens-shaped; stigmas 2. Flowering May–Aug.

HABITAT open floating mats around lakes and ponds, fens, conifer swamps, interdunal hollows.

Carex capillaris

STATUS MW-OBL | NCNE-OBL | GP-OBL

NOTE The creeping stems which root at each flowering stem are distinctive.

CAREX SECTION **Cyperoideae**
One wetland species in Upper Midwest region.

Carex sychnocephala Carey
MANY-HEAD SEDGE

DESCRIPTION Clumped, native grasslike perennial (sometimes annual), from fibrous roots. Stems many and crowded, rounded 3-angled, 0.5–6 dm long. Leaves 1.5–4 mm wide; sheaths tight, white-translucent. Spikes with both staminate and pistillate flowers, pistillate flowers borne above staminate, densely clustered in ovate heads 1.5–3 cm long; bracts leaflike, 2–4 per head, the longest bracts much longer than the heads; pistillate scales thin and translucent with a green midvein, 2/3 length of perigynia, tapered to a tip or with a short sharp point. Perigynia green to straw-colored, flat, lance-shaped, 5–7 mm

Carex chordorrhiza

long and to 1 mm wide, narrowly wing-margined, spongy at base when mature, tapered to a finely toothed, notched beak 3–5 mm long. Achenes lens-shaped, 1–1.5 mm long; stigmas 2. Flowering June–Aug.
HABITAT wet meadows, sandy lake and stream shores, marshes.
STATUS MW-FACW | NCNE-FACW | GP-FACW

CAREX SECTION Deweyanae

Two members of the section in Upper Midwest wetlands. Plants tufted; rhizomes mostly short; bases brown. Inflorescence slender, open, at least the lowest spike(s) distinct; bracts setaceous. Spikes mostly gynecandrous, lateral spikes sometimes pistillate, mixed, or (rarely) staminate. Perigynia appressed to ascending, ovate to lanceolate, plano-convex, slender; base spongy; beak distinct, margins serrate, tip bidentate. Achenes mostly filling the perigynium body. Usually in moist to wet shaded places.

CAREX SECTION *Dewyanae*

1 Perigynia to 1.2 mm wide and 4–5x as long as wide, usually strongly nerved on both sides . . . . *C. bromoides*
1 Perigynia 1.3–1.6 mm wide and 3–4x as long as wide, nerves faint or absent . . . . *C. deweyana*

Carex bromoides Willd. *Deweyanae*
BROME HUMMOCK SEDGE

DESCRIPTION Densely clumped, native grasslike perennial. Stems very slender, 3–8 dm long. Leaves 1–2 mm wide. Spikes 3–7, narrowly oblong, 1–2 cm long, terminal spike with both staminate and pistillate flowers, the staminate below pistillate; lateral spikes all pistillate or with a few staminate flowers at base, the spikes clustered or overlapping; pistillate scales obovate, about as long as perigynia body, pale brown or orange-tinged with translucent margins, tapered to tip or short-awned. Perigynia lance-shaped, flat on 1 side and convex on other, light green, 4–6 mm long and 1–1.5 mm wide, nerved on both sides, gradually tapered to a finely sharp-toothed beak, the beak 1/2–2/3 as long as body. Achenes lens-shaped, to 2 mm long, in upper

part of perigynium body; stigmas 2. Flowering April–July.
HABITAT floodplain forests, old river channels, swamps.
STATUS MW-FACW | NCNE-FACW

Carex deweyana Schwein. *Deweyanae*
DEWEY'S HUMMOCK SEDGE

DESCRIPTION Large, loosely clumped, native grasslike perennial from short rhizomes. Stems weak and spreading, 2–12 dm long, rough-to-touch below the head. Leaves shorter than stems, yellow-green to waxy blue-green, soft, flat, 2–5 mm wide; sheath tight. Spikes 2–6, the lower separate, the upper grouped, forming a head 2–6 cm long and often drooping near tip; terminal spike with staminate flowers at base, lateral spikes usually pistillate, the perigynia upright; pistillate scales ovate, blunt to short-awned at tip, thin and translucent with green center, slightly shorter than perigynia.

Perigynia flat on 1 side and convex on other, 4–6 mm long and 1–2 mm wide, oblong lance-shaped, pale-green, very spongy at base, the beak 2–3 mm long, finely toothed and weakly notched. Achenes lens-shaped, nearly round, yellow-brown, 2 mm long; stigmas 2. Flowering May–Aug.
HABITAT thickets, swamps, and moist to dry woods.
STATUS MW-FACU | NCNE-FACU | GP-FACU

CAREX SECTION Dispermae

One wetland species in Upper Midwest region.

Carex disperma Dewey
♦TWO-SEEDED SEDGE

DESCRIPTION Small, loosely clumped native grasslike perennial, from slender rhizomes. Stems slender, weak, 3-angled, 1–4 dm long, shorter to longer than leaves. Leaves soft and spreading, 1–2 mm wide; sheaths tight, translucent.

Spikes 2–5, with both staminate and pistillate flowers, staminate flowers borne above pistillate, few flowered and small, with 1–6 perigynia and 1–2 staminate flowers, to 5 mm long, stalkless, separate or upper spikes grouped in interrupted heads 1.5–2.5 cm long; bracts sheathlike and resembling the pistillate scales, or threadlike and to 2 cm long; pistillate scales white, translucent except for the darker midrib, tapered to tip or short-awned, 1–2 mm long. Perigynia convex on both sides to nearly round in section, oval, 2–3 mm long, strongly nerved and rounded on the margins, beak tiny. Achenes lens-shaped, oval, 1–2 mm long; stigmas 2. Flowering May–July.

HABITAT hummocks in conifer swamps and alder thickets, usually where shaded, wetland margins.

STATUS MW-OBL | NCNE-OBL | GP-FACW

CAREX SECTION **Divisae**

One wetland species in Upper Midwest region.

Carex praegracilis W. Boott
◆EXPRESSWAY SEDGE

DESCRIPTION Colony-forming perennial, from long black rhizomes. Stems single or few together, 3-angled, 1–7 dm long, longer than the leaves. Leaves on lower part of stems, 2–3 mm wide; sheaths white-translucent. Spikes with both staminate and pistillate flowers, staminate flowers above pistillate, or spikes nearly all staminate or pistillate, 4–8 mm long, upper spikes crowded, lower spikes separated, in narrowly ovate heads 1–4 cm long; bracts absent; pistillate scales brown, shiny, shorter or equal to perigynia. Perigynia green-brown, turning dark brown, flat on 1 side and convex on other, ovate to lance-shaped, 3–4 mm long and 1 mm wide, sharp-edged, spongy at base, tapered to a finely toothed beak 2 mm long, unequally notched. Achenes lens-shaped, 1–2 mm long; stigmas 2. Flowering May–June.

HABITAT wet to moist meadows, shores, streambanks and ditches; in s part of region, most common along salted highways..

STATUS MW-FACW | NCNE-FACW | GP-FACW; considered native in Minn, introduced and adventive in Wisc and Mich.

CAREX SECTION **Glareosae**

Tufted sedges of wetlands, soils often peaty. Spikes distinct, mostly non-overlapping (except *Carex arcta* which has spikes overlapping, the upper not separated),, mostly or all gynecandrous, lateral spikes sometimes pistillate. Perigynia ascending to spreading; margins rounded in most species, smooth or finely serrate, often finely papillose.

ADDITIONAL SPECIES Hudson Bay sedge (***Carex heleonastes*** L. f.), native perennial,

Carex disperma

Carex praegracilis

in small clumps from long rhizomes. Stems slender and stiff, 2–5 dm long, sharply 3-angled, rough on upper stem angles, longer than leaves. Leaves 4–8 on lower stem, 5–12 cm long and 1–2 mm wide. Spikes 2–4, with both staminate and pistillate flowers, the staminate below the pistillate, grouped into a head 1–2 cm long and to 1 cm wide; pistillate scales brown with pale margins, about as long as perigynia. Perigynia 5–10 per spike, oval, flat on one side and convex on other, 2–3 mm long, with sharp margins, tapered to a short beak to 0.5 mm long. Achenes lens-shaped, filling the perigynium; stigmas 2. Rare in UP of Mich (END) in calcareous fens, where known from a single 1980 collection; a nearly circumboreal species, s in North America to n Mich.

CAREX SECTION *Glareosae*

1 Lowest bract bristle-like, several times as long as its spike; perigynia mostly 3–4 mm long, including very short smooth beak; spikes widely separated, containing 1–5 perigynia each............ *C. trisperma*
1 Lowest bract absent or at most about twice as long as its spike (if rarely prolonged, the perigynia smaller and often with serrulate beak); perigynia and spikes various **2**
2 Perigynia broadest near the base of the body, with a conspicuous beak 0.7–1.1 mm long; spikes mostly 7–15, usually ± overlapping or crowded into an ovoid to narrowly pyramidal head 2–4.5 cm long ...
.... *C. arcta*
2 Perigynia broadest at or near the middle of the body; beak essentially absent or less than 0.6 mm long; spikes 2–8, at least the lower spikes well separated or, if crowded, the inflorescence only 0.6–2 cm long.... **3**
3 Spikes 2–4, crowded into a short inflorescence 0.6–2 cm long; perigynia 2.5–3.5 mm long, beak often smooth-margined
.... *C. tenuiflora*
3 Spikes 4–8, remote or ± crowded, but total inflorescence over 2 cm long; perigynia 1.7–

2.6 mm long, beak serrulate usually minutely or scabrous.................. **4**
4 Perigynia 3–9 per spike (occasionally one or two spikes on a plant, especially terminal one, with as many as 15), loosely spreading, becoming rich brown in age; largest leaves 1–2 mm wide; foliage and perigynia green when fresh.............. *C. brunnescens*
4 Perigynia mostly 10–many per spike, appressed-ascending, greenish or dull brown in age; largest leaves 2–2.7 (–3.7) mm wide; foliage and perigynia glaucous or gray-green at least when fresh. ... *C. canescens*

Carex arcta F. Boott *Glareosae*
NORTHERN CLUSTERED SEDGE

DESCRIPTION Loosely to densely clumped, native grasslike perennial from very short, thick rhizomes. Stems 2–8 dm long, soft, sharply triangular, very rough-to-touch above, usually shorter than the leaves. Leaves clustered near base, light green, flat, 2–4 mm wide, long-tapered to a tip, very rough; sheaths loose, purple-dotted. Spikes 5–15, each with both staminate and pistillate flowers, the staminate small and below the pistillate; flowers crowded in oblong heads, 1.5–3 cm long, upper spikes densely packed, lower spikes slightly separate; pistillate scales ovate, acute, translucent with a brown-tinged center, shorter than the perigynia. Perigynia flat on 1 side and convex on other, ovate, 2–3 mm long and 1–1.5 mm wide, green to straw-colored or brown when mature, covered with white dots, widest near the broad base, tapered to a sharp-toothed, notched beak 0.5–1.5 mm long. Achenes lens-shaped, brown, 1–2 mm long; stigmas 2. Flowering June–Aug.

HABITAT floodplain forests, old river channels, swamps and wetland margins.

STATUS MW-OBL | NCNE-OBL

Carex brunnescens (Pers.) Poiret
Glareosae
GREEN BOG SEDGE

DESCRIPTION Densely clumped, native grasslike perennial from a short, fibrous rootstock.

Stems slender, sharply 3-angled, 0.5–5 dm long, smooth or slightly rough-to-touch below the head, mostly longer than the leaves. Leaves 1–3 mm wide; sheaths tight, thin and translucent. Spikes 5–10 in a head 2–5 cm long, all with pistillate flowers borne above staminate, 4–8 mm long, each spike with 5–15 perigynia, lower spikes separated; lowermost bract bristlelike, shorter or longer than lowermost spike; pistillate scales ovate, rounded or acute at tip, shorter than the perigynia. Perigynia 3-angled, not winged or sharp-edged, 2–3 mm long, faintly nerved on both sides, not spongy-thickened at base, tapered at tip to a short, minutely notched beak, the beak and upper body finely toothed and white-dotted. Achenes lens-shaped, 1–1.5 mm long; stigmas 2. Flowering June–Aug.

HABITAT wet forests and swamps, peatland margins.

STATUS MW-FACW | NCNE-FACW | GP-FAC

Carex canescens L. *Glareosae*
♦ GRAY BOG SEDGE

DESCRIPTION Clumped, native grass-like perennial. Stems 2–6 dm long. Leaves waxy blue- or gray-green, 2–4 mm wide, mostly near base of plant and shorter than stems. Spikes 4–8, silvery green or grayish, with both staminate and pistillate flowers, the staminate below the pistillate, ovate to cylindric, 5–10 mm long, the lower spikes ± separate, each spike with 10–30 perigynia. Perigynia flat on one side and convex on other, 2–3 mm long and 1–2 mm wide, with a beak to 0.5 mm long, not noticeably finely toothed on the margins; pis-

tillate scales shorter than perigynia. Achenes lens-shaped; stigmas 2. Flowering May–July.

HABITAT peatlands (including hummocks in patterned fens), tamarack swamps, floating mats, alder thickets, wet forest depressions.

STATUS MW-OBL | NCNE-OBL | GP-OBL

NOTE similar to **green bog sedge** (*Carex brunnescens*) but leaves waxy blue-green rather than green and spikes somewhat larger and silver-green vs. brown.

Carex tenuiflora Wahlenb. *Glareosae*
SMALL-HEADED BOG SEDGE

DESCRIPTION Loosely clumped, delicate native grasslike perennial, spreading from long, slender rhizomes. Stems very slender, 2–6 dm long. Leaves 1–2 mm wide. Spikes 2–4, with both staminate and pistillate flowers, the staminate below the pistillate, stalkless, clustered into a head 8–15 mm long; pistillate scales white-translucent with green center, covering most of the perigynium. Perigynia 3–15, oval, flat on 1 side and convex on other, 3–4 mm long, dotted with small white depressions, sharp-edged, beakless. Achenes lens-shaped, nearly filling the perigynia; stigmas 2. Flowering June–Aug.

HABITAT hummocks in peatlands, floating mats, conifer swamps; mostly confined to tamarack swamps in s part of our region.

STATUS MW-OBL | NCNE-OBL | GP-OBL

NOTE easily overlooked as plants are similar to the more common **two-seeded bog sedge** (*C. disperma*) and **three-seeded bog sedge** (*C. trisperma*) and often growing with these species.

Carex trisperma Dewey *Glareosae*
THREE-SEEDED BOG SEDGE

DESCRIPTION Loosely clumped, native grass-like perennial, with short, slender rhizomes. Stems very slender and weak, 2–7 dm long. Leaves 1–2 mm wide. Spikes 1–3 (usually 2), stalkless, 1–4 cm apart

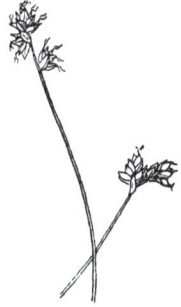

PATRICK ALEXANDER

Carex canescens

in a slender, often zigzagged head, each spike with 2-5 perigynia and a few staminate flowers at the base; lowest spike subtended by a bristlelike bract 2-6 cm long; pistillate scales ovate, translucent with a green center, shorter or equal to the perigynia. Perigynia flat on 1 side and convex on other, oval, 3-4 mm long, finely many-nerved, tapered near tip to a short, smooth beak 0.5 mm long. Achenes oval-oblong, filling the perigynia; stigmas 2. Flowering May-Aug.

HABITAT forested wetlands and conifer swamps, alder thickets, true bogs; a dominant species of forested peatlands in Red Lake peatland of Minn.

STATUS MW-OBL | NCNE-OBL | GP-OBL

CAREX SECTION **Granulares**

Plants tufted or shoots arising singly from elongate rhizomes. Pistillate spikes oblong to narrowly oblong, densely packed with perigynia. Pistillate scales and perigynia dotted or finely streaked with red. Perigynia more than 25 per pistillate spike; veins 25-40, raised.

CAREX SECTION *Granulares*

1 Staminate spike long-peduncled, elevated above summit of uppermost pistillate spikes; lowest pistillate spike usually on a separate basal peduncle; stems mostly solitary from elongate rhizomes; widest leaves 1.5-4 mm broad........*C. crawei*

1 Staminate spike sessile or nearly so; lowest pistillate spike not on a basal peduncle; stems clumped, without elongate rhizomes; widest leaves 4.5-10 mm broad ....*C. granularis*

Carex crawei Dewey *Granulares*
EARLY FEN SEDGE

DESCRIPTION Native grasslike perennial, from long-creeping rhizomes. Stems single or several together, faintly 3-angled, 0.5-4 dm long. Leaves 1-4 mm wide; sheaths tight,

translucent. Spikes either staminate or pistillate, cylindric, densely flowered, 1-3 cm long, terminal spike staminate; lateral spikes pistillate, 2-5, separate, the lowest spike near base of plant; bract leaflike with well-developed sheath, its blade shorter than terminal spike; pistillate scales redbrown with a pale or green midrib, shorter and narrower than the perigynia. Perigynia green to brown, ovate, 2-3.5 mm long, many-nerved; beak absent or very short, entire to notched. Achenes 3-angled, 1-2 mm long; stigmas 3. Flowering May-July.

HABITAT wet to moist meadows and prairies, marly lakeshores, ditches, especially where calcium-rich.

Carex granularis Muhl. *Granulares*
◊PALE SEDGE

DESCRIPTION Clumped, native grasslike perennial, from short rhizomes. Stems rounded 3-angled, 1-5 dm long. Leaves often longer than stems, 3-13 mm wide; sheaths membranous on front, divided-with small swollen joints on back. Spikes either all staminate or pistillate, the terminal spike staminate, stalkless; the lateral spikes pistillate, clustered around the staminate spike; bracts longer than the head; pistillate scales brown, tapered to tip or with a short, sharp point, half as long as perigynia. Perigynia crowded in several rows, green or olive to brown, oval to obovate, 2-3 mm long and 1-2 mm wide, 2-ribbed, strongly nerved;

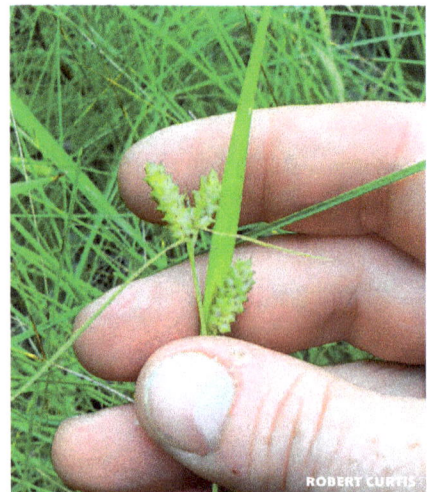

Carex granularis

beak tiny or absent, entire to slightly notched. Achenes 3-angled, 1–2 mm long; stigmas 3. Flowering May–July.

HABITAT wet to moist meadows and swales, streambanks and pond margins, especially where calcium-rich.

STATUS MW-FACW | NCNE-FACW | GP-OBL

CAREX SECTION **Griseae**

One wetland species in Upper Midwest region.

Carex conoidea Schk.
PRAIRIE GRAY SEDGE

DESCRIPTION Clumped, native grasslike perennial. Stems 1–7 dm long, much longer than leaves. Leaves 2–4 mm wide. Spikes either staminate or pistillate; staminate spike on a long stalk and overtopping pistillate spikes, linear, 1–2 cm long; pistillate spikes 2–4, widely spaced or upper 2 grouped, short cylindric, 1–2 cm long, on short, rough stalks; bract leaflike with a rough sheath; pistillate scales ovate and much shorter than perigynia, with a green midvein prolonged into an awn. Perigynia oval, 3–4 mm long and 1–2 mm wide. Achenes 3-angled; stigmas 3. Flowering May–July.

HABITAT wet meadows and prairies.

STATUS MW-FACW | NCNE-FACW | GP-FAC

CAREX SECTION **Heleoglochin**

Plants densely tufted; bases brown. Stems narrowing toward the tip, typically arching at maturity. Inner band of the leaf sheaths smooth, pigmented toward the summit. Leaf blades less than 3 mm wide (ours). Spikes androgynous, the lower branched. Perigynia plano-convex to biconvex, darkening at maturity, mostly less than 3 mm long; beak short-triangular, scabrous on the margin, bidentate. Primarily in wet, peaty soils.

CAREX SECTION *Heleoglochin*

1 Leaf sheaths whitish or pale ventrally except for purplish dots; inflorescence ± crowded, the lowermost spike (or branch) usually at least slightly overlapping the next above it (occasionally separated by a distance no more than its total length); perigynia tending to spread at maturity, therefore not concealed by the scales. . . . .
. . . .*C. diandra*

1 Leaf sheaths strongly tinged with copper color toward their summits ventrally; inflorescence ± interrupted, the lowermost spikes (or branches) often well separated or even peduncled; perigynia ± appressed at maturity, nearly or completely concealed by the large scales. . . . .*Carex prairea*

Carex diandra Schrank *Heleoglochin*
◆LESSER PANICLED SEDGE

DESCRIPTION Densely clumped, native grasslike perennial. Stems sharply 3-angled, 3–8 dm long, usually longer than leaves. Leaves 1–3 mm wide; sheaths white with fine pale lines, translucent on front or slightly copper-colored at mouth. Spikes with both staminate and pistillate flowers, staminate flowers borne above pistillate, clustered in ovate heads 1–4 cm long; bracts small and inconspicuous, shorter than the spikes; pistillate scales brown, tapered to tip or with a short sharp point, about equaling the perigynia. Perigynia brown,

MATTI VIRTALA

Carex diandra

shiny, unequally convex on both sides, broadly ovate, 2–3 mm long and 1–2 mm wide, beak finely toothed, entire to notched, 1–2 mm long. Achenes lens-shaped, 1 mm long; stigmas 2. Flowering May–July.

HABITAT wet meadows, ditches, peatlands (especially calcareous fens), floating mats.

STATUS MW-OBL | NCNE-OBL | GP-OBL

Carex prairea Dewey *Heleoglochin*
PRAIRIE SEDGE

DESCRIPTION Densely clumped, native grass-like perennial, from short rootstocks. Stems sharply 3-angled, 5–10 dm long, longer than the leaves. Leaves 2–3 mm wide; sheaths translucent, yellow-brown or bronze-colored. Spikes with both staminate and pistillate flowers, staminate flowers borne above pistillate, ovate, 4–7 mm long, lower spikes usually separate, in linear-oblong heads 3–8 cm long; bracts small; pistillate scales red-brown, tapered to tip, as long as and covering most of perigynia. Perigynia dull brown, flat on 1 side and convex on other, lance-shaped to ovate, 2–3 mm long and 1–2 mm wide, tapered to a finely toothed, unequally notched beak 1–2 mm long. Achenes lens-shaped, 1 mm long; stigmas 2. Flowering May–July.

HABITAT wet meadows, calcareous fens, marshes, tamarack swamps and peaty lakeshores.

STATUS MW-OBL | NCNE-FACW | GP-OBL

CAREX SECTION **Holarrhenae**
One wetland species in Upper Midwest region.

Carex sartwellii Dewey
♦RUNNING MARSH SEDGE

DESCRIPTION Colony-forming native grasslike perennial, from long black rhizomes. Stems single or few together, stiff, sharply 3-angled, 3–8 dm long, longer than the leaves. Leaves 2–4 mm wide, few per stem, the lowest leaves small and without blades; sheaths with green lines on front, and a translucent ligule around stem. Spikes with both staminate and pistillate flowers, staminate flowers above pistillate, or upper spikes staminate; clustered or lower spikes

separate, 5–10 mm long, in cone-shaped heads, 3–6 cm long; bracts small, the lower bracts sometimes bristlelike and longer than the spike; pistillate scales brown with a prominent green midvein, acute or with a short sharp point, about equal to perigynia. Perigynia tan to brown, flat on 1 side and convex on other, ovate, 2.5–3.5 mm long and 1–2 mm wide, finely nerved on both sides, sharp-edged, tapered to a short, finely toothed beak. Achenes lens-shaped, 1–1.5 mm long; stigmas 2. Flowering May–July.

HABITAT wet to moist meadows, marshes, fens and shores, often where calcium-rich.

STATUS MW-FACW | NCNE-OBL | GP-FACW

CAREX SECTION
Hymenochlaenae
Includes nearly all of our forest understory sedges with long, nodding pistillate spikes (some of which not found in wetlands). Superficially similar to section Gracillimae but plants more delicate. Terminal spike wholly staminate. Perigynia 8–45 per spike, narrow and long-tapering to the beak.

CAREX SECTION *Hymenochlaenae*

1 Terminal spike gynecandrous; sheaths ± softly pubescent or perigynia essentially beakless (except *C. prasina*)........... **2**

1 Terminal spike staminate; sheaths glabrous (except *C. castanea*) and perigynia conspicuously beaked.................. **3**

2 Perigynia strongly angled, gradually tapering into a beak 1–1.5 mm long; bract of lowest pistillate spike sheathless or with sheath up to 1.2 cm long; terminal spike mostly staminate, with at most a few perigynia at apex............. *C. prasina*

MATT LAVIN

Carex sartwellii

2 Perigynia obscurely angled or nearly terete, essentially beakless or beak less than 0.5 mm long; bract of lowest spike with sheath 1.5–8 cm or more in length; terminal spike staminate at base, pistillate toward apex.. . . . .*C. davisii*

3 Leaf sheaths and blades (at least toward the base) ± hairy; pistillate spikes 1–2.5 cm long . . . .*C. castanea*

3 Leaf sheaths and blades glabrous (at most the lowermost bladeless sheaths minutely hispidulous); pistillate spikes mostly (2–) 2.5–6.5 cm long 4

4 Basal sheaths reddish purple for at least several cm above the base; perigynia clearly nerved between the 2 ribs . . . . . .*C. debilis*

4 Basal sheaths brown, lacking any trace of reddish purple color (at most a small trace); perigynia 2-ribbed, but otherwise nerveless or faintly nerved . . . .*C. prasina*

Carex castanea Wahl. *Hymenochlaenae*
◆ CHESTNUT-COLOR SEDGE

DESCRIPTION Clumped, native grasslike perennial. **Stems** 3–10 dm long, purple-tinged at base, longer than leaves. **Leaves** 3–6 mm wide, softly hairy. **Spikes** either staminate or pistillate; the **terminal spike** staminate, upright atop a long stalk; **lateral spikes** pistillate, usually 3, on slender, drooping stalks, short cylindric, 1–2.5 cm long and 6–8 mm wide; pistillate scales ovate, brown-tinged, about as long as perigynia. **Perigynia** ovate, 4–6 mm long, somewhat 3-angled, strongly 2-ribbed with several faint nerves, tapered to a notched

ROB ROUTLEDGE

Carex castanea

beak. **Achenes** 3-angled; stigmas 3. **Flowering** June–July.

HABITAT swamps, moist openings, wetland margins and ditches.

STATUS MW-FACW | NCNE-FACW | GP-FACW

Carex davisii Schwein. & Torr.
Hymenochlaenae
AWNED GRACEFUL SEDGE

DESCRIPTION Clumped, native grasslike perennial. **Stems** 3–10 dm long, purple at base. **Leaves** 4–8 mm wide, hairy on underside; sheaths hairy. **Terminal spike** staminate with pistillate flowers near tip; **pistillate spikes** 2–3, the upper 2 overlapping, cylindric, 2–4 cm long and 5–6 mm wide, upright to nodding on short stalks; pistillate scales obovate, white or translucent with green center, tipped with a long awn, shorter or longer than perigynia. **Perigynia** ovate, dull orange when mature, 4–6 mm long and 2–3 mm wide, somewhat 3-angled, tapered to a notched beak to 1 mm long. **Achenes** 3-angled; stigmas 3. **Flowering** May–June.

HABITAT floodplain forest and moist woods.

STATUS MW-FAC | NCNE-FAC | GP-FAC; Minn (THR).

Carex debilis Michx. *Hymenochlaenae*
SOUTHERN WEAK SEDGE

DESCRIPTION Clumped, native grasslike perennial. **Stems** 6–10 dm long, purple-tinged at base. **Leaves** 2–4 mm wide. **Staminate spike** linear, 2–4 cm long, sometimes with a few pistillate flowers near tip; **pistillate spikes** 2–4, separate along stem, spreading or nodding, 3–6 cm long and 3–5 mm wide, flowers loose in spikes; pistillate scales oblong, half the length of perigynia with translucent or brown margins and a green midrib. **Perigynia** lance-shaped, somewhat 3-angled, 2-ribbed, 5–8 mm long, narrowed to a beak. **Achenes** 3-angled; stigmas 3. **Flowering** May–Aug.

HABITAT wet woods (usually under conifers), swamp margins, wet sandy ditches.
STATUS MW-FACW | NCNE-FACW | GP-OBL

Carex prasina Wahlenb.

Hymenochlaenae
DROOPING SEDGE

DESCRIPTION Clumped, native grasslike perennial. Stems 3–8 dm long, brown or green at base. Leaves 3–5 mm wide. Terminal spike staminate or with a few pistillate flowers at tip; pistillate spikes 2–4, widely separated, cylindric, 2–5 cm long and 5 mm wide, curved or nodding, lower spikes on long stalks, the upper stalks much shorter; upper bract ± sheathless; pistillate scales ovate to obovate, shorter than the perigynia, tipped with a short point. Perigynia 3–4 mm long, ovate, 3-angled, tapered to beak. Achenes 3-angled; stigmas 3. Flowering May–June.

HABITAT low areas in deciduous woods, streambanks.
STATUS MW-OBL | NCNE-OBL; uncommon in Wisc (THR).

CAREX SECTION
Leptocephalae

One wetland species in Upper Midwest region.

Carex leptalea Wahlenb.

SLENDER SEDGE

DESCRIPTION Densely clumped, native grasslike perennial. Stems slender, rounded 3-angled, 1–7 dm long, equal or longer than leaves. Leaves narrow, 0.5–1.5 mm wide; sheaths tight, white, translucent on front. Spikes single on the stems, few-flowered, 5–15 mm long, with both staminate and pistillate flowers, the staminate flowers borne above pistillate; bracts absent; pistillate scales rounded or with a short sharp point, shorter than the perigynia (or the tip of lowest scale sometimes longer than the perigynium). Perigynia yellow-green, nearly round in section to slightly flattened, oblong to oval, 3–5 mm long,

finely many-nerved, beakless or with a short beak. Achenes 3-angled, obovate, 1–2 mm long; stigmas 3. Flowering May–July.

HABITAT swamps, alder thickets, open bogs, calcareous fens; usually in partial shade.
STATUS MW-OBL | NCNE-OBL | GP-OBL

CAREX SECTION
Leucoglochin

One wetland species in Upper Midwest region.

Carex pauciflora Lightf.
◆ FEW-FLOWERED BOG SEDGE

DESCRIPTION Native grasslike perennial, from long slender rhizomes. Stems single or several together, longer than leaves. Leaves 1–2 mm wide, lower stem leaves reduced to scales; bract absent. Spike single, to 1 cm long, with both staminate and pistillate flowers, the staminate above the pistillate; staminate scales infolded to form a slender terminal cone; pistillate scales lance-shaped, 4–6 mm long, pale brown, soon deciduous. Perigynia 1–6, soon turned downward, slender, spongy at base, nearly round in section, straw-colored or pale brown, deciduous when mature, 6–8 mm long. Achenes 3-angled, not filling the perigynium; stigmas 3. Flowering June–July.

HABITAT open peatlands and floating mats in sphagnum moss, true bogs.
STATUS MW-OBL | NCNE-OBL | GP-OBL

JEAN-CLAUDE BOUZAT

Carex pauciflora

CAREX SECTION **Limosae**

Plants loosely tufted or stems arising singly, strongly rhizomatous; bases reddish. Roots covered in a dense yellow felt-like tomentum. Vegetative shoots becoming decumbent, behaving like stolons, producing shoots at the nodes. Pistillate spikes pendulous on slender stalks. Perigynia pale, short-beaked, papillose. Stigmas 3. Common in northern bogs and fens.

CAREX SECTION *Limosae*

1 Pistillate scales nearly or quite as broad as the perigynia and often only slightly if at all longer; staminate spike 15–30 (–50) mm long; plants strongly stoloniferous . . . . *C. limosa*

1 Pistillate scales distinctly narrower than perigynia, generally with narrowly acuminate tips much exceeding them; staminate spike 5–15 mm long; plants loosely clumped . . . . *C. magellanica*

Carex limosa L. *Limosae*
⬥MUD SEDGE

DESCRIPTION Loosely clumped native grasslike perennial, from long scaly rhizomes. **Stems** sharply 3-angled, 3–5 dm long, longer than leaves, usually rough-to-touch above. **Leaves** 1–3 mm wide; sheaths translucent, shredding

Carex limosa

into threadlike fibers near base. **Spikes** either all staminate or pistillate, the terminal spike staminate, 1–3 cm long; the lower 1–3 spikes pistillate, drooping on lax, threadlike stalks 1–3 cm long, ovate to short-cylindric, 1–2 cm long; pistillate scales brown, rounded or with a short, sharp point, about same size as perigynia. **Perigynia** waxy blue-green, ovate, flattened except where filled by achene, 2.5–4 mm long and 1–2 mm wide, strongly 2-ribbed with a few faint nerves on each side; beak tiny. **Achenes** 3-angled, 2 mm long; stigmas 3. **Flowering** May–July. **HABITAT** open bogs and floating mats; fairly common northward in Upper Midwest region, less common in s where mostly confined to calcareous fens and tamarack swamps. **STATUS** MW-OBL | NCNE-OBL | GP-OBL **NOTE** Poor sedge (*Carex magellanica*) is similar but has scales much narrower than the perigynia.

Carex magellanica Lam. *Limosae*
POOR SEDGE

DESCRIPTION Loosely clumped native perennial, from slender branching rhizomes. **Stems** slender, 1–8 dm high, longer than the leaves, red-brown at base. **Leaves** 3–12 on lower half of stem, flat but with slightly rolled under margins, 2–4 mm wide, the dried leaves of previous year conspicuous; sheaths red-dotted. **Terminal spike** staminate (or sometimes with a few pistillate flowers at tip), on a long stalk, linear, 4–12 mm long and 2–4 mm wide, usually upright; **pistillate spikes** 1–4 (rarely with several staminate flowers at base), clustered, usually drooping on slender stalks, 4–20 mm long and 4–8 mm wide, nearly round to oblong; lowest bract leaflike, equal or longer than the head. Female scales lance-shaped to ovate, tapered to a tip, narrower but usually longer than the perigynia, brown or green in center,

margins brown. Perigynia broadly ovate or oval, 2–3 mm long and 1.5–2.5 mm wide, flattened and 2-ribbed, with several evident nerves, pale or somewhat waxy blue-green, covered with many small bumps, the tip rounded and barely beaked. Achenes 3-angled, obovate, 2 mm long; stigmas 3. Flowering July–Aug.

SYNONYM *Carex paupercula*.

HABITAT open bogs, partly shaded peatlands, floating mats, swamps and thickets, usually in sphagnum moss.

STATUS MW-OBL | NCNE-OBL | GP-OBL

CAREX SECTION **Lupulinae**

Distinctive sedges of wet forests; recognized by the strongly inflated, ribbed perigynia, 1–2 cm long.

CAREX SECTION *Lupulinae*

1　Pistillate spikes spherical or nearly so, scarcely if at all longer than wide; sheath of uppermost leaf absent or less than 1.5 (–2.5) cm; style straight or sinuous or contorted (especially in *C. intumescens*) just below or at the middle; beak of perigynium much shorter than the body **2**

1　Pistillate spikes cylindrical or short-oblong, usually definitely longer than broad; sheath of uppermost leaf usually 1.7 cm or longer; style strongly bent and contorted immediately above the body of the achene; beak of perigynium nearly or quite as long as the body **3**

2　Perigynia (7–) 10–31 per spike, radiating in all directions, narrowed at the base to a ± broad cuneate stalk, sometimes hispidulous basally; pistillate spikes 1–2 (–3) . . . . *C. grayi*

2　Perigynia 2–8 (–12) per spike, mostly spreading-ascending, rounded at the base, glabrous (and often very shiny); pistillate spikes (1–) 2–**5** . . . . *C. intumescens*

3　Body of achene with broadly diamond-shaped sides, mostly 2.4–3.4 mm wide, at most 0.5 mm longer than wide, the angles each with a prominent swollen knob . . . . *C. lupuliformis*

3　Body of achene with somewhat diamond-shaped to ± elliptic or ovate sides, 1.7–2.6 (–2.8) mm wide, usually 1 mm or more longer than wide, the angles obscurely if at all knobbed . . . . *C. lupulina*

Carex grayi Carey *Lupulinae*
COMMON BUR SEDGE

DESCRIPTION Native grasslike perennial, rhizomes absent. Stems single or forming small clumps, 3–9 dm long, rough on upper stem angles, sheaths at base of stem persistent, red-purple. Leaves 5–12 mm wide. Spikes either staminate or pistillate; terminal spike staminate, 1–6 cm long, stalked; pistillate spikes 1–2, rounded, stalked; bracts leaflike; pistillate scales ovate, body shorter than perigynia but sometimes tipped with an awn to 7 mm long. Perigynia 10–30 per spike, spreading in all directions, not shiny, 10–20 mm long, strongly nerved, tapered from widest point to a notched beak 2–3 mm long. Achenes 3–5 mm long, with a persistent, withered style; stigmas 3. Flowering June–Sept.

HABITAT floodplain forests and backwater areas (as along Miss River).

STATUS MW-FACW | NCNE-FACW | GP-FACW

Carex intumescens Rudge *Lupulinae*
♦SHINING BUR SEDGE

DESCRIPTION Native grasslike perennial, rhizomes absent. Stems single or in small clumps, 3–9 dm long, rough on upper stem angles; sheaths at base of stem persistent, red-purple. Leaves 4–12

DOUG MCGRADY

Carex intumescens

mm wide, bracts leaflike. Spikes either staminate or pistillate, or sometimes staminate spikes with a few pistillate flowers; terminal spike staminate, 1–6 cm long, stalked; pistillate spikes 1–4, grouped, 1–3 cm long and wide, rounded, on stalks to 1.5 cm long; pistillate scales narrowly ovate, shorter and narrower than perigynia. Perigynia 1–12 per spike, spreading in all directions, satiny (not dull), 10–17 mm long, tapered to a beak 2–4 mm long. Achenes 3–6 mm long, flattened; stigmas 3. Flowering May–Aug.

HABITAT mixed and deciduous moist forests, kettle wetlands in woods, swamps and alder thickets.

STATUS MW-FACW | NCNE-FACW | GP-OBL

Carex lupuliformis Sartwell
Lupulinae
KNOBBED HOP SEDGE

DESCRIPTION Loosely clumped, native grasslike perennial, from rhizomes. Stems stout, 3-angled, 4–10 dm long. Leaves longer than head, 5–15 mm wide. Spikes either all staminate or pistillate, or pistillate spikes sometimes with few staminate flowers at tip; the uppermost spike staminate, 3–8 cm long, stalked or stalkless; pistillate spikes 3–6, clustered or overlapping, 3–6 cm long and 2–3 cm wide; bracts leaflike and much longer than head; pistillate scales lance-shaped, shorter than the perigynia, tipped with a short awn. Perigynia ascending to spreading, yellow-brown when mature, lance-shaped, inflated, 10–20 mm long and 4–8 mm wide, tapered to a beak 5–10 mm long. Achenes 3–5 mm long, 3-angled with concave sides, each angle with a hard knob near the middle; stigmas 3.

HABITAT low areas in forests (floodplains and seasonally wet depressions); swamps and marshes, often in shallow water.

STATUS MW-OBL | NCNE-OBL | GP-OBL; uncommon in Wisc (END) and Mich (THR).

NOTE similar to **common hop sedge** (*Carex lupulina*) but less common.

Carex lupulina Muhl. *Lupulinae*
♦ COMMON HOP SEDGE

DESCRIPTION Loosely clumped, native grasslike perennial, from rhizomes. Stems stout, 3-angled, 3–12 dm long. Leaves much longer than head, 4–15 mm wide; upper sheaths white and translucent, the lower sheaths brown. Spikes either all staminate or pistillate, the upper spike staminate, short-stalked, 2–5 cm long; pistillate spikes 2–6, clustered or overlapping, the lowermost sometimes separate, 2.5–6 cm long and 1.5–3 cm wide; bracts leaflike and spreading, much longer than head; pistillate scales narrowly ovate, tapered to tip or with a short awn, much shorter than the perigynia. Perigynia many, upright, dull green-brown, lance-shaped, inflated, 10–20 mm long and 4–7 mm wide, many-nerved, tapered to a finely toothed beak 5–10 mm long, the beak teeth 1–2 mm long. Achenes 3-angled, 3–4 mm long; stigmas 3. Flowering June–Aug.

HABITAT wet woods and floodplain forests, swamps, wet meadows and marshes, ditches and shores.

STATUS MW-OBL | NCNE-OBL | GP-OBL

NOTE **Shining bur sedge** (*Carex intumescens*) is similar but differs from common hop sedge by having fewer, uncrowded perigynia which are olive-green and glossy.

CAREX SECTION **Multiflorae**
One wetland species in Upper Midwest region.

Carex vulpinoidea Michx.
♦ BROWN FOX SEDGE

DESCRIPTION Densely clumped, native grasslike perennial, from

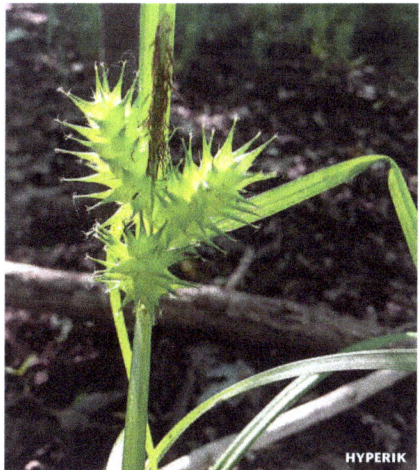

Carex lupulina

short rootstocks. Stems stiff, sharply 3-angled, 3–9 dm long, shorter to longer than the leaves. Leaves 2–4 mm wide; sheaths tight, cross-wrinkled and translucent on front, mottled green and white on back. Spikes with both staminate and pistillate flowers, staminate flowers borne above pistillate; heads oblong to cylindric, 3–9 cm long, with several spikes per branch at lower nodes; bracts small and bristlelike, longer than the spikes; pistillate scales awn-tipped, the awns equal or longer than the perigynia. Perigynia yellow-green, becoming straw-colored or brown when mature, flat on 1 side and convex on other, ovate to nearly round, 2–3 mm long and 1–2 mm wide, abruptly contracted to a notched, finely toothed beak 1 mm long. Achenes lens-shaped, 1–2 mm long; stigmas 2. Flowering May–Aug.

SYNONYM *Carex annectens.*

HABITAT wet to moist meadows, marshes, lakeshores, streambanks, roadside ditches.

STATUS MW-FACW | NCNE-OBL | GP-FACW

CAREX SECTION **Ovales**

In general, Ovales are characterized by a tufted habit, brownish basal sheaths, and sterile shoots with both nodes and internodes; this is in contrast to the sterile shoots of most species of *Carex,* where the stem-like portion is formed only of overlapping leaf sheaths, and nodes and internodes are absent. Mature perigynia (and often a dissecting microscope) may be needed to accurately identify species in this large group.

Carex vulpinoidea

CAREX SECTION *Ovales*

1 Achenes narrow, mostly 0.5–0.8 mm wide; perigynia usually more than 2.5x as long as wide, not oval-shaped or broadest above the middle **2**

1 Achenes wider, 0.9–1.5 mm wide (or if a little narrower, then the perigynium body obovate), perigynia to about 2.5x as long as wide **9**

2 Perigynia 7–10 mm long . . . . *C. muskingumensis*

2 Perigynia 2–6 mm long **3**

3 Perigynia 2–4 mm long and 1–1.5 mm wide, to 3x as long as wide **4**

3 Perigynia either at least 4 mm long, or more than 3x as long as wide, or both . . . **5**

4 Perigynia beak somewhat spreading, the perigynia often not winged to their base . . . . *C. cristatella*

4 Perigynia with a stiff, upright beak, the perigynia winged to their base . . *C. bebbii*

5 leaves 1–3 mm wide **6**

5 leaves 7 mm or more wide **8**

6 Perigynia 1.5–2.5x longer than wide . . . . *C. straminea*

6 Perigynia 2.5–5x longer than wide **7**

7 Perigynia 3–5 mm long and 1.5–2 mm wide, strongly flattened . . . . *C. crawfordii*

7 Perigynia 4–6 mm long and to 1 mm wide, flat on 1 side, convex on other *C. scoparia*

8 Spikes overlapping and crowded, 8–12 mm long, with more than 30 perigynia in each spike . . . . *C. tribuloides*

8 Lowermost spikes separate, spikes 5–8 mm long, with 15–30 perigynia in each spike . . . . *C. projecta*

9 Perigynia less than 4 mm long and less than 2 mm wide **10**

9 Perigynia more than 4 mm long or more than 2 mm wide, or both **13**

10 Perigynia oval and broadest above middle, achene to 1 mm wide **11**

10 Perigynia oval, oblong, or circular; achene often more than 1 mm wide **12**

11 Perigynia beak broad, the winged margins extending to tip of beak; pistillate scales boat-shaped (keeled) and blunt-tipped. . . . . *C. longii*

11 Perigynia beak slender, the winged margins not reaching the tip; pistillate scales flat, tapered to a pointed tip . *C. albolutescens*

12 Heads compact; leaves mostly 2.5–6 mm wide . . . . *C. normalis*

12 Heads loose, the lower spikelets widely spaced; leaves mostly 1.5–2.5 mm wide ... . *C. tenera*

13 Portion of leaf sheath long-translucent; perigynia oval, widest at about one-third of total length................. *C. normalis*

13 Leaf sheath green-veined almost to its top, with only a short translucent area; perigynia round to obovate, widest at or above middle.......................... **14**

14 Female scales lance-shaped, much narrower and somewhat shorter than perigynia **15**

14 Female scales inconspicuous, blunt-tipped or gradually tapered to a tip **16**

15 Spikes widely separated.... *C. straminea*

15 Spikes crowded and overlapping. *C. alata*

16 Perigynia 4–5 mm long and 2–2.6 mm wide, broadest at a third to two-fifths of its total length .... *C. suberecta*

16 Perigynia broadest at two-fifths to half its length **17**

17 Perigynia beak broad, its winged margins extending to its tip; pistillate scales boat-shaped and mostly blunt-tipped, the scale midvein not reaching the tip .... *C. longii*

18 Perigynia beak slender, its winged margins not reaching the tip; pistillate scales flat, gradually tapered to tip, the scale midvein reaching the tip .... *C. albolutescens*

Carex alata T. & G. *Ovales*
WINGED OVAL SEDGE

DESCRIPTION Clumped, native grasslike perennial. Stems 3–10 dm long, longer than leaves. Leaves 2–4 mm wide, sheaths green-veined on inner side. Spikes 4–8, with both staminate and pistillate flowers, pistillate flowers above the few staminate flowers at base of spike, silvery green or silvery brown, round to ovate, 8–12 mm long, grouped into heads 2–4 cm long; pistillate scales narrowly ovate, awn-tipped, translucent but for the narrow green center, shorter than perigynia. Perigynia body obovate, very flat, 4–5 mm long and 2–3 mm wide, several-nerved on both sides, narrowed to a slightly notched beak about 1 mm long. Achenes lens-shaped, 1.5–2 mm long; stigmas 2. Flowering June–July.
HABITAT swamps, peatlands, marshes, sandy or peaty shores.
STATUS MW-OBL | NCNE-OBL | GP-OBL

Carex albolutescens Schwein. *Ovales*
LONG-FRUITED OVAL SEDGE

DESCRIPTION Densely clumped, native grasslike perennial. Stems stiff, 3–7 dm long, longer than leaves. Leaves 2–4 mm wide, sheaths green-veined on inner side. Spikes 4–7, terminal spike pistillate, lateral spikes pistillate with few staminate flowers, 8–12 mm long, densely flowered and clustered into a head or lower spikes somewhat separate; pistillate scales ovate, as long as perigynia but narrower, silvery translucent with a darker midrib extending to scale tip. Perigynia body obovate, flat on 1 side and slightly convex on other, widest above top of achene, 3–4 mm long and 2–3 mm wide, finely nerved on both sides, abruptly narrowed to a short beak. Achenes lens-shaped, 1.5–2 mm long; stigmas 2. Flowering June–Aug.
HABITAT swamps and wet woods, thickets.
STATUS MW-FACW | NCNE-FACW | GP-FACW; uncommon in Mich (THR).

Carex bebbii (L. H. Bailey) Fernald *Ovales*
♦BEBB'S OVAL SEDGE

DESCRIPTION Clumped, native grasslike perennial. Stems sharply 3-angled, 2–8 dm long. Leaves shorter to slightly longer than the stems, 2–5 mm wide; sheaths white, thin and translucent. Spikes

5 - 1 0 ,   *Carex bebbii*

with both staminate and pistillate flowers, pistillate flowers above staminate, 5–8 mm long, clustered in an ovate head 1.5–3 cm long; pistillate scales tapered to tip, narrower and slightly shorter than the perigynia. Perigynia green to brown, flat on 1 side and convex on other, ovate, 2.5–3.5 mm long, finely nerved on back, nerveless on front, wing-margined, with a finely toothed beak 1/3–1/2 the length of the body, shallowly notched at tip. Achenes lens-shaped, 1–1.5 mm long; stigmas 2. Flowering June–Aug.
HABITAT wet to moist meadows, marshes, streambanks, ditches and other wet places; calcareous fens in south.
STATUS MW-OBL | NCNE-OBL | GP-OBL

Carex crawfordii Fernald *Ovales*
CRAWFORD'S OVAL SEDGE

DESCRIPTION Densely clumped, native grass-like perennial. Stems 1–8 dm long, stiff. Leaves 3–4 on each stem, longer or shorter than the stems, 1–4 mm wide. Spikes 3–15, with both staminate and pistillate flowers, the staminate below the pistillate, grouped into a narrowly oblong, sometimes drooping head, 1–3 cm long and 4–15 mm wide; pistillate scales ovate, light brown with green center, shorter and about as wide as perigynia. Perigynia flattened except where enlarged by the achenes, lance-shaped, 3–4 mm long and 0.5–1 mm wide, brown, narrowly winged nearly to the base, finely toothed above the middle, tapered to a long, slender, toothed, notched beak. Achenes brown, lens-shaped, 1 mm long; stigmas 2. Flowering July–Sept.
HABITAT moist openings and wetland margins, sandy shorelines.
STATUS MW-FAC | NCNE-FACW | GP-OBL

Carex cristatella Britton *Ovales*
♦CRESTED OVAL SEDGE

DESCRIPTION Clumped, native grasslike perennial, from short rhizomes. Stems sharply 3-angled, 3–10 dm long, slightly shorter to longer than the leaves. Leaves 3–7 mm wide; sheaths loose, with fine green lines. Spikes with both staminate and pistillate flowers, pistillate flowers borne above staminate; spikes 5–12, 4–8 mm long,

crowded in an ovate to oblong head 2–3.5 cm long; bracts much reduced; pistillate scales tapered to tip, shorter than the perigynia. Perigynia widely spreading when mature, green to pale brown, flat on 1 side and convex on other, ovate to lance-shaped, 2.5–4 mm long and 1–2 mm wide, faintly nerved on both sides, strongly winged above the middle, tapered to a finely toothed, notched beak 1–2 mm long. Achenes lens-shaped, 1–1.5 mm long; stigmas 2. Flowering June–Aug.
HABITAT wet meadows, ditches, floodplains, marshy shores and streambanks.
STATUS MW-FACW | NCNE-FACW | GP-FACW

Carex longii Mackenzie *Ovales*
ROUND-SHOULDERED OVAL SEDGE

DESCRIPTION Densely clumped, native grass-like perennial. Stems stiff, 3–12 dm long, longer than the leaves. Leaves 2–3 mm wide, lower stem leaves reduced to scales. Spikes 3–10, with both staminate and pistillate flowers, the staminate below the pistillate, stalkless, crowded in a head or the lower separate, 5–15 mm long; lateral spikes mostly pistillate; pistillate scales ovate, shorter or nearly as long as the perigynia. Perigynia upright to spreading, obovate, flat on 1 side and convex on other, 3–5 mm long and 1.5–2.5 mm wide, the body broadest above the top of enclosed achene, tapered to a beak. Achenes lens-shaped, to 1.5 mm long; stigmas 2. Flowering May–Aug.

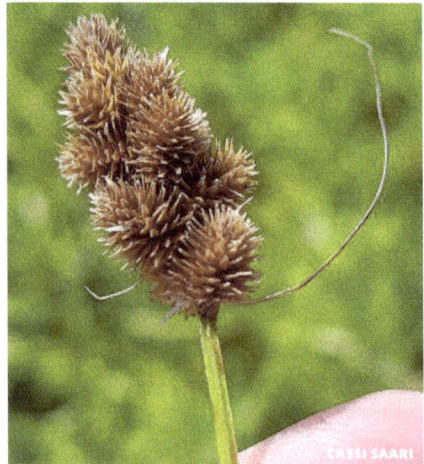

Carex cristatella

HABITAT open wetlands and wetland margins, often where sandy.

STATUS MW-OBL | NCNE-OBL | GP-OBL; historical record from Wisc.

Carex muskingumensis Schwein.
Ovales
SWAMP OVAL SEDGE

DESCRIPTION Clumped, native grasslike perennial. Stems stout, 5–10 dm long, with many leafy sterile stems present. Leaves 3–5 mm wide. Spikes 5–10, with both staminate and pistillate flowers, the staminate below the pistillate, pointed at both ends, 15–25 mm long and 4–6 mm wide, in a dense head 4–8 cm long; pistillate scales lance-shaped, pale brown with translucent margins, half as long as the perigynia. Perigynia upright, lance-shaped, 6–10 mm long, finely nerved on both sides, tapered to a finely toothed, deeply notched beak half as long as the body. Achenes lens-shaped, 2 mm long; stigmas 2. Flowering June–Aug.

HABITAT floodplain forests (as along Miss River), wet woods.

STATUS MW-OBL | NCNE-OBL

Carex normalis Mackenzie *Ovales*
SPREADING OVAL SEDGE

DESCRIPTION Clumped, native grasslike perennial. Stems 3–8 dm long, longer than the leaves. Leaves 2–6 mm wide, lower stem leaves reduced to scales. Spikes 5–10, with both staminate and pistillate flowers, the staminate below the pistillate, round in outline, 6–9 mm long, stalkless, loosely grouped in heads 3–5 cm long; pistillate scales ovate, translucent, lightly brown-tinged, with green midvein, tapered to a point or blunt-tipped, shorter than the perigynia. Perigynia upright, ovate, flat on 1 side and convex on other, green or pale green-brown, 3–4 mm long, finely nerved, tapered to a finely toothed beak. Achenes lens-shaped, 2 mm long; stigmas 2. Flowering June–Aug.

HABITAT moist to wet deciduous woods, floodplain forests, alder thickets, marshes, pond margins.

STATUS MW-FACW | NCNE-FACW | GP-OBL

Carex projecta Mackenzie *Ovales*
♦ NECKLACE SEDGE

DESCRIPTION Clumped, native grasslike perennial from short rhizomes. Stems slender and weak, 3-angled, 4–10 dm long, longer than leaves, upper stems rough. Leaves stiff, 3–7 mm wide; sheaths loose. Spikes 7–15, with both staminate and pistillate flowers, pistillate flowers above staminate in each spike, obovate to nearly round, 5–10 mm long and about as wide, straw-colored, distinct and ± separated (at least the lower spikes) in a somewhat lax and zigzagged inflorescence 3–5 cm long; bracts inconspicuous; pistillate scales narrowly ovate, straw-colored, narrower and shorter than the perigynia. Perigynia ascending to spreading when mature, lance-shaped, 3–5 mm long and 1–2 mm wide, dull brown, flattened except where filled by the achene, winged on margin, the wing gradually narrowing from middle to base, tapered to a notched, finely toothed beak 1-2 mm long. Achenes lens-shaped, 1–2 mm long; stigmas 2. Flowering June–Aug.

HABITAT floodplain forests, swamps, thickets, wet openings, shaded slopes.

STATUS MW-FACW | NCNE-FACW | GP-FACW

NOTE similar to awl-fruited oval sedge (*Carex tribuloides*) but the perigynia tips spreading rather than erect as in *Carex tribuloides*.

BENOIT RENAUD

Carex projecta

Carex scoparia Schk. *Ovales*
♦LANCE-FRUITED OVAL SEDGE

DESCRIPTION Densely clumped, native grass-like perennial, sometimes spreading by surface runners. **Stems** 2–10 dm long, sharply 3-angled, usually longer than the leaves. **Leaves** 1–3 mm wide; sheaths tight, white-translucent. **Spikes** 4–10, with both staminate and pistillate flowers, pistillate flowers borne above staminate, ovate to broadest at middle, 6–12 mm long, clustered or separate, in a narrowly ovate head 1–5 cm long; bracts small, the lowest often bristle-like; pistillate scales tapered to tip, slightly shorter than perigynia.
Perigynia green-white, flat, narrowly lance-shaped, 3–7 mm long and 1–2 mm wide, margins narrowly winged, tapered to a finely toothed, slightly notched beak 1–2 mm long. **Achenes** lens-shaped, 1–1.5 mm long; stigmas 2. **Flowering** May–July.
HABITAT wet meadows and openings, low prairie, swamps and sandy lakeshores.
STATUS MW-FACW | NCNE-FACW | GP-FACW

Carex straminea Willd. *Ovales*
AWNED OVAL SEDGE

DESCRIPTION Densely clumped, native grass-like perennial. **Stems** 4–10 dm long, longer than the leaves. **Leaves** 2–3 mm wide, lower stem leaves reduced to scales. **Spikes** 4–8, stalkless, separated in a head 3–6 cm long, with both staminate and pistillate flowers, the staminate in a slender cluster below the pistillate; pistillate scales lance-shaped, narrower and shorter than the perigynia, translucent and brown-tinged, with a paler midvein, tapered to a tip or short point. **Perigynia** flat and thin, tapered to a flattened, finely toothed beak to half as long as the body. **Achenes** lens-shaped, 1.5 mm long; stigmas 2. **Flowering** May–July.
SYNONYM *Carex richii.*
HABITAT marshes and wetland margins.
STATUS MW-OBL | NCNE-OBL; historical record from Mich.

Carex suberecta (Olney) Britton *Ovales*
WEDGE-FRUITED OVAL SEDGE

DESCRIPTION Clumped, native grasslike perennial. **Stems** 3–7 dm long, shorter or longer than the leaves. **Leaves** 2–3 mm wide, lower stem leaves reduced to scales. **Spikes** 2–5, stalkless, loosely grouped into a head, with both staminate and pistillate flowers, the staminate below the pistillate, 7–12 mm long; pistillate scales shorter and narrower than the perigynia, yellow-brown with a pale midvein and narrow translucent margins, tapered to a tip. **Perigynia** numerous, conspicuously swollen over the achene, 4–5 mm long, abruptly contracted to the flat, finely toothed beak. **Achenes** lens-shaped, 1.5 mm long; stigmas 2. **Flowering** May–July.
HABITAT calcareous fens, swamps, marshes, wet meadows, low prairie, shores.
STATUS MW-OBL | NCNE-OBL

Carex tenera Dewey *Ovales*
NARROW-LEAVED OVAL SEDGE

DESCRIPTION Clumped, native grasslike perennial, from short rhizomes. **Stems** slender, sharply 3-angled, 3–8 cm long, rough-to-touch above, longer than leaves. **Leaves** 0.5–3 mm wide; sheaths white-translucent on front, mottled green and white on back. **Spikes** 4–8, with both staminate and pistillate flowers, pistillate flowers borne above staminate, ovate to round, 4–10 mm long, loose in nodding heads 2.5–5 cm long; bracts small, sometimes bristlelike, longer than the spike; pistillate scales tapered to tip, slightly shorter than perigynia. **Perigynia** ovate, flat on 1 side

DOUG MCGRADY
Carex scoparia

and convex on other, straw-colored when mature, 2.5–4 mm long and 1–2 mm wide, wing-margined, tapered to a notched, finely toothed beak 1–2 mm long. Achenes lens-shaped, 1–2 mm long; stigmas 2. Flowering June–Aug.

HABITAT wet to moist meadows, streambanks, floodplains and moist woods.

STATUS MW-FACW | NCNE-FAC | GP-FACW

Carex tribuloides Wahlenb. *Ovales*
AWL-FRUITED OVAL SEDGE

DESCRIPTION Clumped, native grasslike perennial from short rhizomes. Stems sharply 3-angled, 3–9 dm long, longer than leaves. Leaves stiff, 3–7 mm wide; sheaths loose, with green lines. Spikes 5–15, with both staminate and pistillate flowers, pistillate flowers borne above staminate, obovate, 6–13 mm long, densely to loosely clustered into an ovate or oblong head 2–5 cm long; bracts inconspicuous; pistillate scales tapered to tip, shorter than the perigynia. Perigynia light green to pale brown, flattened except where filled by the achenes, lance-shaped, 3–6 mm long and 1–2 mm wide, broadly winged near middle, tapered to a notched, finely toothed beak 1–2 mm long. Achenes lens-shaped, 1.5 mm long; stigmas 2. Flowering June–July.

HABITAT floodplain forests, shady low areas in woods, pond and lake margins, marshes, low prairie.

STATUS MW-OBL | NCNE-FACW | GP-OBL

CAREX SECTION Paludosae

Mostly slender, long-rhizomatous plants, with red basal leaf sheaths (and ladder-fibrillose) and pubescent perigynia. *Carex lacustris* is somewhat different, having glabrous perigynia.

CAREX SECTION *Paludosae*

1   Perigynia pubescent 2
1   Perigynia glabrous 3
2   Leaf blades involute to triangular-channeled, 0.7–2 mm wide, those of vegetative shoots especially long-prolonged into a curled, filiform tip; leaves and lowermost bracts with the midvein low, rounded, and forming an inconspicuous keel (at least proximally) . . . . *C. lasiocarpa*

2   Leaf blades flat or folded into an M-shape except at the base and near the tip, 2.2–4.5 mm wide, not prolonged into a long filiform tip; leaves and lowest bract with the midvein forming a prominent and sharply pointed keel for much of the length . . . . *C. pellita*

3   Longest ligules 13–40 mm long, much longer than wide; culms lateral, with reddish bladeless basal sheaths; perigynia usually strongly many-nerved *C. lacustris*

3   Longest ligules 2–10 mm long, less than twice as long as wide; culms central, with persistent dead brownish remains of previous years leaves at base; perigynia nerveless or delicately nerved . . . . *C. hyalinolepis*

Carex hyalinolepis Steudel *Paludosae*
SHORELINE SEDGE

DESCRIPTION Large native perennial, from thick, scaly rhizomes. Stems stout, 5–15 dm long, with downward-pointing teeth, single or few together; lower leaf sheaths not breaking into fibers. Leaves 8–15 mm wide, leaves at base of stem well-developed (not reduced to scales), old leaf-bases persistent around stems; sheaths white or pale brown. Spikes either staminate or pistillate; staminate spikes 2–4, slender; pistillate spikes 2–4, separate, cylindric, 3–10 cm long and to 1.5 cm wide, upright, densely flowered, stalkless or short-stalked; bracts leaflike and longer than inflorescence. Body of pistillate scales ovate, shorter than perigynia, with a green midrib prolonged into an awn to 3 mm long. Perigynia 5–8 mm long, very faintly nerved. Achenes 3-angled; stigmas 3. Flowering May–July.

HABITAT wet meadows, marshes and swamps.

STATUS MW-OBL | NCNE-OBL | GP-OBL; Mich.

NOTE similar to **common lakeshore sedge** (*Carex lacustris*) and sometimes treated as a variety of it (*Carex lacustris* var. *laxiflora*).

Carex lacustris Willd. *Paludosae*
◆ COMMON LAKESHORE SEDGE

DESCRIPTION Large, clumped, native perennial, from scaly rhizomes. Stems erect, 3-angled, 6–13 dm long,

rough-to-touch. Leaves equaling or slightly longer than the stem, 6–15 mm wide; sheaths often red-tinged, the lower disintegrating into a network of fibers. Spikes either staminate or pistillate, the upper 2–4 staminate, stalkless, 4–7 cm long; the lower 2–4 spikes pistillate, erect, usually separate, stalkless or short-stalked, cylindric, 3–10 cm long and 3–15 mm wide; bracts leaflike, some or all longer than the head; pistillate scales awned or tapered to tip, the body shorter than the perigynia, the sides thin and translucent to pale brown. Perigynia olive, flattened to nearly round in section, narrowly ovate, 5–7 mm long, with more than 10 raised nerves, tapered to a smooth beak about 1 mm long. Achenes 3-angled, 2.5 mm long; stigmas 3. Flowering May–Aug.

HABITAT swamps, marshes, kettle wetlands, wetland margins, usually in shallow water; low areas in tamarack swamps.
STATUS MW-OBL | NCNE-OBL | GP-OBL

Carex lasiocarpa Ehrh. *Paludosae*
♦ SLENDER SEDGE

DESCRIPTION Colony-forming native grasslike perennial, from long, scaly rhizomes. Stems loosely clumped, 3-an-

gled, 3–10 dm long. Leaves elongate and inrolled, 1–2 mm wide; sheaths tinged with yellow-brown. Spikes either all staminate or pistillate, usually the upper 2 staminate; the staminate spikes slender, on a long stalk; the lower 1–3 spikes pistillate, widely separate, ± stalkless, cylindric, 1–4 cm long; bracts leaflike, the lowest usually longer than the stem; pistillate scales purple-brown with a green center, narrowly ovate, narrower and shorter or longer than the perigynia. Perigynia dull brown green, obovate, nearly round in section, 3–5 mm long, densely soft hairy, contracted to a beak about 1 mm long, the beak teeth erect. Achenes yellow-brown, 3-angled with concave sides, to 2 mm long; stigmas 3. Flowering June–Aug.

HABITAT wet peaty soils, open bogs, pond margins (where a pioneer mat-former), hollows in Red Lake peatland of Minn.
STATUS MW-OBL | NCNE-OBL | GP-OBL

Carex pellita Muhl. ex Willd. *Paludosae*
♦ WOOLLY SEDGE

DESCRIPTION Colony-forming native perennial, from scaly rhizomes. Stems 3-angled, 2–10 dm long. Leaves 2–5 mm wide; sheaths thin and translucent, lower sheaths often purple-tinged on back, shredding into a loose network of fibers. Spikes either all staminate or pistillate, the upper 1–3 staminate, 2–6 cm long; the lower 1–3 spikes pistillate, separate, stalkless or nearly so, cylindric, 1–4 cm long; bracts leaflike, the lowest usually longer than the head; pistillate scales brown to purple-brown, tapered to a tip to awned, shorter or longer than the perigynia. Perigynia brown to yellow-green to gray-brown, nearly round in section, obovate, 2.5–5 mm long, densely hairy, many-nerved, contracted to a finely toothed beak 1–2 mm long, the beak teeth

MATT LAVIN

Carex lasiocarpa

MATT LAVIN

Carex pellita

spreading. Achenes 3-angled with concave sides, 1.5–2 mm long; stigmas 3. Flowering June–Aug.

SYNONYM *Carex lanuginosa, Carex lasiocarpa* var. *latifolia*.

HABITAT wet to moist meadows and swales, marshes, shores, streambanks and other wet places.

STATUS MW-OBL | NCNE-OBL | GP-OBL

CAREX SECTION **Paniceae**

Plants colonial, shoots arising singly or few together; rhizomes elongate; bases brown to maroon. Leaf blades typically stiff. Terminal spike staminate, typically raised above the uppermost pistillate spike. Lateral spikes generally cylindrical, ascending (except *Carex vaginata*). Perigynia several-veined, mostly short-beaked, papillose (except *C. vaginata*). Calciphiles, growing mostly in wet soils. The section is fairly distinctive and easy to recognize, apart from *C. vaginata,* which is morphologically distinct.

CAREX SECTION *Paniceae*

1 Perigynium with a beak 1–2 mm long . . . .
    . . . . *C. vaginata*
1 Perigynium beakless, indistinctly beaked, or contracted to beak less than 0.5 mm . . **2**
2 Plants waxy blue-green; pistillate scales white-translucent on margins . . . *C. livida*
2 Plants green; pistillate scales purple-brown on margins . . . . *C. tetanica*

Carex livida (Wahlenb.) Willd. *Paniceae*
⧫ LIVID SEDGE

DESCRIPTION Native grasslike perennial, forming small clumps from long slender rhizomes. Stems slender, erect, 0.5–6 dm long, shorter or longer than the leaves, light brown at base. Leaves 6–12 on lower third of stem, strongly waxy blue-green, channeled, 0.5–4 mm wide, dried leaves of the previous year conspicuous; sheaths very thin. Ter-

QUINTEN WIEGERSMA

minal spike staminate (or rarely with both staminate and pistillate flowers, the staminate below the pistillate), linear, 1–3 cm long and 3–4 mm wide; pistillate spikes 1–3, the lowest ± separate, sometimes long-stalked, the upper grouped, stalkless or short-stalked, oblong, 1–2 cm long and 5 mm wide, with 5–15 upright perigynia; bracts leaflike, sometimes longer than the head; pistillate scales ovate, rounded to somewhat acute, shorter than the perigynia, light purple with broad green center and white translucent margins. Perigynia obovate, slightly flattened and rounded 3-angled, 2–5 mm long and 1–2 mm wide, strongly waxy blue-green, with small dots, two-ribbed and with many fine nerves, tapered to a beakless tip. Achenes ovate, 3-angled with prominent ribs, 2.5 mm long, brown-black; stigmas 3. Flowering July–Aug.

SYNONYM *Carex grayana*.

HABITAT wet meadows and fens, especially where calcium-rich.

STATUS MW-OBL | NCNE-OBL | GP-OBL

Carex tetanica Schk. *Paniceae*
COMMON STIFF SEDGE

DESCRIPTION Clumped perennial from slender rhizomes. Stems 3-angled, 1–6 dm long, rough-to-touch above.

Leaves 1–5 mm wide; sheaths tight, white or yellow and translucent. Spikes either all staminate or pistillate, terminal spike staminate, 1–3 cm long; lateral spikes pistillate, usually widely separated, the lower spikes short-cylindric, stalked, 6–30 mm long and 3–5 mm wide, loosely flowered with perigynia in 3 rows; bracts shorter than the head; pistillate scales purple-brown on margins, rounded to acute or short-awned, as wide as but shorter than the perigynia. Perigynia green, faintly 3-angled, obovate, 2–4 mm long and about 1–2 mm wide, 2-ribbed; beak tiny, bent. Achenes 3-angled with concave sides, 2 mm long; stigmas 3. Flowering May–July.

HABITAT wet meadows and openings, low prairies, marshy areas.

STATUS MW-FACW | NCNE-FACW | GP-FACW

Carex vaginata Tausch *Paniceae*

◆SHEATHED SEDGE

DESCRIPTION Native grasslike perennial, from long rhizomes. **Stems** 2–6 dm long, several together. **Leaves** 2–5 mm wide, leaves not scalelike at base of stem. **Terminal spike** staminate, 1–2 cm long; **pistillate spikes** 1–3, sometimes staminate at tip, loosely spreading, widely separated, the lower stalks long, the upper shorter; bracts with loose sheaths and blades shorter than the spikes; pistillate scales shorter and narrower than the perigynia, purple-brown, sometimes with a narrow green center, tapered to a tip. **Perigynia** usually in 2 rows, the lower separate, the upper overlapping, 3–5 mm long, narrowly obovate, with a curved beak 1 mm long. **Achenes** 3-angled, nearly filling the perigynia; stigmas 3. **Flowering** June–Aug.

SYNONYM *Carex saltuensis*.

HABITAT swamps and thickets, especially where calcium-rich.

STATUS MW-OBL | NCNE-OBL | GP-OBL

CAREX SECTION **Phacocystis**

Plants often clumped; rhizomes short or long. Lower leaf sheaths brown to red, fibrous in some species. Terminal spike typically staminate, ascending. Lateral spikes pistillate or androgynous, ascending to nodding or drooping, elongate.

Carex vaginata

Perigynia biconvex with distinct marginal veins. Stigmas 2. Mostly common in Upper Midwest wetlands, ranging from floodplains and wet forests (*Carex crinita, C. gynandra, C. emoryi*), to sedge meadows (*C. stricta*), wet prairies (*C. haydenii*), bogs and marshes (*C. aquatilis*), and wet roadsides and ditches.

ADDITIONAL SPECIES Smooth black sedge [*Carex nigra* (L.) Reichard], rare in Wisc wetlands in Douglas and Manitowoc counties, and from several Mich locations in sandy soils of meadows and swales (END); distinguished by dark-spotted perigynia 2–3.5 mm long and black pistillate scales.

CAREX SECTION *Phacocystis*

1 Pistillate spikes on ± lax peduncles, at length drooping, the scales prominently awned; body of achene with an irregular notch, constriction, or wrinkle on one side **2**

1 Pistillate spikes erect or strongly ascending, often sessile, the scales acute or acuminate, not awned; body of achene smooth and ± regular **3**

2 Sheaths smooth; bodies of most if not all pistillate scales shallowly lobed at summit (on each side of base of the awn) *C. crinita*

2 Sheaths scabrous-hispidulous; bodies of most or all pistillate scales on lower part of spike truncate or tapered at summit .... *C. gynandra*

3 Fertile stems of current year with conspicuous bladeless sheaths at base, not surrounded by dried-up bases of the previous year's leaves but arising laterally; lowest bract usually shorter than to approximately equaling the inflorescence **4**

3 Fertile stems of current year mostly lacking bladeless sheaths at base, arising centrally from tufts of dried-up bases of previous years leaves; lowest bract usually conspicuously longer than the inflorescence **6**

4 Perigynia suborbicular to obovoid, 2–2.3 mm long at maturity, broadest at or slightly above middle, rather abruptly contracted to a minute apiculus, at least the lower ones in a spike much exceeded by the spreading scales; lower leaf sheaths not or only slightly tearing to form a ladder like arrangement of fibers, the intact sheaths

smooth ventrally; ligule longer than width of leaf blade; plants with short, ascending rhizomes . . . . *C. haydenii*

4   Perigynia elliptic to ovate, mostly 2.2–2.7 (–3.3) mm long at maturity, broadest at or slightly below the middle, ± tapered to apex, as long as or longer than the scales (rarely exceeded by scales); ventral surface of lower leaf sheaths tearing to form a ladder-like arrangement of fibers (ladder-fibrillose) or if not, then the ligule shorter than width of leaf blade; plants with long horizontal rhizomes **5**

5   Ligule longer than width of leaf blade (deeply inverted V-shaped); ventral surface of lower leaf sheaths tearing to form a ladder-like arrangement of fibers and usually minutely scabrous and red-dotted, especially near the tip . . . . *C. stricta*

5   Ligule shorter than width of leaf blade (often nearly horizontal); ventral surface of lower leaf sheaths not tearing to form a ladder like arrangement of fibers, smooth and whitish . . . . *C. emoryi*

6   Perigynia essentially nerveless, except sometimes at the base only; staminate spikes usually 2 or more . . . . . *C. aquatilis*

6   Perigynia conspicuously few-ribbed on both sides; staminate spike usually 1 . . . . **7**

7   Plants densely tufted, without long rhizomes; scales with a broad central green portion about as wide as the darker margins; leaves mostly overtopping spikes . . . . *C. lenticularis*

7   Plants colonial from elongated rhizomes; scales with very narrow green portion much narrower than the broad, dark margins, scarcely if at all broader than the midrib; leaves mostly shorter than stems . . . . . . *C. nigra* (see note preceeding page)

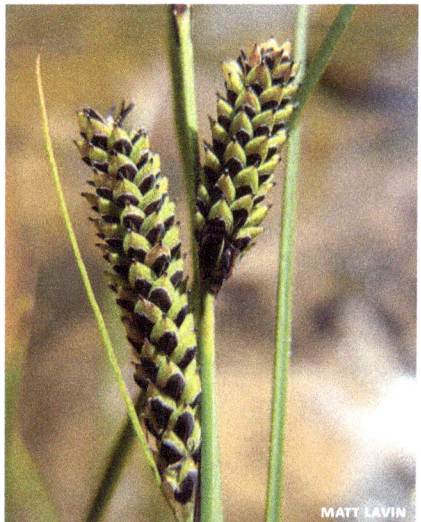

Carex aquatilis Wahlenb. *Phacocystis*
♦ WATER SEDGE

DESCRIPTION Native grasslike perennial, forming clumps or turfs; spreading by many slender rhizomes. Stems 3–12 dm long and longer than the leaves, 3-angled, usually rough-to-touch below the spikes. Leaves waxy blue-green, 2–7 mm wide; sheaths white or purple-dotted. Spikes 3–5, the upper spikes staminate, the middle and lower spikes pistillate or often with staminate flowers borne above pistillate, 2–5 cm long; pistillate scales tapered to tip. Perigynia pale green to yellow-brown or red-brown, obovate, broadest near tip, not inflated, 2–3 mm long; beak tiny. Achenes lens-shaped, 1–2 mm long; stigmas 2. Flowering May–Aug.

HABITAT wet meadows, marshes, shores, streambanks, kettle lakes, ditches, fens.

STATUS MW-OBL | NCNE-OBL | GP-OBL

Carex crinita Lam. *Phacocystis*
♦ FRINGED SEDGE

DESCRIPTION Large, densely clumped native grasslike perennial. Stems 5–15 dm long and longer than leaves. Leaves 7–13 mm wide, lowest stem leaves reduced to scales; sheaths smooth. Spikes staminate or pistillate, drooping on slender stalks; staminate spikes 1–3, above pistillate spikes, 4–10 cm long; pistillate spikes 2–5, narrow cylindric, 4–12 cm long; bract leaflike, without a sheath; pistillate scales rounded and notched at tip with pale midvein prolonged into a toothed awn to 10 mm long, scale edges copper-brown. Perigynia green, 2-ribbed, nerves faint or absent, round in cross-section, abruptly tapered to a tiny beak. Achenes lens-shaped; stigmas 2. Flowering May–July.

HABITAT swamps and alder thickets, wet openings, ditches and potholes.

STATUS MW-OBL | NCNE-OBL | GP-OBL

NOTE similar to *Carex gynandra* but with

Carex aquatilis

smooth sheaths, lower pistillate scales rounded at tip, and perigynia inflated.

Carex emoryi Dewey *Phacocystis*
RIVERBANK SEDGE

DESCRIPTION Loosely clumped, native grass-like perennial from scaly rhizomes. **Stems** 3-angled, 4–12 dm long. **Leaves** 3–7 mm wide, lowest leaves reduced to red-brown sheaths; upper sheaths white or yellow-tinged and translucent, lower sheaths red-brown. **Spikes** 3–7, the terminal 1 or 2 all staminate, 2–4.5 cm long, the lateral spikes all pistillate or with staminate flowers above the pistillate, 2–10 cm long; lowest bract leaflike; pistillate scales blunt or tapered to tip, narrower than the perigynia. **Perigynia** light green, becoming straw-colored at maturity, convex on both sides, oval or ovate, 1.5–3 mm long, stigmas 2. **Flowering** May–July.

HABITAT shores, streambanks, wet meadows and floodplain forests, sometimes forming pure stands.

STATUS MW-OBL | NCNE-OBL | GP-OBL

Carex gynandra Schwein. *Phacocystis*
FRINGED SEDGE

DESCRIPTION Large, clumped native grass-like perennial. **Stems** 5–15 dm long, longer than leaves. **Leaves** 7–14 mm wide, lowest leaves reduced to scales; sheaths finely hairy; bracts leaflike, lowest bract 1–3.5 dm long. **Spikes** either staminate or pistillate, spreading or drooping and often curved, stalked; **staminate spikes** 1–3, above

Carex crinita

pistillate, 5–9 cm long; **pistillate spikes** 2–5, long-cylindric, 5–12 cm long; lower pistillate scales 5–6 mm long, with a pale midrib, tapered to an awned tip about 5 mm long. **Perigynia** green, ovate to oval, not inflated, 3–4 mm long. **Achenes** lens-shaped, stigmas 2. **Flowering** June–July.

SYNONYM *Carex crinita* var. *gynandra*.

HABITAT wet openings and swamps.

STATUS MW-FACW | NCNE-OBL

NOTE similar to *Carex crinita*, but with finely hairy sheaths, lower pistillate scales tapered to an awned tip, and perigynia somewhat flattened and not inflated.

Carex haydenii Dewey *Phacocystis*
LONG-SCALED TUSSOCK SEDGE

DESCRIPTION Loosely clumped, native grass-like perennial, from short rhizomes. **Stems** arising from previous year's clumps of leaves (which are persistent at the base of the new leaves), 3-angled, 3–10 dm long, usually longer than the leaves, rough-to-touch above. **Leaves** green, 2–5 mm wide, the lower leaves bladeless and sheathlike; sheaths white to yellow on front, green on back, translucent. **Spikes** 3–6, the upper 1–3 staminate, the terminal one largest, 2–5 cm long, the others smaller; the lower 2–3 pistillate or with staminate flowers above pistillate, 1–3 cm long; lowest bract leaflike, usually shorter than the head; pistillate scales tapered to tip, longer than the perigynia. **Perigynia** pale brown when mature, often with dark brown spots, convex on both sides, obovate, inflated at tip, 2–3 mm long; beak tiny. **Achenes** lens-shaped, 1 mm long; stigmas 2. **Flowering** May–July.

HABITAT wet to moist meadows and swales, marshes and streambanks; often with dark-scaled sedge (*Carex buxbaumii*).

STATUS MW-OBL | NCNE-OBL | GP-OBL; historically known from Mich LP.

Carex lenticularis Michx. *Phacocystis*
SHORE SEDGE

DESCRIPTION Densely clumped, native grass-like perennial. **Stems** 1–6 dm long, upright, slender, usually shorter than leaves, brown at base. **Leaves** clustered on lower one-third of stem, upright, long-

tapered to tip, 1–2 mm wide; sheaths dotted with yellow-brown on front. Staminate spike single, sometimes with a few pistillate flowers, stalked, linear, 8–30 mm long and 2–3 mm wide; pistillate spikes 3–5, upright, the upper stalkless, the lower stalked, the upper grouped, the lower separate, linear, 1–5 cm long and 3–4 mm wide; lowest bract leaflike, erect, much longer than the head, the upper bracts shorter; pistillate scales ovate, red or red-brown, with a 3-veined, green center, the margins translucent near tip, narrower and usually shorter than the perigynia. Perigynia upright, soon deciduous, obovate, flattened, convex on both sides and sharply two-edged, 2–3 mm long and 1–1.5 mm wide, waxy blue-green, with a few yellow glandular dots or bumps, tapered or rounded at the abruptly pointed tip; the beak small, to 0.2 mm long. Achenes lens-shaped, 2 mm long, brown; stigmas 2. Flowering June–Sept. HABITAT rocky and sandy lakeshores, rock pools along Lake Superior, shallow ponds, sedge mats. STATUS MW-OBL | NCNE-OBL | GP-OBL; Wisc (THR).

Carex stricta Lam. *Phacocystis*
⬧COMMON TUSSOCK SEDGE

DESCRIPTION Densely clumped, native grass-like perennial from long scaly rhizomes, forming large, raised hummocks to 1 m tall. Stems 3-angled, 3–10 dm long, rough-to-touch, longer than leaves. Leaves 2–6 mm wide, the lower leaves reduced to sheaths around the base of stem;

Carex stricta

sheaths white to red-brown on front, green on back, the lower sheaths breaking into ladderlike thin strands. Spikes mostly all staminate or pistillate (sometimes mixed), the upper 1–3 spikes staminate, the terminal spike 1.5–5 cm long, the lower 2–5 spikes pistillate or some with staminate flowers borne above pistillate, 2–8 cm long; lowest bract leaflike; pistillate scales rounded or tapered to tip, equal or longer than the perigynia but narrower. Perigynia green at tip and margins, golden to yellow-brown in middle, with white or brown bumps, convex on both sides to nearly flat, ovate to oval, 2–3 mm long and 0.5–2 mm wide, 2-ribbed with a few faint nerves on both sides; beak short and tubelike, to only 0.3 mm long. Achenes lens-shaped, 1.5 mm long; stigmas 2. Flowering May–July. HABITAT sedge meadows, marshes, fens, shores, streambanks, ditches; common, and a dominant component of sedge meadow communities.. STATUS MW-OBL | NCNE-OBL | GP-OBL

CAREX SECTION **Physoglogin**
One wetland species in Upper Midwest region.

Carex gynocrates Wormsk.
⬧NORTHERN BOG SEDGE

DESCRIPTION Small, native perennial, from long, slender rhizomes. Stems single or few together, 0.3–3 dm long, smooth, usually longer than the leaves, brown at base. Leaves clustered near base of plant, blades inrolled and threadlike, to 1 mm wide. Spikes only 1 per stem, all staminate or all pistillate, or with both staminate and pistillate flowers and with the staminate flowers borne above the pistillate, 0.5–2 cm long; the staminate spike or portion of spike narrowly cylindric, the pistillate spike or portion short-cylindric; bract absent; pistillate scales brown or red-brown, obovate, tapered to tip, shorter but wider than perigynia. Perigynia 4–10, widely spreading, yellow to dark brown, shiny, plump, obovate, 2–4 mm long and 1–2 mm wide, spongy at base, abruptly

contracted to the beak; beak nearly entire to unequally notched, 0.5 mm long. Achenes lens-shaped, 1–2 mm long; stigmas 2. Flowering June–July.

SYNONYM *Carex dioica.*

HABITAT conifer swamps and open peatlands, usually in sphagnum and wet, peaty soils.

STATUS MW-OBL | NCNE-OBL | GP-OBL

CAREX SECTION **Racemosae**

Plants loosely to densely clumped; rhizomes variable in length; bases dark red, generally fibrous; roots not clothed with yellow felt. Terminal spike gynecandrous (in our species). Pistillate scales dark, often black. Perigynia pale, often greenish, very short-beaked to beakless, smooth or papillose, 2-ribbed, inconspicuously veined (in our species). Stigmas 3.

CAREX SECTION *Racemosae*

1 Terminal spike pistillate, androgynous or all staminate . . . . *C. hallii*
1 Terminal spike gynecandrous 2
2 Pistillate scales mostly awned or narrowly acuminate, exceeding the perigynia; ventral surface of lower leaf sheaths tearing into fibers . . . . *C. buxbaumii*
2 Pistillate scales obtuse or acute, equaling or shorter than the perigynia; ventral surface of lower sheaths not tearing to form fibers . . . . *C. media*

Carex buxbaumii Wahlenb.
Racemosae
◆ DARK-SCALED SEDGE

DESCRIPTION Loosely clumped, grasslike perennial, from long rhizomes. Stems single or few together, 3-angled, 3–10 dm long, rough-to-touch above, red-tinged near base. Leaves 1–3 mm wide, the lowest leaves without blades; lower sheaths shredding into thin strands, the upper sheaths membranous and purple-dotted. Spikes 2–5, 1–3 cm long, terminal spike with pistillate flowers above staminate and larger than the lateral spikes, lateral spikes pistillate, short-cylindric, stalkless or nearly so; bracts leaflike, the lowest shorter than the head; pistillate scales dark brown, tapered to an awn at tip. Perigynia light green, golden brown near base, oval, 2.5–3.5 mm

long, 2-ribbed, with 6–8 faint nerves on each side; beak tiny, notched. Achenes 3-angled, 1–2 mm long; stigmas 3. Flowering May–Aug.

HABITAT wet meadows and fens, shallow marshes, low prairie, hollows in patterned peatlands.

STATUS MW-OBL | NCNE-OBL | GP-OBL

Carex hallii Olney *Racemosae*
DEER SEDGE

DESCRIPTION Loosely clumped, native perennial, from short scaly rhizomes. Stems 1–6 dm tall, obtusely triangular. Leaves 2–4 mm wide. Spikes unisexual, terminal spikes usually staminate, 1.5–2.5 cm long, lateral spikes pistillate, pistillate scales obtuse to mucronate. Perigynia planoconvex to obtusely triangular, green or white, 2-ribbed, 2–3 mm long; beak 0.5 mm long, bidentate. Achenes triangular; stigmas 3. Flowering June–Aug.

HABITAT low prairies, sandy sloughs.

STATUS MW-FACW | GP-FACW

Carex media R. Br. *Racemosae*
SCANDINAVIAN SEDGE

DESCRIPTION Loosely clumped, grasslike perennial, from short rhizomes. Stems slender, not stiff, 2–8 dm long, smooth or slightly rough-to-touch

Carex buxbaumii

above, sharply triangular above, much longer than the leaves, red-tinged at base. Leaves 7–15 and mostly near base of stem, pale-green, flat or margins slightly rolled under, 2–3 mm wide, rough-to-touch on margins, the dried leaves of previous year conspicuous; sheaths translucent. Spikes usually 3, densely flowered, the terminal spike with both staminate and pistillate flowers, the staminate below the pistillate, clustered, upright, oblong to nearly round in outline per spike, 4–8 mm long and 3–5 mm wide, stalkless; the lateral spikes pistillate, on short stalks; lowest bract usually shorter than the head; pistillate scales ovate, 2–3 mm long, purple-black, acute to rounded, margins white-translucent, nearly as wide as perigynia but much shorter. Perigynia obovate, 2–4 mm long and 1.5 mm wide, rounded 3-angled, slightly inflated, yellow-green to brown, two-ribbed, otherwise without nerves, tip rounded and abruptly beaked, the beak short (0.5 mm long), red-tinged, with a small notch. Achenes obovate, 1–2 mm long, 3-angled, yellow-brown; stigmas 3. Flowering July–Aug.

SYNONYM *Carex norvegica.*

HABITAT rocky streambanks, rocky Lake Superior shores, talus slopes.

STATUS MW-FACW |NCNE-FACW | GP-FACW; Wisc (END), Mich (THR, Keweenaw Peninsula and Isle Royale).

CAREX SECTION Rostrales

Plants clumped; rhizomes elongate; bases brown, not reddish or purplish. Staminate spike solitary, the base lower than or roughly equaling the apex of the uppermost pistillate spike. Pistillate spikes 2–6, approximately as long as thick. Perigynia ≤20 per spike, divergent or the lowermost reflexed, somewhat inflated, narrow, tapering continuously to the apex, generally 4–7x as long as wide, beakless, subtly bidentate. Species in this section superficially resemble the more common *Carex intumescens,* but differ in their narrower perigynia.

CAREX SECTION *Rostrales*

1 Broadest leaf blades 5–17 mm wide; sheaths of bracts with a ± prolonged lobe at mouth; staminate spike usually peduncled, its tip projecting well above the pistillate spikes.
..........................*C. folliculata*

1 Broadest leaf blades 1.5–3.5 mm wide; sheaths of bracts concave at mouth; staminate spike sessile or very short-peduncled, scarcely if at all projecting above the pistillate spikes......*C. michauxiana*

Carex folliculata L. *Rostrales*
FOLLICLE SEDGE

DESCRIPTION Large, clumped, native grass-like perennial. Stems to 1 m long. Leaves 5–15 mm wide. Staminate spike single, 10–25 mm long, long-stalked; pistillate spikes 2–5, widely separated, upright, 15–30 mm long and as wide, stalked; bracts leaflike, longer than stems; pistillate scales ovate, translucent or brown-tinged, green in center, much shorter than perigynia. Perigynia lance-shaped, many-nerved, 10–15 mm long, tapered to a long, finely toothed beak, the teeth upright. Achenes 3-angled; stigmas 3. Flowering June–Aug.

HABITAT wet woods and cedar swamps.

STATUS MW-OBL | NCNE-OBL

Carex michauxiana Boeckeler
Rostrales
♦ MICHAUX'S SEDGE

DESCRIPTION Clumped, native grasslike perennial. Stems 2–6 dm long. Leaves 2–4 mm wide. Spikes either staminate or pistillate; terminal spike staminate, 0.5–1.5 cm long,

REUVEN MARTIN

Carex michauxiana

stalkless or short stalked; pistillate spikes 2–4, broadly ovate, 1.5–2.5 cm long, upright, the lower spikes stalked, the upper on shorter stalks; bracts leaflike, 1–3 mm wide, longer than the stems; pistillate scales ovate, shorter than the perigynia, margins translucent or brown, with a green midrib, tapered to a tip. Perigynia narrowly lance-shaped, 8–13 long and to 2 mm wide, round in section, long-tapered to a beak with upright teeth 1 mm long. Achenes rounded 3-angled; stigmas 3. Flowering June–Aug.
HABITAT wet meadows, sphagnum peatlands, ditches and swales.
STATUS NCNE-OBL; Wisc (THR).

CAREX SECTION Squarrosae

The obconic perigynia of this section are highly distinctive, widest at the apex and abruptly narrowed to the beak; perigynia in all of the other "bladder" and "bottle-brush" sections taper more gradually to the beak. *Carex typhina* is the only one of our "bladder" and "bottlebrush" sedges to occasionally produce plants with a single spike.

CAREX SECTION *Squarrosae*

1 Pistillate scales with a long awn exceeding the beak of the perigynium; achenes ca. 1.5 mm long .... *C. frankii*

1 Pistillate scales awnless or short-awned, much shorter than the beak, usually ± hidden among the dense perigynia; achenes ca. 2.2–3 mm long **2**

2 Achenes slenderly ellipsoid, slightly more than twice as long as wide, terminated by a strongly sinuous style; pistillate scales very sharp-tipped or short-awned ....
.... *C. squarrosa*

2 Achenes broadly ellipsoid, about or slightly less than twice as long as wide, terminated by a ± straight style; pistillate scales ± acute in outline but blunt at the very tip ....
.... *C. typhina*

Carex frankii Kunth *Squarrosae*
BRISTLY CAT-TAIL SEDGE

DESCRIPTION Native grasslike perennial, forming small clumps. Stems 2–8 dm long. Leaves 6–10 mm wide, lower stem leaves reduced to scales. Spikes

either all staminate or pistillate; terminal spike staminate, short-stalked, 0.5–3 cm long; pistillate spikes 3–6, grouped or separate, cylindric and rounded at each end, 1.5–4 cm long and 1 cm wide, with many flowers; bracts leaflike, longer than head; body of pistillate scales small, tapered to a long awn as long or longer than perigynia beak. Perigynia oblong, inflated, the body 2–4 mm long, 2-ribbed, very abruptly tapered to a notched beak 1–2 mm long. Achenes 3-angled, 2 mm long, with a long, persistent style; stigmas 3. Flowering June–Sept.
HABITAT low openings in forests.
STATUS MW-OBL | NCNE-OBL | GP-OBL

Carex squarrosa L. *Squarrosae*
NARROW-LEAVED CAT-TAIL SEDGE

DESCRIPTION Densely clumped, native grass-like perennial. Stems 3–9 dm long. Leaves 3–6 mm wide, lower stem leaves reduced to scales. Spikes 1 (or sometimes 2–3), with both staminate and pistillate flowers, the staminate below the pistillate; pistillate portion oval, 1–3 cm long and 1–2 cm wide; lateral spikes (if present) pistillate, on upright stalks; bract of the terminal spike short and narrow; pistillate scales tapered to a tip or short-awned, smaller than the perigynia. Perigynia numerous and crowded, spreading, obovate, inflated, the body 3–6 mm long, abruptly tapered to a long notched beak 2–3 mm long. Achenes 2–3 mm long, with a persistent, strongly bent style; stigmas 3. Flowering June–Aug.
HABITAT swamps, floodplain forests, alder thickets, forest depressions.
STATUS MW-OBL | NCNE-OBL

Carex typhina Michx. *Squarrosae*
♦ COMMON CAT-TAIL SEDGE

DESCRIPTION Clumped, native grass-like perennial. Stems 3–8 dm long and usually shorter than upper leaves. Leaves 5–10 mm wide. Spikes 1–6, the terminal spike mostly pistillate with a short staminate base, the pistillate portion cylindric, 2–4 cm long and 1–1.5 cm wide, subtended by a short narrow bract; lateral spikes pistillate, smaller, upright or spreading on short stalks; pistillate scales hidden by

the perigynia, blunt or tapered to a tip. Perigynia obovate, crowded, body 3-5 mm long, abruptly narrowed to a notched beak 2-3 mm long. Achenes 2-3 mm long; stigmas 3. Flowering June–Sept.

HABITAT marshy areas and floodplain forests of large rivers (especially Miss and St. Croix), often occurring with **swamp oval sedge** (*Carex muskingumensis*) and **common bur sedge** (*Carex grayi*).

STATUS MW-OBL | NCNE-OBL | GP-OBL; historical records from Mich.

CAREX SECTION **Stellulatae**

Plants clumped; rhizomes short; bases brown, not fibrous. Inflorescence mostly open, spikes readily distinguished from each other, the lowest in our more common species not overlapping; bracts inconspicuous or lacking. Spikes 2-10 (solitary in *Carex exilis*), gynecandrous (unisexual in *C. sterilis*). Perigynia spreading to reflexed, typically plano-convex, widest at the base, generally chestnut brown to dark brown or even blackish at maturity; margins acute; base spongy; beak generally bidentate, margins finely serrate. Achenes much smaller than the perigynia.

NOTE The distinctions between species in this section are subtle; however, the species have habitat preferences that help with field identification. When examining

Carex typhina

perigynia, view the lowest 2-3 perigynia in the spike; the upper perigynia are very similar in all of our species.

CAREX SECTION *Stellulatae*

1 Spikes solitary; leaves involute; anthers 2-3.6 mm long . . . . *C. exilis*

1 Spikes 2-8; leaves flat or plicate; anthers 0.6-2.2 (-2.3) mm long **2**

2 Perigynium beak smooth-margined . . . . *C. seorsa*

2. Perigynium beak at least sparsely serrulate on margins **3**

3 Widest leaves 2.8-5 mm wide . . . . *C. wiegandii*

3 Widest leaves 0.8-2.7 mm wide **4**

4 Terminal spikes entirely staminate . . . . *C. sterilis*

4 Terminal spikes partly or wholly pistillate **5**

5 Terminal spikes without a distinct clavate base of staminate scales, staminate portion less than 1 mm long; anthers (1-) 1.2-2.2 (-2.3) mm long . . . . *C. sterilis*

5 Terminal spikes with a distinct clavate base of staminate scales 1-8 (-16) mm long; anthers 0.6-1.6 (-2) mm long **6**

6 Lower perigynia in the spikes 2-3 mm wide . . . . *C. atlantica*

6 Lower perigynia in the spikes 0.9-1.9 mm wide **7**

7 Lower perigynia mostly 2.9-3.6 (-4) mm long, 1.8-3.6 times as long as wide; beaks 0.9-2 mm long, mostly 0.5-0.8 times as long as the body . . . . *C. echinata*

7 Lower perigynia mostly 1.9-3 mm long, 1-2 times as long as wide; beaks 0.4-0.9 mm long, mostly 0.2-0.5 times as long as body **8**

8 Perigynia mostly nerveless over achene on ventral surface; perigynium beaks conspicuously setulose-serrulate; perigynia often ± convexly tapered from widest point to beak, forming a shoulder . . . . *C. interior*

8 Perigynia 1-10-nerved over achene on ventral surface; perigynium beaks more sparsely serrulate with definite spaces between the often single teeth; perigynia cuneately or even concavely tapered from widest point to beak . . . . *C. atlantica*

Carex atlantica L. H. Bailey *Stellulatae*
ATLANTIC STAR SEDGE

DESCRIPTION Densely clumped, native grass-like perennial. **Stems** 3–7 dm long, rough-to-touch on upper stem angles. **Leaves** 0.5–4 mm wide, mostly near base of stem. **Terminal spike** with staminate flowers at base, 12–15 mm long; **lateral spikes** pistillate, 8–12 mm long; spikes grouped or separate; pistillate scales broadly ovate, shorter than perigynia. **Perigynia** green, ovate to nearly round, 2–3.5 mm long and 1–3 mm wide, several-nerved on both sides, finely toothed on margin and beak; beak sharply notched, 0.5–1 mm long; stigmas 2. **Flowering** June–Aug.

SYNONYMS *Carex howei, Carex incomperta.*
HABITAT thickets and swamps, moist sand, sphagnum peat.
STATUS MW-FACW | NCNE-FACW | GP-OBL
NOTE similar to **inland sedge** (*Carex interior*) but less common in Upper Midwest region.

Carex echinata Murray *Stellulatae*
♦ LARGE-FRUITED STAR SEDGE

DESCRIPTION Clumped, native grasslike perennial. **Stems** 1–6 dm long, rough above. **Leaves** scalelike at base of stem; leaves with blades 3–6 on lower stem, 1–3 mm wide, shorter to as long as the stems. **Spikes** 3–7, stalkless, few-flowered;

Carex echinata

terminal spike with a slender staminate portion near its base; **lateral spikes** usually all pistillate; bract small; pistillate scales ovate, shorter than perigynia, yellow-tinged with green midvein. **Perigynia** 5–15 and crowded in each spike, spreading or curved downward, green or light brown, flat on 1 side and convex on other, narrowly ovate, spongy-thickened at base, 3–4 mm long, tapered to a toothed, notched beak 1–2 mm long. **Achenes** lens-shaped; stigmas 2. **Flowering** July–Sept.

SYNONYMS *Carex angustior, Carex cephalantha.*
HABITAT swamp margins, wet sandy lakeshores, hummocks in peatlands (as at Red Lake peatland in Minn).
STATUS MW-OBL | NCNE-OBL | GP-OBL

Carex exilis Dewey *Stellulatae*
COAST SEDGE

DESCRIPTION Densely clumped, native grass-like perennial. **Stems** stiff, 2–7 dm long and longer than the leaves. **Leaves** narrow and rolled inward. **Spikes** usually 1, either staminate or pistillate, or with both staminate and pistillate flowers, the staminate below the pistillate, 1–3 cm long; **lateral spikes** (if present) 1 or 2 and much smaller than terminal spike; lower 2 scales empty and upright; pistillate scales ovate, red-brown with translucent margins, about as long as perigynia. **Perigynia** spreading or drooping, ovate, flat on 1 side and convex on other, 3–5 mm long, spongy-thickened at base, tapered to a toothed beak to 2 mm long. **Achenes** lens-shaped; stigmas 2. **Flowering** June–Aug.

HABITAT sphagnum moss peatlands, interdunal wetlands near Great Lakes.
STATUS NCNE-OBL | GP-OBL; rare in n Minn (but locally common in Red Lake peatland), Wisc (THR).

Carex interior L. Bailey *Stellulatae*
♦ INLAND SEDGE

DESCRIPTION Densely clumped, native grass-like perennial. **Stems** slender, sharply 3-angled, 1–6 dm long, equal or longer than the leaves. **Leaves** 1–2 mm wide; sheaths tight, thin and translucent on

front. Spikes 2-4, the terminal spike with pistillate flowers borne above staminate (or rarely all staminate), the lateral spikes pistillate (or rarely with pistillate flowers borne above staminate), round in outline, about 5 mm wide, ± overlapping in heads 1-2.5 cm long; bracts small or absent; pistillate scales blunt-tipped, much shorter than the perigynia. Perigynia green-brown to brown, ovate, filled to margins by the achenes, sharp-edged but not wing-margined, 2-3 mm long and 1-2 mm wide, the base spongy so that achene fills upper perigynium body, tapered to a finely toothed beak to 1 mm long; the beak teeth small, not longer than 0.3 mm. Achenes lens-shaped, 1-1.5 mm long; stigmas 2. Flowering May-Aug.

HABITAT swamps, tamarack bogs, alder thickets, wet meadows and wetland margins.

STATUS MW-OBL | NCNE-OBL | GP-OBL

Carex seorsa Howe *Stellulatae*
SWAMP STAR SEDGE

Densely clumped, native grasslike perennial. Stems 2-7 dm long, longer than the leaves. Leaves 2-4 on lower third of stem, 1-4 mm wide, lower stem leaves reduced to scales. Spikes 4-8, stalkless, with both staminate and pistillate flowers, the staminate below the pistillate, or the lateral spikes often all pistillate; pistillate scales ovate, much shorter than perigynia. Perigynia 5-25, crowded, widely spreading,

green, broadly ovate, flat on 1 side and convex on other, spongy-thickened at base, 2-3 mm long and 1-2 mm wide, tapered to a short beak to a third as long as the body. Achenes lens-shaped; stigmas 2. Flowering May-July.

HABITAT moist to wet deciduous woods, swamps and open bogs.

STATUS MW-FACW | NCNE-FACW; Mich (THR).

Carex sterilis Willd. *Stellulatae*
FEN STAR SEDGE

DESCRIPTION Clumped, native grasslike perennial. Stems stiff, 1-7 dm long, longer than the leaves, rough-to-touch on the upper stem angles. Leaves 3-5 from lower part of stem, 1-4 mm wide, rough, lower stem leaves reduced to scales. Spikes 3-8, 3-12 mm long, stalkless, clustered or the lower separate; staminate and pistillate flowers mostly on separate plants; pistillate scales red-brown with green midvein and translucent margins, tapered to a tip or short point, about as long as body of perigynia. Perigynia 5-25, the lower spreading, red-brown, flat on 1 side and convex on other, broadly ovate, spongy-thickened at base, 2-4 mm long tapered to a finely toothed, notched beak 0.5-1.5 mm long, the beak teeth sharp, to 0.5 mm long. Achenes lens-shaped; stigmas 2. Flowering April-June.

SYNONYM Carex muricata var. sterilis.

HABITAT occasional in spring-fed calcareous fens and calcium-rich wet meadows; often near Great Lakes.

STATUS MW-OBL | NCNE-OBL | GP-OBL; Minn (THR).

NOTE similar to the more common inland sedge (*Carex interior*).

Carex wiegandii Mackenzie *Stellulatae*
WIEGAND'S SEDGE

DESCRIPTION Clumped, native grasslike perennial. Stems 3-6 dm long, longer than leaves, sometimes elongating to 1-2 m. Leaves 3-8 on lower stem, fan-folded, 2-5 mm wide, rough-to-touch on upperside, lower stem leaves reduced to scales. Spikes 4-6, with both staminate and pistillate flowers, the staminate below the pistillate, 4-12 mm long, stalkless, grouped

MATT LAVIN

Carex interior

in a cylindric head 1–3 cm long, the 2 lower spikes sometimes separate; pistillate scales red-brown with green midvein and translucent margins. Perigynia 5–25, the lower ones spreading or turned downward, green to red-brown, flat on 1 side and convex on other, broadly ovate, spongy-thickened at base, 2–4 mm long, tapered to a finely toothed, notched beak about 1 mm long. Achenes lens-shaped; stigmas 2. Flowering June–Aug.

HABITAT sphagnum moss peatlands, marshes, wet meadows.

STATUS NCNE-OBL

CAREX SECTION **Vesicariae**

Includes typical bottlebrush sedges of former section Pseudocypereae, with pistillate spikes tightly packed with perigynia, and pistillate scales with scabrous awns conspicuous between the perigynia; and also former section Vesicariae, with pistillate spikes often narrower, longer, less densely packed with perigynia in some species, and pistillate scales mostly not awned, hidden by the perigynia.

CAREX SECTION *Vesicariae*

1 Pistillate scales with a prominent, scabrous awn; often the body also ciliate 2
1 Pistillate scales smooth-margined, obtuse to acuminate, awnless (rarely the lowermost awned in *C. rostrata* and *C. utriculata*) 6
2 Perigynia ± reflexed at maturity, hard-walled, uninflated, flattened-triangular in cross-section, strongly and closely nerved with most nerves separated by less than three times their width; longest beak teeth 0.7–2.2 mm long 3
2 Perigynia spreading to ascending, thin-textured, ± inflated, ± round in cross-section; many nerves separated by more than three times their width; longest beak teeth 0.3–0.7 mm long 4
3 Spikes 12–18 mm thick; beak teeth strongly outcurved, the longest 1.3–2.1 mm long ... . *C. comosa*
3 Spikes 9–12 mm thick; beak teeth straight or slightly outcurved, the longest 0.7–1.2 mm .... *C. pseudocyperus*
4 Staminate scales (except sometimes the lowermost) acute to acuminate, essentially smooth-margined except at the very tip;

plants extensively colonial from elongate, creeping rhizomes; perigynia 7–11-nerved .... *C. schweinitzii*
4 Staminate scales (at least some) with a distinct, scabrous awn and sometimes also ciliate-margined; plants densely to loosely tufted, rhizomes connecting individual stems in a clump not more than 10 cm long; perigynia 7–25-nerved 5
5 Perigynia 15–20-nerved, the nerves (except for the two prominent lateral nerves) fusing together and becoming indistinguishable from about the middle of the beak to the apex, bodies ellipsoid, 1.4–2.2 mm wide; achenes smooth .... *C. hystericina*
5 Perigynia 7–12-nerved, the nerves separate nearly to the beak tip, the bodies broadly ellipsoid to ± spherical, 2–3.5 mm wide; achenes rough-papillate .... *C. lurida*
6 Leaf blades and bracts involute-filiform, wiry, 1–3 mm wide; stems round or obtusely 3-angled in cross-section, smooth; pistillate spikes 3–15-flowered, nearly spherical or short-oblong (not over 2 cm long); staminate spike usually solitary .... *C. oligosperma*
6 Leaf blades and bracts flat, U-, V-, or W-shaped in cross-section, 1.5–12 mm wide; stems round to 3-angled, often scabrous-angled; pistillate spikes usually more than 15-flowered, oblong to long-cylindric; staminate spikes normally 2 or more (often only 1 in *C. retrorsa*).................... 7
7 Perigynia 4–7 mm thick; achenes with a deep notch or constriction on one angle .. .. *C. tuckermanii*
7 Perigynia 2.5–3.5 mm thick; achenes symmetrical, not notched on one angle . 8
8 Lowest pistillate bract 3–9 times as long as the entire inflorescence; mature perigynia 7–12 mm long, at least the lower reflexed or widely spreading; staminate spike often 1, its base (or base of lowest staminate spike if more than one) slightly if at all elevated above summit of the crowded pistillate spikes (rarely lower spike remote) .... *C. retrorsa*
8 Lowest pistillate bract less than 3 times as long as inflorescence; perigynia 4–7.5 mm long, ascending or spreading; staminate spikes mostly 2–4, generally well elevated above the pistillate spikes 9
9 Leaves strongly papillose on upper surface, U-shaped in cross-section, glaucous, widest leaves 1.5–4.5 (-7.5) mm wide; stems round or very obtusely triangular, smooth below inflorescence .... *C. rostrata*

9 Leaves smooth or scabrous on upper surface, flat or folded, pale to dark green, widest leaves 3–12mm wide; stems triangular, often scabrous below the inflorescence.........................**10**

10 Colonial from long-creeping rhizomes; widest leaves 5–12 mm wide; ligules about as long as wide; basal sheaths usually spongy-thickened with little or no red tingeing; perigynia (at least those on lower portion of fully mature spike) ± widely spreading; stems bluntly triangular and sparsely and irregularly scabrous below the inflorescence..............*C. utriculata*

10 Tufted; widest leaves 3–6) mm wide; ligules longer than wide; basal sheaths not spongy-thickened and often tinged with reddish purple; perigynia ascending; stems sharply triangular and scabrous-angled below the inflorescence .....*C. vesicaria*

Carex comosa F. Boott *Vesicariae*
BRISTLY SEDGE

DESCRIPTION Native grasslike perennial, often forming large clumps. **Stems** stout, sharply 3-angled, 5–15 dm long. **Leaves** 5–12 mm wide; sheaths translucent on front, with small swollen joints on back. **Spikes** either staminate or pistillate; **terminal spike** staminate, 3–7 cm long; **lateral spikes** pistillate, 3–5, cylindric, 3–8 cm long and 9–12 mm wide, the lower spikes longer stalked and drooping when mature; bracts leaflike, much longer than the head; pistillate scales with translucent margins, tapered into a long, rough awn. **Perigynia** numerous, spreading outward when ripe, flattened 3-angled, lance-shaped, 5–8 mm long, shiny, strongly nerved, gradually tapered to the 2–3 mm long beak, the beak with curved teeth 1–2 mm long. **Achenes** 3-angled, 1.5–2 mm long; stigmas 3. **Flowering** June–Aug.
HABITAT marshes, wetland margins, floating mats, ditches.
STATUS MW-OBL | NCNE-OBL | GP-OBL

Carex hystericina Muhl. *Vesicariae*
◆ PORCUPINE SEDGE

DESCRIPTION Native grasslike perennial, from short rhizomes, often forming large clumps. **Stems** upright or leaning, 3-angled, 2–10 dm long, usually longer than the leaves. **Leaves** yellow-green, 3–8 mm wide; sheaths white, thin and translucent on front, green to yellow or red on back, the lower sheaths breaking into threadlike fibers. **Spikes** either all staminate or pistillate, the **terminal spike** staminate, 1–5 cm long, usually short-stalked and often with a bract; **lateral spikes** pistillate or occasionally with staminate flowers above pistillate, 1–4, short-cylindric, 1–5 cm long and 1–1.5 cm wide, separate or clustered, the lower spikes usually nodding on slender stalks, the upper spikes short-stalked and upright; pistillate scales small, narrow and much shorter than the perigynia, tipped with a rough awn. **Perigynia** spreading or upright, green to straw-colored, ovate, round in section when mature, 5–8 mm long, strongly nerved, abruptly tapered to a slender, toothed beak 3–4 mm long; the beak teeth to 1 mm long. **Achenes** 3-angled with concave sides, 1.5 mm long; stigmas 3. **Flowering** May–July.
HABITAT swamps, alder thickets, wet meadows and ditches; often in calcareous fens in s part of our region.
STATUS MW-OBL | NCNE-OBL | GP-OBL

Carex lurida Wahlenb. *Vesicariae*
BOTTLEBRUSH SEDGE

DESCRIPTION Clumped native grasslike perennial. **Stems** 3–10 dm long, shorter than the leaves, rounded 3-an-

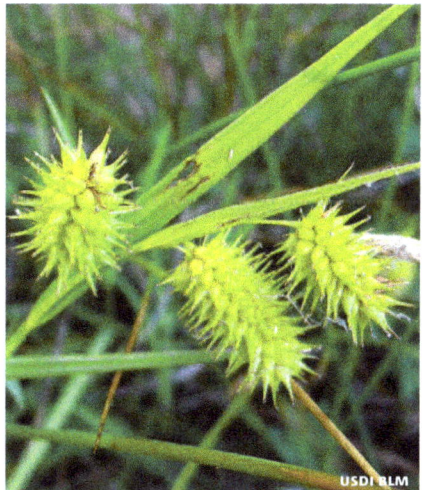

Carex hystericina

gled and ± smooth, purple-tinged at base. Leaves flat, 3–7 mm wide, lower stem leaves reduced to scales. Spikes either staminate or pistillate, terminal spike staminate, 1–7 cm long; pistillate spikes 1–4 (usually 2), many-flowered, grouped or the lower separate, stalkless and erect or the lower short-stalked and sometimes drooping, 2–7 cm long and 15–20 mm wide; bracts leafy, longer than the head; staminate scales with the midrib prolonged into an awn, pistillate scales awned or sharp-pointed. Perigynia in many rows, broadly ovate, somewhat inflated, 6–9 mm long, pale, smooth and shining, strongly nerved, tapered to a notched beak half to as long as the body. Achenes 3-angled, loosely enclosed in the lower part of the perigynium, the style persistent and twisted; stigmas 3. Flowering June–Aug.

HABITAT river floodplains, swamps, open bogs, fens and wet meadows.

STATUS MW-OBL | NCNE-OBL | GP-OBL

Carex oligosperma Michx. *Vesicariae*
♦RUNNING BOG SEDGE

DESCRIPTION Native grasslike perennial, forming colonies from creeping rhizomes. Stems slender, 4–10 dm long, purple-tinged at base. Leaves stiff, rolled inward, 1–3 mm wide. Spikes either all staminate or pistillate; staminate spike usually single; pistillate spikes l (or 2–3 and widely separated), stalkless or nearly so, ovate to short-cylindric, 1–2 cm long, lowest bract leaflike. Perigynia 3–15, ovate, somewhat inflated, compressed, 4–7 mm long, strongly several-nerved, abruptly tapered to a beak 1–2 mm long. Achenes 3-angled, 2–3 mm long; stigmas 3. Flowering June–Aug.

HABITAT open bogs and swamps, floating mats, pioneer mat-former along pond margins; common or dominant sedge in peatlands of northern part of region, less common southward.

STATUS MW-OBL | NCNE-OBL | GP-OBL

Carex pseudocyperus L. *Vesicariae*
♦FALSE BRISTLY SEDGE

DESCRIPTION Large, clumped grasslike native perennial. Stems stout, 3–10 dm long, 3-angled, rough-to-touch. Leaves 5–15 mm wide; sheaths translucent, yellow-tinged on back. Spikes either all staminate or pistillate, the terminal spike staminate, 1.5–7 cm long; lateral spikes pistillate, 2–6, cylindric, 3–8 cm long and 1 cm wide, lower spikes drooping on slender stalks; bracts much longer than the head; pistillate scales tipped by an awn, the awn shorter or longer than the perigynia. Perigynia spreading, ovate, 3-angled, 4–6 mm long, shiny, strongly nerved, tapered to a toothed beak, the beak teeth 0.5–1 mm long. Achenes 3-angled, 1.5 mm long; stigmas 3. Flowering June–Aug.

HABITAT marshy lake margins, swamps, fens, wet ditches; Red Lake peatland (Minn) where

Carex oligosperma

Carex pseudocyperus

indicator of mineral-rich fens.
STATUS MW-OBL | NCNE-OBL | GP-OBL

Carex retrorsa Schwein. *Vesicariae*
DEFLEXED BOTTLEBRUSH SEDGE

DESCRIPTION Densely clustered, native grasslike perennial. Stems 4–10 dm long. Leaves 3–4 dm long and 4–10 mm wide, flat and soft; sheaths dotted with small bumps. Spikes either all staminate or pistillate, or the terminal 1-2 spikes with both staminate and pistillate flowers, the staminate above the pistillate, stalkless or lowest spike on a slender stalk; lower spikes 3-8, pistillate, 1.5–5 cm long and 1.5–2 cm wide; pistillate scales conspicuous, shorter and narrower than the perigynia. Perigynia crowded in rows, spreading or the lowest perigynia angled downward, smooth and shiny, 6–13-nerved, 7–10 mm long, somewhat inflated, tapered to a long, smooth beak 2–4 mm long, the beak teeth short, to 1 mm long. Achenes dark brown, 3-angled, 2 mm long, loose in the lower part of the perigynium; stigmas 3. Flowering June–Aug.
HABITAT floodplain forests, swamps, thickets and marshes.
STATUS MW-OBL | NCNE-OBL | GP-OBL

Carex rostrata J. Stokes *Vesicariae*
BEAKED SEDGE

DESCRIPTION Native grasslike perennial, with short to long, creeping rhizomes. Stems round or bluntly 3-angled, 3–10 dm long, smooth below inflorescence. Leaves waxy blue, with many fine bumps on upper surface, to 4 mm wide, inrolled or channeled in section. Spikes either staminate or pistillate, the upper 2-5 staminate; lower 2-5 spikes pistillate or sometimes 1 or 2 with staminate flowers above the pistillate, cylindric, 2-10 cm long and 8-12 mm wide. Perigynia upright when young, becoming widely spreading when mature, yellow-green to brown, shiny, ovate, nearly round in section, inflated, 2–6 mm long, narrowed to a beak about 1 mm long. Achenes 3-angled, 1-2 mm long; stigmas 3. Flowering July–Sept.
HABITAT peat mats or shallow water.
STATUS MW-OBL | NCNE-OBL | GP-OBL

NOTE similar to *Carex utriculata* but much less common in Upper Midwest region, and with the leaves waxy blue and dotted with fine bumps on upper surface, v-shaped in section or inrolled, and only 2-4 mm wide.

Carex schweinitzii Dewey *Vesicariae*
SCHWEINITZ' SEDGE

DESCRIPTION Loosely clumped, native grasslike perennial. Stems 3–7 dm long, single or few together from rhizomes, sharply 3-angled. Leaves 4–10 mm wide, rough-to-touch near tip, lower stem leaves reduced to scales. Spikes either all staminate or pistillate; terminal spike staminate, on a slender stalk and usually with a bract; pistillate spikes 2-5, grouped or the lowest separate, cylindric, 3-8 cm long and to 1.5 cm wide, spikes ascending or spreading; pistillate scales translucent or brown-tinged, the midvein prolonged into a finely toothed awn often longer than the perigynium. Perigynia spreading or upright, ovate, inflated, round in section, 5-7 mm long, with 7-9 nerves, abruptly tapered to a beak, the beak teeth upright or spreading. Achenes 3-angled, loosely enclosed in the perigynia; stigmas 3. Flowering May-July.
HABITAT shaded streambanks.
STATUS MW-OBL | NCNE-OBL; historical record from Wisc.

Carex tuckermanii F. Boott *Vesicariae*
TUCKERMAN'S SEDGE

DESCRIPTION Clumped, native grasslike perennial, from short rhizomes. Stems 4–8 dm long. Leaves 3-6 mm wide and 2-4 dm long, soft and flat. Spikes either staminate or pistillate; staminate spikes usually 2, separated, raised above pistillate spikes; pistillate spikes 2-4, separated, cylindric, 2-5 cm long. Perigynia overlapping and ascending in 6 rows, broadly ovate, 7-10 mm long and 4-7 mm wide, inflated, tapered to a notched beak 2 mm long. Achenes 3-angled, obovate, 3-4 mm long, with a deep indentation near the middle of 1 angle; stigmas 3. Flowering June–Aug.
HABITAT swamps, alder thickets, low areas in forests, pond margins.
STATUS MW-OBL | NCNE-OBL | GP-OBL

Carex utriculata F. Boott *Vesicariae*
♦BEAKED SEDGE

DESCRIPTION Large, densely clumped, native grasslike perennial from short rootstocks, also forming turfs from long rhizomes. **Stems** bluntly 3-angled, 3–12 dm long, spongy at base. **Leaves** strongly divided with swollen joints 4–12 mm wide; sheaths white-translucent on front, divided with swollen joints on back. **Spikes** either staminate or pistillate, the upper 2–5 staminate, held well above the pistillate spikes, the **terminal spike** 3–6 cm long; lower 2–5 spikes pistillate or sometimes 1 or 2 with staminate flowers above the pistillate, usually separate, cylindric, 2–10 cm long and 8–12 mm wide, the **upper spikes** stalkless or short-stalked, **lower spikes** stalked, upright; bracts shorter to slightly longer than the head; pistillate scales acute to awn-tipped, body of scale shorter than perigynia. **Perigynia** upright at first to widely spreading when mature, in many rows, yellow-green to brown, shiny, ovate, nearly round in section, inflated, 3–8 mm long and 2–4 mm wide, strongly 7-9-nerved, contracted to a toothed beak 1–2 mm long, the teeth mostly straight, 0.5 mm long. **Achenes** 3-angled, 2 mm long; stigmas 3. **Flowering** June–Aug. **HABITAT** wet meadows, marshes, fens, swamps and lakeshores. **STATUS** MW-OBL | NCNE-OBL | GP-OBL

NOTE long confused with *Carex rostrata* (a boreal species uncommon in Upper Midwest region) which has waxy blue leaves to only 4 mm wide and with numerous small bumps on upper leaf surface.

Carex vesicaria L. *Vesicariae*
♦INFLATED SEDGE

DESCRIPTION Clumped, native grasslike perennial, from stout, short rhizomes. **Stems** sharply 3-angled and rough-to-touch below the head, 3–10 dm long, not spongy at base (as in *Carex utriculata*). **Leaves** 2–7 mm wide; sheaths white-translucent on front, not conspicuously divided with small swollen joints on back, the lowest sheaths often shredding into ladder-like fibers. **Spikes** either all staminate or pistillate, the upper 2–4 staminate, held well above the pistillate, 2–4 cm long; lower 1–3 spikes pistillate, separate, cylindric, 2–8 cm long and 4–15 mm wide, stalkless or short-stalked, erect; lowest bract usually longer than the head; pistillate scales acute to awn-tipped, shorter to as long as perigynia. **Perigynia** upright and overlapping in rows, dull yellow-green to brown, ovate to round, inflated, 3–8 mm long and 3 mm wide, strongly nerved, abruptly tapered to a toothed beak 1–2 mm long, the teeth 0.5–1 mm long. **Achenes** 3-angled, 2–3 mm long; stigmas 3. **Flowering** June–Aug. **HABITAT** wet meadows, marshes, forest depressions and shores. **STATUS** MW-OBL | NCNE-OBL | GP-OBL

CAREX SECTION Vulpinae

Plants clumped; bases generally pale. Inner band of the leaf sheaths hyaline, in other regards various: corrugated or

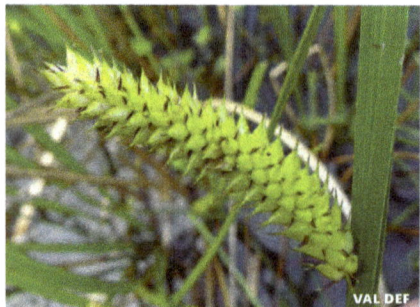

Carex utriculata

Carex vesicaria

smooth, thickened or fragile at the summit, sparsely purple-dotted or lacking pigmentation, and combinations of the above. Stems thick, spongy, weak, the angles narrowly winged, scabrous. Inflorescence longer than wide (ours), ovate to cylindrical. Bracts setaceous. Spikes densely flowered, the lower branched, mostly or all androgynous (the terminal always androgynous). Perigynia planoconvex, bases spongy (not spongy in *Carex alopecoidea*). Wetlands. The thick, spongy stems, branched lower spikes, and spongy perigynium bases (except in *C. alopecoidea*) are characteristic.

CAREX SECTION *Vulpinae*

1 Perigynia 6.5–8 mm long, enlarged below with a spongy disc-like area much broader than the rest of the body; beak twice as long as body of perigynium, or longer; thin ventral surface of leaf sheaths with many tiny purplish dots ..... *Carex crus-corvi*

1 Perigynia smaller, 3–6.2 mm long, corky below but without so distinct a disc-like area; beak slightly longer than the body of perigynium, or shorter; thin ventral surface of leaf sheaths dotted or not **2**

2 Perigynia somewhat contracted or ± cuneately tapered into the beak (this then difficult to define, but about equaling or slightly exceeding the body, if the latter is measured from the base of perigynium to summit of achene), 4–6.2 mm long, with at least a few nerves ventrally; ventral surface of leaf sheaths not dotted with purplish . **3**

2 Perigynia contracted into a beak no longer than the body, (2.6–) 3–4.5 (–4.8) mm long, essentially nerveless ventrally; ventral surface of leaf sheaths sparsely to strongly dotted with purplish, especially toward the apex **4**

3 Sheaths thickened (or even ± cartilaginous) at the concave or truncate mouth, smooth and unwrinkled ventrally; perigynia 4.7–6.2 mm long .... *C. laevivaginata*

3 Sheaths thin (usually broken) at the prolonged (when intact) mouth, rather strongly puckered or cross-wrinkled ventrally, very rarely nearly or quite smooth; perigynia 4–5 (–5.5) mm long ... *C. stipata*

4 Sheath fronts not cross-wrinkled ventrally; achenes about as long as wide; perigynia nerveless or with 3–5 very faint veins dorsally .... *C. alopecoidea*

4 Sheath fronts clearly cross-wrinkled ventrally, at least near the apex; achenes distinctly longer than wide; perigynia with 3–5 prominent veins dorsally .. *C. conjuncta*

Carex alopecoidea Tuckerman *Vulpinae*
♦ BROWN-HEADED FOX SEDGE

DESCRIPTION Clumped, native grasslike perennial. **Stems** soft, 4–10 dm long, 3-angled and sharply winged. **Leaves** 3–8 mm wide; sheaths purple-dotted, not cross-wrinkled. **Spikes** with both staminate and pistillate flowers, staminate flowers above pistillate, in heads 1.5–5 cm long; pistillate scales tapered to tip or with a short sharp tip. **Perigynia** yellow-brown when mature, ovate, flat on 1 side and convex on other, 3–5 mm long, spongy-thickened at base, narrowed to a beak half to as long as the body. **Achenes** lens-shaped, 1–2 mm long; stigmas 2. **Flowering** May–July.

HABITAT swamps and floodplain forests, streambanks, swales, moist fields.

STATUS MW-FACW | NCNE-FACW | GP-FACW

Carex conjuncta Boott *Vulpinae*
GREEN-HEADED FOX SEDGE

DESCRIPTION Clumped, native grasslike perennial; plants light green. **Stems** slender, 3-angled, 4–9 dm long, as long or longer than leaves, somewhat roughened above. **Leaves** soft, 5–10 mm wide; margins rough; sheaths somewhat cross-wrinkled on inner side. **Spikes** 6–10, either staminate or pistillate, the staminate above the pistillate, in a narrow head 2–5 cm long; bract small and bristlelike or absent; pistillate scales ovate, about as long as perigynia, tapered to a sharp tip or short awn. **Perigynia** ovate, green or yellow-green, 3–4 mm long and 1–2 mm wide, slightly spongy at base, tapered to a rough, 2-toothed beak. **Achenes** flattened; stigmas 2. **Flowering** June–July.

HABITAT low prairie, streambanks, wet meadows and thickets.

STATUS MW-FACW | NCNE-FACW | GP-FAC; Minn (THR), Mich (THR).

Carex crus-corvi Shuttlew. *Vulpinae*
CROWFOOT FOX SEDGE

DESCRIPTION Clumped, native grasslike perennial. Stems stout, 4–8 dm long, sharply 3-angled, shorter than leaves. Leaves 5–10 mm wide; sheaths not cross-wrinkled. Spikes in large head, 8–20 cm long, upper spikes grouped, lower spikes separate. Spikes with both staminate and pistillate flowers, staminate above pistillate; pistillate scales triangular-ovate, shorter or equal to perigynia. Perigynia ovate, flat on 1 side and convex on other, 5–9 mm long, with a broad, spongy base to 2.5 mm wide, tapered to a notched beak much longer than body. Achenes lens-shaped; stigmas 2. Flowering June–July.

HABITAT floodplains, marshes, edges of seasonally wet woodland depressions.

STATUS MW-OBL | NCNE-OBL | GP-OBL; Wisc (END), Mich (END); historical records for Minn.

Carex laevivaginata (Kuk.) Mack.
Vulpinae
SMOOTH-SHEATHED FOX SEDGE

DESCRIPTION Densely clumped, native grasslike perennial. Stems stout, 3-angled, 3–10 dm long. Leaves 3–10 mm wide; sheaths not cross-corrugated (as in *Carex stipata*). Spikes with both staminate and pistillate flowers, the staminate above the pistillate, numerous, grouped into a dense head 2–5 cm long and 1–1.5 cm wide, green or straw-colored when mature; bracts short or reduced to bristles sometimes longer than the spikes; pistillate scales shorter than the perigynia, tapered to a tip or short awn. Perigynia green or straw-colored, spreading, broadly lance-shaped, 5–6 mm long, long-tapered to the tip, flat on 1 side and convex on other, spongy-thickened at base. Achenes lens-shaped; stigmas 2. Flowering May–July.

HABITAT swamps, marshy areas, streambanks.

STATUS MW-OBL | NCNE-OBL; Minn (THR), Wisc (END).

Carex stipata Muhl. ex Willd. *Vulpinae*
♦ COMMON FOX SEDGE

DESCRIPTION Densely clumped, native grasslike perennial. Stems 3-angled and slightly winged, 2–12 dm long. Leaves 4–8 mm wide; sheaths cross-wrinkled on front, divided with small swollen joints on back. Spikes with both staminate and pistillate flowers, staminate flowers borne above pistillate, clustered or the lowest spikes often separate, in oblong heads 3–10 cm long; bracts small and sometimes bristlelike, longer than the spike; pistillate scales tapered to a tip or with a short, sharp point, half to 3/4 as long as the perigynia. Perigynia yellow-green to dull brown, flat on 1 side and convex on other, narrowly ovate, 3–5 mm long and 1–2 mm wide, strongly several-nerved on both sides, tapered to a finely toothed, notched beak 1–3 mm long. Achenes lens-shaped, 1.5–2 mm long; stigmas 2. Flowering May–July.

HABITAT floodplain forests and swamps, thickets, wet meadows, wetland margins and ditches; usually not in sphagnum bogs.

STATUS MW-OBL | NCNE-OBL | GP-OBL

Cladium TWIG-RUSH

Cladium mariscoides (Muhl.) Torr.
♦ TWIG-RUSH

DESCRIPTION Grasslike native grasslike peren-

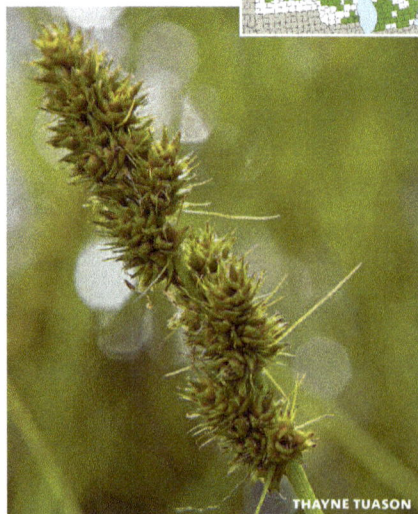

THAYNE TUASON

Carex stipata

nial, spreading by rhizomes and forming colonies. Stems single or in small groups, stiff, slender, smooth, 0.3–1 m tall. Leaves 1–3 mm wide, upper portion round in section, middle portion flattened. Flowers in lance-shaped spikelets, 3–5 mm long, in branched clusters (umbels) at end of stem and also with 1–2 clusters on slender stalks from leaf axils; uppermost flower perfect, the style 3-parted; middle flowers staminate; lowest scale of each spikelet empty; scales overlapping, ovate, brown; bristles absent. Achenes dull brown, 2–3 mm long, pointed at tip; tubercle absent. Flowering June–Aug.

SYNONYM *Mariscus mariscoides.*

HABITAT shallow water, sandy or mucky shores, floating bog mats, calcium-rich wet meadows, seeps, fens and low prairie.

STATUS MW-OBL | NCNE-OBL | GP-OBL

NOTE *Cladium mariscus* subsp. *jamaicense* (sawgrass) is a dominant species of the Florida Everglades.

Cyperus FLATSEDGE

Small to medium, annual or perennial grasslike plants. Stems often clumped, unbranched, sharply 3-angled. Leaves mostly from base of plants, with 1 or more leaflike bracts near top of stems, the blades flat or folded along midvein. Flower heads in umbels at ends of stems; the spikelets many, grouped in 1 to several rounded or cylindric spikes. Flowers per-

Cladium mariscoides

fect; bristlelike sepals and petals absent; stamens 1–3; styles 2–3-parted. Achenes lens-shaped or 3-angled, beakless.

Cyperus FLATSEDGE

1 Achenes lens-shaped; stigmas 2 . . . . **2**
1 Achenes 3-angled; stigmas 3 . . . . **4**
2 Scales straw-colored . . . . *C. flavescens*
2 Scales red-purple or deep brown **3**
3 Style cleft to slightly below middle . . . .
   . . . . *C. bipartitus*
3 Style cleft almost to base . . . . *C. diandrus*
4 Plants perennial **5**
4 Plants annual. **6**
5 Scales 2–3 mm long, only slightly keeled . .
   . . . . *C. esculentus*
5 Scales 3–5 mm long, keeled . . *C. strigosus*
6 Scales curved outward at tip; stamens 1 . **7**
6 Scales not curved outward at tip; stamens 3
   . . . . **8**
7 Scales 3-nerved, not awn-tipped . . . .
   . . . . *C. acuminatus*
7 Scales 7–9-nerved, tipped with a short awn
   . . . . *C. squarrosus*
8 Scales to 2 mm long, with 3–5 veins near
   center of scale . . . . *C. erythrorhizos*
8 Scales 2–5 mm long, with 7 or more well-
   spaced veins. . . . . *C. odoratus*

Cyperus acuminatus Torr. & Hook.
SHORT-POINT FLATSEDGE

DESCRIPTION Clumped, grasslike native annual. Stems 3-angled, 5–30 cm tall. Leaves light green, as long as stems or longer, to 2 mm wide; bracts to 3 mm wide, longer than the head. Spikelets flat, 3–7 mm long, crowded in 1–5 round clusters (spikes), 1 spike stalkless, the other spikes on stalks 1–3 cm long; scales 1–2 mm long, ovate, pale green, becoming tan when mature, strongly 3-nerved; stamens 1; style 3-parted. Achenes tan to pale brown, 3-angled, 0.5–1 mm long. Aug–Sept.

HABITAT muddy or sandy shores, stream-banks and flats.

STATUS MW-OBL | NCNE-OBL | GP-OBL; Minn (THR); historical record for Mich.

Cyperus bipartitus Torr.
SHINING FLATSEDGE

DESCRIPTION Clumped, grasslike native annual. Stems 3-angled, 1–3 dm tall. Leaves usually shorter than stems; leaves and bracts 0.5–2 mm wide, the bracts usually 3, longer than the spikes. Spikelets linear, 10–15 mm long and 2–3 mm wide, in clusters (spikes) of 3–10, the spikes stalkless or on stalks to 10 cm long; scales overlapping, ovate, shiny, purple-brown on margins; stamens 2 or 3; style 2-parted, lower third not divided. Achenes lens-shaped, 1–2 mm long, hidden by the scales. Flowering July–Sept.
SYNONYM *Cyperus rivularis.*
HABITAT wet, sandy, gravelly or muddy shores, streambanks, wet meadows, ditches.
STATUS MW-OBL | NCNE-FACW | GP-FACW

Cyperus diandrus Torr.
UMBRELLA FLATSEDGE

DESCRIPTION Clumped, grasslike native annual. Stems 3-angled, 5–30 cm tall. Leaves about as long as stems, 1–3 mm wide; bracts usually 3, longer than the spikes. Spikelets 5–10, linear, 5–20 mm long and 2–3 mm wide; in 1–3 loose, rounded spikes, the spikes on stalks to 6 cm long; scales loosely overlapping, ovate, 2–3 mm long, not shiny, purple-brown on margins; stamens 2; style 2-parted, divided nearly to the base, persistent. Achenes lens-shaped, pale brown, 1 mm long, visible between the scales. Flowering July–Sept.
HABITAT sandy or muddy shores, streambanks, wet meadows.
STATUS MW-FACW | NCNE-OBL | GP-FACW

Cyperus erythrorhizos Muhl.
RED-ROOT FLATSEDGE

DESCRIPTION Clumped, stout or slender native annual, roots red. Stems 3-angled, 1–7 dm long. Leaves mostly near base of plant, shorter to longer than stems, 2–8 mm wide; bracts 3–7, to 9 mm wide, usually much longer than the spikes. Spikelets linear, 2–10 mm long and 1–2 mm wide; grouped in a pinnate manner along a stalk (rachilla), in cylindric clusters, the terminal cluster stalkless, the others on stalks to 8 cm long; scales ovate, satiny brown, 1–2 mm long, overlapping; stamens 3; style 3-parted. Achenes ivory white, sharply 3-angled, ovate, 0.5–1 mm long. Flowering July–Sept.
HABITAT sandy or muddy shores, streambanks, exposed mud flats, ditches; often with **rusty flatsedge** (*Cyperus odoratus*).
STATUS MW-OBL | NCNE-OBL | GP-OBL

Cyperus esculentus L.
⧫ CHUFA, YELLOW NUTSEDGE

DESCRIPTION Introduced, grasslike perennial, with rhzomes ending in small tubers. Stems single, 3-angled, erect, 2–7 dm long. Leaves light green, mostly from base of plant, about as long as stems, 3–10 mm wide, with a prominent midvein; the bracts 3–6, usually much longer than the spikes. Spikelets linear, 3–12 cm long and 1–2 mm wide; pinnately arranged on a stalk, forming loose cylindrical spikes, the spikes to 5 cm long and 1–2 mm wide; scales straw-colored, 2–3 mm long, overlapping; stamens 3; style 3-parted. Achenes pale brown, 3-angled, 1–2 mm long. Flowering July–Sept.
HABITAT sandy or muddy shores, streambanks, marshes, ditches and other wet places; weedy in wet or moist cultivated fields.
STATUS MW-FACW | NCNE-FACW | GP-FACW; listed noxious weed in Mich; common lawn weed in se USA.

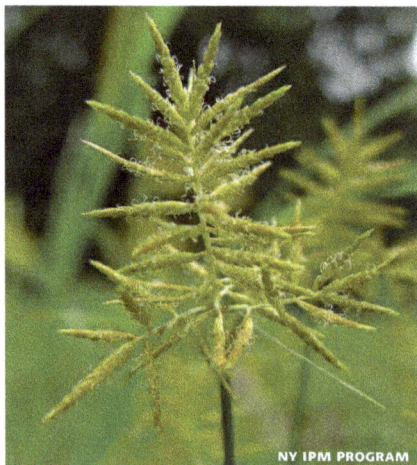

Cyperus esculentus

Cyperus flavescens L.
YELLOW FLATSEDGE

DESCRIPTION Clumped, grasslike native annual. **Stems** very slender, 3-angled, 5–30 cm tall. **Leaves** mostly at base of plant, 1–3 mm wide, mostly shorter than the stems; bracts 2–4, much longer than the spikes. Spikelets linear, 10–15 cm long and 2–3 mm wide, stalkless or on stalks to 5 cm long; scales ovate, straw-colored, 2–3 mm long; stamens mostly 3; style 2-parted. **Achenes** lens-shaped, black and shiny, 1 mm long. **Flowering** Aug–Sept.

HABITAT wet, sandy or mucky shores and wet meadows.

STATUS MW-OBL | NCNE-OBL | GP-OBL

Cyperus odoratus L.
RUSTY FLATSEDGE

DESCRIPTION Stout, grasslike, fibrous-rooted, native annual. **Stems** clumped or single, 3-angled, 2–7 dm long. **Leaves** mostly from base of plant, shorter to longer than flowering stems, the blades 2–8 mm wide; the involucral bracts much longer than the spikes. **Spikelets** linear, 1–2 cm long, pinnately arranged along a stalk, forming several to many cylindrical spikes, the spikes stalkless or stalked; scales red-brown, 2–3 mm long, overlapping; stamens 3; style 3-parted. **Achenes** brown, 3-angled, 1–2 mm long. **Flowering** July–Sept.

SYNONYMS *Cyperus engelmannii, Cyperus speciosus, Cyperus ferruginescens.*

HABITAT sandy or muddy shores, floating mats, ditches, wet cultivated fields.

STATUS MW-FACW | NCNE-OBL | GP-FACW

Cyperus squarrosus L.
◆ AWNED FLATSEDGE

DESCRIPTION Small, clumped, sweet-scented, grasslike annual; native. **Stems** very slender, 3-angled, 3–15 cm long. **Leaves** few, all at base of plant, 1–2 mm wide; bracts 2–3, longer than the spikes. **Spikelets** linear, flattened, 3–10 mm long, in 1–4 dense, rounded spikes, 1 spike stalkless, the other spikes on stalks to 3 cm long; scales 1–2 mm long, tipped by an awn to 1 mm long, pale brown; stamens 1; style 3-parted. **Achenes** brown, 3-angled, 0.5–1 mm long. **Flowering** July–Sept.

SYNONYMS *Cyperus aristatus, Cyperus inflexus.*

HABITAT wet, sandy or muddy shores and streambanks, mud and gravel bars, wet meadows.

STATUS MW-OBL | NCNE-OBL | GP-OBL

Cyperus strigosus L.
FALSE NUTSEDGE, STRAW-COLORED CYPERUS

DESCRIPTION Grasslike native perennial, from tuberlike corms. **Stems** single or few, slender, sharply 3-angled, 1–8 dm long. **Leaves** mostly at base of plants, the blades 2–12 mm wide, margins rough-to-touch; the bracts mostly longer than the spikes. **Spikelets** flat, linear, 6–20 mm long and 1–2 mm wide, golden-brown, pinnately arranged and spreading, in several to many cylindric spikes, the spikes often bent downward, on stalks 1–12 cm long, the stalks sometimes branched; scales straw-colored, 3–5 mm long; stamens 3; style 3-parted. **Achenes** brown, 3-angled, 1–2 mm long. **Flowering** July–Sept.

HABITAT wet, sandy or muddy shores, streambanks, marshes, wet meadows, ditches, cultivated fields.

STATUS MW-FACW | NCNE-FACW | GP-FACW

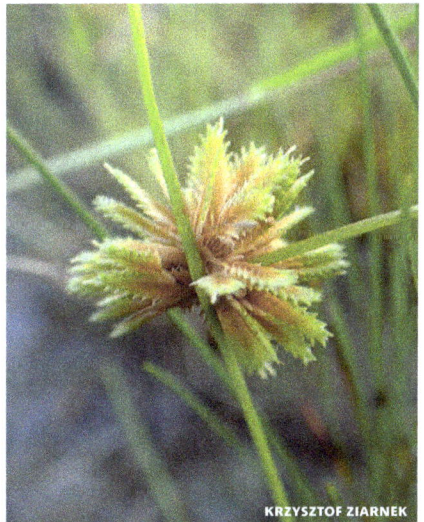

KRZYSZTOF ZIARNEK

Cyperus squarrosus

Dulichium
THREE-WAY SEDGE

Dulichium arundinaceum (L.) Britton
♦ THREE-WAY SEDGE, DULICHIUM

DESCRIPTION Grasslike native perennial, spreading by rhizomes and often forming large colonies. **Stems** stout, erect, 3–10 dm long, jointed, hollow, rounded in section. **Leaves** 3-ranked, flat, short, 4–15 cm long and 3–8 mm wide; lower leaves reduced to sheaths. **Flower heads** from leaf axils, in linear clusters of 5–10 spikelets, the clusters 1–2.5 cm long; scales lance-shaped, green to brown, 5–8 mm long. **Flowers** perfect; sepals and petals reduced to 6–9 downwardly barbed bristles; stamens 3; style 2-parted. **Achenes** light brown, oblong, 2–4 mm long, beaked by the persistent, slender style. **Flowering** July–Sept.

HABITAT shallow marshes, wet meadows, shores, bog margins.

STATUS MW-OBL | NCNE-OBL | GP-OBL

Eleocharis SPIKE-RUSH

Small to medium rushlike plants, mostly perennial from rhizomes (annual in several species), often forming large, matlike colonies. Stems round, flattened, or angled in section. Leaves reduced to sheaths at base of stems. Flower head a single spikelet at tip of stem; scales of the spikelets spirally arranged and overlapping. Flowers perfect; sepals and petals bristlelike or absent, the bristles usually 6 if present; stamens 3; styles 2-3-parted, the base of style swollen and persistent as a projection (tubercle) atop the achene, or sometimes joined with the achene body. Achenes rounded on both sides or 3-angled.

ADDITIONAL SPECIES
Eleocharis atropurpurea (Retz.) J. Presl & C. Presl; reported from sandy shores of lakes and ponds with fluctuating water levels in the w LP of Michigan (END); known elsewhere from scattered locations in the s USA. *Eleocharis atropurpurea* is a tiny annual plant, similar to *E. geniculata* in having shiny black achenes, but even smaller.

Eleocharis mamillata (H.Lindb.) H. Lindb., a boreal species reported from Ashland, Bayfield, and Douglas counties in n Wisc; found between wet meadow and shrub-carr communities and on sphagnum-sedge bog mats; often included within *E. palustris*.

Eleocharis tenuis (Willd.) Schultes; perennial, reported from Pierce County in w Wisc; stems slender, to only 0.5 mm wide (*see key*).

Eleocharis SPIKE-RUSH

1   Tubercle joined with achene and not forming a distinct cap 2
1   Tubercle forming a distinct cap atop the achene 4
2   Small plants mostly less than 1 dm tall; scales 1–2 mm long; achenes to 1 mm long . . . . *E. parvula*
2   Plants larger, more than 1 dm tall; scales 3–5 mm long; achenes 2 mm or more long . **3**
3   Stems 1–4 dm long, not rooting at tips; spikelets with 3–9 flowers *E. quinqueflora*
3   Stems 4 dm or more long, often rooting at tips; spikelets with 10 or more flowers . . . . *E. rostellata*
4   Spikelets slender, about as wide as stems; the scales persistent 5
4   Spikelets wider than stems; scales deciduous 7
5   Stems round in section and divided by cross-walls (appearing jointed) . . . . *E. equisetoides*

KRZYSZTOF ZIARNEK

Dulichium arundinaceum

5 Stems 3–4-angled and not divided by cross-walls **6**

6 Stems 3-angled, 1 to 2 mm wide; scales 10 or less .... *E. robbinsii*

6 Stems 4-angled, 2–6 mm wide; scales 10 or more................. *E. quadrangulata*

7 Styles 3-parted; achenes 3-angled to ± round in section **8**

7 Styles 2-parted; achenes lens-shaped . . **17**

8 Body of achene flat across top, tubercle flattened, its base nearly as wide as top of achene................. *E. melanocarpa*

8 Achene body rounded at top; tubercle long and narrow or constricted at base **9**

9 Achene round in section, 2x longer than wide **10**

9 Achene 3-angled, nearly as wide as long **12**

10 Stems flattened, 1 mm or more wide; rare. .... *E. wolfii*

10 Stems ± round in section, less than 0.5 mm wide **11**

11 Scales 1.5–3 mm long; achene grayish or yellow-white; common ..... *E. acicularis*

11 Scales to 1.3 mm long; achene dark-yellow to orange; rare in northern portion of region........................ *E. nitida*

12 Achenes pitted or rough (use hand lens) **13**

12 Achenes smooth...................... **15**

13 Stems flattened............ *E. compressa*

13 Stems 4–8-angled **14**

14 Stems usually 6- to 8-angled; achenes yellow to orange-yellow, somewhat persistent after the scales are shed, with 12–20 horizontal ridges through the length of the achene; widespread species .... *E. elliptica*

14 Stems 4- or 5-angled; achenes green to brown-green (rarely yellow), deciduous with the falling scales, with 6–10 (–14) horizontal ridges through the length of the achene; reported for w Wisc .... *E. tenuis* .... (see *Additional Species*, previous page)

15 Annual; achenes gray to nearly white..... .... *E. microcarpa*

15 Perennial; achenes olive to brown **16**

16 Achene angles not winged; bristles present .... *E. intermedia*

16 Achene angles narrowly winged; bristles absent.................... *E. tricostata*

17 Plants clumped annuals with soft, easily compressed stems.................... **18**

17 Plants perennial with stiff stems; spreading by short to long rhizomes **19**

18 Tubercle very depressed, not over 1/4 of the total length of the achene, nearly or quite as wide as the truncate body, on which it appears as a flattish cap.......... .... *E. engelmannii*

18 Tubercle ± broadly triangular, more than 1/4 the total length of the achene and about 3/4 to nearly as wide as the broadest part of the body. .... *E. obtusa*

19 Plants small, 3–15 cm tall; stems clumped .... *E. flavescens*

19 Plants larger, 2–10 dm tall; stems scattered and single, or in small clusters .... *E. palustris*

Eleocharis acicularis (L.) Roemer & Schultes ◆LEAST SPIKE-RUSH

DESCRIPTION Small, clumped, mat-forming native perennial, from slender rhizomes. Stems threadlike, 3–15 cm long and to 0.5 mm wide, somewhat 4-angled and grooved; sheaths membranous, usually red at base. Spikelets narrowly ovate, 3–6 mm long and 1–1.5 mm wide; scales with a green midvein and chaffy margins; sepals and petals reduced to 3–4 bristles or absent; style 3-parted. Achenes gray, round-ed 3-angled, ridged, to 1 mm long; tubercle cone-shaped, constricted at base. Flowering May–Sept.

HABITAT shallow water, exposed muddy or sandy shores, marshes, streambanks.

STATUS MW-OBL | NCNE-OBL | GP-OBL

Eleocharis compressa Sullivant FLAT-STEM SPIKE-RUSH

DESCRIPTION Clumped native perennial, from stout black rhizomes. Stems flattened and of-ten twisted, 1.5–4 dm

Eleocharis acicularis

long and 0.5–1 mm wide, shallowly grooved; sheaths red or purple at base. Spikelets ovate, 4–10 mm long and 3–4 mm wide; lowest scale sterile and encircling the stem; fertile scales with a green midvein, purple-brown on sides, and white translucent margins; sepals and petals absent or reduced to 1–5 bristles; style 3-parted. Achenes yellow-brown, covered with small bumps, somewhat 3-angled, 1–1.5 mm long; tubercle small, constricted at base. Flowering May–Aug.

SYNONYM *Eleocharis acuminata.*

HABITAT low calcareous prairie, wet meadows, swamps, ditches, dolomite limestone crevices in e UP of Mich.

STATUS MW-FACW | NCNE-FACW | GP-FACW; Mich (THR).

Eleocharis elliptica Kunth
ELLIPTIC SPIKE-RUSH

DESCRIPTION Mat-forming native perennial, with long rhizomes. Stems subterete to sometimes compressed, often with 5–10 ridges, to 90 cm long and 0.3–0.8 mm wide, spongy; sheaths persistent, not splitting, dark red at base, usually red-brown at tip, tooth to 0.5 mm long (usually present on some stems). Spikelets ovoid, 3–8 cm long and 2–3 mm wide, scales spreading in fruit, 10–30, brown, midrib region often paler, entire or shallowly notched; perianth bristles absent or rarely 1–3, pale brown, to 1/2 of achene length, sparsely retrorsely spinulose; style 3-parted. Achenes yellow, orange, or medium brown, obpyriform, angles evident to prominent, ca. 1 mm long, rugulose (visible at 10x), with 12–20 horizontal ridges in a vertical series; tubercle brown to whitish, depressed, apiculate.

HABITAT wet, often marly ground: fens, beaches, stony and often marshy shores, interdunal pools and flats, sandy swales and ditches; occasionally in bogs and along trails in conifer swamps; sometimes in shallow water (and if so, sometimes with stems over 5 dm long, although stems normally much shorter).

STATUS MW-OBL | NCNE-OBL | GP-OBL

NOTE the yellow achenes of *E. elliptica* often persist after the scales are shed, making this species easily recognized late in the season.

Eleocharis engelmannii Steud.
ENGELMANN'S SPIKE-RUSH

DESCRIPTION Native annual. Stems 2–40 cm long and 0.5–1.5 mm wide; sheaths with a small tooth to 0.3 mm long. Spikelets cylindric or ovoid, 5–10 mm long and 2–3 mm wide; floral scales 25–100(–200), orange brown to stramineous, midribs mostly keeled; perianth bristles present or often absent, 5–8, brown, rudimentary to slightly exceeding tubercle; styles 2–3-parted. Achenes ca. 1 mm long; tubercle depressed, triangular, to nearly 1/2 as high as wide, nearly as wide as achene.

HABITAT moist, open, usually sandy ground.

STATUS MW-FACW | NCNE-FACW | GP-FACW

Eleocharis equisetoides (Elliott) Torr.
HORSETAIL SPIKE-RUSH

DESCRIPTION Medium to large native perennial, spreading by stout rhizomes. Stems round, hollow, to 1 m long and 5 mm wide; sheaths membranous, brown or green. Spikelets cylindric, 1.5–3 cm long, not wider than the stem; scales ovate to rounded, with narrow chaffy margins; stamens and petals reduced to 5–6 small bristles; style 3-parted. Achenes rounded on sides, golden-brown, shiny, 2 mm long; tubercle black, flattened, 3-angled, 1 mm long.

SYNONYM *Eleocharis elliottii, Eleocharis interstincta.*

HABITAT sandy or mucky lakeshores and pond margins, sometimes in shallow water.

STATUS MW-OBL | NCNE-OBL | GP-OBL

Eleocharis flavescens (Poir.) Urban
BRIGHT GREEN SPIKE-RUSH

DESCRIPTION Small, clumped, mat-forming native perennial, spreading by slender rhizomes. Stems bright green, flattened, 3–15 cm long. Spikelets ovate, 2–7 mm long and much wider than stem; scales ovate, red-brown, with a green midvein; sepals and petals reduced to 6–8 barbed bristles; style 2-parted (rarely 3-parted). Achenes lens-shaped, brown, 1 mm

long; tubercle pale, cone-shaped, constricted at base. Flowering July–Aug.

SYNONYM *Eleocharis olivacea.*

HABITAT shallow water, sandy or muddy lakeshores; sometimes where calcium-rich.

STATUS NCNE-OBL | GP-OBL; Minn (THR).

Eleocharis intermedia (Muhl.) Schultes
MATTED SPIKE-RUSH

DESCRIPTION Small, densely clumped, native annual. Stems threadlike, grooved, of unequal lengths, 5–20 cm long; sheaths toothed on 1 side. Spikelets long-ovate, wider than stem; scales oblong lance-shaped, purple-brown, with a green midvein and white, translucent margins; sepals and petals reduced to barbed bristles or sometimes absent; style 3-parted. Achenes light brown to olive, 3-angled, 1 mm long; tubercle cone-shaped, constricted at base. Flowering June–Sept.

SYNONYM *Eleocharis macounii.*

HABITAT wet, sandy or mucky shores, streambanks, mud flats.

STATUS MW-OBL | NCNE-OBL | GP-OBL

Eleocharis melanocarpa Torr.
BLACK-FRUIT SPIKE-RUSH

DESCRIPTION Densely clumped, native perennial. Stems flattened, wiry, 2–6 dm long; sheaths 1-toothed. Spikelets cylindric to narrowly ovate, 5–15 mm long, wider than stem; scales ovate, rounded at tip, brown with a lighter midvein; sepals and petals reduced to 3–4 rough bristles; style 3-parted. Achenes oblong pyramid-shaped, rounded 3-angled, dark brown to black, 1 mm long; tubercle light brown, depressed across top of achene and pointed in middle. Flowering June–July.

HABITAT wet, sandy or mucky lakeshores, moist sandy prairie, mud flats; disjunct in Upper Midwest region from Atlantic coast from Mass to Fla and Tex.

STATUS MW-FACW | NCNE-FACW | GP-FACW

Eleocharis microcarpa Torr.
HAIR SPIKE-RUSH

DESCRIPTION Densely clumped, native annual. Stems slender, somewhat 4-angled, to

3 dm long (or longer when underwater). Spikelets ovate or oblong, 2–6 mm long, wider than stems; scales ovate, brown-red, with a green midvein and paler margins, all but lowest scale soon deciduous; sepals and petals reduced to 3–6 slender bristles; style 3-parted. Achenes pale white, 3-angled, to 1 mm long; tubercle small, pyramid-shaped. Flowering July–Aug.

SYNONYM *Eleocharis torreyana.*

HABITAT marshes and wet meadows.

STATUS MW-OBL | NCNE-OBL | GP-OBL; uncommon in LP of Mich (END); disjunct from main range along Atlantic coast from Mass to Fla and Tex.

Eleocharis nitida Fernald
NEAT SPIKE-RUSH

DESCRIPTION Small, mat forming, native perennial, from matted or creeping purplish rhizomes. Stems round in section to somewhat 4-angled, to 15 cm long, very thin and delicate, to only 0.3 mm wide. Spikelets ovoid, small, to 4 mm long and 2 mm wide; fertile scales 1–1.3 mm long, brown to dark brown, midrib usually pale or greenish, the scales often early-deciduous; bristles absent; style 3-parted. Achenes persistent after scales fall, dark yellow-orange or brown, 3-angled (the angles evident), covered with small bumps (under magnification); tubercle brown, flattened and saucer-like, with a tiny central tip. Flowering May–June.

HABITAT wet soil in openings in alder thickets and marshes, sometimes in shallow water, usually where little competing vegetation; disturbed moist places such as ditches and wheel ruts; crevices in rocks along Lake Superior shoreline.

STATUS NCNE-OBL; rare in Minn (THR), Wisc (END), Mich (END); more common in boreal regions.

Eleocharis obtusa (Willd.) J.A. Schultes
♦ BLUNT SPIKE-RUSH

DESCRIPTION Clumped, fibrous-rooted native annual. Stems slender, round in section, ribbed, 0.5–5 dm long and 1–2 mm wide; sheaths green. Spikelets ovate to cylindric, 4–15 mm long and 2–4

mm wide; scales purple-brown, with a green midvein and pale margins; sepals and petals reduced to 6–7 brown bristles, or absent; styles 2- or 3-parted. Achenes lens-shaped, light to dark brown or olive, shiny, 1–1.5 mm long; tubercle flattened-triangular, about as wide as the broad top of achene. Flowering June–Sept.

SYNONYM *Eleocharis ovata*.

HABITAT wet, sandy or muddy shores, marshes, ditches, mud flats, temporary ponds.

STATUS MW-OBL | NCNE-OBL | GP-OBL

Eleocharis palustris L.
♦ CREEPING SPIKE-RUSH

DESCRIPTION Native perennial, spreading by rhizomes. Stems single or in small clusters, slender to stout, round in section, 1–8 dm long and 1–3 mm wide; sheaths red or purple at base. Spikelets long-ovate, 5–30 mm long and 2–4 mm wide, wider than stems; lowest scale sterile, encircling the stem; fertile scales lance-shaped to ovate, 2–5 mm long, brown or red-brown, with a green or pale midvein; sepals and petals reduced to usually 4, pale brown, barbed bristles; style 2-parted. Achenes lens-shaped, yellow to brown, 1–2 mm long; tubercle flattened-triangular, constricted at base. Flowering May–Aug.

SYNONYMS a variable species known by a number of synonyms including *Eleocharis erythropoda*, *Eleocharis macrostachya*,

Eleocharis obtusa

Eleocharis smallii.

HABITAT shallow water of marshes, wet meadows, muddy shores, bogs, ditches, streambanks and swamps.

STATUS MW-OBL | NCNE-OBL | GP-OBL

Eleocharis parvula (Roemer & Schultes)
Link. SMALL SPIKE-RUSH

DESCRIPTION Very small, clumped, native perennial, spreading by slender rhizomes and forming dense mats; rhizomes often with tubers. Stems threadlike, mostly 2–6 cm long, less than 1 mm wide. Spikelets ovate, 2–5 mm long and 1–2 mm wide; scales ovate, green, straw-colored or brown, 1–2 mm long; sepals and petals absent or reduced to small bristles; style 3-parted. Achenes pale brown, 3-angled, about 1 mm long; tubercle small, not forming a distinct cap on top of achene body. Flowering July–Sept.

HABITAT wet saline or alkaline flats and shores.

STATUS MW-OBL | NCNE-OBL | GP-OBL; uncommon in Mich LP (END), reported from Wisc.

Eleocharis quadrangulata (Michx.)
Roemer & Schultes
SQUARE-STEM SPIKE-RUSH

DESCRIPTION Medium to large, clumped, native perennial, spreading by rhizomes. Stems stout, sharply 4-angled, to 1 m long and 2–5 mm wide. Spikelets cylindric, 2–5 cm long, about as thick as stem; scales in 4 rows, brown, oval to obovate,

Eleocharis palustris

5–6 mm long, margins chaffy; sepals and petals reduced to bristles; style 2- or 3-parted. Achenes rounded on sides, brown, 2–3 mm long; tubercle dark brown, flattened cone-shaped.

HABITAT shallow water and wet, sandy or mucky shores; sedge meadows.

STATUS MW-OBL | NCNE-OBL | GP-OBL; historical records from Wisc.

Eleocharis quinqueflora (F.X. Hartmann) Schwarz
FEW-FLOWER SPIKE-RUSH

DESCRIPTION Small, clumped, perennial, spreading by rhizomes. Stems threadlike, grooved, 1–3 dm long and less than 1 mm wide. Spikelets ovate, 4–8 mm long and 2–3 mm wide; scales ovate, brown, chaffy on margins, 2–5 mm long; sepals and petals reduced to bristles or absent; style 3-parted. Achenes gray-brown or brown, 3-angled, 1–3 mm long; tubercle slender, joined to the achene and beaklike. Flowering June–Aug.

SYNONYM *Eleocharis pauciflora.*

HABITAT wet, sandy or gravelly shores and flats marshes and fens; often where calcium-rich.

STATUS MW-OBL | NCNE-OBL | GP-OBL

ANDREY ZHARKIKH

Eleocharis quinqueflora

Eleocharis robbinsii Oakes
ROBBINS' SPIKE-RUSH

DESCRIPTION Clumped, native perennial, spreading by rhizomes. Stems slender, 3-angled, 2–6 dm long and 1–2 mm wide; when underwater, plants often with numerous sterile stems from base; sheaths brown. Spikelets lance-shaped, 1–2 cm long and 2–3 mm wide, barely wider than stems; scales narrowly ovate, margins chaffy; sepals and petals reduced to 6 barbed bristles; style 3-parted. Achenes rounded on both sides, light brown, 2–3 mm long; tubercle flattened and cone-shaped, with a raised ring at base. Flowering July–Aug.

HABITAT wet, sandy or mucky lake and pond shores, marshes, exposed flats.

STATUS MW-OBL | NCNE-OBL

Eleocharis rostellata (Torr.) Torr.
BEAKED SPIKE-RUSH

DESCRIPTION Clumped, native perennial, without creeping rhizomes. Stems flattened, wiry, 3–10 dm long and 1–2 mm wide; the fertile stems upright, the sterile stems often arching and rooting at tip; sheaths brown. Spikelets oblong, tapered at both ends, 5–15 mm long and 2–5 mm wide, wider than the stem; scales ovate, 3–5 mm long, green to brown with a darker midvein and translucent margins; sepals and petals reduced to 4–8 barbed bristles; style 3-parted. Achenes olive to brown, rounded 3-angled, 2–3 mm long; tubercle cone-shaped, joined with the achene body and beaklike. Flowering June–Aug.

HABITAT shores, wet meadows, calcareous fens and mud flats; sites typically calcium-rich, often associated with mineral springs.

STATUS MW-OBL | NCNE-OBL | GP-OBL; Minn (THR), Wisc (THR).

Eleocharis tricostata Torr.
THREE-RIBBED SPIKE-RUSH

DESCRIPTION Native perennial, spreading by short rhizomes. Stems single or in small clusters, slender, somewhat flattened, grooved, 2–6 dm long; sheaths 1-toothed at tip. Spikelets cylindric, 5–15 mm

long; scales overlapping in many rows, ovate, brown with a green midvein; bristles absent; style 3-parted. Achenes brown to olive, 1 mm long, 3-angled, narrowly winged on the angles; tubercle very small, nearly flat, light brown.

HABITAT sandy shores.

STATUS NCNE-OBL; rare in Mich (THR); disjunct from main range along Atlantic coast.

Eleocharis wolfii A. Gray
WOLF'S SPIKE-RUSH

DESCRIPTION Clumped, native perennial, from slender rhizomes. Stems flattened, 2-edged, often twisted, 1-3 dm long and 1–2 mm wide; sheaths often purple at base, membranous at tip. Spikelets narrowly ovate, 4–10 mm long and 2–3 mm wide, wider than stem; scales green, tinged with purple, narrowly ovate, 2–3 mm long, with chaffy margins; bristles absent; style 3-parted. Achenes gray, ± round in section, about 1 mm long; tubercle cone-shaped, constricted at base where joins achene. Flowering June–July.

HABITAT wet meadows and low prairie.

STATUS MW-OBL | NCNE-OBL | GP-OBL; Minn (END), Wisc (END).

Eriophorum COTTON-GRASS

Grasslike perennials. Stems clumped or single, round to rounded 3-angled in section. Leaves mostly at base of plant, the blades flat, folded or inrolled; upper leaves often reduced to bladeless sheaths. Flower heads at ends of stems, with 1 or several spikelets; scales many, spirally arranged, chaffy on margins; involucral bracts leaflike in species with several spikelets in the head, or reduced to scales in species with 1 spikelet at end of stems (*E. chamissonis, E. vaginatum*). Flowers perfect; sepals and petals numerous, reduced to long, cottony, persistent white to tawny brown bristles; stamens 3; styles 3-parted. Achenes brown, ± 3-angled, sometimes with a short beak formed by the persistent style.

NOTE *Eriophorum* spikelets resemble "cottonballs" when mature; the cottonballs are composed of the bristle-like sepals and petals.

Eriophorum COTTON-GRASS

1   Head a single spikelet at end of stem; leaflike bracts absent **2**

1   Head of 2 or more spikelets; leaflike bracts present **3**

2   Plants forming colonies from rhizomes . . . . *E. chamissonis*

2   Plants densely clumped, rhizomes absent . . . . *E. vaginatum*

3   Leaves 1–2 mm wide; leaflike bract 1, erect, the head appearing lateral from side of stem **4**

3   Leaves 3 mm or more wide; leaflike bracts 2 or 3, the head appearing terminal **5**

4   Blade of uppermost stem leaf much shorter than its sheath . . . . *E. gracile*

4   Blade as long or longer than its sheath . . . . *E. tenellum*

5   Scales 3–7-nerved, copper-brown on sides . . . . *E. virginicum*

5   Scales with 1 nerve, sides olive-green to nearly black **6**

6   Midvein of scale slender, fading before reaching tip of scale . . . . *E. angustifolium*

6   Midvein of scale widening toward tip of scale and reaching scale tip . . . . *E. viridicarinatum*

Eriophorum angustifolium Honck.
♦THIN-SCALE COTTON-GRASS

DESCRIPTION Grasslike native perennial, spreading by rhizomes and forming colonies.

UDO SCHMIDT

Eriophorum angustifolium

Stems mostly single, 2–8 dm long and 2–3 mm wide, ± round in section, becoming 3-angled below the head. Leaves few, flat or folded along midrib, 3–8 mm wide, often dying back from the tips; sheaths sometimes red, dark-banded at tip. Spikelets 3–10, clustered in heads 1–3 cm wide when mature, the heads drooping on weak stalks; involucral bracts leaflike, often black at base, the main bract upright and usually longer than the head; scales lance-shaped, brown or purple-green, 4–6 mm long, the midvein not extending to tip of scale; bristles bright white, 2–3 cm long. Achenes brown to nearly black, 2–3 mm long. Flowering May–July.

SYNONYM *Eriophorum polystachion.*

HABITAT bogs, calcareous fens, wet meadows.

STATUS MW-OBL | NCNE-OBL | GP-OBL

Eriophorum chamissonis C. A. Meyer
⬧RUSTY COTTON-GRASS

DESCRIPTION Grasslike native perennial, spreading by rhizomes and forming colonies. Stems single or in small groups, stout, ± round in section, 2–6 dm long and 1–3 mm wide. Leaves few, mostly from base of plant and shorter than stems, the uppermost leaves from near middle of stem and often without blades, lower leaves round in section to 3-angled and channeled, 1–2 mm wide. Spikelets single, erect at end of stems, clustered in a ± round head 2–3

Eriophorum chamissonis

cm wide; involucral bracts not leaflike, reduced to black scales; flower scales narrowly ovate, black-green, with broad white margins and tips; bristles white to bright red-brown. Achenes dark brown, beaked, 2–3 mm long. Flowering June–July.

SYNONYM *Eriophorum russeolum.*

HABITAT bogs.

STATUS MW-OBL | NCNE-OBL | GP-OBL

NOTE sometimes combined into *Eriophorum angustifolium.*

Eriophorum gracile Koch
SLENDER COTTON-GRASS

DESCRIPTION Grasslike native perennial, spreading from rhizomes. Stems single, spreading or reclining, slender, ± round in section, 2–6 dm long and 1–2 mm wide. Leaves few, channeled on upper side, 1–2 mm wide, the basal leaves often withered by flowering time, blades of uppermost leaves small. Spikelets in clusters of 2–5 at ends of stems, on spreading to nodding stalks 2–3 cm long; involucral bract leaflike and erect, shorter than spikelet cluster; scales ovate, pale to black-brown with a prominent midvein; bristles bright white. Achenes light brown, 3–4 mm long. Flowering May–July.

HABITAT fens and bogs.

STATUS MW-OBL | NCNE-OBL | GP-OBL

Eriophorum tenellum Nutt.
CONIFER COTTON-GRASS

DESCRIPTION Grasslike native perennial, with rhizomes and forming colonies. Stems single, slender, erect, 3–8 dm long, rounded 3-angled, rough-to-touch on upper angles. Leaves linear, 1–2 mm wide, channeled, not reduced and bladeless on upper stem. Spikelets 3–6, in short-stalked clusters at ends of stems, or with 1–2 rough, drooping stalks to 5 cm long; involucral bract leaflike, stiff and erect, usually shorter than the spikelet cluster; scales ovate, straw-colored to red-brown; bristles white. Achenes brown, 2–3 mm long.

HABITAT bogs and conifer swamps.

STATUS MW-OBL | NCNE-OBL | GP-OBL

Eriophorum vaginatum L.

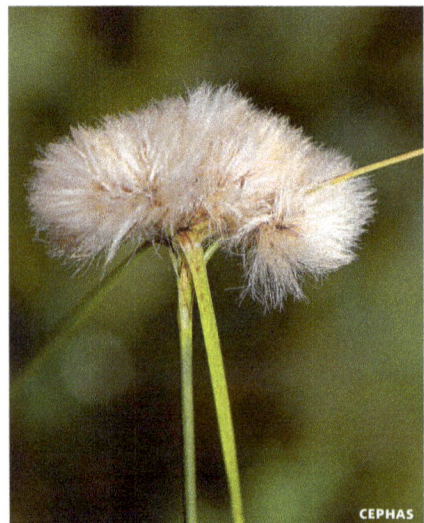

♦TUSSOCK COTTON-GRASS

DESCRIPTION Densely clumped, grasslike native perennial, forming large hummocks. **Stems** stiff, rounded 3-angled, 2–7 dm long. **Leaves** at base of stems, mostly shorter than stems, only 1 mm wide, with 1–3 inflated, bladeless sheaths on stem. **Spikelets** clustered in a single head at end of stems; involucral bracts absent; scales narrowly ovate, purple-brown to black, with white margins, spreading when mature; bristles usually white (rarely red-brown). **Achenes** obovate, 3–4 mm long. June.

SYNONYM *Eriophorum callitrix, Eriophorum spissum.*

HABITAT sphagnum bogs and tamarack swamps.

STATUS MW-OBL | NCNE-OBL | GP-OBL

Eriophorum virginicum L.

♦TAWNY COTTON-GRASS

DESCRIPTION Large, grasslike native perennial, with slender rhzomes. **Stems** single or in small groups, stiff, erect, to 1 m long, leafy, mostly smooth. **Leaves** flat, 2–4 mm wide, the uppermost often longer than the head. **Spikelets** in dense clusters of several to many at ends of stems, on short stalks of ± equal lengths, the clusters wider than long; involucral bracts 2–3, leaflike, spreading or bent downward, unequal, much longer than the head; scales ovate, thick, copper-brown with a green center; bristles tawny or copper-brown. **Achenes** light brown, 3–4 mm long. **Flowering** July–Aug.

HABITAT Sphagnum moss peatlands.

STATUS MW-OBL | NCNE-OBL | GP-OBL

NOTE the widely spreading, leaf-like involucral bracts and copper-colored, cottony bristles are distinctive.

Eriophorum viridicarinatum (Engelm.) Fern.

DARK-SCALE COTTON-GRASS

DESCRIPTION Grasslike native perennial, forming colonies from spreading rhizomes. **Stems** mostly single, ± round in section, 3–7 dm long. **Leaves** flat except at tip, the uppermost leaves 10–15 cm long; sheaths green. **Spikelets** usually 20–30, clustered in heads at ends of stems, on short to long, finely hairy stalks; involucral bracts 2–4, not black at base, longer or equal to head; scales narrowly ovate, black-green, the midvein pale, extending to tip of scale; bristles white. **Achenes** brown, 3–4 mm long. **Flowering** May–July.

HABITAT bogs and open conifer swamps.

STATUS MW-OBL | NCNE-OBL | GP-OBL

NOTE similar to **thin-scale cotton-grass** (*Eriophorum angustifolium*), but usually with more spikelets, the scale midvein extending

Eriophorum vaginatum

KRZYSZTOF ZIARNEK

Eriophorum virginicum

CEPHAS

to the tip of scale, and the leaf sheaths not dark-banded at tip.

Fimbristylis FIMBRY

Annual or perennial grasslike plants. Stems slender, clumped or single. Leaves mostly at base of plants, narrowly linear, flat to inrolled. Spikelets many-flowered, in umbel-like clusters at ends of stems; involucral bracts 2–3, short and leaflike; scales spirally arranged and overlapping. Flowers perfect, sepals and petals absent, stamens 1–3; styles 2–3-parted, swollen at base, deciduous when mature. Achenes lens-shaped or 3-angled.

Fimbristylis FIMBRY

1   Plants annual; achenes 3-angled; styles 3-parted .... *F. autumnalis*
1   Plants perennial; achenes lens-shaped; styles 2-parted .... *F. puberula*

Fimbristylis autumnalis (L.) Roemer & Schultes ♦ AUTUMN SEDGE

DESCRIPTION Clumped, grasslike native annual, with shallow fibrous roots. Stems flattened, slender, sharp-edged, 0.5–3 dm long. Leaves shorter than the stems, flat, 1–2 mm wide. Spikelets usually many in an open umbel-like cluster, the spikelets lance-shaped, 3–8 mm long, single or several at ends of threadlike, spreading stalks; involucral bracts 2–3, leaflike, usually shorter than the head; scales ovate, golden-brown with a prominent green midvein, 1–2 mm long; style 3-parted. Achenes ivory to tan, 3-angled and ribbed on the angles, to

Fimbristylis autumnalis

0.5 mm long. Flowering July–Sept.
HABITAT sandy or mucky shores (especially where seasonally flooded and then later drying), streambanks, wet meadows, ditches.
STATUS MW-OBL | NCNE-FACW | GP-OBL

Fimbristylis puberula (Michx.) Vahl
CHESTNUT SEDGE

DESCRIPTION Grasslike native perennial, with short rhizomes. Stems single or in small clumps, slender, stiff, 2–7 dm long, round to oval in section, sometimes swollen at base. Leaves shorter than the stems, usually inrolled, 1–3 mm wide, often hairy. Spikelets few to many in an umbel-like cluster; the spikelets ovate, 5–10 mm long, the central spikelet stalkless, the others on slender stalks; involucral bracts 2–3, leaflike, the longest equal or longer than the head; scales ovate, brown with a lighter midvein, 3–4 mm long, usually finely hairy, often tipped with a short awn; style 2-parted, the style branches finely hairy. Achenes light brown, lens-shaped, 1–2 mm long. Flowering June–Sept.

HABITAT wet meadows, shores, and low prairie, often where sandy and calcium-rich; also in drier prairies.
STATUS MW-OBL | NCNE-OBL | GP-OBL; Minn (END), Wisc (END); historical records from Mich.

Fuirena UMBRELLA-SEDGE

Fuirena pumila (Torr.) Sprengel
UMBRELLA-SEDGE

DESCRIPTION Clumped, grasslike native annual. Stems smooth, 1–6 dm long. Leaves flat, to 10 cm long and 5 mm wide, the margins hairy; lower sheaths hairy. Spikelets stalkless, in clusters of mostly 1–3, the spikelets ovate, 6–15 mm long; scales ovate, brown, 2–4 mm long, hairy, 3-nerved, tipped with a spreading awn almost as long as scale; bristles 3, downwardly barbed. Achenes smooth, 3-angled. Flowering July–Sept.
SYNONYM *Fuirena squarrosa*.
HABITAT sandy or mucky shores, especially where seasonally flooded; floating sedge mats.

Lipocarpha
HALFCHAFF SEDGE

Small, tufted, grasslike annuals, the hair-like stems bearing 2 hair-like leaves at the base, only the upper leaves with a short blade. Spikelets very small, 1–3 in a stalkless cluster surpassed by 2 or 3 hairlike involucral bracts. flowers of 1 stamen, a 2-cleft style and no bristles; in addition to the larger, pointed, spirally overlapping, outer scales, a minute inner scale occurs on the side of the flower next to the axis of the spikelet.

Lipocarpha HALFCHAFF SEDGE

1 Main outer scales with a long curved awn almost equaling the length of the scale itself; small inner scale equaling and partly inclosing the achene; mature achene obovate, compressed, black. . . . .
. . . . *L. drummondii*

1 Main outer scales with a short extended tip much shorter than the length of the scale; small inner scale much shorter than the achene, often absent; mature achene cylindrical, brown. . . . . *L. micrantha*

Lipocarpha drummondii (Nees)
G. Tucker
DRUMMOND'S HALFCHAFF SEDGE

DESCRIPTION Small, tufted, grasslike annual. Stems to 12 cm long and to 0.5 mm wide. Leaves 1–3 cm long and  to 0.6 mm wide; longest involucral leaf erect or nearly so. Spikelets in mostly 1–2 ovoid heads, not bristly; bracts 1–2, longest erect, to 2 cm long; scales 2; 1st scale light brown to reddish brown, with greenish midvein, widest above mid-length; 2nd scale reddish brown near tip, with 2–4 reddish veins. Achenes black, obovoid, 0.5–0.75 mm long. Flowering Aug–Sept.
SYNONYMS *Cyperus hemidrummondii, Hemicarpha drummondii*
HABITAT Wet sandy margins of ponds and streams.
STATUS MW-FACW | NCNE-FACW | GP-FACW

Lipocarpha micrantha (Vahl) G. Tucker
✦SMALL FLOWER HALFCHAFF SEDGE

DESCRIPTION Small, densely clumped, grasslike annual; native.  Stems compressed, 3–15 cm long. Leaves slender, 2 per stem, to 10 cm long and 0.5 mm wide, mostly shorter than the stems. Spikelets in stalkless clusters of 1–3; the spikelets many-flowered, ovate, 2–5 mm long; involucral bracts 2–3, leaflike, the main bract upright and longer than the spikelets (the head appearing lateral); scales brown with a green midvein, 1–2 mm long, tipped with an awn; bristles absent; style 2-parted, not swollen at base. Achenes brown, oblong, to 1 mm long. Flowering Aug–Sept.
SYNONYMS *Cyperus subsquarrosus, Hemicarpha micrantha*.
HABITAT sandy or muddy shores and stream-banks, usually where seasonally flooded.
STATUS MW-OBL | NCNE-OBL | GP-FACW

Rhynchospora BEAK-RUSH

Grasslike perennials (annual in *Rhynchospora scirpoides* and sometimes in *Rhynchospora macrostachya*), clumped or spreading by rhizomes. Stems erect, leafy, usually 3-angled or sometimes round. Leaves flat or rolled inward. Spikelets clustered in dense heads, the heads open to crowded; scales overlapping in a spiral. Flowers perfect, or sometimes upper flowers staminate only; sepals and petals reduced to usually 6 (1–20) bristles or sometimes absent; stamens usually 3; styles 2-parted, swollen at base and persistent on the achene as a tubercle. Achenes lens-shaped.

DOUG MCGRADY

Lipocarpha micrantha

ADDITIONAL SPECIES *Rhynchospora recognita* (Gale) Kral (Mich state endangered), formerly considered a variety of *Rhynchospora globularis,* is taller, stiffer, broader leaved, with wider and more bristly spikelet clusters, and distinctly orange tinted in comparison to the darker and less bristly spikelet clusters of *Rhynchospora globularis*. In Mich, the species is reported on pond shores and moist, sandy swales in the sw Lower Peninsula.

Rhynchospora **BEAK-RUSH**

1 Plants annual; spikelets with many perfect flowers, scales not empty; bristles subtending the achene absent . . . . *R. scirpoides*

1 Plants mostly perennial; spikelets few-flowered, the lower scales empty; bristles present **2**

2 Leaves 5 mm or more wide; achenes flattened, with a tubercle more than 10 mm long; annual or short-lived perennial . . . . *R. macrostachya*

2 Leaves 1–4 mm wide; achenes rounded or lens-shaped, tubercle less than 3 mm long; perennials. **3**

3 Surface of achene covered with conspicuous ridges or wrinkles; bristles subtending achene much shorter than achene. . . . .*R. recognita* . . . . (see *Additional Species*)

3 Surface of achene smooth or only slightly wrinkled; bristles longer than achene . . . **4**

4 Spikelets white to tan; bristles 8 or more . . . . *R. alba*

4 Spikelets brown, dark olive-green or nearly black; bristles 5–6 . . . . **5**

5 Scales dark olive-green to black; bristles with upward-pointing barbs, at least some of the bristles longer than the tubercle. . . . .*R. fusca*

5 Scales brown; bristles with downward pointing barbs (rarely smooth), the bristles shorter to as long as the tubercle **6**

6 Stems narrow and threadlike; achene margins not translucent, achene body less than half as wide as long . . .*R. capillacea*

6 Stems stout; achene with translucent margins, body more than half as wide as long . . . .*R. capitellata*

Rhynchospora alba (L.) Vahl
♦ WHITE BEAK-RUSH

DESCRIPTION Clumped, grasslike native perennial. Stems slender, erect, 1–6 dm long. Leaves bristlelike, 0.5–3 mm wide, shorter than the stems. Spikelets in 1–3 rounded heads, 5–20 mm wide, at or near ends of stems, the lateral heads usually long-stalked; the spikelets oblong, narrowed at each end, 4–5 mm long, white, becoming pale brown; bristles 8–15, downwardly barbed, about equaling the tubercle. Achenes lens-shaped, brown-green, 1–2 mm long; tubercle triangular, about half as long as achene. Flowering June–Sept.

HABITAT bogs, fens, open conifer swamps of black spruce and tamarack.

STATUS MW-OBL | NCNE-OBL | GP-OBL

Rhynchospora capillacea Torr.
♦ NEEDLE BEAK-RUSH

DESCRIPTION Small, clumped, grasslike native perennial. Stems slender, 0.5–4 dm long. Leaves threadlike, rolled inward, to only 0.5 mm wide, much shorter than the stem. Spikelets in 1–2 small, separated clusters, each cluster subtended by 1 to several short, bristlelike bracts; the spikelets ovate, 3–7 mm long; scales overlapping, ovate, brown with a paler, sharp-tipped midvein; bristles 6, downwardly

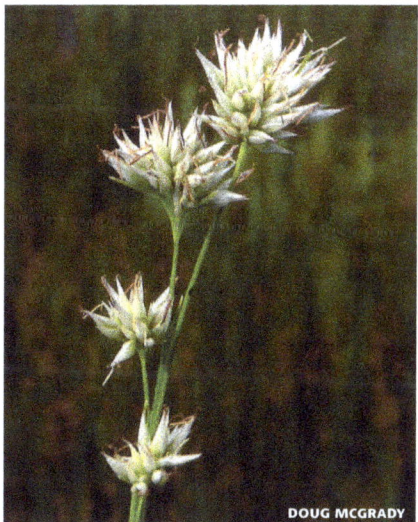

Rhynchospora alba

barbed, longer than the achenes; style 2-parted. Achenes lens-shaped, satiny yellow-brown, 2 mm long; tubercle dull brown, narrowly triangular, about 1 mm long. Flowering June–Aug.

HABITAT calcareous fens, interdunal flats, wet sandy or gravelly shores, seeps; usually where calcium-rich.

STATUS MW-OBL | NCNE-OBL | GP-OBL; Minn (THR).

Rhynchospora capitellata (Michx.) Vahl BROWN BEAK-RUSH

DESCRIPTION Clumped, grasslike native perennial. Stems erect, 3-angled, 3–8 dm long. Leaves flat, 2–4 mm wide, rough on margins, shorter than stems. Spikelets 3–5 mm long, several to many in 2–7 rounded, ± loose clusters 1–1.5 cm wide, the lateral clusters often in pairs on slender stalks; bristles 6, 2–3 mm long, usually downwardly barbed, about equaling the achene; style 2-parted. Achenes lens-shaped, dark brown, 1–2 mm long; tubercle triangular, about as long as achene. Flowering June–Sept.

SYNONYM *Rhynchospora glomerata*.

HABITAT wet sandy or mucky shores and flats, wet meadows, bogs, calcareous fens, ditches.

STATUS MW-OBL | NCNE-OBL | GP-OBL

Rhynchospora fusca (L.) Aiton f.
◆ GRAY BEAK-RUSH

DESCRIPTION Clumped, grasslike native perennial, spreading by short rhizomes and forming colonies. Stems slender, 3-angled, l–3 dm long. Leaves very slender, rolled inward, mostly shorter than the stems. Spikelets spindle-shaped, dark brown, 4–6 mm long, in 1–4 loose clusters, the lower clusters on long stalks, each cluster subtended by an erect, leafy bract, the bract longer than the cluster; bristles 6, upwardly barbed; style 2-parted. Achenes light brown, 1–1.5 mm long; tubercle flattened-triangular, nearly as long as achene.

HABITAT wet, sandy shores, interdunal wetlands, sedge meadows, bog mats.

STATUS MW-OBL | NCNE-OBL | GP-OBL

Rhynchospora macrostachya Torr.
TALL BEAK-RUSH

DESCRIPTION Large, grasslike, native annual (sometimes a short-lived perennial), rhizomes absent. Stems single or few, 3-angled, 0.5–2 m long. Leaves flat, to 1 cm wide. Spikelets 15–20 mm long, in a single, loose cluster at end of stem; scales lance-shaped, light brown; bristles 6 (3 on each side of achene), about 1 cm long; style 2-parted. Achenes flattened, dark brown, 4–6 mm long; tubercle long and beaklike, 15–20 mm long.

HABITAT sandy or mucky shores, especially on wet, exposed flats.

STATUS MW-OBL | NCNE-OBL | GP-OBL

Rhynchospora scirpoides (Vahl) Griseb. LONG-BEAKED BEAK-RUSH

DESCRIPTION Clumped, grasslike, native annual. Stems 1–5 dm long. Leaves 1–3 mm wide, sometimes longer than stems. Spikelets ovate, 3–6 mm long, in open clusters at ends of stems; scales overlapping, ovate, brown, 2–3 mm long; bristles absent; style 2-parted. Achenes lens-shaped, dark brown, to 1 mm long and about as wide; tubercle flattened-triangular, about as long as achene. Flowering Aug–Sept.

SYNONYM *Psilocarya scirpoides*.

HABITAT wet sandy or mucky shores and mudflats.

STATUS MW-OBL | NCNE-OBL | GP-OBL; Wisc (THR).

R. capillacea   R. fusca

Schoenoplectiella
CLUB-RUSH

This genus, recently segregated from *Schoenoplectus,* includes our annual, tufted, rush species.

Schoenoplectiella CLUB-RUSH

1 Achene covered with prominent transverse ridges; perianth bristles none; rare in Wisc and LP of Mich . . . . *S. hallii*

1 Achene smooth or obscurely pitted; perianth bristles present or absent; widespread species **2**

2 Taller shoots with stems (base of plant to inflorescence) more than 3/4 as long as height of the plant, including the involucral bract; achenes thickly and asymmetrically biconvex (inner face slightly but clearly convex, outer faces forming a clear angle) . . . . *S. purshiana*

2 Taller shoots with stems to 3/4 as long as height of the plant; achenes flattened-plano-convex (inner face essentially flat, the outer faces gently rounded) *S. smithii*

Schoenoplectiella hallii (Gray) Lye
SHARPSCALE-BULRUSH

DESCRIPTION Small, clumped, native annual. Stems slender, 1–3 dm long, rounded 3-angled, finely grooved. Leaves 1 to several near base of stem, reduced to sheaths or upper leaves with a narrow blade to 10 cm long; main bract erect, 2–10 cm long, resembling a continuation of stem. Spikelets cylindric, 4–12 mm long and 2–3 mm wide, in heads of 3–8 spikelets, ± stalkless, or some spikelets in clusters of 2–3 on stalks to 1 cm long, the head appearing lateral from side of stem; scales 2–3 mm long, light brown with a green midvein, tipped by an awn to 1 mm long; bristles several or absent; style 3-parted. Achenes 3-angled, light to dark brown, 1–2 mm long, cross-ridged (under 10x magnification), tipped with a short beak. Flowering July–Sept.

SYNONYMS *Schoenoplectus hallii, Scirpus hallii, Scirpus saximontanus.*

HABITAT sandy and muddy lakeshores and flats, often where seasonally flooded.

STATUS MW-OBL | NCNE-OBL | GP-OBL; Wisc (END), Mich (THR).

Schoenoplectiella purshiana (Fern.)
Lye ♦ WEAK-STALK CLUB-RUSH

DESCRIPTION Clumped, native annual. Stems often arching (to decumbent), cylindric, to 1 m long and to 2 mm wide. Leaves 1, to as long as the stem; blade absent, or if present, C-shaped in cross-section, 0.5–1 mm wide; bract erect or often divergent, to 15 cm long. Spikelets 1–12; scales straw-colored to orange-brown, midrib often greenish, broadly obovate, 2.5–3 mm long, margins ciliolate at tip and with a small sharp point; perianth members 6, brown, bristle-like, equaling to slightly exceeding achene, densely retrorsely spinulose. Achenes biconvex, brown, turning blackish, 1.6–2.2 mm long, rounded at base to a distinct stipelike constriction; beak 0.1–0.3 mm long. Flowering July–Aug.

SYNONYMS *Schoenoplectus purshianus, Scirpus purshianus.*

HABITAT sandy to mucky, sometimes marly, shores, especially where water levels have receded.

STATUS MW-OBL | NCNE-OBL | GP-OBL

Schoenoplectiella smithii (Gray) Hayas.
SMITH'S CLUB-RUSH

DESCRIPTION Clumped, native annual. Stems slender, smooth, round or rounded 3-angled, to 6 dm long. Leaves re-

SHAUN POGACNIK

Schoenoplectiella purshiana

duced to sheaths, or some with short blades; bract narrow, upright, 2–10 cm long, appearing to be a continuation of stem. Spikelets ovate, 5–10 mm long, in a single cluster of 1–12 spikelets; scales yellow-brown with a green midvein; bristles 4–6, barbed or smooth, longer than achene, sometimes smaller or absent; style 2-parted. Achenes lens-shaped or flat on 1 side and convex on other, glossy brown to black, 1–2 mm long. Flowering July–Aug.

SYNONYMS *Schoenoplectus smithii, Scirpus smithii.*

HABITAT sandy, gravelly or mucky shores, floating mats, bogs.

STATUS MW-OBL | NCNE-OBL | GP-OBL

Schoenoplectus CLUB-RUSH

Perennial or annual, tufted or rhizomatous herbs. Stems cylindric to strongly 3-angled, smooth, spongy with internal air cavities. Leaves basal, rarely 1(–2) on stem; sheaths tubular; ligules membranous; blades well-developed to rudimentary. Inflorescences terminal, head-like to openly paniculate; spikelets 1–100 or more; involucral bracts 1–5, leaflike, proximal bract erect to spreading. Spikelets terete; scales deciduous, spirally arranged, each subtending a flower, or proximal scale empty, midrib usually prolonged into short awn, margins ciliate. Flowers bisexual; perianth of 0–6(–8) spinulose bristles shorter than to somewhat longer than the achene; stamens 3. Achenes biconvex to trigonous, with apical beak; rugose or with transverse wavy ridges.

Schoenoplectus CLUB-RUSH

1 Spikelets (at least several of them) distinctly pediceled (sometimes congested in *S. acutus*); stems terete, often over 1 m tall. 2

1 Spikelets 1-few, crowded, sessile or nearly so (rarely one on a short pedicel); stems 3-angled or terete (if terete, then slender, soft, and not over 1 m tall) 4

2 Styles 3-cleft; achenes 3-sided; perianth bristles 2–4 (–5); scales glabrous on the back; mature achenes ca. 2.5 mm long, including short apiculus; stems firm . . . . .
. . . . *S. heterochaetus*

2 Styles 2-cleft; achenes plano-convex (flat on one side and rounded on the other) or biconvex; perianth bristles mostly 6; scales puberulent on back; achenes shorter, or stems soft and easily compressed 3

3 Stems firm and dark olive-green when fresh; spikelets ovoid to cylindrical (often 2.5 or more times as long as wide), usually in a stiffer, sometimes condensed, inflorescence; scales dull, pale or whitish brown, the midrib not strongly contrasting, the margins often more copiously ciliate than in S. tabernaemontani, and the backs copiously flecked with shiny red dots, often puberulent; mature achenes ca. 2.2–2.7 mm long, including apiculus, completely hidden by the scales . . . . *S. acutus*

3 Stems rather soft and easily compressed, pale blue-green when fresh; spikelets ovoid (about twice as long as wide, or shorter), in an open, lax inflorescence; scales ± shiny, rich orange-brown, often with prominent greenish midrib, the margins ciliate but the backs essentially glabrous (puberulence and swollen red flecks, if any, limited to region of midrib); mature (dark gray or lead-colored) achenes ca. 1.6–2.1 (–2.4) mm long, including apiculus, barely covered by the scale . . . . *S. tabernaemontani*

4 Spikelet 1, strongly ascending, the involucral bract surpassing its tip by not more than 15 (–20) mm; leaves normally many, hair-like, submersed; stem seldom over 1 mm thick; anthers (2.1–) 2.5–3.5 mm long; achenes 3-sided, the body ca. 2.5–3 mm long. . . . . *S. subterminalis*

4 Spikelets usually more than 1 and the involucral bract surpassing them by more than 15 mm (except in smallest plants of some populations); leaves stiff and stems thicker; anthers and achenes various. . . . 5

5 Midrib of scale ± greenish, excurrent as a short (not over 0.5 mm) tip extending beyond the tapered (sometimes very slightly notched) apex of the scale; bristles slightly exceeding body of achene; rhizome soft; achene with apiculus 0.5 mm or more in length; styles 3-cleft and achenes 3-sided; leaves more than half as tall as the stems . . . . *S. torrey*

5 Midrib of scale brown, excurrent as a long (0.5–1 mm) tip equaling or exceeding lobes; bristles shorter than body of achene; rhizome firm and hard; achene with apiculus shorter than 0.5 mm; styles usually 2-cleft and achenes biconvex to

plano-convex (occasionally some styles 3-cleft and achenes 3-sided in a spikelet); leaves less than half as tall as the stems…
.... *S. pungens*

Schoenoplectus acutus (Muhl.) A. & D. Löve ♦ HARDSTEM CLUB-RUSH

DESCRIPTION Native perennial, from stout rhizomes and often forming large colonies. Stems round in section, 1–3 m long. Leaves reduced to 3–5 sheaths near base of stem, blades absent, or upper leaves with blades to 25 cm long; main bract erect, appearing as a continuation of stem, 2–10 cm long, eventually turning brown. Spikelets 5–15 mm long and 3–5 mm wide, in clusters of mostly 3–7, the clusters grouped into a branched head of up to 60 spikelets, the head appearing lateral from side of stem, the branches stiff and spreading; scales chaffy, mostly translucent, 3–4 mm long, often with red-brown spots, usually tipped with an awn to 1 mm long; bristles 6, unequal, usually shorter than achene; style 2-parted (rarely 3-parted). Achenes light green to dull brown, flat on 1 side and convex on other, 2–3 mm long, the style beak small, to 0.5 mm long. Flowering May–Aug.
HABITAT usually emergent in shallow to deep water (1–2 m deep) of marshes, ditches, ponds and lakes; sometimes where brackish.
STATUS MW-OBL | NCNE-OBL | GP-OBL
NOTE Hardstem-bulrush stems are difficult to flatten between fingers in contrast to the otherwise similar softstem-bulrush (*S. tabernaemontani*), in which the stems are easily compressed.

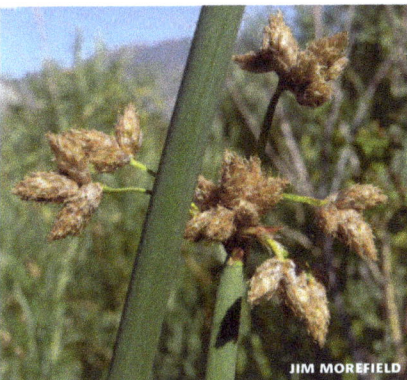

Schoenoplectus heterochaetus
(Chase) Soják **PALE GREAT CLUB-RUSH**

DESCRIPTION Native perennial, spreading by stout rhizomes. Stems slender, round in section, 1–2 m long. Leaves reduced to 3–4 sheaths at base of stem, upper sheaths with blades 6–8 cm long; main bract erect, 1–10 cm long, shorter than head. Spikelets mostly single at ends of stalks, the spikelets 5–15 mm long and 3–6 mm wide, in open, lax heads; scales chaffy, brown, 3–4 mm long, tipped with an awn to 2 mm long; bristles 2–4, unequal, about as long as achene; style 3-parted. Achenes 3-angled, light green to brown, 2–3 mm long, with a beak about 0.5 mm long. Flowering June–Aug.
SYNONYM *Scirpus heterochaetus*
HABITAT emergent in shallow to deep water (1–2 m deep) of marshes, ponds and lakes, ditches
STATUS MW-OBL | NCNE-OBL | GP-OBL
NOTE Slender bulrush is similar to **hardstem club-rush** (*S. acutus*) but much less common; slender bulrush is more slender, the head is more open, and the achene is 3-angled.

Schoenoplectus pungens (Vahl) Palla
♦ COMMON THREESQUARE

DESCRIPTION Native perennial, from slender rhizomes and forming colonies. Stems erect to somewhat curved, 2–12 dm long, 3-angled, the sides concave to slightly convex. Leaves mostly 1–3 near base of stem, usually folded, or channeled near tip, reaching to about middle of stem and 1–3 mm wide; main bract erect, sharp-tipped,

JIM MOREFIELD

Schoenoplectus acutus

MATT LAVIN

Schoenoplectus pungens

resembling a continuation of the stem, 2–15 cm long. Spikelets 5–20 mm long and 3–5 mm wide, clustered in heads of 1–6 stalkless spikelets, the head appearing lateral; scales brown and translucent, 3–5 mm long, notched at tip, with a midvein extended into a short awn 1–2 mm long; bristles 4–6, unequal, shorter than achene; style 2–3-parted. Achenes light green or tan to dark brown, 3-angled or flat on 1 side and convex on other, 2–3 mm long, the beak to 0.5 mm long. Flowering May–Sept.

SYNONYMS *Scirpus pungens*; here, includes *Scirpus americanus* and *Scirpus olneyi*.

HABITAT shallow water, wet sandy, gravelly or mucky shores, streambanks, wet meadows, ditches, seeps and other wet places.

STATUS MW-OBL | NCNE-OBL | GP-OBL

Schoenoplectus subterminalis

(Torr.) Soják WATER-BULRUSH

DESCRIPTION Aquatic native perennial, spreading by rhizomes. Stems slender, weak, round in section, to 1 m or more long, floating or slightly emergent from water surface near tip. Leaves many, threadlike, channeled, from near base of stem and extending to just below water surface; bract 1–6 cm long, appearing to be a continuation of stem. Spikelets single at ends of stems, with several flowers, light brown, narrowly ovate, tapered at each end, 7–12 mm long; scales thin, 4–6 mm long, light brown with a green midvein; bristles shorter to about as long as achene, downwardly barbed; style 3-parted. Achenes 3-angled, brown, 2–4 mm long, tipped with a slender beak to 0.5 mm long. Flowering July–Aug.

SYNONYM *Scirpus subterminalis*.

HABITAT in water to about 1 m deep of lakes, ponds and bog margins.

STATUS MW-OBL | NCNE-OBL | GP-OBL

Schoenoplectus tabernaemontani

(K.C.Gmel.) Palla ♦SOFT-STEM CLUB-RUSH

DESCRIPTION Native perennial, spreading by rhizomes and sometimes forming large colonies. Stems stout, smooth, erect, 1–3 m long, round in section. Leaves reduced to 4–5 sheaths at base of stem, or upper leaves with a blade to 7 cm long; main bract erect, 1–10 cm long, shorter than the head. Spikelets red-brown, 4–12 mm long and 3–4 mm wide, single or in clusters of 2–5 at ends of stalks, the stalks spreading or drooping, the clusters in paniclelike heads; scales ovate, light to dark brown, 2–3 mm long, the midvein usually extended into a short awn to 0.5 mm long; bristles 4–6, downwardly barbed, equal or longer than achene; style 2-parted. Achenes flat on 1 side and convex on other, brown to black, about 2 mm long, tapered to a very small beak to 0.2 mm long. Flowering June–Aug.

SYNONYM *Scirpus validus*.

HABITAT shallow water and shores of lakes, ponds, marshes, streams, ditches.

STATUS MW-OBL | NCNE-OBL | GP-OBL

NOTE similar to **hardstem club-rush** (*S. acutus*) but the stems easily crushed between the fingers, plants generally smaller and more slender, and the head more open.

Schoenoplectus torreyi (Olney) Palla

TORREY'S CLUB-RUSH

DESCRIPTION Native perennial, spreading by rhizomes and often forming colonies. Stems erect, sharply 3-angled, 5–10 dm long. Leaves several, narrow, often longer than the stem; bract erect, 5–15 cm long, appearing to be a continuation of stem. Spikelets ovate, light brown, 8–15

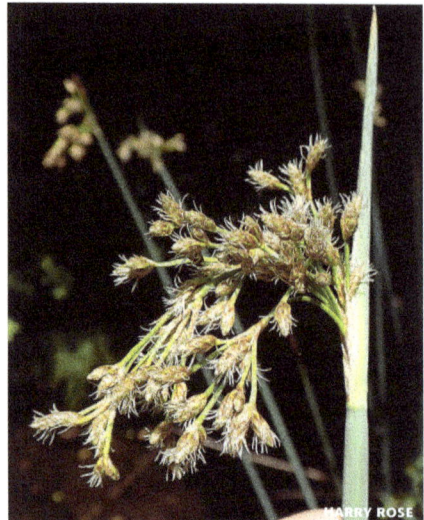

Schoenoplectus tabernaemontani

mm long, in a single head of 1–4 spikelets, the head appearing lateral from side of stem; scales ovate, shiny brown, with a greenish midvein sometimes extended as a short awn to 0.5 mm long; bristles about 6, downwardly barbed, longer than achene; style 3-parted. Achenes compressed 3-angled, shiny, light brown, 3–4 mm long, tipped by a slender beak to 0.5 mm long. Flowering June–Aug. HABITAT shallow water, wet sandy or mucky shores. STATUS MW-OBL | NCNE-OBL | GP-OBL

Scirpus BULRUSH

Stout, rushlike perennials, mostly spreading by rhizomes. Stems unbranched, 3-angled or round in section, solid or pithy. Leaves broad and flat, to narrow and often folded near tip, or reduced to sheaths at base of stems; involucral bracts several and leaflike, or single and appearing like a continuation of the stem. Spikelets single, or in panicle-like or umbel-like clusters at ends of stems, or appearing lateral from the stem; the spikelets stalked or stalkless; scales overlapping in a spiral. Flowers perfect; sepals and petals reduced to 1–6 smooth or downwardly barbed bristles, or sometimes absent; stamens 2 or 3; styles 2-3-parted. Achenes lens-shaped, flat on 1 side and convex on other, or 3-angled, usually tipped with a beak.

Scirpus BULRUSH

1 Lower sheaths red-tinged 2
1 Sheaths green or brown 3
2 Styles 3-cleft; achenes 3-angled; bristles (3–) 6; summit of culm usually scabrous . . . . *S. expansus*
2 Styles 2-cleft; achenes 2-sided; bristles 4 (–5); summit of culm usually smooth . . . . *S. microcarpus*
3 Spikelets many in dense, more or less round heads; bristles about as long as achene or shorter . . . . *S. atrovirens*
3 Spikelets few in open clusters; bristles much longer than achene 4
3 Mature bristles equal or only slightly longer than scales, spikelets not woolly . . . . *S. pendulus*
3 Mature bristles longer than scales, giving spikelets woolly appearance . . . . *S. cyperinus* complex (see desc.)

Scirpus atrovirens Willd.
♦ BLACK BULRUSH

DESCRIPTION Loosely clumped native perennial, with short rhizomes. Stems 3-angled, leafy, 0.5–1.5 m long. Leaves mostly on lower half of stem, blades ascending, usually shorter than the head, 6–18 mm wide; bracts 3–4, leaflike, to 15 cm long, mostly longer than the head. Spikelets many, 2–8 mm long and 1–3 mm wide, crowded in rounded heads at end of stems, the heads on stalks to 12 cm long; scales brown-black, translucent except for the broad green midvein, 1–2 mm long, tipped by an awn to 0.5 mm long; bristles 6, white or tan, shorter or equal to the achene; style 3-parted. Achenes tan to nearly white, compressed 3-angled, about 1 mm long, with a short beak 0.2 mm long. Flowering June–Aug. SYNONYMS *Scirpus atrovirens var. pallidus, Scirpus hattorianus, Scirpus pallidus.* HABITAT wet meadows, shores, ditches, streambanks, swamps, springs. STATUS MW-OBL | NCNE-OBL | GP-OBL

Scirpus cyperinus (L.) Kunth
♦ WOOL-GRASS

DESCRIPTION Coarse, densely tufted, native perennial, rhzomes short. Stems leafy, to 2 m tall, rounded 3-angled to nearly round in section. Leaves flat,

Scirpus atrovirens

3–10 mm wide, rough-to-touch on margins; sheaths brown; bracts 2–4, leaflike, spreading, usually drooping at tip, often red-brown at base. Spikelets numerous, ovate, 3–8 mm long and 2–3 mm wide, appearing woolly due to the long bristles, in clusters of 1 to several spikelets; the spikelet clusters grouped into large, spreading, branched heads at ends of stems; scales ovate, 1–2 mm long; bristles 6, smooth, brown, much longer than achene and scale; styles 3-parted. Achenes white to tan, flattened 3-angled, 0.5–1 mm long, with a short beak. Flowering July–Sept.

HABITAT common in wet meadows, marshes, swamps, ditches, bog margins, thickets; where wet or in very shallow standing water.

STATUS MW-OBL | NCNE-OBL | GP-OBL

NOTE the *Scirpus cyperinus* complex, including this species, *S. atrocinctus* Fern., and *S. pedicellatus* Fern., is often regarded as a single, highly variable species. Alternately, the 3 taxa can be separated as follows:

1 Spikelets all or mostly all sessile in clusters of (2–) 3–7 or more . . . . *S. cyperinus*
1 Spikelets mostly pediceled, the ultimate branches of the inflorescence typically bearing 1 central, sessile spikelet with 2–3 pediceled ones 2
2 Scales and bases of bracts dark blackish green; plants slender with leaves 2–5 mm wide . . . . *S. atrocinctus*
2 Scales and bases of bracts brown or gray-brown; plants more robust with leaves 3–10 mm wide . . . . *S. pedicellatus*

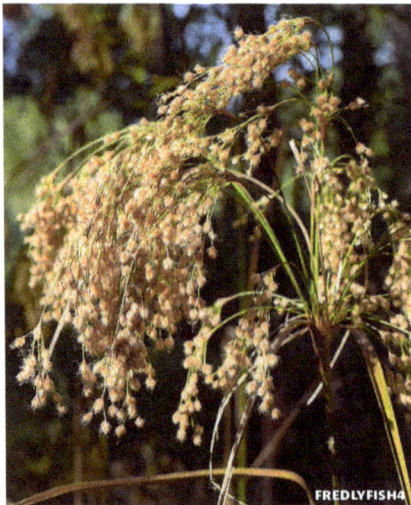

Scirpus cyperinus

Scirpus atrocinctus flowers and fruits earlier than the other two species, often with inflorescences fully developed by late June, and achenes ripe by late July. *S. atrocinctus* readily hybridizes with *S. cyperinus* to form hybrid swarms. Scales of *S. atrocinctus* are usually distinctly blackened, at least near the tip, while those of *S. pedicellatus* have no black pigment or only slightly so. *S. pedicellatus* is paler and larger than *S. atrocinctus,* the spikelets greenish to pale brown.

Scirpus expansus Fern.
WOODLAND BULRUSH

DESCRIPTION Native perennial, with long red rhizomes. Stems single, stout, to 2 m long, 3-angled. Leaves flat, 1–2 cm wide, margins rough-to-touch; lower sheaths red at base; bracts leaflike, the largest 30 cm or more long. Spikelets ovate, 3–5 mm long, clustered in small heads; the heads 2–3x branched; scales 1–2 mm long, oval, brown with a green midvein; bristles 6, light brown, about as long as achene, downwardly barbed, deciduous; style 3-parted. Achenes compressed 3-angled, dull brown to purple-brown, 1–2 mm long, with a short beak. Flowering June–July.

HABITAT Marshes, streambanks, shores and ditches.

STATUS MW-OBL | NCNE-OBL

Scirpus microcarpus J. & K. Presl
◆ RED-TINGED BULRUSH

DESCRIPTION Native perennial, from stout rhizomes. Stems single or few together, 5–15 dm long, weakly 3-angled. Leaves several along stem, flat, ascending, 7–15 mm wide, the upper leaves longer than the head, margins rough-to-touch; sheaths often red-tinged; bracts 3–4, leaflike, to 2–3 dm long. Spikelets numerous, 3–6 mm long and 1–2 mm wide; in a loose, spreading, umbel-like head, the head formed of clusters of 4–20 or more spikelets on stalks to 15 cm long; scales 1–2 mm long, brown and translucent except for green midvein; bristles 4–6, white to tan, downwardly barbed, longer than achene; style 2-parted. Achenes lens-shaped, pale tan to nearly white, about 1 mm long, the beak tiny. Flow-

ering June–July.

HABITAT streambanks, wet meadows, marshes, wet shores, thickets, swamps, springs; not in dense shade.

STATUS MW-OBL | NCNE-OBL | GP-OBL

Scirpus pendulus Muhl.
DROOPING BULRUSH

DESCRIPTION Loosely clumped native perennial, from short, thick rhizomes. **Stems** upright, rounded 3-angled, to 1.5 m long, lower stem covered by old leaf bases. **Leaves** several on stem, flat, 4–10 mm wide, shorter than head; bracts leaflike, 3 or more, shorter than the head, pale brown at base. **Spikelets** many, cylindric, 4–10 mm long and 2–4 mm wide; in an open, umbel-like head at end of stem, the spikelets drooping and clustered in groups of 1 stalkless and several stalked spikelets; scales about 2 mm long, red-brown with a green midvein; bristles 6, brown, smooth, longer than achene and about as long as scale; style 3-parted. **Achenes** compressed 3-angled, light brown, about 1 mm long, with a short, slender beak. **Flowering** June–Aug.

SYNONYM *Scirpus lineatus.*

HABITAT marshes, wet meadows, streambanks, swamp openings and ditches.

STATUS MW-OBL | NCNE-OBL | GP-OBL

Scleria
NUT-RUSH, STONE-RUSH

Annual or sometimes perennial sedgelike herbs, tufted or spreading by short rhizomes. Stems slender, 3-angled. Leaves narrow, shorter than the stem. Flowers either staminate or pistillate, borne in separate spikelets on the same plant; staminate spikelets few-flowered; pistillate spikelets with uppermost flower fertile, the lower scales empty. Flowers in clusters at ends of stems, or with both terminal clusters and clusters from upper leaf axils; sepals and petals absent; stamens 1–3; style 3-parted. Fruit a hard, white achene.

Scleria **NUT-RUSH, STONE-RUSH**
1 Flowers in clusters at ends of stems or from leaf axils .... *S. reticularis*
1 Flowers 1 to few in stalkless heads along a spike .... *S. verticillata*

Scleria reticularis Michx.
RETICULATED NUT-RUSH

DESCRIPTION Native annual or perennial, from short, slender rhizomes. **Stems** slender, 3-angled, to 8 dm long. **Leaves** linear, smooth or sometimes hairy, 2–4 mm wide, shorter than the stem. **Spikelets** in a cluster at end of stem and in 1–3 short-stalked clusters from upper leaf axils; scales lance-shaped to ovate. **Achenes** round, dull white or gray, 1–2 mm wide, covered with a fine cross-hatch of small ridges, tipped with a small sharp point. **Flowering** July–Sept.

HABITAT Sandy marshes, wet sand flats.

STATUS MW-OBL | NCNE-OBL | GP-FACW; Wisc (END), Mich (THR).

Scleria verticillata Muhl.
◆ LOW NUT-RUSH

DESCRIPTION Clumped native annual, roots fibrous. **Stems** slender, smooth, 3-angled, 2–6 dm long. **Leaves** erect,

Scirpus microcarpus

Scleria verticillata

linear, 1 mm wide, shorter than the stems; sheaths often hairy. Spikelets in 2–8 separated heads, each head stalkless, 2–4 mm long, subtended by a small, bristlelike bract 4–6 mm long; scales lance-shaped. Achenes ± round, white, 1 mm wide, covered with horizontal ridges, tipped with a short, sharp point. Flowering July–Sept.

HABITAT sandy or gravelly shores, interdunal flats, wet meadows, marshes.

STATUS MW-OBL | NCNE-OBL | GP-OBL; Minn (THR).

Trichophorum
LEAFLESS-BULRUSH

Tufted perennials. Stems 3-angled or terete. Leaves basal or nearly so; sheaths bladeless or with very short blades less than 1 cm long and to 1 mm wide. Inflorescences terminal; spikelets 1; involucral bracts 1, suberect, scale-like, tip mucronate or awned. Spikelets with 3–9 spirally arranged scales, each subtending a flower. Flowers bisexual; perianth of 0–6 bristles, straight, shorter than to about 20 times as long as the achene, smooth or scabrous; stamens 3. Achenes 3-angled or plano-convex.

Tricophorum **LEAFLESS-BULRUSH**

1 Stems more or less round in section, smooth . . . . *T. cespitosum*
1 Stems 3-angled, rough on angles . . . . *T. alpinum*

Trichophorum alpinum (L.) Pers.
◆ALPINE
COTTON-GRASS
DESCRIPTION Native grasslike perennial,

Trichophorum alpinum

from short rhizomes. Stems single to clustered, slender, 1–4 dm long, sharply 3-angled, rough- to-touch on the angles. Leaves reduced to scales at base of stem, with 1–2 leaves upward on stem, these with short narrow blades 5–15 mm long. Spikelets single at ends of stems, brown, 5–7 mm long, with 10–20 flowers, involucral bract awl-shaped, shorter than spikelet, sometimes absent; scales ovate, blunt-tipped, yellow-brown; bristles 6, white, flattened, longer than the scales, when mature forming a white tuft 1–2 cm longer than the spikelet. Achenes 3-angled, dull brown, 1–4 mm long.

SYNONYMS *Eriophorum alpinum, Scirpus hudsonianus.*

HABITAT open bogs, conifer swamps, wet meadows, wet sandy shores; sometimes where calcium-rich.

STATUS MW-OBL | NCNE-OBL | GP-OBL

NOTE similar to the cottongrasses (*Eriophorum*) and previously placed in that genus as *Eriophorum alpinum* or treated as *Scirpus hudsonianus.*

Trichophorum cespitosum (L.) Hartman ◆TUFTED LEAFLESS-BULRUSH

DESCRIPTION Densely tufted, native grasslike perennial, from short rhizomes. Stems slender, smooth, more or less round in section, 1–4 dm long. Leaves light brown and scalelike at base of stems, and also usually 1 leaf upward on stem, the

Trichophorum cespitosum

blade narrow, short, to 6 mm long. Spikelets 1 at end of stems, brown, 4–6 mm long, several-flowered; scales yellow-brown, deciduous, the lowest scale about as long as spikelet; bristles 6, usually slightly longer than achene; style 3-parted. Achenes brown, 3-angled, 1.5 mm long.

SYNONYM *Scirpus cespitosus*

HABITAT open bogs, cedar swamps, calcareous fens, wet swales between dunes; also Lake Superior rocky shores.

STATUS MW OBL | NCNE OBL | GP OBL; Wisc (THR).

Eriocaulaceae
Pipewort Family

Eriocaulon PIPEWORT

Eriocaulon aquaticum (Hill.) Druce
◆ PIPEWORT

DESCRIPTION Native perennial, spongy at base, with fleshy roots. Stems usually single, leafless, slightly twisted, 5–7-ridged, 5–20 cm long (or reaching 2–3 m long when in deep water). Leaves grasslike, in a rosette at base of plant, thin and often translucent, 2–10 cm long and 2–5 mm wide, 3–9-nerved with conspicuous cross-veins. Flowers either staminate or pistillate, grouped together in a single, ± round head at end of stem, the heads white-woolly, 4–6 mm wide. Fruit a 2–3-seeded capsule.

Eriocaulon aquaticum

Flowering July–Sept.

SYNONYM *Eriocaulon septangulare.*

HABITAT shallow water, sandy or peaty shores.

STATUS MW-OBL | NCNE-OBL

Hydrocharitaceae
Tape-Grass Family

AQUATIC, PERENNIAL HERBS. Stems leafy, the leaves whorled (*Elodea*), plants floating on water surface (*Hydrocharis*), or plants stemless with clusters of long, linear, ribbonlike leaves (*Vallisneria*). Flowers either staminate or pistillate and borne on separate plants,; pollination occurs at water surface. Fruit several-seeded, maturing underwater.

Hydrocharitaceae **TAPE-GRASS FAMILY**

1 Leaves floating, blades orbicular with a ± cordate base; invasive species in Mich. . . . . *Hydrocharis morus-ranae*

1 Leaves submerged, linear; widespread native species **2**

2 Leaves very long and ribbon-like (mostly 3–11 mm wide), in a basal rosette . . . . *Vallisneria americana*

2 Leaves to 6 (–12) cm long, opposite or whorled **3**

3 Leaves whorled, entire . . . . *Elodea*

3 Leaves opposite, minutely denticulate to visibly toothed . . . . *Najas*

Elodea WATER-WEED

Aquatic perennial herbs, rooting from lower nodes or free-floating. Stems slender, leafy, branched. Leaves crowded near tip of stem, mostly in whorls of 3–4, or opposite, stalkless; margins finely sharp-toothed. Flowers either staminate or pistillate and on separate plants, tiny, single from upper leaf axils, subtended by a 2-parted spathe, usually extended to the water surface by a long, threadlike hypanthium, or stalkless and breaking free to float to water surface in staminate flowers of *Elodea nuttallii;* sepals 3; petals 3 or absent, white or purple; staminate flowers with 9 stamens; pistillate flowers with 3 stigmas, the stigmas entire or 2-parted. Fruit a capsule, ripening underwater. *Anacharis* or *Philotria* in older floras.

ADDITIONAL SPECIES **Two-leaf waterweed** (*Elodea bifoliata*) is reported from several locations in Minnesota, at the e edge of this species' main range of w USA; at least some of the leaves of this species are in twos rather than mostly 3–4.

Elodea **WATER-WEED**

1   Leaves mostly 2 mm or more wide; staminate flowers long-stalked in a spathe, the spathe more than 7 mm long, extended to water surface by a long, threadlike hypanthium . . . . *E. canadensis*

1   Leaves to 1.5 mm wide; staminate flowers stalkless in a spathe, the spathe 2–4 mm long, breaking free to float to water surface at flowering time . . . . *E. nuttallii*

Elodea canadensis Michx.
♦ COMMON WATER-WEED

DESCRIPTION Submerged, native perennial herb. Stems round in section, usually branched, 2–10 dm long. Leaves bright green, firm; lower leaves opposite, reduced in size, ovate or lance-shaped; upper leaves in whorls of 3, the uppermost crowded and overlapping, lance-shaped, 5–15 mm long and about 2 mm wide, rounded at tip. Flowers either staminate or pistillate and on separate plants, at ends of threadlike stalks, 2–30 cm long; staminate flowers in spathes from upper leaf axils, the spathes about 10 mm long and to 4 mm wide; sepals green, 3–5 mm long; petals white, 5 mm long; stamens 9. Female flowers in spathes from upper leaf axils, the spathes 10–20 mm long, extended to water surface by a threadlike hypanthium; sepals 2–3 mm long; petals white, 2–3 mm long. Fruit a capsule, 5–6 mm long, tapered to a beak 4–5 mm long. Flowering June–Aug.

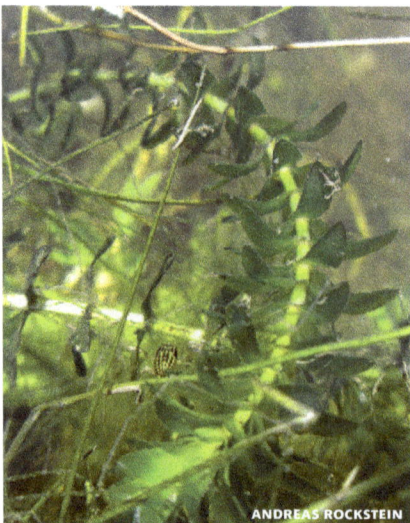

SYNONYM *Anacharis canadensis.*

HABITAT shallow to deep water of lakes (including Great Lakes), streams and ditches.

STATUS MW-OBL | NCNE-OBL | GP-OBL

Elodea nuttallii (Planchon) St. John
♦ FREE-FLOWERED WATER-WEED

DESCRIPTION Submerged, native perennial herb. Stems slender, round in section, usually branched, 3–10 dm long. Lower leaves opposite, reduced in size, ovate to lance-shaped; upper leaves in whorls of 3 (or sometimes 4), not densely overlapping at tip, linear to lance-shaped, 6–13 mm long and 0.5–1.5 mm wide, tapered to a pointed tip. Flowers either staminate or pistillate and on separate plants; staminate flowers in stalkless spathes from middle leaf axils, the spathes ovate, 2–3 mm long, the flowers single and stalkless in the spathe, breaking free and floating to water surface and then opening; sepals green or sometimes red, 2 mm long; petals absent or very short (to 0.5 mm long); stamens 9. Female flowers in cylindric spathes from upper leaf axils, the spathes 1–2.5 cm long, extended to water surface by a threadlike stalk to 10 cm long; sepals green, about 1 mm long; petals white, longer than sepals. Fruit a capsule, 5–7 mm long. Flowering June–Aug.

SYNONYMS *Anacharis nuttallii, Anacharis occidentalis, Philotria nuttallii.*

Elodea canadensis

Elodea nuttallii

HABITAT shallow to deep water of lakes, streams and ditches.

STATUS MW-OBL | NCNE-OBL | GP-OBL

NOTE Similar to **common water-weed** (*Elodea canadensis*) but less common, and plants smaller, with leaves narrower, paler green, and not closely overlapping at stem tips, and the staminate flowers not elevated on a long, slender stalk.

Hydrocharis FROG'S-BIT

Hydrocharis morsus-ranae L.
♦EUROPEAN FROG'S-BIT

DESCRIPTION Invasive perennial aquatic herb; spreading in Great Lakes region. Roots to 30 cm long, not anchored in bottom substrate but free-floating in water. Leaves basal, floating, or sometimes becoming emersed; blade cordate, 3–5 cm wide. green and veined on upper surface; underside purplish-red with a spongy coating. Flowers white, 3-petaled, about 1.5 cm wide; staminate flowers 1–4 per spathe; pistillate flowers solitary in spathe; petals more than 1.5 times the length of sepals; stamens 9–12. Flowering June–Aug.

HABITAT shallow water of slow-moving streams, marshes and ditches.

STATUS MW-OBL | NCNE-OBL

NOTE Stolons from the center of the plant produce new plants, creating dense mats. The stolons also produce vegetative winter buds (*turions*), that break free from floating plants, sink and overwinter on the bottom. In spring, the turions float to the water surface and begin to grow; a single plant can produce an estimated 100 to 150 turions in each season.

KRZYSZTOF ZIARNEK

Hydrocharis morsus-ranae

Najas NAIAD, WATER-NYMPH

Aquatic annual herbs, roots fibrous, rhizomes absent. Stems wavy, with slender branches. Leaves simple, opposite or in crowded whorls, stalkless, abruptly widened at base to sheath the stem; margins toothed to nearly entire, the teeth sometimes spine-tipped. Flowers either staminate or pistillate, separate on same plant or on different plants, tiny, single and stalkless in leaf axils, enclosed by the sheathing leaf bases; staminate flowers a single anther within a membranous envelope (spathe), this surrounded by perianth scales, the scales sometimes joined into a tube; pistillate flowers surrounded by 1–2 spathes, pistils 1, stigmas 2–4, style usually persistent. Fruit a 1-seeded achene.

Najas **NAIAD, WATER-NYMPH**

1   Leaves coarsely toothed and spine-tipped (spines visible without a lens), bright green; midvein of leaf underside and stems between nodes often prickly .. *N. marina*

1   Leaves nearly entire or toothed (if spine-tipped, the spines, except in *Najas minor,* not visible without a lens), often olive green; leaf surface and stems between leaves smooth **2**

2   Base of leaves lobed or clasping stem ... **3**

2   Leaves tapered to base, not lobed or clasping stem.......................... **4**

3   Leaves somewhat stiff, curved downward near tip; base of leaf lobed; seed coat pitted, the pits wider than long and arranged in regular, ladderlike rows .... *N. minor*

3   Leaves slender, not stiff, not curved downward near tip; base of leaf clasping; seed coat pitted, the pits longer than wide .... *N. gracillima*

4   Achenes smooth and glossy, widest above middle .... *N. flexilis*

4   Achenes rough and pitted, widest at middle and tapered to ends ... *N. guadalupensis*

Najas flexilis (Willd.) Rostkov & Schmidt
♦NORTHERN WATER-NYMPH

DESCRIPTION Native annual aquatic herb. Stems branched, 5–40 cm long. Leaves densely clustered at tips of stems, linear, tapered to a long slender point, spreading or ascending, 1–4 cm long and to

0.5 mm wide; margins with tiny sharp teeth. Flowers either staminate and pistillate, separate on same plant. Achenes oval, olive-green to red, the beak 1 mm or more long; seeds straw-colored, shiny, 2–4 mm long. Flowering July–Sept.

HABITAT ponds, lakes and streams.

STATUS MW-OBL | NCNE-OBL | GP-OBL

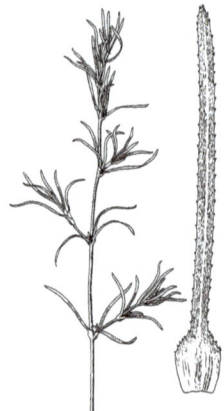

NOTE The newly described *Najas canadensis* Michx., believed to be a hybrid derived from *N. flexilis* and *N. guadalupensis*, is reported from scattered locations in the region.

Najas gracillima (A. Braun) Magnus
◊ SLENDER WATER-NYMPH

DESCRIPTION Native annual aquatic herb; plants light green. Stems very slender, branched, 0.5–5 dm long. Leaves opposite or in groups of 3 or more, bristlelike, 0.5–3 cm long and to 0.5 mm wide, spreading or ascending; margins with very small teeth. Flowers either staminate or pistillate and on the same plant. Achenes cylindric, narrowed at ends; seeds light brown, 2–3 mm long.

HABITAT shallow water of lakes, usually in muck; intolerant of polluted water.

STATUS MW-OBL | NCNE-OBL | GP-OBL

Najas guadalupensis (Sprengel) Magnus
SOUTHERN WATER-NYMPH

DESCRIPTION Native annual aquatic herb. Stems much branched, 1–6 dm long. Leaves

numerous, linear, spreading and often curved downward at tip, 1–3 cm long and 0.5–2 mm wide; groups of smaller leaves also present in leaf axils; margins with very small teeth. Flowers either staminate or pistillate, separate on same plant. Achenes cylindric, the beak to 0.5 mm long; seeds brown or purple, 1–3 mm long. Flowering July–Sept.

HABITAT shallow to deep water of lakes, ponds and sometimes rivers; often with **northern water-nymph** (*Najas flexilis*) but less common.

STATUS MW-OBL | NCNE-OBL | GP-OBL

Najas marina L.
ALKALINE WATER-NYMPH

DESCRIPTION Native annual aquatic herb. Stems stout, 1–5 dm long and 1–4 mm wide, compressed, branched, prickly. Leaves opposite or whorled, linear, 0.5–4 cm long and 1–4 mm wide, sometimes with spines on underside; margins coarsely toothed, the teeth 1–4 mm apart and spine-tipped. Flowers either staminate or pistillate and on different plants. Achenes olive-green, the beak about 1 mm long; seeds dull, 2–5 mm long. Flowering July–Sept.

HABITAT shallow water (to 1 m deep) of lakes and marshes.

STATUS MW-OBL | NCNE-OBL | GP-OBL

MARY KRIEGER

Najas flexilis

SHOW RYU

Najas gracillima

Najas minor Allioni
EUTROPHIC WATER-NYMPH

DESCRIPTION Introduced annual aquatic herb; plants dark green. Stems slender, branched, 1–2 dm long. Leaves opposite or whorled, linear, 0.5–3.5 cm long and to 0.2 mm wide, curved downward at tip; margins sharp-toothed, with 7–15 teeth on each side. Flowers either staminate and pistillate, separate on the same plant. Achenes oval, the beak about 1 mm long; seeds 2–3 mm long, purple-tinged. **HABITAT** marshes, lakes, ponds; introduced and potentially invasive throughout e USA. **STATUS** MW-OBL | NCNE-OBL | GP-OBL

Vallisneria TAPE-GRASS

Vallisneria americana Michx.
♦ **TAPE-GRASS, EEL-GRASS, WATER-CELERY**

DESCRIPTION Submerged native perennial herb, fibrous rooted, spreading by stolons and often forming large colonies. Stems absent. Leaves long and ribbonlike, in tufts from a small crown, to 1 m or more long and 3–10 mm wide, rounded at tip, margins smooth. Flowers either staminate or pistillate and on separate plants. Staminate flowers small, about 1 mm wide, in a many-flowered head, the head within a stalked spathe from base of plant, the stalk 3–15 cm long; sepals 3, petals 1, stamens 2; the staminate flowers released singly from the spathe and floating to water surface where they open. Female flowers single in a spathe, on long slender stalks that extend to water surface, the stalk contracting and coiling after flowering to draw the fruit underwater; sepals 3, petals small, 3; stigmas 3. Fruit a cylindric, curved capsule, 4–10 cm long. Flowering July–Sept. **HABITAT** shallow (to sometimes deep) water of lakes and streams. **STATUS** MW-OBL | NCNE-OBL | GP-OBL

Vallisneria americana

Iridaceae
Iris Family

PERENNIAL HERBS with rhizomes, bulbs, or fibrous roots. Leaves parallel-veined, narrow, 2-ranked, the margins joined to form an edge facing the stem (equitant). Flowers perfect, with 6 petal-like segments, single or in clusters at ends of stem, stamens 3, style 3-parted. Fruit a 3-chambered capsule.

Iridaceae **IRIS FAMILY**

1 Flowers more than 2 cm wide; stems not winged; leaves more than 6 mm wide . *Iris*
1 Flowers to 2 cm wide; stems winged; leaves to 6 mm wide . . . . *Sisyrinchium*

Iris IRIS, FLAG
Perennial herbs, spreading by thick rhizomes. Stems erect. Leaves swordlike, erect or upright, the margins joined to form an edge facing the stem. Flowers 1 or several at ends of stems; yellow or blue-violet; sepals 3, spreading or bent downward, longer and wider than the petals; petals 3, erect or arching; stamens 3; styles 3-parted, the divisions petal-like and arching over the stamens. Fruit an oblong capsule.

Iris **IRIS, FLAG**

1 Flowers yellow . . . . *I. pseudacorus*
1 Flowers blue or violet 2
2 Stems as long or longer than leaves; base of plant often purple-tinged; sepal base unspotted, or with a with a hairless, green-yellow spot . . . . *I. versicolor*
2 Stems shorter than leaves; base of plant usually brown; sepal base with a hairy, bright yellow spot . . . . *I. virginica*

Iris pseudacorus L.
♦ **YELLOW FLAG**

DESCRIPTION Introduced perennial herb, from thick rhizomes. Stems 0.5–1 m long, shorter or equal to the leaves. Leaves sword-shaped, stiff and erect, waxy, 1–2 cm wide. Flowers several at end of stems, yellow, 7–9 cm wide, sepals spreading, upper portion marked with brown; petals

erect, narrowed in middle, 1–2.5 cm long. Fruit a 6-angled, oblong capsule, 5–9 cm long. Flowering May–June.

HABITAT lakeshores, streambanks, marshes, ditches; introduced from Europe and sometimes invasive.

STATUS MW-OBL | NCNE-OBL | GP-OBL

Iris versicolor L.
◊NORTHERN BLUE FLAG

DESCRIPTION Native perennial herb, from thick, fleshy rhizomes and forming colonies. Stems ± round in section, often branched above, 4–9 dm long. Leaves sword-shaped, erect or arching, somewhat waxy, 2–3 cm wide, usually shorter than stem. Flowers several on short stalks at ends of stems, blue-violet, 6–8 cm wide; sepals spreading, unspotted, or with a green-yellow spot near base, surrounded by white streaks and purple veins; petals erect, about half as long as sepals. Fruit an oblong capsule, 3–6 cm long. Flowering June–July.

HABITAT marshes, shores, wet meadows, open bogs, swamps, thickets, forest depressions; often in shallow water.

STATUS MW-OBL | NCNE-OBL | GP-OBL

NOTE *Iris versicolor* is similar to **southern blue flag** (*Iris virginica*), which is sometimes considered a variety of this species, but *Iris versicolor* has a more northerly distribution.

Iris virginica L.
SOUTHERN BLUE FLAG

DESCRIPTION Native perennial herb, from thick rhizomes, often forming large colonies. Stems ± round in section, to 1 m long. Leaves sword-shaped, erect or arching, 2–3 cm wide, usually longer than stems. Flowers several on short-stalks at ends of stems, blue-violet, often with darker veins, 6–8 cm wide; sepals spreading, curved backward at tip, with a hairy, bright yellow spot near base; petals shorter than sepals. Fruit an ovate to oval capsule, 4–7 cm long. Flowering May–July.

SYNONYM *Iris shrevei.*

HABITAT swamps, thickets, shores, streambanks, marshes, ditches.

STATUS MW-OBL | NCNE-OBL | GP-OBL

Sisyrinchium
BLUE-EYED-GRASS

Clumped perennial herbs, from fibrous roots. Stems slender, leafless, flattened or winged. Leaves narrow and linear, from base of plant, the margins joined and turned to form an edge facing the stem. Flowers in an umbel at end of stem, above a pair of erect green bracts (spathe), blue-violet (our species), with 6 spreading segments, the segments joined only at base, the tips rounded but with an small bristle. Fruit a rounded capsule; seeds round, black.

ANDREAS ROCKSTEIN

Iris pseudacorus

DAVID BERGER

Iris versicolor

Sisyrinchium BLUE-EYED GRASS

1   Spathes stalked from axils of leafy bracts, upper stem appearing branched . . . . *S. atlanticum*

1   Spathes stalkless at end of unbranched stem, upper stem not appearing branched **2**

2   Stems flattened and winged, 2–4 mm wide; larger leaves 2–3 mm wide . *S. montanum*

2   Stem slender, barely winged, to 1 mm wide; larger leaves to 1.5 mm wide . . . . *S. mucronatum*

Sisyrinchium atlanticum E. Bickn.
EASTERN BLUE-EYED-GRASS

DESCRIPTION Native perennial herb, spreading and forming small clumps; plants pale green and waxy. **Stems** slender, 0.5–2 mm wide, often angled at the upper nodes, the margins smooth and narrowly winged. **Leaves** mostly from base of plant, shorter than stems, 1–3 mm wide. **Flowers** single in 2–3 spathes, the spathes of 2 bracts, 1–2 cm long, often purple-tinged, stalked from the axils of leaflike bracts; flower segments (tepals) blue-violet, 8–12 mm long, bristle-tipped. **Fruit** a stalked, oval capsule, 3–5 mm long. **Flowering** June–July.

HABITAT wet sandy shores.

STATUS MW-FACW | NCNE-FACW | GP-FACW; Mich (THR).

Sisyrinchium montanum Greene
♦ MOUNTAIN BLUE-EYED-GRASS

DESCRIPTION Clumped, native perennial, from fibrous roots; plants pale-green and waxy. **Stems** stiff and erect, leafless, flattened and winged, 1–5 dm tall and 2–4 mm wide. **Leaves** mostly from base of plant, narrow and grasslike, about half as long as stem, 1–3 mm wide. **Flowers** in head of 1 to several flowers at end of stem, subtended by a spathe, the spathe of 2 bracts, the outer bract 3–7 cm long, the inner bract about half as long; the flower segments (tepals) blue-violet with a yellow center, 5–15 mm long, with a short, slender tip. **Fruit** a ± round, pale brown capsule, 4–7 mm wide, on an erect stalk shorter than the inner

bract. **Flowering** May–July.

SYNONYM *Sisyrinchium angustifolium.*

HABITAT wet meadows, shores, thickets, ditches, swales; also in drier woods and fields.

STATUS MW-FAC | NCNE-FAC | GP-FAC

Sisyrinchium mucronatum Michx.
MICHAUX'S BLUE-EYED-GRASS

DESCRIPTION Clumped, native perennial herb; plants dark green. **Stems** very slender, to 1 mm wide, leafless, margins not or barely winged. **Leaves** from near base of plant, narrow and linear, to 1.5 mm wide. **Flowers** in a single head at end of stem, subtended by a spathe, the spathe of 2 bracts, the bracts often purple-tinged, the outer bract 2–3 cm long, the inner bract shorter, 1–2 cm long; the segments (tepals) deep violet-blue, 8–10 mm long, tipped with a sharp point. **Fruit** a ± round, pale brown capsule, 2–4 mm long, on spreading stalks. **Flowering** May–June.

HABITAT wet meadows, calcareous fens.

STATUS MW-FAC | NCNE-FAC | GP-FAC

Juncaceae
Rush Family

Juncus RUSH

Clumped or rhizomatous rushes, mostly perennial (annual in *Juncus bufonius*). Stems erect and unbranched. Leaves from base of plant or along stem, alternate,

JASON HOLLINGER

Sisyrinchium montanum

round in section, or flat to rolled inward, or reduced to sheaths at base of stem; leaves in some species with cross-partitions at intervals (septate). Flowers perfect, regular, in compact to open clusters of few to many flowers, subtended by 1 or several leaflike involucral bracts; sepals and petals of 6 chaffy, scalelike, green to brown tepals; stamens 6 or 3; stigmas 3, ovary superior, 1 or 3-chambered. Fruit a many-seeded capsule; seeds with a short slender tip or with a tail-like appendage at each end.

NOTE *Juncus* are distinguished from the **club-rushes** and **bulrushes** (*Schoenoplectus, Schoenoplectiella, Scirpus*) by the fruit being a capsule rather than an achene. Correctly identifying species within the genus *Juncus* is difficult, and based, in part, on characteristics of the small seeds within the capsule.

ADDITIONAL SPECIES **Stout rush** (*Juncus nodatus* Coville, subgenus *Septati*) is reported from central Wisc (more common southward in Miss River valley). Plants are large (to 1 m tall), with leaves round in section and conspicuously cross-divided. Inflorescence is terminal, with several hundred heads, each head with 2–10 flowers.

Juncus **Groups (subgenera) RUSHES**

1 Flowers borne singly (each flower atop a stalk); bractlets present (except in *Juncus pelocarpus*) 2
1 Flowers borne in heads (flowers ± stalkless); bractlets absent 4
2 Inflorescences appearing lateral; inflorescence bract round in section, erect, appearing to be continuation of stem; basal leaves without blades, stem leaves absent . . . . GROUP 1 (Subgenus *Genuini*)
2 Inflorescences appearing terminal; inflorescence bract erect or upright, flat, involute or round in section; basal leaves (at least some) usually with blade, stem leaves present or absent 3
3 Leaves round in section, with cross-partitions at intervals (septate); capsules beaked . . . . GROUP 2 (Subgenus *Septati*)
3 Leaves flat, involute, or round in section, not septate; capsules not beaked . . . . GROUP 3 (Subgenus *Poiophylli*)

4 Leaves flat or ensiform (sword-shaped) . **5**
4 Leaves round in section or compressed. . **6**
5 Leaves ensiform, imperfectly septate . . . . GROUP 4 (Subgenus *Ensifolii*)
5 leaves flat, not septate . . . . GROUP 5 . . . . (Subgenus *Graminifolii*)
6 Capsules large; seeds large, long tailed; leaves not noticeably septate . . GROUP 6 . . . . (Subgenus *Alpini*)
6 Capsules smaller; seeds not tailed or if tailed not long; leaves septate or not. . . . . GROUP 2 (Subgenus *Septati*)

Juncus **Group 1 (subgenus *Genuini*)**
1 Stems densely clumped; stamens 3 . . . . *J. effusus*
1 Stems single from rhizomes, the stems often in rows; stamens 6 2
2 Tepals dark brown; capsules red-brown; anthers equal to or longer than filaments . . . . *J. balticus*
2 Tepals greenish; capsules green to light brown; anthers shorter than filaments. . . . . *J. filiformis*

Juncus **Group 2 (subgenus *Septati*)**
1 Seeds tailed, 0.7–2.6 mm long (including tails); seed body with a whitish translucent covering 2
1 Seeds not tailed, 0.3–0.7 mm long; seed body clear yellow-brown 4
2 Seeds 1.1–1.9 mm long; heads ovate, 5–50-flowered; inflorescence branches erect to ascending. . . . . *J. canadensis*
2 Seeds 0.7–1.2 mm; heads 2–8-flowered; inflorescence branches erect to spreading 3
3 Outer tepals obtuse to nearly acute. . . . . *J. brachycephalus*
3 Outer tepals acuminate to acute . . . . *J. brevicaudatus*
4 Flowers 1–2 (–4) at each node, not in heads; capsules fertile only below middle . . . . *J. pelocarpus*
4 Flowers in heads of 3–60; capsules fertile throughout or only below middle 5
5 Heads sphere-shaped or nearly so, 15–60-flowered 6
5 Heads obconic to hemispheric, 3–15-flowered 11
6 Stamens 3 7
6 Stamens 6 10

7 Plants clumped; tepals lance-shaped . . . . *J. acuminatus*

7 Plants rhizomatous; tepals narrowly lance-shaped **8**

8 Capsules shorter than (and hidden by) tepals . . . . *J. brachycarpus*

8 Capsule longer to slightly shorter than tepals **9**

9 Capsules remaining joined at tip when mature . . . . *J. scirpoides*

9 Capsule valves separating at tip as seeds mature . . . . *J. nodosus*

10 Outer tepals 2.4–4.1 mm long, equaling inner tepals; auricles 0.5–1.7 mm . . . . *J. nodosus*

10 Outer tepals 4–6 mm, outer and inner tepals unequal in length; auricles 1–4 mm . . . . *J. torreyi*

11 Stamens 3 . . . . **12**

11 Stamens 6 . . . . **13**

12 Heads 5–50; tepals 2.6–3.5 mm long, nearly equal; capsules 2.8–3.5 mm long . . . . *J. acuminatus*

12 Heads 30–250; tepals 1.7–2.9 mm long, inner tepals shorter than outer tepals; capsules 2–3 mm . . . . *J. nodosus*

13 Stems sometimes creeping or floating; submersed leaves may be threadlike and formed before flowering; inflorescences with 1–9 heads . . . . *J. articulatus*

13 Stems erect; threadlike submersed leaves not formed except in *Juncus militaris*; heads 1–60 or more **14**

14 Lower stem leaves overtopping inflorescence, uppermost stem leaf usually reduced to an inflated, bladeless sheath . . . . *J. militaris*

14 Lower stem leaves shorter than inflorescence, uppermost stem leaf usually with a blade, its sheath not inflated **15**

15 Inner tepals blunt-tipped; inflorescence stiffly erect . . . . *J. alpinoarticulatus*

15 Inner tepals tapered to a sharp tip; inflorescence spreading. . . . *J. articulatus*

Juncus Group 3 (subgenus *Poiophylli*)

1 Plants annual, to 10 cm tall . . . *J. bufonius*

1 Plants perennial, more than 10 cm tall . . **2**

2 Auricles at summit of leaf sheath 3–6 mm long, membranous, transparent . *J. tenuis*

2 Auricles absent or very short membranous or hardened projections less than 2 mm long **3**

3 Leaf blade flat . . . . *J. compressus*

3 Leaf blade round in cross-section or channeled and closed for ± entire length . . . . *J. vaseyi*

Juncus Group 4 (subgenus *Ensifolii*)

One species in subgenus in Upper Midwest region; leaves folded and joined near tip, with one edge facing the stem (similar to *Iris*); nw Wisc and n Mich . . . . *J. ensifolius*

Juncus Group 5 (subgenus *Graminifolii*)

1 Stems single from long rhizomes; stamens 6, pale yellow; n Minn . *Juncus longistylis*

1 Stems single or in small clumps from short rhizomes; stamens 3, red-brown; occasional across Upper Midwest . . . . . *J. marginatus*

Juncus Group 6 (subgenus *Alpini*)

One species (rare) in subgenus in northern portions of Upper Midwest; leaf blades without cross-partitions at regular intervals . . . . *J. stygius*

Juncus acuminatus Michx.

♦ TAPER-TIP RUSH

DESCRIPTION Clumped, native perennial rush. Stems erect, slender, 2–8 dm tall, with 1–2 leaves. Leaves from stem and at base of plant, round to compressed in section, 5–40 cm long and 1–3 mm wide; auricles rounded, 1–2 mm long; bract erect, round, 1–4 cm long, shorter than the head. Flowers in an open, pyramid-shaped inflorescence, 5–12 cm long and less than half as wide, composed of 5–50 rounded heads 6–10 mm wide, each head with 5–30 flowers, the branches spreading, 1–10 cm long; tepals lance-shaped, green or straw-colored, 3–4 mm long; stamens 3, shorter than the tepals. Capsules oval, straw-colored to light brown, 3–4 mm long, about as long as the tepals, tipped with a short, blunt point. Flowering June–Aug.

HABITAT wet sandy shores, streambanks and ditches; not in open bogs.

STATUS MW-OBL | NCNE-OBL | GP-OBL

Juncus alpinoarticulatus Chaix.
ALPINE RUSH

DESCRIPTION Native perennial rush, spreading by rhizomes. Stems in small clumps, 1.5–4 dm long. Leaves mostly from base of plant and with 1–2 stem leaves, round in section, hollow, with small swollen joints, 2–12 cm long and 0.5–1 mm wide; sheaths green to red, auricles rounded, 0.5–1 mm long; bract round in section, 2–6 cm long and shorter than the head. Flowers in an open panicle of 5–25 heads, 2–15 cm long and 1–5 cm wide, the heads oblong pyramid-shaped, 2–6 mm wide, mostly 2–5-flowered, the branches upright, 1–7 cm long; tepals green to brown, 2–3 mm long, the inner tepals shorter, the margins chaffy; stamens 6. Capsules oblong, 3-angled, straw-colored to chestnut brown, satiny, 2–3 mm long, slightly longer than the tepals, tapered to a rounded tip. Flowering June–Sept.

SYNONYM *Juncus alpinus.*

HABITAT sandy or gravelly shores, streambanks, fens; often where calcium-rich.

STATUS MW-OBL | NCNE-OBL | GP-OBL

Juncus articulatus L. JOINTED RUSH

DESCRIPTION Native perennial rush, with coarse white rhizomes. Stems usually clumped, 2–6 dm long. Leaves from stem and at base of plant, ± round in section, hollow, with small swollen joints, 4–12 cm long and 1–3 mm wide; sheaths green or sometimes red, auricles rounded, about 1 mm long; bract erect, round in section, 1–4 cm long, shorter than the head. Flowers in open panicles, 4–10 cm long and 3–6 cm wide, composed of 3–30 heads, the heads rounded, 6–8 mm wide, 3–10-flowered, panicle branches erect to widely spreading, 1–4 cm long; tepals green to dark brown, 2–3 mm long; stamens 6. Capsules oval, dark brown, shiny, 3–4 mm long, longer than the tepals, tapered to a tip. Flowering July–Sept.

HABITAT sandy, gravelly or mucky shores, streambanks and springs.

STATUS MW-OBL | NCNE-OBL | GP-OBL

Juncus balticus Willd.
♦ **BALTIC RUSH, WIRE-RUSH**

DESCRIPTION Native perennial rush, spreading by stout, brown to black rhizomes. Stems slender and tough, dark green, 3–9 dm long, in rows from the rhizomes. Leaves reduced to red-brown sheaths at base of stem; bract erect, round in section, 1–2 dm long, longer than the head and resembling a continuation of stem. Flowers single on stalks, in dense to spreading heads, the heads appearing lateral, extending outward from stem 1–7 cm; tepals lance-shaped, dark brown, 3–5 mm long, margins chaffy; stamens 6. Capsules ovate, somewhat 3-angled, red-brown, 3–4 mm long, shorter to slightly longer than the tepals, tapered to a sharp point. Flowering May–Aug.

SYNONYM *Juncus arcticus.*

HABITAT wet sandy or gravelly shores, interdunal wetlands near Great Lakes, meadows, ditches, marshes, seeps.

STATUS MW-OBL | NCNE-OBL | GP-FACW

Juncus acuminatus

Juncus balticus

Juncus brachycarpus Engelm.
WHITE-ROOT RUSH

DESCRIPTION Native perennial rush, spreading by stout white rhizomes. Stems erect, round in section, 3–8 dm long. Leaves from stem and base of plant, round in section, 3–50 cm long and 1–2 mm wide, cross-divided; auricles rounded, 0.5–2 mm long; bract erect, channeled, 1–3 cm long, shorter than the head. Flowers in an open or crowded raceme or panicle of 3–10 or more heads, 2–8 cm long and 1–3 cm wide, the heads round, 8–10 mm wide, with 30 or more flowers, the branches upright, 1–4 cm long; tepals narrowly lance-shaped, straw-colored, 2–4 mm long; stamens 3, shorter than the tepals. Capsules ovate, brown, 2–3 mm long, shorter than the tepals, abruptly tapered to a small point. Flowering June–Aug.

HABITAT wetland margins, sandy swales and prairies.

STATUS MW-FACW | NCNE-FACW | GP-FACW; Mich (THR).

Juncus brachycephalus (Engelm.)
Buchenau **SMALL-HEAD RUSH**

DESCRIPTION Densely clumped, native perennial rush. Stems erect, round in section, 3–7 dm long. Leaves from stem and base of plant, round in section, 2–20 cm long and 1–2 mm wide, often spreading; auricles rounded, to 1 mm long; bract erect, round in section, 1–5 cm long, shorter than the head. Flowers in an open raceme or panicle of 10–80 heads, 5–25 cm long and 2–12 cm wide, the heads oval, 2–5 mm wide, 2–6-flowered, branches upright to spreading, 1 5 cm long; tepals lance-shaped, green to light brown, 3-nerved, 2–3 mm long, margins chaffy; stamens 3 or sometimes 6. Capsules ovate, ± 3-angled, light brown, 3–4 mm long, longer than the tepals, abruptly narrowed to a short beak. Flowering June–Sept.

HABITAT sandy or gravelly shores, streambanks, open bogs, calcium-rich springs.

STATUS MW-OBL | NCNE-OBL | GP-OBL

Juncus brevicaudatus (Engelm.) Fernald
♦ NARROW-PANICLE RUSH

DESCRIPTION Densely clumped, native perennial rush. Stems erect, round in section, 1.5–5 dm long. Leaves from stem and base of plant, round in section, hollow, with small swollen joints, 3–20 cm long and 1–2 mm wide; sheaths green or sometimes red, auricles rounded, 1–2 mm long; bract erect, round in section, 2–7 cm long, shorter to longer than the head. Flowers in a raceme or panicle of 3–35 heads, 3–12 cm long and 1–4 cm wide, the heads oval, 2–6 mm wide, 2–7-flowered, branches upright, 0.5–3.5 cm long; tepals green to light brown, often red-tinged near tip, 3-nerved, 3–4 mm long, margins chaffy; stamens 3. Capsules oval, 3-angled, dark brown, 3–5 mm long, longer than the tepals, tapered to a sharp point. Aug–Sept.

HABITAT wet meadows, marshes, fens, sandy lakeshores, streambanks, rocks along Lake Superior.

STATUS MW-OBL | NCNE-OBL | GP-OBL

Juncus bufonius L. ♦ TOAD-RUSH

DESCRIPTION Small native annual rush. Stems clumped, erect to spreading, 5–20 cm long. Leaves from stem and at base of plant, flat or channeled, 1–7 cm long and to 1 mm wide, usually shorter than stem; sheaths green to red or brown, auricles absent; bract erect, 1–10 cm long, shorter than the head. Flowers single, mostly stalkless, with 1–7 flowers along each branch of the inflorescence, the inflorescence comprising half or more of the entire length of

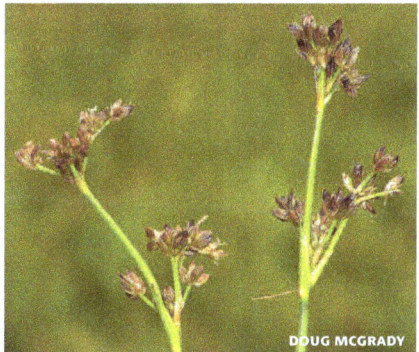

Juncus brevicaudatus

plant; tepals lance-shaped, green to straw-colored, 4–6 mm long, margins chaffy; stamens 6. Capsules ovate, brown or green, 3–4 mm long, rounded at tip, shorter than the tepals. Flowering June–Aug.
HABITAT sandy or silty shores, mud flats, streambanks, wet compacted soil of trails and wheel ruts.
STATUS MW-FACW | NCNE-FACW | GP-OBL

Juncus canadensis J. Gay
⬩CANADA RUSH

DESCRIPTION Clumped, native perennial rush. Stems erect, rigid, round in section, 3–9 dm long. Leaves from stem and at base of plant, round in section, hollow, with small swollen joints, 3–20 cm long and 1–3 mm wide; sheaths green to red, auricles rounded, 1–2 mm long; bract erect, round in section, 3–7 cm long and shorter than the head. Flowers in an open or crowded raceme or panicle of few to many heads, 2–20 cm long and 1–10 cm wide, the heads ± round, 3–8 mm wide, with 5–40 or more flowers, the branches upright, 1–10 cm long; tepals narrowly lance-shaped, green to brown, 3–5 mm long; stamens 3. Capsules ovate, 3-angled, light to dark brown, 3–5 mm long, equal or longer than the tepals, rounded to a short tip. Flowering July–Sept.
HABITAT sandy, muddy or mucky shores, marshes, streambanks, thickets, ditches.
STATUS MW-OBL | NCNE-OBL | GP-OBL

Juncus compressus Jacq.
BLACK GRASS

DESCRIPTION Clumped, introduced perennial rush. Stems erect, flattened, 2–7 dm long.

Leaves from plant base and 1 or 2 along stem, flat or channeled, 5–20 cm long and to 1.5 mm wide; auricles rounded, to 1 mm long; bract erect, somewhat bent, flat or folded, 2–8 cm long, often longer than head. Flowers on short stalks 1–5 mm long, with 1–2 flowers along each branch of the inflorescence, the inflorescence 3–7 cm long and 1–3 cm wide, branches upright; tepals ovate, light to dark brown, 1–3 mm long, margins translucent; stamens 6. Capsules nearly round, light brown, 2–3 mm long, longer than tepals. Flowering June–Aug.
HABITAT wet meadows, disturbed wet areas, ditches along highways where forming dark green colonies; often where salty.
STATUS MW-OBL | NCNE-FACW | GP-FACW

Juncus effusus L. ⬩SOFT RUSH

DESCRIPTION Densely clumped, native perennial rush. Stems erect, round in section, to about 1 m long. Leaves reduced to bladeless sheaths at base of stem, the sheaths to 2 dm long, mostly red-brown; bract round in section, 10–30 cm long, appearing like a continuation of stem, longer than the head. Flowers in a many-flowered inflorescence, with 2–4 flowers along each branch of the inflorescence, the inflorescence appearing lateral, the branches upright to spreading or bent downward, 2–6 cm long; tepals lance-shaped, green to straw-colored, 2–3 mm long; stamens 3. Capsules broadly

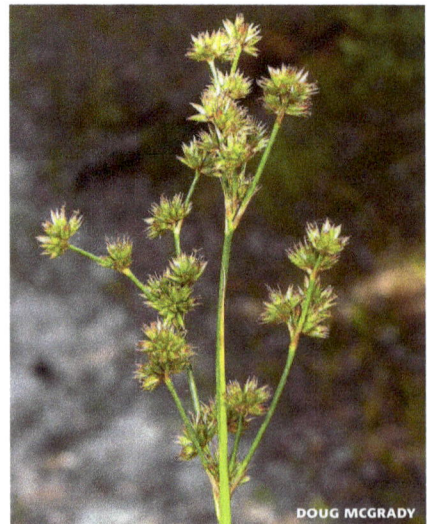

Juncus bufonius

Juncus canadensis

ovate, olive-green to brown, 2–3 mm long, about as long as tepals, sometimes tipped with a short point. Flowering June–July.

HABITAT marshes, shores, streambanks, bog margins, wet meadows.

STATUS MW-OBL | NCNE-OBL | GP-OBL

Juncus ensifolius Wikström
DAGGER-LEAF RUSH

DESCRIPTION Perennial rush, spreading by rhizomes. Stems single or in loose clumps, erect, flattened and narrowly winged, 2–6 dm long. Leaves from stem and at base of plant, the blade folded along midrib, the margins joined, with one edge turned toward the stem, 5–12 cm long and 2–6 mm wide; sheaths green or red, with broad chaffy margins, auricles absent; bract erect, 1–4 cm long, shorter than the head. Flowers in an open panicle of 3–11 heads, 4–7 cm long and 3–5 cm wide, the heads ± round, 8–10 mm wide, with 15 or more flowers, the branches upright, 1–10 cm long; tepals lance-shaped, straw-colored to red-brown, 3–4 mm long; stamens 3 (or 6). Capsules oval, dark brown, 3–4 mm long, tapered to a short beak, shorter to longer than the tepals. Flowering July–Sept.

HABITAT margins of streams, ponds and springs.

STATUS NCNE-FACW | GP-FACW; disjunct from main range of w USA in Ashland Co., Wisc, where probably introduced; also known from n Mich.

Juncus effusus

Juncus filiformis L. THREAD RUSH

DESCRIPTION Native perennial rush, with short or long rhizomes. Stems clumped or in rows from the rhizomes, erect, round in section, 1–5 dm long. Leaves reduced to bladeless sheaths at base of stem, the sheaths pale brown, to 6 cm long; bract erect, round in section, 6–20 cm long, appearing to be a continuation of stem, longer than the head. Flowers in an branched inflorescence, 1–3 cm long, with 1–3 flowers along each branch of the inflorescence, the inflorescence appearing lateral, the branches erect to spreading, to 1 cm long; tepals lance-shaped, green to straw-colored, 2–3 mm long, margins chaffy; stamens 6. Capsules broadly ovate, light brown, 2–3 mm long, slightly longer than the tepals, tipped by a short beak.

HABITAT sandy, mucky, or gravelly shores, streambanks, thickets.

STATUS MW-FACW | NCNE-FACW | GP-OBL

Juncus longistylis Torr.
LONG-STYLE RUSH

DESCRIPTION Native perennial rush, spreading by rhizomes. Stems 3–7 dm long. Leaves mostly at base of plant, flat and grasslike, 4–15 cm long and 1–3 mm wide, smaller upward; sheaths green, with broad membranous margins, auricles rounded, 1–2 mm long. Flowers in an inflorescence of 2–6 stalked, rounded heads, 5–7 cm long and 2–4 cm wide, the heads rounded, satiny chestnut brown, 1–2 cm wide, with 3–10 flowers, the branches erect, 1–4 cm long; tepals lance-shaped, green to brown, 4–6 mm long, margins translucent; stamens 6. Capsules oblong, brown, 2–3 mm long, slightly shorter than the tepals, rounded at tip, with a short beak. Flowering June–Aug.

HABITAT shores, wet meadows, springs.

STATUS MW-FACW | NCNE-FACW | GP-FACW

Juncus marginatus Rostk.
GRASS-LEAF RUSH

DESCRIPTION Native perennial rush, spreading by rhizomes. Stems single or in small clumps, erect, com-

pressed, 2–5 dm long, bulblike at base. Leaves from base of plant and on stem, flat, grasslike, 2–30 cm long and 1–3 mm wide; sheaths green, membranous on margins, auricles rounded, to 0.5 mm long; bract erect to spreading, flat, 1–8 cm long, shorter to slightly longer than the head. Flowers in an open panicle, 2–8 cm long and 1–6 cm wide, composed of 5–15 heads, the heads rounded, 3–6 mm wide, 6–20-flowered, branches upright, 0.5–2.5 cm long; tepals lance-shaped, green with red spots, 2–3 mm long, margins chaffy; stamens 3. Capsules ± round, brown with red spots, 2–3 mm long, slightly longer than the tepals, rounded at tip. Flowering June–Aug.

SYNONYM *Juncus biflorus*.

HABITAT sandy shores and streambanks, wet meadows, marshes, low prairie, springs.

STATUS MW-FACW | NCNE-FACW | GP-FACW; historical record from Minn.

Juncus militaris Bigelow
BAYONET RUSH

DESCRIPTION Native perennial rush, spreading by rhizomes, the rhizomes, when underwater, often producing long threadlike leaves 3–4 dm long. Stems stout, erect, round in section, 3–12 dm long. Leaves mostly 2 from near middle of stem, round in section, 5–7 dm long and about 5 mm wide, longer than the head, upper leaf often reduced to a sheath; sheaths often inflated, auricles rounded, only to 0.2 mm long; bract an inflated sheath with or without a blade, the blade (if present) round in section, 1–2 cm long, shorter than the head. Flowers in an open panicle of 50 or more heads, 4–15 cm long and 4–10 cm wide, the heads oblong pyramid-shaped, 3–6 mm wide, with 5–15 flowers, the branches upright, 1–7 cm long; tepals lance-shaped, straw-colored to red-brown, 2–3 mm long, margins chaffy; stamens 6. Capsules narrowly ovate, 3-angled, straw-colored to brown, satiny, 2–4 mm long, about equal to the tepals, tapered to a conspicuous beak.

HABITAT lakeshores, marshes, open bogs.

STATUS MW-OBL | NCNE-OBL; Mich (THR).

Juncus nodosus L. ◆ KNOTTED RUSH
DESCRIPTION Native perennial rush, spreading by rhizomes. Stems erect, slender, round

in section, 1.5–6 dm long. Leaves on stem and one at base of plant, round in section, hollow, with small swollen joints, 3–30 cm long and 1–2 mm wide, upper leaves usually longer than the head; sheaths green, their margins green, becoming yellow and membranous toward tip, auricles rounded, yellow, 0.5–1 mm long; bract erect to spreading, round in section, 2–12 cm long, usually much longer than head. Flowers in a raceme or panicle of several heads, 1–6 cm long and 1–3 cm wide, the heads ± round, 6–10 mm wide, 6–20-flowered, the branches erect to spreading, 0.5–3 cm long; tepals narrowly lance-shaped, green to light brown, 3–4 mm long; the margins narrowly translucent; stamens 6. Capsules awl-shaped, brown, 4–5 mm long, longer than the tepals, tapered to a sometimes curved beak. Flowering July–Sept.

HABITAT sandy, gravelly or clayey shores and streambanks, wet meadows, fens, ditches, springs; often where calcium-rich.

STATUS MW-OBL | NCNE-OBL | GP-OBL

Juncus pelocarpus E. Meyer
BROWN-FRUIT RUSH

DESCRIPTION Native perennial rush, spreading by rhizomes and forming colonies. Stems erect, round in section, 1–4 dm long. Leaves from stem and at base of plant, round in section, very slen-

Juncus nodosus

der, 2–10 cm long and about 1 mm wide; auricles absent or short and straw-colored; bract erect, round in section, 2–4 cm long, shorter than the head. Flowers single or paired in a much-branched inflorescence, 5–15 cm long and 4–10 cm wide, the flowers on mostly 1 side of each branch, the branches upright to widely spreading, 1–4 cm long, with at least some of the flowers usually replaced by clusters of awl-shaped leaves; tepals ovate, dark brown, about 2 mm long, margins chaffy; stamens 6. Capsules narrowly ovate, dark brown, satiny, 2–3 mm long, equal or slightly longer than tepals, tapered to a slender beak. Flowering July–Aug.

HABITAT shallow water, sandy or mucky shores, bog margins.

STATUS MW-OBL | NCNE-OBL | GP-OBL

Juncus scirpoides Lam.
ROUND-HEADED RUSH

DESCRIPTION Native perennial rush, from coarse, pale rhizomes. Stems single or clumped, erect, round in section, 1–7 dm long. Leaves mostly 2–3 per stem, basal leaves absent or single, blades round in section, hollow, with small swollen joints, 2–20 cm long and 1–2 mm wide, uppermost leaf shorter than the head; sheaths green or the lower red, membranous on upper margins, auricles pointed, 1–2 mm long; bract erect, round in section, 1–6 cm long, shorter to slightly longer than head. Flowers in an open to crowded raceme or panicle of 2–15 heads, 3–12 cm long and 1–5 cm wide, the heads round 5–10 mm wide, with 20–60 flowers, branches upright, 1–7 cm long; tepals lance-shaped, green to brown, 2–3 mm long; stamens 3. Capsules awl-shaped, 3-angled, straw-colored to brown, 3–4 mm long, equaling to slightly longer than the tepals. Flowering June–Sept.

HABITAT wet sandy shores, wet meadows, streambanks.

STATUS MW-FACW | NCNE-FACW | GP-FACW; Mich (THR).

Juncus stygius L. MOOR RUSH

DESCRIPTION Native perennial rush, from slender rhizomes. Stems single or few to-

gether, erect, round in section, 1–4 dm long. Leaves 1–3 from near base of plant, with 1 leaf above middle of stem, round in section or somewhat flattened, 3–15 cm long and 0.5–2 mm wide; auricles short and rounded or absent; bract erect, round in section, 1–2 cm long, shorter than the head. Flowers in an inflorescence of 1–3 heads, the heads obovate, 5–10 mm wide, 1–4-flowered, branches erect, to 1 cm long; tepals lance-shaped, straw-colored to red-brown, 4–5 mm long, margins chaffy; stamens 6, nearly as long as the tepals. Capsules oval, 3-angled, green-brown, 6–8 mm long, longer than the tepals, tipped with a distinct point.

HABITAT open bogs, marshes, and shallow water.

STATUS NCNE-OBL | GP-OBL; Wisc (END), Mich (THR).

Juncus tenuis Willd. ♦ PATH-RUSH

DESCRIPTION Clumped, native perennial rush. Stems erect, round in section to slightly flattened, 1–6 dm long. Leaves near base of stem, flat to broadly channeled, 10–15 cm long and to 1 mm wide; sheaths green, the margins yellow and glossy, auricles triangular, 1–3 mm long; bracts 1–3 (usually 2), the lowest erect, flat, 6–10 cm long, longer than the head. Flowers stalkless or on short stalks to 3 mm long, on branches with 1–7 flowers, in a crowded to spreading head 2–5 cm long; tepals lance-shaped, green

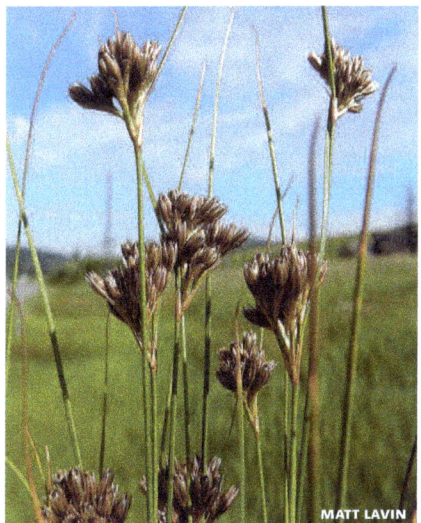

Juncus tenuis

to straw-colored or light brown, 3–5 mm long, margins narrowly translucent; stamens 6. Capsules ovate, green to straw-colored, 2–5 mm long, shorter or equaling the tepals, rounded at tip. Flowering June–July.
HABITAT wet meadows, shores, streambanks, springs, common in disturbed places (often where soils compacted) such as trails, roadsides and ditches; also in drier woods and meadows.
STATUS MW-FAC | NCNE-FAC | GP-FAC

Juncus torreyi Coville ♦ TORREY'S RUSH
DESCRIPTION Native perennial rush, from tuber-bearing rhizomes. Stems single, erect, round in section, 4–8 dm long. Leaves from stem and base of plant, round in section, hollow, with small swollen joints, 15–30 cm long and 1–2 mm wide, the upper leaves often longer than the head; sheaths green, the margins white and translucent, auricles 1–3 mm long; bract erect or spreading, round in section, 4–12 cm long, longer than the head. Flowers in a crowded, rounded raceme or panicle of 3–23 heads, 2–5 cm long and as wide, the heads round, 10–15 mm wide, with 25 to many flowers, branches erect to spreading 1–4 cm long; tepals narrowly lance-shaped, green to brown, 3–5 mm long, margins narrowly translucent; stamens 6. Capsules awl-shaped, brown, 4–6 mm long, equal or longer than the tepals, tapered to a short beak. Flowering June–Sept.
HABITAT sandy shores, streambanks, wet meadows, marsh borders, springs, ditches.
STATUS MW-FACW | NCNE-FACW | GP-FACW

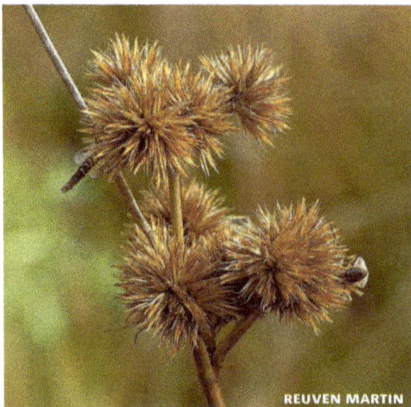

REUVEN MARTIN
Juncus torreyi

NOTE A hymenopteran larva is responsible for the galls which are sometimes present on *Juncus torreyi* and *J. nodosus*. The galls appear in the head as clusters of overlapping, bract-like leaves, yellow and red in color.

Juncus vaseyi Engelm. VASEY'S RUSH
DESCRIPTION Clumped, native perennial rush. Stems erect, 2–6 dm long. Leaves all at base of plant, round in section, solid, narrowly channeled on upper surface, to 3 dm long and to 1 mm wide, usually shorter than stem; sheaths green or red, the margins membranous, auricles short or absent; bract upright, usually shorter than the head. Flowers single, stalkless or on short stalks, in a crowded inflorescence, 1–4 cm long; tepals lance-shaped, green to light brown, 4–6 mm long, margins narrowly translucent; stamens 6. Capsules cylindric, 1–2 mm long, equal or slightly longer than the tepals, blunt-tipped. Flowering July–Aug.
HABITAT wet meadows, sandy shores.
STATUS MW-FACW | NCNE-FACW | GP-FACW; Mich (THR).

Juncaginaceae
Arrow-Grass Family

Triglochin ARROW-GRASS
Grasslike perennial herbs, clumped from creeping rhizomes, often in brackish habitats. Stems slender, leafless. Leaves all from base of plant, slender, linear, round or somewhat flattened in section, sheathing at base. Flowers perfect, regular, on short stalks in a spikelike raceme at end of stem; flower segments (tepals) 6; stigmas 3 or 6, styles short or absent; stamens 6, anthers stalkless, nearly as large as tepals. Fruit of 3 or 6 carpels, these splitting when mature into 1-seeded segments.

Triglochin ARROW-GRASS
1 Plants generally small and slender; stigmas 3; fruits linear, clublike toward tip . . . . *T. palustris*
1 Plants larger, usually 3 dm or more tall; stigmas 6; fruits short-cylindric . . . . *T. maritima*

Triglochin maritima L.
♦ COMMON *or* SEASIDE ARROW-GRASS

DESCRIPTION Clumped, native perennial herb, from a thick crown and spreading by rhizomes. **Stems** 2–8 dm long. **Leaves** upright to spreading, somewhat flattened, to 5 dm long and 1–3 mm wide. **Flowers** 2–3 mm wide, in densely flowered, spike-like racemes 1–4 dm long; the flowers on upright stalks 4–6 mm long, the stalks extending downward on the stem as a wing; tepals 6, 1–2 mm long; stigmas 6; stamens 6. **Fruit** of 6 ovate carpels, 2–5 mm long and 1–3 mm wide, the carpel tips curved outward. **Flowering** June–Aug.

HABITAT Sandy, gravelly, or marly lakeshores and streambanks; marshes, brackish wetlands.

STATUS MW-OBL | NCNE-OBL | GP-OBL

NOTE plants are larger than those of **marsh arrow-grass** (*Triglochin palustris*) and the fruit ovate rather than linear.

Triglochin palustris L.
♦ MARSH ARROW-GRASS

DESCRIPTION Small, clumped, native perennial herb. **Stems** slender, 2–4 dm long. **Leaves** erect, round in section, to 3 dm long and 1–2 mm wide. **Flowers** small, 1–2 mm wide, in loosely flowered racemes, 10–25 cm long; the flowers on erect stalks, 2–5 mm long; tepals 6, 1–2 mm long; stigmas 3; stamens 6. **Fruit** of 3 narrow, clublike carpels, 5–8 mm long and 1 mm wide, splitting upward from base into 3 segments. **Flowering** June–Sept.

HABITAT Sandy, gravelly, or marly lakeshores and streambanks, calcareous fens, marshes, interdunal swales; often where calcium-rich.

STATUS MW-OBL | NCNE-OBL | GP-OBL

Liliaceae
Lily Family

PERENNIAL HERBS, from corms, bulbs or rhizomes. Stems leafy or leafless. Leaves linear to ovate, usually from base of plant, sometimes along stem, alternate to opposite or whorled. Flowers perfect (with both staminate and pistillate parts), regular; sepals and petals of 4 or 6 petal-like tepals; stamens 4 or 6; ovary superior or inferior, 3-chambered. Fruit a capsule or round and berrylike.

NOTE Under the angiosperm Phylogeny Group III system (APG III), most genera in the Liliaceae have been placed into various new familes. However, the family designations are still in a state of flux, and may change in the future. As a convenience, our small number of closely related genera are retained within the traditional Lily Family grouping, with the proposed new family name noted after each genus name.

Liliaceae **LILY FAMILY**

1   Leaves from stem.....................**2**
1   leaves all from base of plant; flowers at ends of leafless stalks (scapes).........**3**
2   Plants large, 3–8 dm tall; flowers showy, orange ....*Lilium*
2   Plants smaller, mostly 0.5–4 dm tall; flowers small, white .....*Maianthemum*
3   Flowers yellow; ovary inferior ....*Hypoxis hirsuta*
3   Flowers white or greenish; ovary superior **4**
4   Stems sticky-hairy... *Triantha glutinosa*
4   Stems smooth and waxy, not sticky-hairy. ....*Anticlea elegans*

Triglochin maritima

FLORENT BECK

Anticlea DEATH CAMAS
Melanthiaceae

Anticlea elegans (Pursh) Rydb.
♦ WHITE CAMAS

DESCRIPTION Native perennial herb, from an ovate bulb; plants waxy, especially when young. **Stems** erect, 2–6 dm long. **Leaves** mostly from base of plant, linear, 2–4 dm long and 4–12 mm wide; stem leaves much smaller. **Flowers** green-yellow or white, in a raceme or panicle, 1–3 dm long, the branches upright, subtended by large, lance shaped, green or purplish bracts; tepals 6, obovate, 7–12 mm long, usually purple-tinged near base; stamens 6. **Fruit** an ovate capsule, 10–15 mm long; seeds 3 mm long. **Flowering** July–Aug.

SYNONYMS *Zigadenus elegans, Z. glaucus.*

HABITAT sandy or rocky shores of Great Lakes, open bogs, calcareous fens.

STATUS MW-FAC | NCNE-FACW | GP-FACW

CAUTION potentially fatal if eaten.

Hypoxis STAR-GRASS
Hypoxidaceae

Hypoxis hirsuta (L.) Cov.
♦ YELLOW STAR-GRASS

DESCRIPTION Low native perennial herb, from a small, shallow corm. **Stems** leafless, lax, 1 to several, silky-hairy in upper part, shorter than leaves when flowering, to 4 dm long when mature. **Leaves** from base of plant, linear, hairy, to 6 dm long and 2–10 mm wide. **Flowers** 1–6 (usually 2), yellow, 1–2.5 cm wide, in racemes at ends of stems, tepals hairy on outside, 5–12 mm long, spreading in flower, closing and turning green after flowering, persistent. **capsule** oval, 3–6 mm long; seeds black. **Flowering** May–July.

HABITAT wet meadows, shores, moist prairie; often where calcium-rich.

STATUS MW-FAC | NCNE-FAC | GP-FACW

Lilium LILY *Liliaceae*

Lilium **LILY**

1　Flowers erect; tepals narrowed at the base to a slender claw; leaves to 8 (–14) mm wide ....*L. philadelphicum*

1　Flowers nodding (fruit becoming erect); tepals narrowed gradually toward base, not clawed; widest leaves to 20 mm wide ....*L. michiganense*

Lilium michiganense Farw.
♦ MICHIGAN LILY

DESCRIPTION Native perennial herb from a scaly bulb. **Stems** stout and erect, to 2.5 m tall. **Leaves** whorled (the upper stem leaves and those of the inflorescence alternate); blades lance-shaped, tapering to both ends, smooth, the larger commonly 8–12 cm long and to 2 cm wide. **Flowers** occasionally solitary, usually several or many, partly in an umbel from the uppermost leaf-whorl and partly in a terminal raceme,

GERTJAN VAN NOORD

Anticlea elegans

STEPHEN HORVATH

Hypoxis hirsuta

nodding from long, erect or ascending pedicels; perianth segments strongly recurved, lance-shaped, 6–9 cm long, orange or orange-red, spotted with purple, bright green at the base within. Flowering July-Aug.

HABITAT wet meadows, along streams, and in floodplain forests.

STATUS MW-FACW | NCNE-FACW | GP-FACW

NOTE the orange flower is nodding instead of upright as in *Lilium philadelphicum*.

Lilium philadelphicum L.
WOOD-LILY

DESCRIPTION Native perennial herb from a scaly bulb. Stems erect, 3–8 dm long. Leaves all from stem, narrowly lance-shaped, 4–10 cm long and 3–9 mm wide, parallel-veined; lower leaves alternate, upper leaves opposite or whorled; petioles absent. Flowers 1–5, erect, large and showy, on stalks 1–8 cm long at ends of stem; tepals orange-red, yellow and dark-spotted toward base, lance-shaped, 4–8 cm long and 0.8–2.8 cm wide, stamens and pistil about as long as tepals; stigma 3-parted; ovary superior. capsule oblong, 2.5–4 cm long; seeds flat. Flowering June-July.

HABITAT wet meadows, low prairie, fens and open bogs, seeps, ditches; also in drier meadows, prairies and woods.

STATUS MW-FAC | NCNE-FAC | GP-FACU

Maianthemum *Asparagaceae*
WILD LILY-OF-THE-VALLEY

Perennial herbs, from rhizomes. Stems unbranched, arching or upright. Leaves 2–3 along the stem, clasping stem or with short petioles. Flowers white, with 4 or 6 tepals and stamens; in a raceme at end of stem. Fruit berrylike, mottled when young, red when mature.

Maianthemum
WILD LILY-OF-THE-VALLEY
1 Tepals 4; leaves usually 2. . *M. canadense*
1 Tepals 6; leaves usually 3 . . . *M. trifolium*

Maianthemum canadense Desf.
♦ CANADA MAYFLOWER,
WILD LILY-OF-THE-VALLEY

DESCRIPTION Small native perennial herb, spreading by rhizomes. Stems erect, 5–20 cm long. Leaves usually 2 along stem, ovate, heart-shaped at base, 3–10 cm long; petioles short or absent. Flowers small, white, 4–6 mm wide, stalked, in a short raceme at end of stem, the raceme 3–6 cm long; tepals 4, spreading; stamens 4; style 2-lobed. Fruit a pale red berry, 3–4 mm wide; seeds 1–2. Flowering May-July.

HABITAT on hummocks in swamps, open bogs and thickets; also common in moist to dry woods.

STATUS MW-FAC | NCNE-FACU | GP-FACU

AARON CARLSON
Lilium michiganense

JOSHUA MAYER
Maianthemum canadense

Maianthemum trifolium (L.) Sloboda
FALSE SOLOMON'S SEAL

DESCRIPTION Native perennial herb, from long rhizomes. **Stems** erect, 1–5 dm long at flowering time. **Leaves** alternate, smooth, usually 3 (2–4), oval or oblong lance-shaped, 6–12 cm long and 1–4 cm wide; petioles absent. **Flowers** small, white, 8 mm wide, stalked, 3–8 in a raceme; tepals 6, spreading; stamens 6. **Fruit** a dark red berry, 3–5 mm wide; seeds 1–2. **Flowering** May–June.

SYNONYM *Smilacina trifolia.*

HABITAT open bogs, conifer swamps, thickets.

STATUS MW-OBL | NCNE-OBL | GP-OBL

Triantha FALSE ASPHODEL
Tofieldiaceae

Triantha glutinosa (Michx.) Baker
◆STICKY FALSE ASPHODEL

DESCRIPTION Native perennial herb, from a bulb. **Stems** erect, nearly leafless, 2–5 dm long, covered with sticky hairs. **Leaves** 2–4 from base of plant, linear, hairy, 8–20 cm long and to 8 mm wide, sometimes with 1 bractlike leaf near middle of stem. **Flowers** white, on sticky-hairy stalks 3–6 mm long, in a raceme 2–5 cm long when in flower, becoming longer when fruiting,

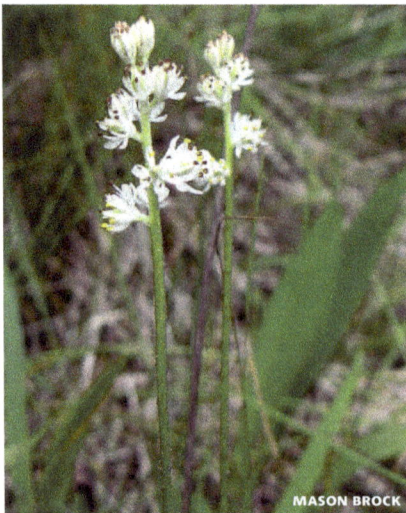

Triantha glutinosa

2–3 at each node of the raceme, upper flowers opening first; tepals 6, oblong lance-shaped, 4 mm long; stamens 6. **Fruit** an oblong capsule, 5–6 mm long; seeds about 1 mm long, with a slender tail at each end. **Flowering** June–Aug.

SYNONYM *Tofieldia glutinosa.*

HABITAT sandy or gravelly shores, interdunal wetlands, calcareous fens, rocky shores of Lake Superior.

STATUS MW-OBL | NCNE-OBL | GP-OBL; Wisc (THR).

Orchidaceae
Orchid Family

PERENNIAL HERBS, from fleshy or tuberous roots, corms or bulbs. Leaves simple, along the stem and alternate, or mostly at base of plant, stalkless and usually sheathing the stem, parallel-veined, often somewhat fleshy. Flowers perfect (with both staminate and pistillate parts), irregular, showy in many species, in heads of 1 or 2 flowers at ends of stems, or with several to many flowers in a spike, raceme or panicle, each flower usually subtended by a bract; sepals 3, green or colored, sometimes resembling the lateral petals, the lateral sepals free, or joined to form an appendage below the lip (as in *Cypripedium*), or joined with the lateral petals to form a hood over the lip (*Spiranthes*); petals 3, white or colored, the 2 lateral petals alike, the lowest petal different and called the lip; stamens 1–2, attached to the style and forming a stout column; ovary inferior. Fruit a many-seeded capsule, opening by 3 or sometimes 6 longitudinal slits, but remaining closed at tip and base; seeds miniscule.

NOTE The Orchidaceae is one of the world's largest flowering plant families (along with the Asteraceae), with over 900 genera and an estimated 28,000 species, most of which occur in the Tropics.

Orchidaceae ORCHID FAMILY

1　Plants without leaves; stems yellowish (chlorophyll absent) *Corallorhiza trifida*

1　Plants with one or more leaves; stems green **2**

2　Leaves whorled ......*Isotria verticillata*

Arethusa DRAGON'S MOUTH

Arethusa bulbosa L.

♦ SWAMP-PINK, DRAGON'S MOUTH

DESCRIPTION Native perennial herb; roots few, fibrous, from a corm. Stems leafless, smooth, 1–4 dm long. Leaves 1, linear, small and bractlike at flowering time, later expanding to 2 dm long and 3–8 mm wide; lower stem with 2–4 bladeless sheaths. Flowers single at ends of stems, sepals rose-purple, oblong, 2.5–5 cm long; petals joined and ± hoodlike over the column, lip pink, streaked with rose-purple, 2.5–4 cm long, curved downward near middle. Flowering June–July.

HABITAT open bogs and conifer swamps (in sphagnum moss), floating mats around bog lakes, calcareous fens; often with grass-pink (*Calopogon tuberosus*) and rose pogonia (*Pogonia ophioglossoides*).

STATUS MW-OBL | NCNE-OBL | GP-OBL

Calopogon GRASS-PINK

Calopogon tuberosus (L.) BSP.

♦ GRASS-PINK

DESCRIPTION Native perennial herb, from a corm. Stems leafless, smooth, 2–7 dm long. Leaves 1 near base of plant, linear, 1–4 dm long and 2–15 mm wide. Flowers pink to purple, 2–15 in a loose raceme, 3–12 cm long; sepals ovate, 1–2.5 cm long; petals oblong, 1–2.5 cm long, the lip located above the lateral petals, 1–2 cm long, bearded on inside with yellow-tipped bristles.

SYNONYM *Calopogon pulchellus*.

HABITAT open bogs and floating mats, openings in conifer swamps, calcareous fens near Great Lakes shoreline.

STATUS MW-OBL | NCNE-OBL | GP-OBL

NOTE Grass-pink is distinguished from *Arethusa bulbosa* and *Pogonia ophioglossoides* by having a raceme of several flowers vs. single flowers in *Arethusa* and *Pogonia*.

ADDITIONAL SPECIES Oklahoma grass-pink (*Calopogon oklahomensis* D. H. Goldman), reported from several locations in se Minn; typically in somewhat drier situ-

Arethusa bulbosa

ations than the more common *C. tuberosus.* The two species can be separated as follows:

1 Flowers opening sequentially; dilated upper portion of middle lip lobe usually much wider than long, typically anvil shaped; corms globose to elongate, not forked. .... *C. tuberosus*

1 Flowers opening nearly simultaneously; dilated upper portion of middle lip lobe usually much narrower than long, triangular to broadly rounded; corms elongate, forked. .... *C. oklahomensis*

Calypso CALYPSO

Calypso bulbosa (L.) Oakes
◆ CALYPSO, FAIRY SLIPPER

DESCRIPTION Native perennial herb, from a corm. Stems 0.5–2 dm long, with 2–3 bladeless sheaths on lower portion. Leaves single from the corm, ovate, 3–5 cm long and 2–3 cm wide, petioles 1–5 cm long. Flowers 1, nodding at end of stem; sepals and lateral petals similar, pale purple to pink, lance-shaped, 1–2 cm long and 3–5 mm wide, lip white to pink, streaked with purple, 1.5–2 cm long and 5–10 mm wide, the lip extended to form a white "apron" with several rows of yellow bristles. Flowering May–June.

HABITAT mature conifer forests or mixed forests of conifers and deciduous trees (such as balsam fir, hemlock, and paper birch), usually in shade; soils rich in woody humus.

STATUS MW-FACW | NCNE-FACW | GP-FACW; Wisc (THR), Mich (THR).

NOTE the single leaf of calypso appears in late August or September, persists through the winter, and withers after flowering in spring. Between fruiting in June and July and the emergence of the new leaf in late summer of fall, no aboveground portions of the plant may be visible.

Corallorhiza CORALROOT

Corallorhiza trifida Chat.
◆ YELLOW CORALROOT

DESCRIPTION Native perennial saprophytic herb, roots absent. Stems yellow-green, smooth, 1–3 dm long, single or in clusters from the coral-like rhizome. Leaves reduced to 2–3 overlapping sheaths on lower stem. Flowers yellow-green, 5–15 in a raceme 3–8 cm long; sepals and lateral petals yellow-green, linear, 3–5 mm long, lip white, sometimes with purple-spots, obovate, 3–5 mm long and 2–3 mm wide. Capsules drooping, 1–1.5 cm long and 3–7 mm wide. Flowering May–June.

HABITAT Moist to wet, mostly conifer woods, swamps (often under cedar) and thickets; usually where shaded.

STATUS MW-FACW | NCNE-FACW | GP-FAC

NOTE Yellow coralroot flowers earlier than our other species of coralroot (and which are typically found in drier forests).

Calypso bulbosa

Corallorhiza trifida

Cypripedium
LADY'S-SLIPPER

Erect perennial herbs, from coarse, fibrous roots. Stems unbranched, often clumped, hairy. Leaves 2 or more at base of plant or along stem, broad. Flowers 1 or 2, large and mostly showy at ends of stems, white, pink or yellow; lateral sepals similar to lateral petals, the sepals joined to form a single appendage below the lip; lateral petals free and spreading, lip inflated and pouchlike, projecting forward; stamens 2, 1 on each side of column. Fruit a many-seeded capsule.

CAUTION the plant's hairs can be irritating if touched.

Cypripedium **LADY'S-SLIPPER**

1 Lip pouch pink to purple; leaves 2 at base of stem .... *C. acaule*
1 Lip pouch yellow or whitish; leaves 3 or more on stem 2
2 Pouch yellow, sometimes brown- or purple-dotted 3
2 Pouch white to pink, or pink with white patches 4
3 Sepals and petals red-brown; lateral petals strongly twisted, brown-purple; pouch less than 4 cm long.......................... .... *C. parviflorum* var. *makasin*
3 Sepals and petals yellow to brown-green; lateral petals wavy, green with red-brown streaks; pouch more than 4 cm long .... *C. parviflorum* var. *pubescens*
4 Pouch projected downward into a cone-shaped spur .... *C. arietinum*
4 Pouch not spurred 5
5 Sepals and lateral petals white; lip 3-5 cm long .... *C. reginae*
5 Sepals and lateral petals green; lip 1.5-2 cm long .... *C. candidum*

Cypripedium acaule Aiton
♦MOCCASIN-FLOWER,
PINK LADY'S-SLIPPER

DESCRIPTION Native perennial herb, from coarse rhizomes; roots long and cordlike. Stems leafless, 2-4 dm long, glandular-hairy. Leaves 2 at base of plant, opposite, oval to obovate, 1-2 dm long and 3-10 cm wide, thinly hairy, stalkless. Flowers 1, nodding at end of stem; sepals and lateral petals yellow-green to green-brown, the 2 lower sepals joined to form a single sepal below the lip; lip drooping, pink with red veins, 3-5 cm long, cleft along the upper side and hiding the opening. Flowering May–June.

HABITAT hummocks in conifer swamps; sites typically shaded, acidic and nutrient-poor, sometimes fairly dry; southward in region also found on hummocks in open bogs.
STATUS MW-FACW | NCNE-FACW | GP-FACW

Cypripedium arietinum R. Br.
RAM'S-HEAD LADY'S-SLIPPER

Native perennial herb, from a coarse rhizome; roots long and cordlike. Stems slender, 1-4 dm long, thinly hairy. Leaves 3-5, above middle of stem, stalkless, oval, often folded, 5-10 cm long and 1.5-3 cm wide, finely hairy. Flowers 1 or sometimes 2 at ends of stems; sepals and lateral petals similar, green-brown; lip an inflated pouch, 1.5-2.5 cm long, white or pink-tinged, with prominent red-veins, extended downward to form a conical pouch. Flowering late May–June.

HABITAT conifer swamps, wet forest openings (often with white cedar); also in drier, sandy, conifer and mixed conifer-deciduous forests, and on low dunes under conifers near shores of Great Lakes.
STATUS MW-FACW | NCNE-FACW | GP-FACW; Minn (THR), Wisc (THR).

JUDY GALLAGHER

Cypripedium acaule

Cypripedium candidum Muhl.
◆WHITE LADY'S-SLIPPER

DESCRIPTION Native perennial herb, from a rhizome; roots long and cordlike. Stems 1.5–3 dm long, hairy. Leaves 2–4, upright, alternate along upper stem, oval, 5–15 cm long and 2–5 cm wide, sparsely glandular-hairy, stalkless; reduced to overlapping sheathing scales below. Flowers 1 at end of stems, the subtending bract leaflike, erect, 3–8 cm long; sepals and lateral petals green-yellow, often streaked with purple, the lateral sepals joined below lip, notched at tip; lateral petals linear lance-shaped, green-yellow, sometimes twisted, 2–4 cm long; lip a small inflated pouch, 1.5–2 cm long, white with faint purple veins. Flowering May–June.

HABITAT calcium-rich wet meadows, low prairie, wet shores along Great Lakes, calcareous fens (often with shrubby cinquefoil, *Potentilla fruticosa*); usually where sunny.

STATUS MW-OBL | NCNE-OBL; Wisc (THR), Mich (THR).

Cypripedium parviflorum Salisb.
var. **makasin** (Farw.) Sheviak
SMALL YELLOW LADY'S-SLIPPER

DESCRIPTION Native perennial herb, from rhizomes; roots long, numerous. Stems 1.5–6 dm long, glandular-hairy. Leaves 2–5, alternate along stem, ascending, oval, 5–18 cm long and 2–7 cm wide, sparsely hairy, stalkless. Flowers 1 (rarely 2) at ends of stems; sepals purple-brown, the lateral sepals joined below the lip, notched at tip; lateral petals linear, purple-brown, spirally twisted, 2–5 cm long; lip an inflated pouch, 1.5–3 cm long, yellow, often with purple veins and spots near opening. Flowering May–July.

SYNONYM *Cypripedium calceolus* var. *parviflorum*.

HABITAT conifer swamps, wet meadows, fens, and moist forests (often under cedar); sphagnum mosses are usually sparse; sites are shaded or sunny, with organic or mineral, often calcium-rich soil; s in region also in open, calcium-rich swales.

STATUS MW-FACW | NCNE-FACW | GP-FACW

Cypripedium parviflorum Salisb.
var. **pubescens** (Farw.) Sheviak
◆LARGE YELLOW LADY'S-SLIPPER

DESCRIPTION Native perennial herb, from a rhizome; roots long and numerous. Stems 1.5–6 dm long, glandular-hairy. Leaves 3–6, alternate along stem, ascending, ovate to oval, 8–20 cm long and 3–8 cm wide, sparsely hairy. Flowers 1 (rarely 2) at ends of stems; sepals yellow-green, the lateral sepals joined below the lip, notched at tip; lateral petals linear, yellow-green, often streaked with red-brown, usually spirally twisted, 4–8 cm long; lip an inflated pouch, 3–6 cm long, yellow, often with purple veins near opening. Flowering May–July.

SYNONYM *Cypripedium calceolus* var. *pubescens*.

HABITAT conifer swamps, bogs, fens, prairies and thickets, especially where soils derived from limestone; also in moist hardwood forests.

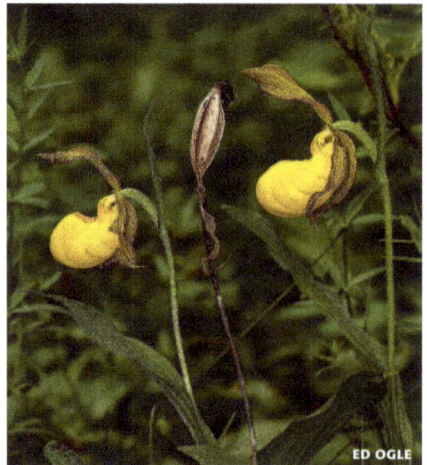

Cypripedium candidum

Cypripedium parviflorum var. *pubescens*

Cypripedium reginae Walter
◆SHOWY LADY'S-SLIPPER

DESCRIPTION Native perennial herb, from a coarse rhizome; roots many, long and cordlike. **Stems** 4–10 dm long, strongly glandular-hairy. **Leaves** 4–12, alternate along stem, spreading or ascending, broadly oval, 10–25 cm long, 4–12 cm wide, abruptly tapered to tip, nearly smooth to hairy, stalkless; reduced to sheaths at base. **Flowers** 1 or often 2 at ends of stems, the subtending bract leaflike, 6–12 cm long; sepals and lateral petals white, the lateral sepals joined to form an appendage under the lip, rounded at tip; lip an inflated pouch, 3–5 cm long, white, often infused with pink or purple. **Flowering** June–July.

HABITAT conifer and hardwood swamps (especially balsam fir-cedar-tamarack swamps), bogs, calcareous fens, sedge meadows, floating mats, wet openings, wet clayey slopes, ditches; especially where open and sunny; most abundant in openings in wet forests and swamps not dominated by sphagnum mosses.

STATUS MW-FACW | NCNE-FACW | GP-FACW

Galearis ORCHIS

Galearis rotundifolia (Banks ex Pursh) R.M. Bateman ◆ROUND-LEAF ORCHID

DESCRIPTION Native perennial herb, roots few from a slender rhizome. **Stems** leafless, smooth, 15–30 cm long.

Leaves single from near base of plant, oval, 4–15 cm long and 2–8 cm wide; usually with 1–2 bladeless sheaths below. **Flowers** 4 or more, in a raceme 3–8 cm long; sepals white to pale pink; petals white to pink or purple-tinged, the 2 lateral petals joined with the upper sepal to form somewhat of a hood over the column; lip white, with purple spots, 6–10 mm long and 4–7 mm wide, 3-lobed, the terminal lobe largest and notched at tip; spur about 5 mm long, shorter than lip. **Flowering** June–July.

SYNONYMS *Amerorchis rotundifolia, Orchis rotundifolia, Platanthera rotundifolia.*

HABITAT conifer swamps (on moss under cedar, tamarack, or black spruce); s in region in cold conifer swamps of balsam fir, black spruce and cedar; usually found over limestone and where sphagnum mosses not predominant.

STATUS MW-OBL | NCNE-OBL; Wisc (THR), Mich (END).

NOTE sometimes included within the rein-orchids as *Platanthera rotundifolia* (Banks ex Pursh) Lindl.

Hammarbya
BOG ADDER'S MOUTH

Hammarbya paludosa (L.) Kuntze
◆BOG ADDER'S MOUTH

DESCRIPTION Small, inconspicuous native perennial (our smallest orchid), from a bulb-

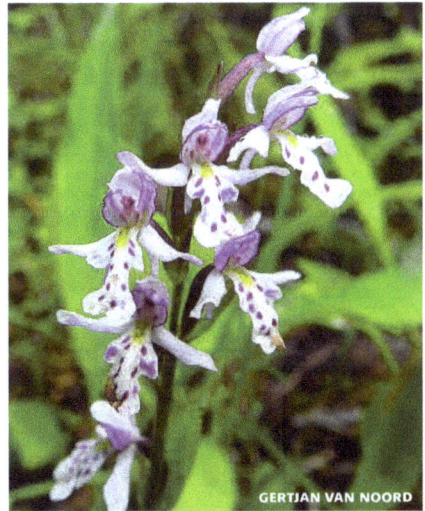

Cypripedium reginae

Galearis rotundifolia

like base; roots few, fibrous. Stems leafless, smooth, 7–15 cm long. Leaves 2–5 from base of plant, obovate, 1–2 cm long and 0.5–1 cm wide, clasping stem at base. Flowers small, yellow-green, 10 or more in a slender, spikelike raceme 3–9 cm long and about 5 mm wide, the flowers evenly spaced in the raceme, twisted so that lip is uppermost in the flowers; lip very small, ovate, 1–1.5 mm long and 0.5 mm wide. Flowering July–Aug.

SYNONYM *Malaxis paludosa.*

HABITAT sphagnum moss hummocks in black spruce swamps, usually where somewhat open.

STATUS MW-OBL | NCNE-OBL | GP-OBL; Minn (END); main range northward in Canada, s in USA to n Minn.

Isotria WHORLED POGONIA
Isotria verticillata (Willd.) Raf.
♦ LARGER WHORLED POGONIA

DESCRIPTION Native perennial herb; roots long, fleshy, covered with hairs. Stems smooth, red-brown, hollow, 1–4 dm long. Leaves 5–6 in a whorl

at top of stem, oblong lace-shaped, upright when young, drooping with age, 3–9 cm long and 2–5 cm wide. Flowers 1–2 at end of stems, on stalks 3–5 cm long; sepals green, purple-tinged, narrowly lance-shaped, 4–6 cm long and 3–5 mm wide; petals yellow-green, oval, shorter and wider than sepals, 2–3 cm long and 4–6 mm wide; lip green-white, 3-lobed near tip, 1.5–2.5 cm long and 1 cm wide. Capsules erect, oval, 2–3 cm long. Flowering May–June.

SYNONYM *Pogonia verticillata.*

HABITAT in sphagnum and in partial shade in tamarack and black spruce bogs; also in moist, sandy woods.

STATUS MW-FAC | NCNE-FAC | GP-FAC; Mich (THR); more common in e USA.

NOTE easily overlooked because of the tendency for colonies of larger whorled pogonia to produce few flowers and its resemblance to **Indian cucumber-root** (*Medeola virginiana*) with which it often occurs in upland situations.

Liparis TWAYBLADE
Liparis loeselii (L.) Rich.
♦ FEN-ORCHID, LOESEL'S TWAYBLADE

DESCRIPTION Small, smooth, native perennial herb, from a bulb-like base. Stems erect, 1–2.5 dm long, upper stem somewhat angled in section. Leaves 2 from base of plant, ascending, sheathing at

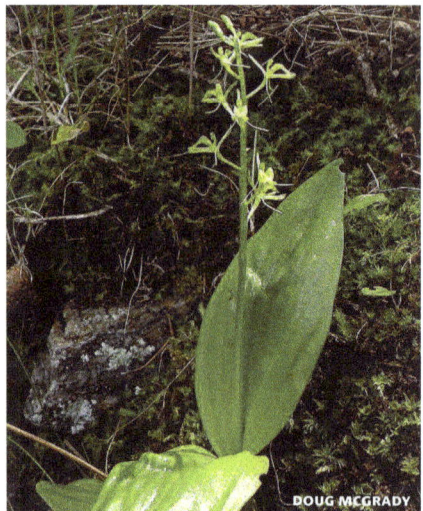

JASON HOLLINGER
Isotria verticillata

DOUG MCGRADY
Liparis loeselii

base, shiny, lance-shaped to oval, 4–15 cm long and 1–4 cm wide. Flowers 2–15, yellow-green, small, upright, in an open raceme 2–10 cm long and 1–2 cm wide; sepals narrowly lance-shaped, 4–6 mm long and 1–2 mm wide; petals linear, 3–5 mm long, often twisted and bent forward under the lip; lip yellow-green, obovate, 4–5 mm long and 2–3 mm wide, tipped with a short point. capsules persistent, short-cylindric, 8–12 mm long. Flowering June–Aug.

HABITAT conifer swamps, fens, floating mats, streambanks, sandy shores, ditches; soils peaty to mineral, acid to calcium-rich.

STATUS MW-FACW | NCNE-FACW | GP-OBL

Malaxis ADDER'S MOUTH

Small perennial herbs. Leaves 1–5 from base of plant or single along stem. Flowers green-white, spaced or crowded in slender or cylindric racemes at ends of stems.

Malaxis **ADDER'S MOUTH**

1 Flowers evenly spaced in a raceme 5–11 cm long . . . . *M. monophyllos*
1 Flowers crowded near top of raceme, the raceme 2–5 cm long . . . . *M. unifolia*

Malaxis monophyllos Swartz
◆WHITE ADDER'S MOUTH

DESCRIPTION Native perennial herb, from a bulblike base; roots few, fibrous. Stems smooth, 1–2 dm long. Leaves single, appearing to be attached well above base of stem, the leaf base clasping stem, ovate to oval, 3–7 cm long and 1.5–4 cm wide. Flowers small, green-white, 14–30 or more, in a long, slender, spikelike raceme 4–11 cm long and to 1 cm wide; on stalks 1–2 mm long, the flowers evenly spaced in the raceme; lip heart-shaped, bent downward, 2–3 mm long and 1–2 mm wide, narrowed at middle to form a long, lance-shaped tip, with a pair of lobes at base. Flowering June–Aug.

SYNONYM *Malaxis brachypoda.*

HABITAT conifer swamps (cedar-balsam fir-spruce), especially in wet depressions and where soils are marly; sphagnum moss hummocks in conifer swamps, wet hardwood forests.

STATUS MW-FACW | NCNE-FACW | GP-FACW

Malaxis unifolia Michx.
GREEN ADDER'S MOUTH

DESCRIPTION Small native perennial herb, from a bulblike base; roots few, fibrous. Stems smooth, 1–3 dm long. Leaves single, attached near middle of stem, ovate, 2–7 cm long and 1–4 cm wide. Flowers small, green, numerous in a cylindric raceme 1.5–6 cm long and 1–2 cm wide, the upper flowers crowded, the lower flowers more widely spaced; lowermost lip very small, 1–2 mm long, with 3 teeth at tip. Flowering June–Aug.

HABITAT sphagnum moss hummocks in swamps, sedge meadows, thickets; also in moist to dry forests.

STATUS MW-FAC | NCNE-FAC | GP-FAC

Neottia TWAYBLADE

Perennial herbs. Stems with a pair of opposite leaves near middle, stems smooth below leaves, hairy above. Leaves broad, stalkless. Flowers small, green to purple, in a raceme at end of stem, the lip 2-lobed or deeply parted. Formerly considered part of genus *Listera.*

Listera **TWAYBLADE**

1 Lip 3–5 mm long, divided to about middle into 2 narrow segments . *Neottia cordata*
1 Lip 7–12 mm long, shallowly notched or divided 1/3 of length, the segments broad **2**

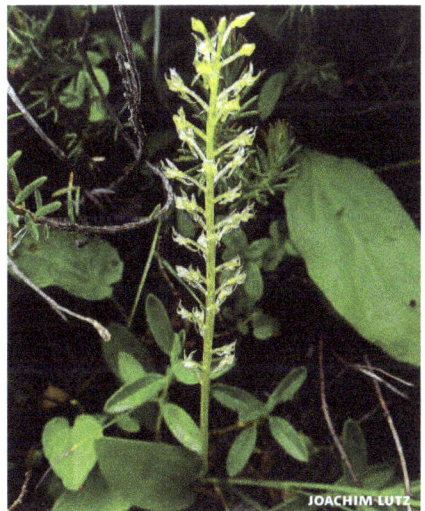

JOACHIM LUTZ

Malaxis monophyllos

2   Lip wide at base, with a pair of auricles . . .
. . . . *Neottia auriculata*

2   Lip narrowed to base, auricles absent . . . .
. . . . *Neottia convallarioides*

Neottia auriculata (Wieg.) Szlach.
AURICLED TWAYBLADE

DESCRIPTION Native perennial herb; roots fibrous. **Stems** 1–2 dm long, smooth below leaves, hairy above. **Leaves** 2 near middle of stem, opposite, ovate, 2–5 cm long and 2–4 cm wide. **Flowers** pale green, 8–15 in a raceme 4–8 cm long and 2–3 cm wide, on stalks 2–5 mm long; lip oblong, 6–10 mm long and 2–5 mm wide, the base with a pair of small clasping auricles, the tip cleft for about 1/4–1/3 of its length. **Flowering** June–Aug.
SYNONYM *Listera auriculata.*
HABITAT alluvial sand along rivers, often under alders, occasionally in moist conifer or mixed conifer and deciduous forests; usually where shaded.
STATUS NCNE-FACW; Minn (END), Wisc (END).

Neottia convallarioides (Sw.) Rich.
BROAD-LEAVED TWAYBLADE

DESCRIPTION Native perennial herb; roots fibrous. **Stems** 1–3 dm long, glandular-hairy above leaves, smooth below. **Leaves** 2, opposite near middle of stem, broadly ovate, 3–6 cm long and 2–5 cm wide, stalkless. **Flowers** yellow-green, 6–20 in a raceme 4–10 cm long and 2–3 cm wide; lip wedge-shaped, 9–11 mm long and to 6 mm wide at tip, usually with a small tooth on each side near the base, the tip shallowly 2-lobed. **Flowering** July–Aug.
SYNONYM *Listera convallarioides.*
HABITAT Seeps in forests, cedar swamps, wet, mixed conifer-deciduous woods, streambanks.
STATUS NCNE-FACW | GP-FACW; Wisc (THR).

Neottia cordata (L.) Rich.
◆HEART-LEAVED TWAYBLADE

DESCRIPTION Native perennial herb; roots fibrous. **Stems** 1–3 dm long, glandular-hairy above leaves, smooth below. **Leaves** 2, opposite near middle of stem, 1–4 cm long and 1–3 cm wide, stalkless. **Flowers** green to red-purple, 6–20 in a raceme 3–12 cm long and 1–2 cm wide; lip slender, 3–5 mm long, with 2 teeth on side near base, the tip cleft halfway or more into spreading linear lobes. **Flowering** June–July.
SYNONYM *Listera cordata.*
HABITAT open bogs and conifer swamps, where usually on sphagnum moss hummocks; hemlock groves.
STATUS MW-FACW | NCNE-FACW | GP-FACU

Platanthera REIN-ORCHID
Perennial herbs, from a cluster of fleshy roots. Stems erect, smooth. Leaves mostly along the stem, upright, reduced to sheaths at base and upward on stem; leaves basal in **large round-leaf orchid** (*Platanthera orbiculata*). Flowers white or green, several to many in a spike or raceme; upper sepal joined with petals to form a hood over the column; lateral sepals spreading; lip linear to ovate or 3-lobed, entire, toothed or fringed, extended backward into a spur, the spur commonly curved; stamens 1, the anther attached to the top of the short column. Fruit a many-seeded capsule.

Platanthera REIN-ORCHID

1   Lip prominently ciliate or fringed 2

1   Lip entire or toothed, but not fringed 6

Neottia cordata

2 Lip simple, not deeply divided (except for fringe) **3**

2 Lip deeply 3-parted in addition to fringe. **4**

3 Flowers white; longest cilia (usually lateral) of fringe about half as long as the undivided portion of the lip, or even shorter............... *P. blephariglottis*

3 Flowers yellow to orange; longest cilia of fringe distinctly more than half the length of the undivided portion of the lip........ .... *P. ciliaris*

4 Flowers pink-purple; divisions of the lip broadly fan-shaped, copiously lacerate-fringed, but the fringe usually cut less than half the distance to the base of the division of the lip .... *P. psycodes*

4 Flowers creamy, greenish, or white; at least the lateral divisions of the lip more narrowly cuneate, mostly cut into a long fringe more than half their length....... **5**

5 Flowers creamy or green-yellow, in narrow compact spikes to 3 cm wide; lip fringed nearly to base .... *P. lacera*

5 Flowers white, large, in spikes more than 3 cm wide; lip less deeply fringed .... *P. praeclara*

6 Flowers pink-purple .... *P. psycodes*

6 Flowers white or green-white........... **7**

7 Stems leafless; leaves 1–2 at base of stem **8**

7 Stems leafy; leaves 1 or more **9**

8 Leaves 1, ascending; stems to 3 dm long .. .. *P. obtusata*

8 Leaves 2, prostrate; stems 3–6 dm long ... . *P. orbiculata*

9 Stem leaf 1, sometimes with several small leaves on upper stem........ *P. clavellata*

9 Stem leaves 2 or more................. **10**

10 Lip wide, 2-lobed and fringed at base..... .... *P. flava*

10 Lip narrow, neither lobed nor fringed .. **11**

11 Flowers white; lip widened at base .... *P. dilatata*

11 Flowers green-white; lip narrowly lance-shaped .... *P. huronensis*

Platanthera blephariglottis (Willd.)

Lindl. **WHITE FRINGED ORCHID**

DESCRIPTION Native perennial herb; roots fleshy. Stems 2–10 dm long. Leaves 1–3, alternate along stem, narrowly lance-shaped, 5–30 cm long and 1–5 cm wide; upper several leaves smaller and bractlike. Flowers bright white, in a compact, densely flowered raceme, 5–15 cm long and 4–5 cm wide; sepals nearly round, 5–11 mm long; lateral petals narrowly oblong; lip oblong lance-shaped, fringed, 8–11 mm long; spur slender 1.5–4 cm long. Flowering July-Aug.

SYNONYM *Habenaria blephariglottis.*

HABITAT open sphagnum bogs, occasional in wet, open sandy areas.

STATUS MW-OBL | NCNE-OBL

Platanthera ciliaris (L.) Lindl.

♦ YELLOW FRINGED ORCHID

DESCRIPTION Native perennial herb; roots long and fleshy. Stems 3–10 dm long. Leaves 1–3, alternate along stem, oblong lance-shaped, 6–30 cm long and 4–6 cm wide; upper leaves abruptly reduced in size and bractlike. Flowers orange, many, in a densely flowered cylindric raceme 5–15 cm long and 5 cm wide; sepals broadly oval to obovate, 6–8 mm long, the lateral sepals spreading; lip oblong, 10–16 mm long, long-fringed, some fringes branched; spur slender, 2–3 cm long. Flowering July-Aug.

SYNONYM *Habenaria ciliaris.*

HABITAT open sphagnum bogs, moist sandy meadows; often with white fringed orchid (*Platanthera blephariglottis*), and hybrids of the two are common.

STATUS MW-FACW | NCNE-FACW | GP-FACW; Mich (END).

Platanthera ciliaris

Platanthera clavellata (Michx.) Luer
◆CLUB-SPUR ORCHID

DESCRIPTION Native perennial herb; roots fleshy. Stems slender, 1-4 dm long. Leaves 1, near or just below middle of stem, oblong to lance-shaped, 5-15 cm long and 1-3 cm wide, usually with 1-3 bractlike leaves above. Flowers 5-20, green-yellow, spreading, in a short raceme, 2-6 cm long and 1.5-3 cm wide; sepals and lateral petals broadly ovate, 3-5 mm long; lip oblong, 3-5 mm long, shallowly 3-lobed or toothed at tip; spur curved, widened at tip, 8-12 mm long. Flowering June–Aug.

SYNONYM *Habenaria clavellata.*

HABITAT in sphagnum moss of open bogs and floating mats, black spruce and tamarack swamps; also colonizing wet ditches.

STATUS MW-OBL | NCNE-FACW | GP-OBL

Platanthera dilatata (Pursh) Lindl.
◆TALL WHITE BOG-ORCHID,
BOG CANDLES

DESCRIPTION Native perennial herb, clove-scented; roots fleshy. Stems stout or slender, to 1 m long. Leaves 3-6, alternate along stem, upright, lance-shaped, to 10-20 cm long and 1-3 cm wide, with 1-2 small, bractlike leaves above and 1 bladeless sheath at base of stem. Flowers 10-60, bright white, upright, in a raceme 1-2.5 dm long; lateral sepals lance-shaped, 4-9 mm long and 1-3 mm wide; lateral petals similar but joined with upper sepal to form somewhat of a hood over the column; lip lance-shaped, widened at base, 6-8 mm long; spur slender, 4-8 mm long. Flowering June–July.

SYNONYMS *Habenaria dilatata, Piperia dilatata.*

HABITAT wet, open bogs and floating mats, conifer swamps, streambanks, shores and seeps; often where sandy or calcium-rich (as in calcareous fens), not in deep sphagnum moss.

STATUS MW-FACW | NCNE-FACW | GP-FACW

Platanthera flava (L.) Lindl.
PALE GREEN ORCHID

DESCRIPTION Native perennial herb, roots fleshy. Stems 3-7 dm long. Leaves 2-4, alternate along stem, lance-shaped or oval, to 5-15 cm long and 2-5 cm wide, with 1-3 bractlike leaves above. Flowers 15 or more, green-yellow or green, stalkless, in a raceme 5-15 cm long and 2-4 cm wide; sepals ovate, 2-3 mm long; lip bent downward, 3-6 mm long, the margin irregular, with a tooth near base on each side; spur 4-6 mm long. Flowering June–Aug.

SYNONYM *Habenaria flava.*

HABITAT wet depressions in hardwood swamps, alder thickets, sedge meadows, moist sand prairies; often where calcium-rich, sometimes where disturbed.

DOUG MCGRADY
Platanthera clavellata

JOSHUA MAYER
Platanthera dilatata

STATUS MW-FACW | NCNE-FACW | GP-FACW; Minn (END), Wisc (THR).

Platanthera huronensis (Nutt.) Lindl.
NORTHERN BOG-ORCHID

DESCRIPTION Native perennial herb; roots fleshy. Stems 2–8 dm long. Leaves 2–7, alternate on stem, linear to oblong, 5–30 cm long and 2–5 cm wide, with 1–3 smaller leaves above. Flowers small, green, erect, many in a raceme 4–25 cm long; lateral sepals ovate and spreading; lateral petals lance-shaped, curved upward and joined with upper sepal to form a loose hood over column; lip lance-shaped, 3–7 mm long, not abruptly widened at base; spur curved forward under the lip, about as long as lip, 3–7 mm long. Flowering June–Aug.

SYNONYMS *Habenaria hyperborea, Platanthera hyperborea.*

HABITAT moist to wet forests and swamps, thickets, streambanks, wet meadows, wet sand along Great Lakes shoreline, ditches.

STATUS MW-FACW | NCNE-FACW | GP-OBL

NOTE this species and *Platanthera aquilonis* (here included within this complex taxon) were formerly grouped within *Platanthera hyperborea.*

Platanthera lacera (Michx.) G. Don
♦RAGGED FRINGED ORCHID

DESCRIPTION Native perennial herb; roots fleshy. Stems 3–8 dm long. Leaves 3–7, alternate on stem, lance-shaped to oval, to 5–15 cm long and 1–4 cm wide; upper leaves much smaller. Flowers white or green-white, in a usually compact, many-flowered raceme, 5–20 cm long and 2–5 cm wide; sepals broadly oval, 4–7 mm long, the lateral ones deflexed behind the lip; lateral petals linear, entire; lip 10–16 mm long and 5–20 mm wide, deeply 3-lobed, each lobe fringed with a few long segments; spur curved, 1–2 cm long. Flowering June–Aug.

SYNONYM *Habenaria lacera.*

HABITAT hummocks in open sphagnum bogs, conifer bogs, swamps, wet meadows, sandy prairie, thickets, ditches.

STATUS MW-FACW | NCNE-FACW | GP-FACW

Platanthera leucophaea (Nutt.) Lindl.
♦PRAIRIE FRINGED ORCHID

DESCRIPTION Native perennial herb; roots thick, fleshy. Stems 4–8 dm long, smooth. Leaves 5–10, alternate on stem, lance-shaped, 8–15 cm long and 1–4 cm wide, the upper leaves much smaller. Flowers white, large and showy, spreading, in a cylindric raceme 5–15 cm long and 5–9 cm wide; sepals 9–12 mm long, broadly ovate; lateral petals broadly obovate, ragged at tip, 10–15 mm long; lip deeply 3-lobed, the lobes fringed more than half way to base, 1.5–2.5 cm long and about as wide; spur curved, 2.5–5 cm long. Flowering June–July.

SYNONYM *Habenaria leucophaea.*

HABITAT open, calcium-rich wet meadows and low prairie, especially where soils are high in organic matter; occasionally in sedge meadows and on floating bog mats.

STATUS MW-FACW | NCNE-FACW | GP-OBL; Wisc (END), Mich (END).

Platanthera lacera

Platanthera leucophaea

Platanthera obtusata (Banks ex Pursh) Lindl. BLUNT-LEAF ORCHID

DESCRIPTION Native perennial herb; roots fleshy. Stems leafless, slender, 1–3 dm long. Leaves 1 at base of stem, ascending, persistent through flowering, obovate, 5–15 cm long and 1–4 cm wide, blunt-tipped, long-tapered to base. Flowers 4–20, green-white, in a raceme 3–12 cm long and 1–2 cm wide; lateral sepals ovate, spreading; petals ascending, widened below middle; lip lance-shaped, widened at base, 4–6 mm long; spur curved, tapered to a thin tip, 5–8 mm long. Flowering June–Aug.

SYNONYM *Habenaria obtusata.*

HABITAT shaded hummocks in conifer swamps (especially under cedar, black spruce or balsam fir), wet mixed conifer-deciduous forests, alder thickets.

STATUS MW-FACW | NCNE-FACW | GP-FACW

Platanthera orbiculata (Pursh) Lindl.
◊ LARGE ROUND-LEAF ORCHID

DESCRIPTION Native perennial herb; roots fleshy. Stems 2–6 dm long, leafless apart from 1–6 small bracts. Leaves 2, opposite at base of plant, spreading or lying flat on ground, ± round, shiny, 6–15 cm long and 4–15 cm wide. Flowers green-white, several in a raceme 5–20 cm long and 3–6 cm wide; sepals ovate, to 1 cm long; petals ovate, 6–7 mm long; lip entire, rounded at tip, 10–15 mm long and 2 mm wide; spur 2–3 cm long, somewhat widened at tip. Flowering Late June–Aug.

SYNONYM *Habenaria orbiculata.*

HABITAT shaded conifer swamps (white cedar, balsam fir, black spruce), especially where underlain by marl; also in drier conifer forests.

STATUS MW-FAC | NCNE-FAC | GP-FAC

Platanthera praeclara Sheviak & M.L. Bowles
◊ WESTERN PRAIRIE FRINGED ORCHID

DESCRIPTION Native perennial herb; roots thick, fleshy. Stems 4–8 dm long, smooth. Leaves 3–10 or more, upright, scattered and alternate on stem, smooth, lance-shaped to ovate lance-shaped, base of leaf clasping stem; lower leaves 10–15 cm long and 1.5–3.5 cm wide, the upper stem leaves much smaller. Flowers white to creamy white, large and showy, to 2.5 cm wide; up to 24 flowers in an open raceme; petals 3, the two lateral petals ragged at tip; the lip larger and deeply 3-lobed, the lobes fringed more than half way to base, 1.5–2.5 cm long and about as wide; spur 35–55 mm long. Flowering June–Aug.

SYNONYM *Habenaria leucophaea* var. *praeclara.*

Platanthera orbiculata

Platanthera praeclara

HABITAT sedge meadows, low prairie, moist prairie swales; soils usually sandy loams and often calcium-rich.
STATUS MW-FAC | NCNE-FACW | GP-FAC; Minn (END).
NOTE Federally listed as threatened across its entire range (mostly Great Plains region).

Platanthera psycodes (L.) Lindl.
◆ PURPLE FRINGED ORCHID

DESCRIPTION Native perennial herb; roots thick and fleshy. Stems stout, 3–10 dm long. Leaves 4–12, alternate on stem, lance-shaped or oval, the upper much smaller and narrow. Flowers rose-purple, in a densely flowered, cylindric raceme 4–20 cm long and 3–5 cm wide; sepals oval to obovate, 4–6 mm long; petals spatula-shaped, finely toothed on margins; lip broad, 8–14 mm wide, deeply 3-lobed, the lobes fan-shaped, fringed to less than half way to base; spur curved, about 2 cm long. Flowering July–Aug.
SYNONYM *Habenaria psycodes.*
HABITAT wetland margins, shores, wet forests, wet meadows, low prairie, roadside ditches; typically not on sphagnum moss.
STATUS MW-FACW | NCNE-FACW | GP-FACW

Pogonia POGONIA

Pogonia ophioglossoides (L.) Ker
Gawler ◆ ROSE POGONIA, SNAKE-MOUTH
DESCRIPTION Native perennial herb, spreading by surface runners (stolons) which send up a stem every 10 cm or more apart. Stems slender, smooth, 1.5–4 dm long. Leaves single, attached about halfway up stem, narrowly oval, 3–10 cm long and 1–2.5 cm wide, stalkless. Flowers pink to purple, usually 1 at end of stems; sepals widely spreading, petals oval, angled over the column; lip pink with purple veins, 1.5–2 cm long and 5–10 mm wide, fringed at tip, bearded with yellow bristles. Flowering June–July.
HABITAT conifer swamps and open bogs in sphagnum moss, floating sedge mats, sedge meadows, interdunal wetlands.
STATUS MW-OBL | NCNE-OBL | GP-OBL

Spiranthes LADIES' TRESSES
Perennial herbs, from a cluster of tuberous roots. Stems slender, erect. Leaves largest at base of plant, becoming smaller upward on stem, the stem leaves erect and sheathing. Flowers small, white or creamy, spirally twisted in a densely flowered, spikelike raceme; sepals and lateral petals similar, the lateral petals joined with all 3 sepals or with only the upper sepal to form a hood over lip and column; lip folded upward near middle so that margins embrace the column, curved downward beyond the middle, with a pair of bumps or thickenings at base; anthers 1, from back of the short column.

Spiranthes LADIES' TRESSES
1 Lip of flower violin-shaped (constricted near middle and widened near tip) ......
.... *S. romanzoffiana*
1 Lip not violin-shaped 2
2 Lip bright yellow .... *S. lucida*
2 Lip white or yellow-green in center.......
.... *S. cernua*

Platanthera psycodes
AARON CARLSON

Pogonia ophioglossoides
JOSHUA MAYER

Spiranthes cernua (L.) L. C. Rich.
♦ NODDING LADIES' TRESSES

DESCRIPTION Native perennial herb; roots fleshy. Stems 1–5 dm long, upper stem short-hairy, lower stem smooth. Leaves mostly at base of plant, usually present at flowering time, linear to oblong lance-shaped, 6–25 cm long and 5–15 mm wide; upper stem leaves 3–5, much smaller and bractlike. Flowers white, in a spikelike raceme 3–15 cm long, with 2–4 vertical rows of flowers, the rows spirally twisted; sepals and petals hairy on outside; lateral petals joined with upper sepal to form a hood; lip white, yellow-green at center, 6–10 mm long and 3–6 mm wide, slightly narrowed at middle, curved downward, the tip curved inward toward stem, the tip wavy-margined or with small rounded teeth, the base of lip with a pair of backward-pointing bumps. Flowering Aug–Oct.

HABITAT open, usually sandy wetlands such as wet meadows, lakeshores, moist prairies, ditches and roadsides.

STATUS MW-FACW | NCNE-FACW | GP-FACW

Spiranthes lucida (H. Eaton) Ames
SHINING LADIES' TRESSES

DESCRIPTION Small native perennial herb; roots fleshy. Stems slender, smooth to finely hairy above, 1–3 dm

long. Leaves mostly at base of plant, oblong lance-shaped, shiny, 5–10 cm long and 5–15 mm wide; stem leaves usually 2, small and bractlike. Flowers white, nodding, in a spike-like raceme 2–7 cm long, the flowers in 1–2 vertical rows, the rows spirally twisted; upper sepal and petals forming a hood over the column; lip oblong, 5–6 mm long, the outer half bright yellow or yellow-orange, with white margins, bumps at base of lip small, less than 1 mm long. Flowering June–July.

HABITAT streambanks, lakeshores, wet meadows, ditches; especially on calcium-rich soils and limestone gravels, often where somewhat disturbed.

STATUS MW-FACW | NCNE-FACW | GP-FACW

Spiranthes romanzoffiana Cham.
♦ HOODED LADIES' TRESSES

DESCRIPTION Native perennial herb; roots thick and fleshy. Stems 1–4 dm long, upper stem finely hairy. Leaves mostly from base of plant, present at flowering time, upright, linear to narrowly lance-shaped, 5–20 cm long and 3–9 mm wide, the stem leaves becoming smaller and bractlike. Flowers white or cream-colored, in a spikelike raceme 3–10 cm long, with 1–3 vertical rows of flowers, the rows spirally twisted; sepals and lateral petals joined to form a hood over the lip; lip ovate, strongly constricted near middle (violin-shaped), curved downward, the tip ragged and bent

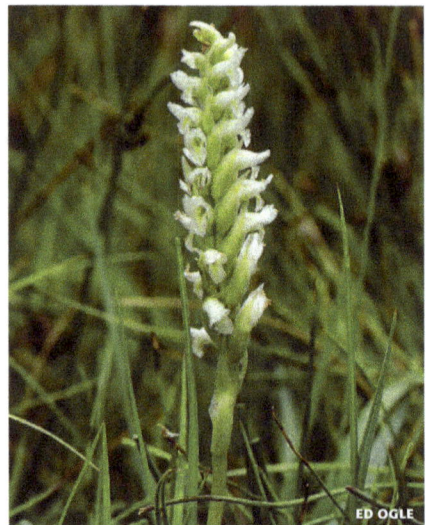

Spiranthes cernua

Spiranthes romanzoffiana

inward toward stem, the bumps at base very small. Flowering July–Sept.

HABITAT open wetlands including wet meadows, fens, lakeshores, open swamps, ditches, seeps; usually in neutral or calcium-rich habitats.

STATUS MW-OBL | NCNE-OBL | GP-OBL

Poaceae
Grass Family

PERENNIAL OR ANNUAL herbaceous plants, clumped or spreading by rhizomes. Stems (culms) usually hollow, with swollen, solid nodes. Leaves long, linear, parallel-veined, alternate in 2 ranks or rows, sheathing the stem, the sheaths usually split vertically, sometimes joined and tubular as in brome (*Bromus*) and mannagrass (*Glyceria*); with a membranous or hairy ring (ligule) at top of sheath between blade and stem, or the ligule sometimes absent; a pair of projecting lobes (auricles) sometimes present at base of blade. Flowers (florets) small, usually perfect (with both staminate and pistillate parts), or sometimes either staminate or pistillate, the staminate and pistillate flowers separate on the same or different plants.

Florets grouped into spikelets, each spikelet with 1 to many florets, the florets stalkless and alternate along a small stem or axis (rachilla), with a pair of small bracts (glumes) at base of each spikelet (the glumes rarely absent); the glumes usually of different lengths, the lowermost (or first) glume usually smaller, the upper (or second) glume usually longer. Within the spikelet, each floret subtended by 2 bracts, the larger one (lemma) containing the flower, the smaller one (palea) covering the flower; the lemma and palea often enclosing the ripe fruit (grain or caryopsis); stamens usually 3 or sometimes 6, usually exserted when flowering; ovary superior, never enclosed in a sac (as in sedges); styles 2-3-parted, the stigmas often feathery.

Spikelets grouped in a variety of heads, most commonly in branching heads (panicles), or stalked along an unbranched stem (rachis) in a raceme, or the spikelets stalkless along an unbranched stem in a spike; spikelets breaking (disarticulating) either above or below the glumes when mature, the glumes remaining in the head if falling above the glumes, or the glumes falling with the florets if disarticulation is below the glumes.

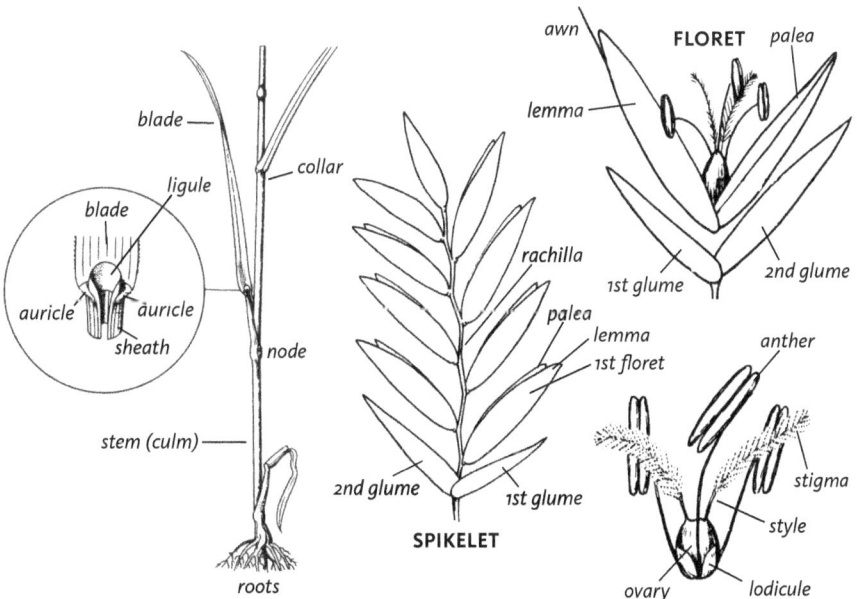

Grass Terminology

Poaceae Groups GRASS FAMILY

1 Stems often 2 m or more tall **2**
1 Stems usually less than 2 m tall **4**
2 Staminate and pistillate flowers separate on same plant . . . . *Zizania aquatica*
2 Flowers perfect **3**
3 Spikelets several-flowered; flowers disarticulating above the glumes. . . . . *Phragmites australis*
3 Spikelets 1-flowered; disarticulating below the glumes *Andropogon gerardii*
4 Spikelets breaking below the glumes, the glumes falling with the florets . GROUP 1
4 Spikelets breaking above the glumes, the glumes remaining in the head **5**
5 Each spikelet with 1 fertile floret . . . . GROUP 2
5 Each spikelet with 2 or more fertile florets. **6**
6 Spikelets in spikes or spikelike racemes. . . . . GROUP 3
6 Spikelets in open or dense panicles. . . . . GROUP 4

Poaceae Group 1

Spikelets falling as entire unit (the glumes falling with the florets).

1 Spikelets falling with attached stalk and bristles. . . . . *Hordeum jubatum*
1 Spikelets falling separately, without attached stalk **2**
2 Each spikelet with 2 florets . . . . *Sphenopholis obtusa*
2 Each spikelet with 1 perfect floret (sterile or staminate florets sometimes present) . . . **3**
3 Spikelets in a spike or raceme. **4**
3 Spikelets in a panicle. **5**
4 Spikelets ± round in outline; first and second glumes of equal lengths . . . . *Beckmannia syzigachne*
4 Spikelets lance-shaped; first and second glumes of different lengths . . . . . *Spartina*
5 Spikelets flattened on back sides. **6**
5 Spikelets flattened along margins **9**
6 Spikelets awned; ligules absent . . . . *Echinochloa*
6 Spikelets not awned; ligules present. **7**
7 Spikelets at least sparsely pubescent towards their margins . . . *Dichanthelium*
7 Spikelets glabrous **8**
8 Terminal panicle 8–40 cm long, (smaller in occasional depauperate individuals of annual species); annuals or perennials, but without clear remnants of overwintering basal rosette leaves; flowering and fruiting summer-fall . . . . *Panicum*
8 Terminal panicle 2.5–8 (–12) cm long; tufted perennials with clear remnants of old, dead leaves from the previous year present at the base, these sometimes formed into a clear overwintering rosette; flowering spring and fruiting in late spring–early summer . . . . *Dichanthelium*
9 Bracts below each spikelet 2; glumes absent . . . . *Leersia*
9 Bracts below each spikelet 3–4; glumes present **10**
10 Panicle open . . . . *Cinna*
10 Panicle dense, cylinder-shaped and spikelike. *Alopecurus*

Poaceae Group 2

Spikelets with 1 fertile floret, disarticulating above glumes (the glumes remaining attached to the head).

1 Florets stiff and shiny . . . . *Phalaris arundinacea*
1 Florets soft and papery, not shiny **2**
2 Glumes much smaller than floret **3**
2 At least 1 glume about same length as spikelet **5**
3 Lemma 5-veined . . . . *Leersia*
3 Lemma 3-veined **4**
4 Lemma not awned, the lemma veins running parallel to a blunt tip . . . . *Catabrosa aquatica*
4 Lemma usually awned, the veins converging to the tip. . . . . . *Muhlenbergia*
5 Florets 3, the 2 lower florets staminate or reduced to scales **6**
5 Florets 1, with both staminate and pistillate parts. **7**
6 Panicle dense and spikelike. . . . . *Phalaris arundinacea*
6 Panicle open . . . . *Anthoxanthum hirtum*
7 Stalk within spikelet (rachilla) elongate and bristlelike behind the palea . . . . *Calamagrostis*
7 Rachilla not elongated **8**
8 Lemmas with 3 pronounced veins and tipped with an awn . . . . *Muhlenbergia*
8 Lemmas with 5 faint veins, awn (if present) from back of lemma . . . . *Agrostis*

Poaceae Group 3

Head a spike.

1   Spikes several to many at ends of stems; 1-sided, the spikelets all in 2 rows on lower side of rachis.... *Leptochloa fascicularis*

1   Spikes single; spikelets on opposite sides of rachis **2**

2   Spikelets in groups of 3 at nodes of spike, the 2 side spikelets reduced and on short stalks about 1 mm long . . . . . *Hordeum jubatum*

2   Spikelets mostly 2 at each node, the spikelets all alike . . . . *Elymus*

Poaceae Group 4

Head a panicle; fertile florets more than 1; spikelets disarticulating above the glumes (the glumes remaining attached to head).

1   Glumes about as long as spikelet **2**

1   Glumes much shorter than spikelet **4**

2   Florets 3 or more; plant of w and sc Minn and w Iowa . . . . *Scolochloa festucacea*

2   Florets 2 . . . . **3**

3   Lemmas awned . *Deschampsia cespitosa*

3   Lemmas awnless . . . . *Graphephorum melicoides*

4   Lemmas with 3 prominent veins **5**

4   Lemmas with 5 or more prominent veins **6**

5   Each spikelet with 2 florets; lemma veins running parallel to a blunt tip . . . . *Catabrosa aquatica*

5   Each spikelet with 3 florets; lemma veins converging at tip . . . . *Eragrostis*

6   Edges of leaf sheaths joined for at least half of their length **7**

6   Edges of leaf sheaths joined only at base **8**

7   Lemmas awned . . . . *Bromus*

7   Lemmas awnless . . . . *Glyceria*

8   Staminate and pistillate flowers on separate plants; plant of brackish areas . . . . *Distichlis spicata*

8   Flowers perfect, with both staminate and pistillate parts **9**

9   Lemmas tapered to a point, the veins converging at tip . . . . *Poa*

9   Lemmas with parallel veins extending to a blunt tip **10**

10   Lemma veins obscure; often in saline habitats . . . . *Puccinellia*

10   Lemma veins 5, prominent; habitats not saline . . . . *Torreyochloa*

Agrostis BENTGRASS

Perennial grasses, clumped or spreading by rhizomes or sometimes by stolons. Leaves soft, flat. Head an open panicle. Spikelets small, 1-flowered, breaking above glumes; glumes ± equal length, 1-veined; floret shorter than glumes; lemma awnless or with a short straight awn; palea small or absent; stamens usually 3.

Agrostis BENTGRASS

1   Plants clumped; palea ± absent . . . . *A. hyemalis*

1   Plants with rhizomes and|or stolons; palea present, about half as long as lemma . . . . *A. stolonifera*

Agrostis hyemalis (Walter) BSP.
TICKLEGRASS

DESCRIPTION Clumped, native perennial grass. **Stems** slender, erect to reclining, 2–6 dm long. **Leaves** mostly at or near base of plant, upright to spreading, flat to inrolled, 1–2 mm wide, smooth or somewhat rough-to-touch; sheaths smooth, the ligule translucent, 1–2 mm long, rounded and usually ragged at tip. **Head** an open panicle, 1–3 dm long, the branches threadlike and spreading, the branches themselves branched and with spikelets only above their middle. **Spikelets** 1-flowered, often purple, 1–3 mm long; glumes lance-shaped, 1–3 mm long; lemma 1–2 mm long, unawned or with a short straight awn; palea absent. **Flowering** June–Aug.

HABITAT wet meadows, bogs, ditches, streambanks, shores; more commonly in dry, sandy places.

STATUS MW-FAC | NCNE-FAC | GP-FACW

Agrostis stolonifera L.
♦REDTOP, SPREADING BENTGRASS

DESCRIPTION Introduced perennial grass, spreading by rhizomes and also sometimes by stolons. **Stems** erect or ± horizontal at base, 3–10 dm or more long. **Leaves** ascending, 2–8 mm wide, rough-to-touch; sheaths smooth, the ligule translucent, usually splitting at tip, 2–5 mm long. **Head** an open panicle, 3–20 cm long, the branches

spreading, branched and with spikelets along their entire length. Spikelets 1-flowered, usually purple, 2–4 mm long; glumes lance-shaped, 1.5–2.5 mm long; lemma 2/3 length of glumes, 1–2 mm long; palea present, about half as long as lemma. Flowering July–Sept.

SYNONYMS *Agrostis alba, Agrostis gigantea, Agrostis palustris.*

HABITAT wet meadows, ditches, streambanks and shores; disturbed areas.

STATUS MW-FACW | NCNE-FACW | GP-FACW; introduced from Europe as a pasture grass, naturalized throughout most of USA and s Canada.

Alopecurus
MEADOW-FOXTAIL

Annual or perennial grasses. Stems erect or more or less horizontal at base. Leaves mostly from lower 1/2 of the stems; sheaths open; auricles absent; ligules membranous, entire to lacerate. Heads densely flowered, cylindric, spike-like panicles. Spikelets 1-flowered, flattened, breaking below the glumes; glumes equal length, 3-nerved, often silky hairy on back, awnless; lemma about as long as glumes or shorter, awned from the back, the awn shorter to longer than the glume tips; palea absent. The narrow panicles resemble those of timothy (*Phleum*).

Alopecurus **MEADOW-FOXTAIL**

1 Spikelets (excluding awns) ca. 4–6.5 mm long; awns mostly exserted ca. 3.5–6 mm beyond tips of glumes; anthers ca. 2.4–3.5 mm long . . . . *A. pratensis*

1 Spikelets not over 3 mm long; awns at most exserted ca. 2–3 mm; anthers less than 2 mm long **2**

2 Awn exserted at most about 1 mm beyond tips of glumes, usually included, inserted about a third or half the distance from base of lemma . . . . *A. aequalis*

2 Awn of most lemmas exserted ca. 2–3 mm, inserted near the base of lemma (on lower 1/5–1/4) **3**

3 Anthers 0.3–0.7 mm long; spikelets 2–2.5 mm long. . . . . *A. carolinianus*

3 Anthers 1.4–1.8 mm long; larger spikelets 2.6–3 mm long . . . . *A. geniculatus*

Alopecurus aequalis Sobol.
♦ SHORT-AWN FOXTAIL

DESCRIPTION Native, annual or short-lived perennial grass. Stems single or in small clumps, slender, erect to ± horizontal, 2–6 dm long, often rooting at the nodes. Leaves 1–5 mm wide, finely rough-to-touch above; ligule membranous, rounded to elongate, 2–7 mm long. Head an erect, spikelike panicle, 2–7 cm long and 3–5 mm wide. Spikelets 1-flowered; glumes 2–3 mm long, blunt-tipped, hairy on the keel and veins; lemma about equaling the glumes, awned from back, the awn straight, to 1.5 mm longer than glume tips. Flowering June–Aug.

HABITAT shallow water or mud of wet meadows, marshes, ditches, springs, open bogs, fens, shores and streambanks; sometimes where calcium-rich.

STATUS MW-OBL | NCNE-OBL | GP-OBL

MATT LAVIN

Agrostis stolonifera

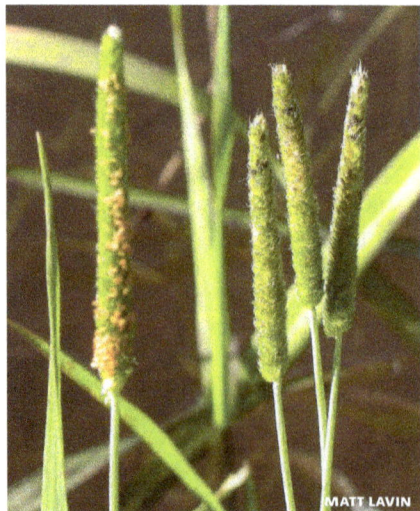

MATT LAVIN

Alopecurus aequalis

Alopecurus carolinianus Walter
CLUMPED FOXTAIL

DESCRIPTION Densely clumped, native annual grass. Stems erect to upright, 1–4 dm long. Leaves 1–3 mm wide, finely rough-to-touch above; ligule membranous, rounded to elongate, 1–5 mm long. Head a cylindric, spikelike panicle, 1–5 cm long and 3–5 mm wide. Spikelets 1-flowered; glumes 2–3 mm long, blunt-tipped, hairy on keel; lemma about as long as glumes, awned from back, the awn bent near middle and 2–3 mm longer than glume tips. Flowering May–July.

HABITAT mud flats, temporary ponds, wet meadows, marshes, low prairie, fallow fields.
STATUS MW-FACW | NCNE-FACW | GP-FACW

Alopecurus geniculatus L.
MARSH-FOXTAIL

DESCRIPTION Introduced, clumped perennial grass. Stems usually decumbent, sometimes rooting at the nodes, 2–8 dm long; auricles absent; ligules 2–5 mm long, obtuse. Leaves mostly 1–4 mm wide; upper sheaths somewhat inflated. Head a spike-like panicle, 3–5 cm long, 4–6 mm wide, often tinged with purple; glumes as in A. carolinianus; awn inserted about halfway between the base and middle of the lemma, exserted 1.5–3.2 mm long.

HABITAT mud and shallow water.
STATUS MW-OBL | NCNE-OBL | GP-OBL

Alopecurus pratensis L.
♦ FIELD MEADOW-FOXTAIL

DESCRIPTION Introduced, perennial grass, shortly rhizomatous. Stems erect or decumbent at base, 4–8 dm long; auricles absent; ligules 1.5–3 mm long, obtuse to truncate; upper sheaths not or scarcely inflated. Head a spike-like panicle, 2–8 cm long, 5–10 mm wide, scarcely tapering. Spikelets 1-flowered; glumes 4–5.5 mm long, the keel narrowly winged, conspicuously ciliate, especially above the middle, with hairs 1–1.5 mm long; awn inserted about halfway between the base and middle of the lemma, exserted 2–6 mm.

HABITAT native of Eurasia; naturalized in moist meadows, fields, and waste places.
STATUS MW-FACW | NCNE-FAC

Andropogon BLUESTEM

Andropogon gerardii Vitman
♦ BIG BLUESTEM

DESCRIPTION Tall, loosely clumped, native perennial grass. Stems stout, 1–3 m tall, forming large bunches or extensive sod. leaf blades usually 5–10 mm wide, the lower ones and the sheaths sometimes villous; ligules membranous. Head a cylindric, spikelike panicle, 1–5 cm long and 3–5 mm wide. Spikelets in 2–6 racemes, subdigitate, on a long-exserted peduncle, 5–10 cm long; joints of the rachis and pedicels equal, sparsely or usually densely ciliate,

Alopecurus pratensis

Andropogon gerardii

densely bearded at the summit. Spikelets of two kinds, in pairs at the joints of the rachis, one sessile and perfect, the other pediceled and staminate, sterile, or abortive; glumes of the fertile spikelet equal or nearly so, lacking a midnerve, often ciliate; fertile lemma shorter than the glumes, narrow, hyaline, usually ending in a long awn 8–15 mm long, twisted below and more or less bent.

HABITAT a dominant species of tallgrass prairie, also occurs in wet grasslands, sedge meadows, and occasionally in calcareous fens, especially in s and w portions of the Upper Midwest region.

STATUS MW-FACW | NCNE-FACU | GP-FACU

Anthoxanthum
SWEETGRASS

Anthoxanthum hirtum (Schrank) Y.
Schouten & Veldkamp ♦ SWEETGRASS, HOLY GRASS, VANILLA GRASS

DESCRIPTION Native perennial grass, from creeping rhizomes; plants sweet-scented, especially when dried. Stems erect, 2–6 dm tall, smooth. **Leaves** flat, 2–6 mm wide, smooth or short-hairy; stem leaves short, 1–4 cm long, leaves on sterile shoots much longer; sheaths smooth or short-hairy at top; ligule membranous, 1–4 mm long. **Head** a pyramid-shaped pan-

icle, 5–10 cm long, the branches spreading to drooping. **Spikelets** 3-flowered, the lower 2 florets staminate, the terminal spikelet perfect, golden brown, or green or purple at base and golden near tips, 5 mm long, breaking above the glumes; glumes ovate, shiny, 4–6 mm long; lemmas 3–4 mm long, the staminate lemma hairy. **Flowering** May–July.

SYNONYM *Hierochloe odorata*.

HABITAT wet meadows, shores, low prairie; often where sandy.

STATUS MW-FACU | NCNE-FACU | GP-FACU

NOTE the fragrance emitted when fresh plants are crushed or burned is from coumarin, an anti-coagulant agent.

Beckmannia SLOUGHGRASS

Beckmannia syzigachne (Steud.)
Fernald ♦ AMERICAN SLOUGHGRASS

DESCRIPTION Stout, native annual grass. **Stems** single or in small clumps, 4–10 dm long. **Leaves** flat, 3–10 mm wide, rough-to-touch; sheaths overlapping, smooth, the upper sheath often loosely enclosing lower part of panicle; ligule membranous, rounded to acute, 3–6 mm long. **Head** of many 1-sided spikes in a narrow panicle 10–30 cm long, the panicle branches erect, overlapping, 1–5 cm long; each spike 1–2 cm long, with several to many spikelets

Anthoxanthum hirtum

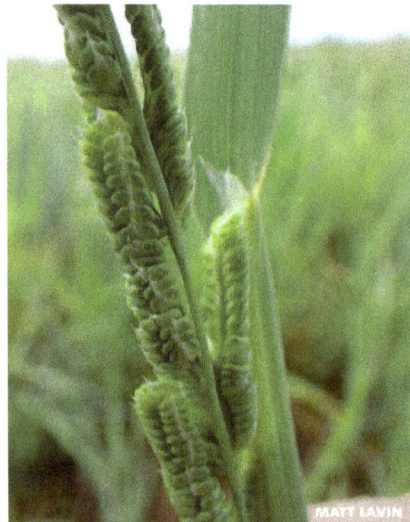

Beckmannia syzigachne

in 2 rows on the rachis. Spikelets 1–2-flowered, overlapping, nearly round, 2–4 mm long, straw-colored when mature, breaking below the glumes; glumes equal, broad, inflated along midvein, with a short, slender tip; lemma about as long as glumes but narrower; palea nearly as long as lemma. Flowering June–Sept.

HABITAT wet meadows, marshes, ditches, shores and streambanks.

STATUS MW-OBL | NCNE-OBL | GP-OBL; rare in Mich (THR).

Bromus BROME

Perennial grasses. Leaves flat; sheaths closed to near top. Head a panicle of drooping spikelets. Spikelets with several to many flowers, breaking above the glumes; glumes shorter than lemmas; lemmas awned (in species included here); stamens usually 3.

Bromus **BROME**

1 Lemmas hairy across back.. *B. latiglumis*
1 Lemmas smooth on back, lemma margins
   fringed with hairs............. *B. ciliatus*

Bromus ciliatus L. ♦ FRINGED BROME

DESCRIPTION Native perennial grass, rhizomes absent. Stems single or few together, smooth or hairy at nodes, 5–12 dm long. Leaves flat, 4–10 mm wide, usually with long, soft hairs mainly on upper surface; sheaths usually with long hairs; ligule membranous, short, to 2 mm long, ragged across tip. Head a loose, open panicle 1–3 dm long, the branches usually drooping. Spikelets large, 4–10-flowered,

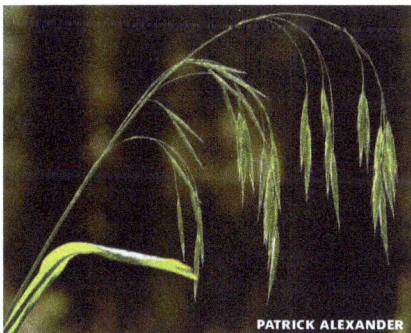

PATRICK ALEXANDER

Bromus ciliatus

1.5–3 cm long and 5–10 mm wide; glumes usually ± smooth, lance-shaped, the first glume 4–9 mm long, the second glume 6–10 mm long, often tipped with a short awn; lemma 10–15 mm long, ± smooth on back, usually long-hairy along lower margins, tipped with an awn 2–6 mm long; palea about as long as body of lemma. Flowering July–Sept.

HABITAT streambanks, shores, thickets, sedge meadows, fens, marshes; also in moist woods.

STATUS MW-FACW | NCNE-FACW | GP-FAC

Bromus latiglumis (Scribn. ex Shear) A.S. Hitchc. EAR-LEAVED BROME

DESCRIPTION Native perennial grass. Stems single or in small clumps, ± smooth, 6–15 dm long. Leaves flat, 8–20 along stem; 10–15 mm wide, with a pair of auricles at base; sheaths with a dense ring of hairs at top. head a panicle 1–2 dm long, the branches spreading or drooping. Spikelets several-flowered, 2–3 cm long, first glume 5–8 mm long, awl-shaped, second glume wider, 6–10 mm long; lemmas 10–12 mm long, hairy, awned, the awn 2–7 mm long.

SYNONYM *Bromus altissimus.*

HABITAT floodplain forests, thickets and streambanks, sometimes in rocky woods.

STATUS MW-FACW | NCNE-FACW | GP-FACW

NOTE similar to **Canada brome** (*Bromus pubescens*), a species of mostly drier woods, but top of sheaths of ear-leaved brome with a ring of dense hairs, and leaf blades with well-developed auricles.

Calamagrostis REEDGRASS

Perennial grasses, spreading by rhizomes. Stems single or in clumps. Leaves flat or inrolled, green or waxy blue-green, smooth or rough-to-touch; sheaths smooth; ligule large, membranous, usually with an irregular, ragged margin. Head a loose and open, or dense and contracted panicle. Spikelets 1-flowered, breaking above glumes; glumes nearly equal, lance-shaped; lemma shorter than glumes, lance-shaped, awned from back, the awn about as long as lemma, the base of lemma (callus) bearded with a tuft of hairs, these shorter to as long as lemma; palea shorter than lemma; stamens 3.

Calamagrostis REEDGRASS

1 Panicle ± loose and open, the branches ascending to spreading; leaves ± lax, flat, 2–6 mm wide . . . . *C. canadensis*
1 Panicle contracted, the branches short, ascending to appressed; leaves stiff, often inrolled, 1–4 mm wide when flattened . . . . *C. stricta*

Calamagrostis canadensis (Michx.)

P. Beauv. ♦BLUEJOINT

DESCRIPTION Native perennial grass, from creeping rhizomes. Stems erect, in small clumps, 6–15 dm long, often rooting from lower nodes when partly underwater. Leaves flat, green to waxy blue-green, 3–8 mm wide, rough-to-touch on both sides; sheaths smooth; ligules 3–7 mm long. Head a ± open panicle, 8–20 cm long, the branches upright or spreading. Spikelets 1-flowered, 2–6 mm long; glumes ± equal, 2–4 mm long, smooth or finely rough-hairy on back; lemma ± smooth, awned from middle of back, the awn straight, base with dense callus hairs about as long as lemma. Flowering June–Aug.

HABITAT wet meadows, shallow marshes, calcareous fens, streambanks, thickets.

STATUS MW-OBL | NCNE-OBL | GP-FACW

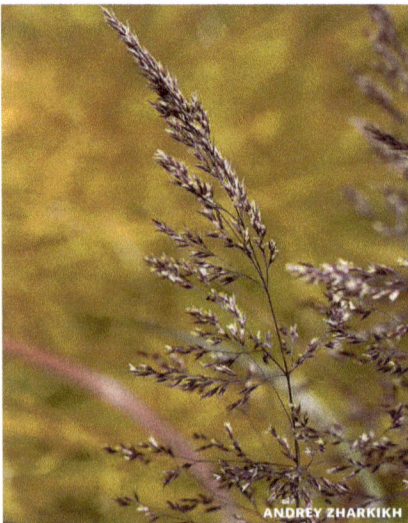

Calamagrostis stricta (Timm) Koeler

♦NARROW-SPIKE REEDGRASS

DESCRIPTION Native perennial grass, spreading by rhizomes; plants waxy blue-green. Stems erect, 3–12 dm long. Leaves stiff, often inrolled, 1–4 mm wide when flattened. Head a narrow panicle, 5–15 cm long, the branches short, upright to erect. Spikelets 1-flowered; glumes 3–6 mm long, smooth or rough-hairy on back; lemma rough-hairy, 2–4 mm long, awned, the awn straight, from near middle of back, base with many callus hairs, half to as long as lemma. Flowering June–Sept.

SYNONYM *Calamagrostis inexpansa, Calamagrostis neglecta.*

HABITAT wet meadows, shallow marshes, shores, streambanks; rocky shore of Lake Superior.

STATUS MW-FACW | NCNE-FACW | GP-FACW

NOTE similar to **bluejoint** (*Calamagrostis canadensis*), but the head narrow and crowded and the leaves often inrolled.

Catabrosa BROOK-GRASS

Catabrosa aquatica (L.) P. Beauv.

BROOK-GRASS

DESCRIPTION Native, loosely clumped or sprawling perennial grass. Stems thick, weak, often horizontal, 2–6 dm long, branching and rooting at nodes in mud or water. Leaves flat, smooth, mostly 10–15 cm long and 3–10 mm wide; sheaths smooth; ligule 1–4 mm long. Head an open,

Calamagrostis canadensis

Calamagrostis stricta

pyramid-shaped or oblong panicle 10–20 cm long. Spikelets mostly 2-flowered or sometimes mostly 1-flowered, the second floret (when present) well above the first, breaking above the glumes, golden brown, 2–4 mm long; glumes unequal, the first glume smaller, 1–2 mm long, the second glume ragged at tip; lemma smooth, ragged at tip, 2–3 mm long; palea similar to lemma. Flowering June–Sept. HABITAT shallow water or mud of streambanks, cold springs and seeps. STATUS MW-OBL | NCNE-OBL | GP-OBL; Wisc (END), more common in w USA.

Cinna WOODREED

Tall, perennial grasses, rhizomes weak or absent. Leaves wide, flat and lax; ligule brown, membranous, with an irregular, jagged margin. Head a large, closed to open panicle, the branches upright to spreading or drooping. Spikelets small, 1-flowered, laterally compressed, breaking below the glumes; glumes nearly equal, lance-shaped, keeled; lemma similar to glumes, with a short awn from just below the tip; palea shorter than lemma; stamens 1.

Cinna WOODREED

1 Panicle ± crowded and narrow, the branches upright; second glume 4–6 mm long.................... *C. arundinacea*
1 Panicle open, the branches spreading to drooping; second glume 2–4 mm long.... .... *C. latifolia*

Cinna arundinacea L.
COMMON WOODREED

DESCRIPTION Native perennial grass, rhizomes weak or absent. Stems 1 or few together, erect, 6–15 dm long, often swollen at base. Leaves 4–12 mm wide, margins rough-to-touch; sheaths smooth; ligule red-brown, 3–10 mm long. Head a narrow panicle, dull gray-green, 1–3 dm long, the branches upright. Spikelets 1-flowered; glumes narrowly lance-shaped, 3–5 mm long, the first glume 1-veined, the second glume 3-veined, usually rough-hairy; lemma 3–5 mm long, rough-hairy on back, usually with an awn to 0.5 mm long, attached just

below tip and mostly shorter than lemma tip. Aug–Sept. HABITAT swamps, floodplain forests, streambanks, pond margins, moist woods. STATUS MW-FACW | NCNE-FACW | GP-FACW

Cinna latifolia (Trev. ex Goepp.) Griseb.
♦ DROOPING REEDGRASS

DESCRIPTION Native perennial grass, with weak rhizomes. Stems single or in small groups, erect, 5–13 dm long, not swollen at base. Leaves 5–15 mm wide, usually rough-to-touch; sheaths smooth to finely roughened; ligule pale, 2–7 mm long. Head a loose, open panicle, pale green, satiny, 1–3.5 dm long, the branches spreading to drooping. Spikelets 1-flowered; glumes narrowly lance-shaped, 1-veined, 2–4 mm long; lemma 2–4 mm long, finely rough-hairy on back, usually with an awn to 1.5 mm long from just below the tip, the awn usually longer than the tip. Flowering July–Aug. HABITAT wet woods, swamps, springs. STATUS MW-FACW | NCNE-FACW | GP-OBL

Deschampsia HAIRGRASS

Deschampsia cespitosa (L.) P. Beauv.
♦ TUFTED HAIRGRASS

DESCRIPTION Densely clumped, native perennial grass. Stems stiff, erect, 3–10 dm long. Leaves mostly from base of plant, usually shorter than head, flat or inrolled, 2–4 mm wide; sheaths smooth; ligule white, translucent, 3–10 mm long. Head a narrow to open panicle, 1–4 dm long,

MATT LAVIN

Cinna latifolia

the panicle branches threadlike, upright to spreading, the lower branches in groups of 2–5, flowers mostly near branch tips. Spikelets 2-flowered, purple-tinged, fading to silver with age, 2–5 mm long, breaking above the glumes; glumes shiny, 2–5 mm long, the first glume slightly shorter than second glume; lemma smooth, 2–4-toothed across the flat tip, awned from near base on back, the awn shorter to about as long as lemma. Flowering June–July.
HABITAT wet meadows, streambanks, shores, calcium-rich seeps and springs, rocky shores of Great Lakes.
STATUS MW-FACW | NCNE-FACW | GP-FACW

Dichanthelium PANICGRASS

Perennial grasses, tufted or sometimes rhizomatous, sometimes with hard, corm-like bases. Stems hollow, usually erect or ascending, sometimes decumbent in the fall, usually branching from the lower stem nodes in summer and fall, terminating in small panicles that are usually partly included in the sheaths. Basal rosettes of winter leaves sometimes present. Stem leaves usually markedly longer and narrower than the rosette blades; ligules of hairs, membranous, or membranous and ciliate, sometimes absent. Flowers in terminal panicles (vernal) developing late spring to early summer, and sometimes lateral panicles (autumnal) in late-summer or fall; disarticulation below the glumes.

    Dichanthelium is often included in genus *Panicum*, the two genera being similar in form. However, molecular data re-inforce the separation of *Dichanthelium* as a distinct genus.

Dichanthelium PANICGRASS
1 Widest leaf blades over 15 mm wide......
    ....*D. clandestinum*
1 Widest leaf blades less than 15 mm wide **2**
2 Spikelets 2–3 mm long; ligule absent or a ring of hairs to 1 mm long .....*D. boreale*
5 Spikelets less than 2 mm long; ligule a ring of hairs 2–3 mm long ....*D. acuminatum*

Dichanthelium acuminatum (Sw.)

Gould & C.A. Clark   ♦ SAND PANIC-GRASS
DESCRIPTION Loosely clumped, native perennial grass. Stems erect, smooth, 3–8 dm long. Leaves firm, upright, 5–10 cm long and 3–5 mm wide, fringed with sparse hairs at base; sheaths ± smooth; ligule a fringe of hairs 2–3 mm long. Head a narrow panicle, 8–12 cm long. Spikelets oval, hairy, 1–2 mm long, with 1 fertile flower; first glume short, about 0.5 mm long; second glume and sterile lemma nearly equal to fruit. Flowering June–Aug.
SYNONYM *Dichanthelium spretum, Panicum acuminatum, Panicum spretum.*
HABITAT moist to wet sandy shores and flats.
STATUS MW-FAC | NCNE-FAC | GP-FAC

Deschampsia cespitosa

Dichanthelium acuminatum

Dichanthelium boreale (Nash) Freckmann NORTHERN PANIC-GRASS

DESCRIPTION Native perennial grass, in small clumps. Stems upright, 2–6 dm long. Leaves upright to spreading, 5–20 cm long and 1–2 cm wide, smooth or sometimes hairy on underside, base of leaf often fringed with hairs; sheaths hairy; ligule absent or a fringe of short hairs. Head an open panicle, 5–12 cm long, the branches spreading or upright. Spikelets oval in outline, finely hairy, about 2 mm long, on long stalks with 1 fertile flower; first glume to half as long as second glume; second glume and lemma purple-tinged, about equal, and as long as fruit. Flowering June–Aug.

SYNONYM *Panicum boreale.*

HABITAT wet prairies, tamarack bogs.

STATUS MW-FAC | NCNE-FAC | GP-FACU

Dichanthelium clandestinum (L.) Gould. DEER-TONGUE GRASS

DESCRIPTION Clumped native perennial grass, may form large colonies. Stems stout, erect, 6–15 dm long, hairy at least at nodes. Leaves flat, spreading, smooth to hairy, 5–20 cm long and 1–3 cm wide, often fringed with hairs at base; sheaths hairy to smooth. Head an open panicle, 5–15 cm long, the branches spreading or upright.

Spikelets oblong, finely hairy, about 3 mm long, with 1 fertile flower; first glume to half as long as second glume; second glume shorter than sterile lemma and fruit. Flowering June–July.

SYNONYM *Panicum clandestinum.*

HABITAT floodplain forests, alder thickets, ditches; especially where sandy.

STATUS MW-FACW | NCNE-FACW | GP-FAC

Distichlis SALTGRASS

Distichlis spicata (L.) Greene
♦INLAND SALTGRASS

DESCRIPTION Short perennial grass, spreading by scaly rhizomes and forming patches; the staminate and pis-tillate flowers on separate plants. Stems stiff, erect, 1–3 dm long. Leaves upright, the upper often longer than the head, mostly inrolled, 5–10 cm long and 0.5–3 mm wide, smooth or with sparse hairs; sheaths overlapping, smooth or sparsely hairy, usually long-hairy at collar; ligule small. Head an unbranched, narrow, spikelike panicle, 3–7 cm long. Spikelets several to many, upright, 8–20 mm long; staminate spikelets straw-colored, pistillate spikelets green-gray, breaking above the glumes; glumes unequal, 1–5 mm long; lemmas ovate, 3–6 mm long. Flowering June–Sept.

SYNONYM *Distichlis stricta.*

HABITAT seasonally wet, brackish flats, shores and disturbed areas.

STATUS MW-FACW | NCNE-FACW | GP-FACW; considered introduced in Wisc; main species range Great Plains and w USA.

Echinochloa
BARNYARD-GRASS

Large, weedy, annual grasses. Stems single or several together, erect to ± horizontal, to 1 m or more long. Leaves flat, wide and smooth; sheaths smooth or hairy; ligules absent. Head a dense panicle, the branches crowded with spikelets forming racemes or spikes. Spikelets with 1 terminal fertile floret and 1 sterile floret, breaking below the glumes, nearly stalkless; glumes unequal, the first glume 3-veined, to half the length of second glume, the second glume 5-veined; sterile lemma similar to second glume, awned or awnless; fertile lemma smooth and shiny.

Distichlis spicata

Echinochloa BARNYARD-GRASS

1 Lower leaf sheaths rough-hairy; spikelets each with 2 awns; southern portion of region . . . . *E. walteri*
1 Leaf sheaths smooth; spikelets with usually 1 awn (from sterile lemma); widespread in region. 2
2 Fertile lemma rounded or broadly tapered to a thin, membranous, withered beak . . . . *E. crus-galli*
2 Fertile lemma tapered to a stiff, persistent beak . . . . *E. muricata*

Echinochloa crus-galli (L.) P. Beauv.
BARNYARD-GRASS

DESCRIPTION Weedy, introduced annual grass. **Stems** 1 m or more long. **Leaves** 7–30 mm wide; sheaths smooth; ligule absent. **Head** an erect, green to purple panicle, 1–2.5 dm long; panicle branches spreading to erect, long-hairy, some of the hairs as long or longer than spikelets (excluding spikelet awns). **Spikelets** 3–5 mm long (excluding awns); glumes awnless; sterile lemma awnless or with an awn to 4 cm or more long; tip of fertile lemma firm, shiny, rounded or broadly tapered to a point, the beak usually green and withered, the lemma body and beak separated by a line of tiny hairs. **Flowering** July–Sept.
HABITAT shores, wet meadows, ditches, streambanks, mud flats, moist disturbed areas.
STATUS MW-FACW | NCNE-FAC | GP-FAC; introduced from Europe, naturalized throughout much of s Canada and USA.

Echinochloa muricata (P. Beauv.)
Fernald ♦BARNYARD-GRASS

DESCRIPTION Weedy, native annual grass. **Stems** 1 m or more long. **Leaves** 5–30 mm wide; sheaths smooth; ligule absent. **head** a green to purple panicle, sometimes strongly purple, 1–3 dm long, panicle branches spreading, hairs on branches absent or to 3 mm long and shorter than spikelets. **Spikelets** 2–4 mm long (excluding awns); glumes awnless; sterile lemma awnless or with an awn 5–10 mm long; tip of fertile lemma firm, shiny, gradually tapered

to the stiff beak, the lemma body and beak not separated by a line of tiny hairs (the beak itself often short-hairy). **Flowering** July–Sept.
SYNONYM *Echinochloa pungens.*
HABITAT shores, streambanks and ditches, where sometimes in shallow water.
STATUS MW-OBL | NCNE-OBL | GP-FACW

Echinochloa walteri (Pursh) Heller
SALTMARSH COCKSPUR GRASS

DESCRIPTION Tall, native annual grass. **Stems** usually erect, 1–2 m long. **Leaves** 10–25 mm wide; lower sheaths usually rough-hairy; ligule absent. **Head** a dense panicle, often nodding, 1–3 dm long. **Spikelets** ± hidden by awns, the awns 1–3 cm long from sterile lemmas and 2–10 mm long from second glume; fertile lemma oval, with a small, withering tip, but not separated by a line of hairs as in *Echinochloa crusgalli.* Aug–Sept.
HABITAT streambanks, lakeshores, ditches; locally common in marshes along Lake Erie.
STATUS MW-OBL | NCNE-OBL | GP-OBL

Elymus WILDRYE

Clumped perennial grasses, mostly of upland habitats. Leaves flat; ligules short. Head a densely flowered spike. Spikelets usually 2 at each node of spike, breaking above glumes (*Elymus riparius*) or below

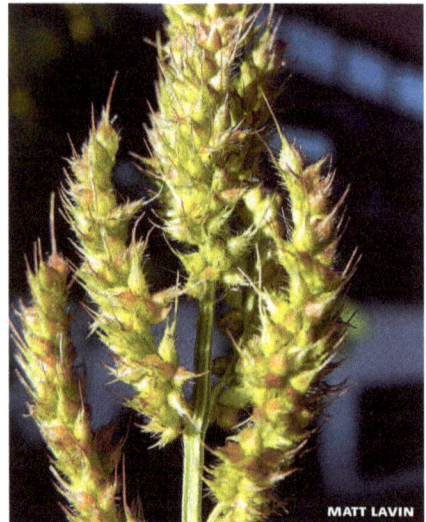

Echinochloa muricata

glumes (*E. virginicus*); glumes narrow and awnlike; lemmas tipped with a long awn; stamens 3.

Elymus **WILDRYE**

1 Spikes somewhat nodding; lemma awned, the awn longer than 1 cm; glumes not bowed-out at base .... *E. riparius*

1 Spikes erect; lemma unawned, or with a short awn to 1 cm long; glumes bowed-out at base .... *E. virginicus*

Elymus riparius Wiegand
STREAMBANK WILDRYE

DESCRIPTION Clumped, native perennial grass. **Stems** 1 m long or more. **Leaves** 8–10 along stem, 5–15 mm wide, upper surface smooth to rough; sheaths smooth; ligule short. **Head** a spike, 6–20 cm long, somewhat nodding. **Spikelets** mostly 2 at each node, 2–4-flowered, finely hairy, breaking above glumes; glumes narrow, to 1 mm wide at middle, not bowed-out at base; lemma finely hairy to smooth, tipped with a straight awn 2–3 cm long.

HABITAT streambanks, floodplain forests.

STATUS MW-FACW | NCNE-FACW | GP-FAC

NOTE similar to **Canada wild rye** (*Elymus canadensis*), a common species of drier, sandy places, but awns in streambank wildrye are straight rather than bent and curved.

Elymus virginicus L.
◆VIRGINIA WILDRYE

DESCRIPTION Clumped, native perennial grass. **Stems** 6–12 dm long. **Leaves** flat, 5–15 mm wide, rough-to-touch on both sides; sheaths smooth. **Head** an erect spike, 5–15 cm long, the base of spike often covered by top of upper sheath. **Spikelets** usually 2 at each node, 2–4-flowered, breaking below glumes; glumes firm, 1–2 mm wide, yellowish, bowed-out at base, tapered to a straight awn about 1 cm long; lemmas 6–9 mm long, smooth to hairy, usually with a straight awn to 3 cm long. **Flowering** July–Aug.

HABITAT floodplain forests, thickets, streambanks.

STATUS MW-FACW | NCNE-FACW | GP-FAC

Eragrostis LOVEGRASS

Annual grasses (those included here), perfect-flowered or with staminate and pistillate flowers on different plants. Stems clumped, or spreading and rooting at lower nodes and with creeping stolons. Leaves with short, flat to folded blades; sheaths hairy near top; ligule a ring of short hairs. Heads usually many, in an open or narrow panicle. Spikelets few- to many-flowered, breaking above glumes, laterally compressed, the florets overlapping; glumes unequal; lemmas 3-veined; palea shorter than lemma, 2-veined.

Eragrostis **LOVEGRASS**

1 Plants mat-forming; base of stems lying along ground, rooting at lower nodes, the nodes bearded with hairs; spikelets 3–10 mm long .... *E. hypnoides*

1 Plants not mat-forming; stems erect, not rooting at lower nodes, the nodes smooth; spikelets to 3 mm long .... *E. frankii*

Eragrostis frankii C. A. Meyer
SANDBAR LOVEGRASS

DESCRIPTION Densely clumped, native annual grass. **Stems** branched, 1–5 dm long. **Leaves** smooth, 1–4 mm wide; sheaths smooth but long-hairy at top; ligule short-hairy. **Head** an open panicle, 5–20 cm

MATT LAVIN

Elymus virginicus

long, the branches mostly ascending. **Spikelets** 3–6-flowered, 2–3 mm long and 1–2 mm wide. **Flowering** Aug–Sept.

HABITAT wet, muddy areas, streambanks, sandbars, roadside ditches, cultivated fields. **STATUS** MW-FACW | NCNE-FACW | GP-FACW

Eragrostis hypnoides (Lam.) BSP.
◆CREEPING LOVEGRASS

DESCRIPTION Mat-forming, native annual grass. **Stems** mostly spreading and rooting at lower nodes, 5–15 cm long, smooth but short-hairy at nodes. **Leaves** flat to folded, 1–5 cm long and 1–3 mm wide, upper surface hairy; sheaths smooth except for hairs at top and sometimes along margins; ligule of short hairs about 0.5 mm long. **Head** a loose panicle, 2–6 cm long. **Spikelets** 10–35-flowered, linear, 3–10 mm long; glumes 1-veined, 0.5–1.5 mm long; lemma smooth and shiny, 1–2 mm long. **Flowering** July–Sept.

HABITAT wet, sandy or muddy shores and streambanks, sand bars, mud flats. **STATUS** MW-OBL | NCNE-OBL | GP-OBL **range** Minn, Wisc, Mich.

Glyceria MANNAGRASS

Perennial grasses, loosely clumped or spreading by rhizomes. Stems upright, or reclining at base and often rooting at lower nodes. Leaves flat; sheaths tubular, the margins mostly closed. Head an open panicle. Spikelets 3-flowered, ovate to linear, ± round in section or somewhat flattened, breaking above the glumes; glumes unequal, shorter than lemmas, 1-veined; lemmas unawned, usually 7-veined; palea about as long as lemma; stamens 3 or 2.

ADDITIONAL SPECIES Reed Mannagrass [*Glyceria maxima* (Hartman) Holmb.], native to Eurasia and an aggressive invader of wetlands, was first discovered in the USA in Racine County in 1975 in Wisc, the species is now known from several additional locations. Spreading by rhizomes, the grass may form large patches and grow to 2.5 m tall. It is similar to large plants of *G. grandis,* but differs in its firmer, larger, and more prow-tipped lemmas as well as its usually larger anthers.

Glyceria **MANNAGRASS**

1   Spikelets linear-cylindric, 10 mm long or longer.................................2
1   Spikelets ovate, 2 to 7 mm long 3
2   Leaves less than 5 mm wide; lemmas ± smooth ....*G. borealis*
2   Leaves 5 mm or more wide; lemmas finely hairy.................*G. septentrionalis*
3   Spikelets 3–4 mm wide; veins of lemma not raised ....*G. canadensis*
3   Spikelets 2–2.5 mm wide; veins of lemma raised.................................4
4   Spikelets 4–7 mm long ....*G. grandis*
4   Spikelets 2–4 mm long ....*G. striata*

Glyceria borealis (Nash) Batchelder
NORTHERN MANNAGRASS

DESCRIPTION Native perennial grass. **Stems** erect or reclining at base, often rooting from lower nodes, 6–12 dm long. **Leaves** flat or folded, 2–5 mm wide, smooth; sheaths smooth; ligule 3–10 mm long. **Head** a panicle, 2–4 dm long, with stiff, erect to ascending, branches to 8–12 cm long, each with several spikelets. **Spikelets** linear, mostly 6–12-flowered, 1–1.5 cm long; glumes rounded at tip, 2–3 mm long; lemmas 3–4 mm long, 7-veined. **Flow-**

Eragrostis hypnoides

ering June–Aug.

HABITAT marshes, ponds, stream, ditches, often in shallow water or mud.

STATUS MW-OBL | NCNE-OBL | GP-OBL

Glyceria canadensis (Michx.) Trin.
⧫RATTLESNAKE-MANNAGRASS

DESCRIPTION Native perennial grass. Stems single or few together, erect, 6–15 dm long. Leaves 3–7 mm wide, upper surface rough; ligule 2–5 mm long. Head an open panicle, 1–3 dm long, the branches drooping, with spikelets mostly near tips. Spikelets ovate, 5–10-flowered, 5–7 mm long, the florets spreading; glumes 2–3 mm long, the first glume lance-shaped, the second glume ovate; lemma veins not raised.

HABITAT marshes, swamps, thickets, open bogs, fens.

STATUS MW-OBL | NCNE-OBL | GP-OBL

Glyceria grandis S. Wats.
⧫AMERICAN MANNAGRASS

DESCRIPTION Loosely clumped, native perennial grass. Stems erect, stout, 1–1.5 m long and

4–6 mm wide. Leaves flat, smooth, 6–12 mm wide; sheaths smooth; the ligule translucent, 3–6 mm long. head a large, open, much-branched panicle, 2–4 dm long, usually nodding at tip, branches lax and drooping when mature. Spikelets ovate, purple, slightly flattened, 5–9-flowered, 4–7 mm long; glumes pale or white, 1–3 mm long; lemmas purple, 2–3 mm long. Flowering June–Sept.

HABITAT marshes, ditches, streams, lakes and ponds, open bogs, fens; usually in shallow water or mud.

STATUS MW-OBL | NCNE-OBL | GP-OBL

Glyceria septentrionalis A. Hitchc.
EASTERN MANNAGRASS

DESCRIPTION Native perennial grass. Stems somewhat fleshy, often ± horizontal at base and rooting from lower nodes, 1–1.5 m long. Leaves 6–10 mm wide; sheaths smooth; ligule large. Head a narrow panicle, 2–4 dm long, the branches to 10 cm long, each with several spikelets. Spikelets 1–2 cm long, 8–14-flowered; glumes 2–4 mm long; lemmas green or pale, 4–5 mm long, spreading when mature; palea often longer than lemma. Flowering June–Aug.

HABITAT swamps, thickets, shallow water of pond margins, wet depressions in forests.

STATUS MW-OBL | NCNE-OBL | GP-OBL

Glyceria striata (Lam.) A. Hitchc.
⧫FOWL-MANNAGRASS

DESCRIPTION Loosely clumped, native perennial grass; plants pale green. Stems erect, slender, 3–10 dm long. Leaves flat or folded, smooth, 2–6 mm wide; sheaths smooth; ligule 1–3 mm long. Head an open, loose panicle, 1–2 dm long, the branches lax, drooping. Spikelets ovate, often purple, 3–7-flowered, 3–4 mm long; glumes 0.5–1.5 mm long; lemma 2 mm long, strongly 7-veined. Flowering June–Aug.

HABITAT swamps, thickets, low areas in forests, wet meadows, springs, streambanks.

STATUS MW-OBL | NCNE-OBL | GP-OBL

MATT LAVIN

Glyceria grandis

Graphephorum FALSE-OAT

Graphephorum melicoides (Michx.)
Desv. PURPLE FALSE-OAT

DESCRIPTION Native, clumped perennial grass. **Stems** smooth or finely hairy, 4–9 dm long. **Leaves** flat, 3–6 mm wide, sparsely long-hairy; sheaths smooth or hairy; ligule membranous, ragged at tip. **Head** a slender, nodding panicle, 10–20 cm long, the branches upright to drooping, to 6 cm long, the spikelets mostly above middle of branch. **Spikelets** 2-flowered, 6–7 mm long, finely hairy; glumes somewhat unequal, 4–7 mm long, the first glume 1-veined, the second glume 3-veined; lemma unawned; stalk within spikelet (rachilla) and base of lemma white-hairy.

SYNONYM *Trisetum melicoides.*

HABITAT Cedar swamps, mixed forests, ridge and swale ecosystems near Lake Michigan, shoreline dolomitic sites, and seepage areas on shoreline bluffs.

STATUS MW-FACW | NCNE-FACW; Wisc (END).

NOTE sometimes retained in genus *Trisetum.*

Hordeum BARLEY

Hordeum jubatum L.
◆FOXTAIL BARLEY

DESCRIPTION Clumped perennial grass; plants smooth to densely hairy. **Stems** erect or reclining at base, 2–7 dm long. **Leaves** usually flat, 2–5 mm wide; ligule less than 1 mm long. **Head** a terminal spike, erect to nodding, 3–10 cm long, appearing bristly due to the long, spreading awns from glumes and lemmas. **Spikelets** 1-flowered, 3 at each node, the center spikelet fertile, stalkless, the 2 lateral spikelets sterile, short-stalked, reduced to 1–3 spreading awns; the 3 spikelets at each node falling as a unit; glumes of fertile spikelet awnlike; lemma lance-shaped, tipped by a long awn; the glume and lemma awns 2–7 cm long. **Flowering** June–Sept.

HABITAT wet meadows, ditches, shores, shallow marshes, disturbed areas; often where brackish.

STATUS MW-FAC | NCNE-FAC | GP-FACW; considered introduced from w USA in Mich.

Leersia CUT-GRASS

Perennial grasses, spreading by long rhizomes. Stems slender, somewhat weak. Leaves flat, smooth to hairy or rough-to-touch; ligules membranous, short. Head an open panicle. Spikelets 1-flowered, laterally compressed, falling as a unit from the stalk; glumes absent; lemmas smooth to bristly hairy, 5-veined; palea narrow, about as long as lemma; stamens 2–3 (in Upper Midwest species).

Leersia CUT-GRASS

1 Spikelets ovate, 3–4 mm wide............
........................*L. lenticularis*
1 Spikelets linear, 1–2 mm wide...........2
2 Stems round in section; leaves very rough-to-touch; spikelets 4–6 mm long ....
..........................*L. oryzoides*
2 Stems flattened in section; leaves smooth or finely roughened; spikelets to 3.5 mm long ....*L. virginica*

Glyceria striata

Hordeum jubatum

Leersia lenticularis Michx.
CATCHFLY GRASS

DESCRIPTION Native perennial grass, from creeping rhizomes. Stems 1–1.5 m long. Leaves lax, smooth to soft-hairy, 1–2 cm wide; sheaths smooth, or hairy at top; ligule flat-topped, 1 mm long. Head a panicle, 1–2 dm long, often drooping, the branches spreading, each branch with 1–4 spikelike racemes 1–2 cm long. Spikelets 1-flowered, pale, flat, nearly round in section, 4–5 mm long, short-stalked, closely overlapping one another; glumes absent; lemma 4–6 mm long, the veins and keel fringed with bristly hairs.

HABITAT river floodplains.

STATUS MW-OBL | NCNE-OBL | GP-OBL

Leersia oryzoides (L.) Swartz
◆ RICE CUT-GRASS

DESCRIPTION Loosely clumped, native perennial grass, from creeping rhizomes. Stems weak and sprawling, rooting at nodes, 1–1.5 m long. Leaves flat, 2–3 dm long and 5–10 mm wide, rough-to-touch, the margins fringed with short spines; sheaths rough-hairy; ligule flat-topped, 1 mm long. Head an open panicle at end of stem and from leaf axils (these often partly enclosed by leaf sheaths), 1–2 dm long, the branches ascending to spreading. Spikelets

STEFAN LEFNAER

Leersia oryzoides

1-flowered, oval, 5 mm long and 1–2 mm wide, compressed, pale green, turning brown with age; glumes absent; lemma covered with bristly hairs. Flowering July–Sept.

HABITAT muddy or sandy streambanks, shores, swales and marshes; sometimes forming large patches.

STATUS MW-OBL | NCNE-OBL | GP-OBL

Leersia virginica Willd.
WHITE GRASS

DESCRIPTION Native perennial grass, spreading by rhizomes. Stems slender and weak, often ± horizontal at base and rooting at nodes, 5–12 dm long. Leaves rough-hairy, especially along margins, 5–20 cm long and 5–15 mm wide; sheaths smooth or finely hairy; ligule short, flat-topped. Head an open panicle, 1–2 dm long, the branches separated along the rachis, stiffly spreading, the spikelets from middle to tip of branches. Spikelets oblong, barely overlapping one another, 3 mm long and 1 mm wide, sparsely hairy; glumes absent; lemma 3–4 mm long, the keel and margins sparsely hairy. Flowering July–Sept.

HABITAT swamps, floodplain forests, shaded forest depressions, streambanks.

STATUS MW-FACW | NCNE-FACW | GP-FACW

Leptochloa SPRANGLETOP

Leptochloa fascicularis (Lam.) A. Gray
SPRANGLETOP

DESCRIPTION Clumped, annual grass. Stems erect to spreading, branched from base, 2–10 dm long, somewhat fleshy. Leaves flat to loosely inrolled, 1–3 mm wide, finely rough-to-touch; sheaths ± smooth, often purple, the upper sheath often partly sheathing the head; ligule 3–5 mm long. Head a ± cylindric panicle, 5–20 cm long and 2–5 cm wide, composed of several to many branches, the branches upright and bearing spikelets in racemes. Spikelets 6–12-flowered, 5–10 mm long, breaking above the glumes; glumes unequal, lance-shaped, 1-veined, the first glume 2–4 mm long, the second glume 4–5 mm long; lemma 4–5 mm long, 3-veined, tipped with an awn 4–5 mm long; palea about as long as lemma. Flow-

ering July–Sept.

SYNONYM *Diplachne acuminata, Diplachne fusca, Leptochloa fusca.*

HABITAT shores, streambanks, muddy or sandy flats, usually where flooded part of year, often where brackish.

STATUS MW-OBL | NCNE-OBL | GP-FACW; considered introduced in Wisc and Mich.

Muhlenbergia MUHLY

Perennial grasses, clumped or with creeping rhizomes. Stems erect or reclining at base, often branching from base. Leaves smooth to hairy, ligules membranous. Head a panicle, usually narrow and spikelike, sometimes open and spreading, at ends of stems and sometimes also from leaf axils. Spikelets 1-flowered, breaking above glumes; glumes usually nearly equal in length, 1-veined, the tip often awned; lemma lance-shaped, 3-veined, sometimes awned, some species with long, soft hairs at lemma base; palea about as long as lemma.

Muhlenbergia MUHLY

1　Panicle open and loose, 4 cm wide or more
　　. . . . .**2**
1　Panicle slender and densely flowered, less than 2.5 cm wide **3**
2　Plants spreading by rhizomes. . . . .
　　. . . . .*M. asperifolia*
2　Plants clumped, rhizomes absent . . . .
　　. . . . .*M. uniflora*
3　Leaf blades usually inrolled, to 1 mm wide; panicles few-flowered, the heads not round in outline . . . .*M. richardsonis*
3　Leaf blades flat, 2–7 mm wide; panicles usually densely flowered, the heads ± round in outline **4**
4　Stems smooth and shiny . . . .*M. frondosa*
4　Stems dull, finely hairy (at least below the nodes) **5**
5　Panicle stiffly erect, 5–10 mm wide; glumes longer than lemma . . . .*M. glomerata*
5　Panicle bent or nodding, less than 5 mm wide; glumes shorter or equal to lemma . **6**
6　Panicle silvery green; lemmas with awn 5 mm or more long . . . .*M. sylvatica*
6　Panicle pale green or purple-tinged; lemmas unawned or with short awn less than 5 mm long. . . . .*M. mexicana*

Muhlenbergia asperifolia (Nees & Meyen) Parodi ALKALI MUHLY

DESCRIPTION Perennial grass, from slender, scaly rhizomes. Stems 1–5 dm long, becoming ± horizontal near base, rooting and branching from lower nodes, the branches spreading, waxy. Leaves upright, flat, 2–6 cm long and 1–3 mm wide, rough-to-touch; sheaths smooth; ligule ragged, to 0.5 mm long. Head an open panicle, 5–15 cm long, the branches threadlike, widely spreading. Spikelets 1-flowered (sometimes 2-flowered), single on the branches, purple or dark gray, 1–2 mm long; glumes nearly equal, half to nearly as long as spikelet; lemma unawned, 1–2 mm long. Flowering July–Sept.

HABITAT wet meadows, seeps, shores and mudflats, often where brackish.

STATUS MW-FACW | NCNE-FACW | GP-FACW range Minn, Wisc, se Mich; considered introduced from w USA in Wisc and Mich.

Muhlenbergia frondosa (Poiret) Fern. ♦WIRESTEM MUHLY

DESCRIPTION Native perennial grass, from stout, scaly rhizomes. Stems 4–10 dm long, unbranched and erect when young, becoming branched and sprawling with age, smooth and shiny between nodes. Leaves lax, smooth, 3–10 cm long and 2–6 mm wide; ligule fringed, 1–2 mm long. Head a narrow panicle, to 10 cm long,

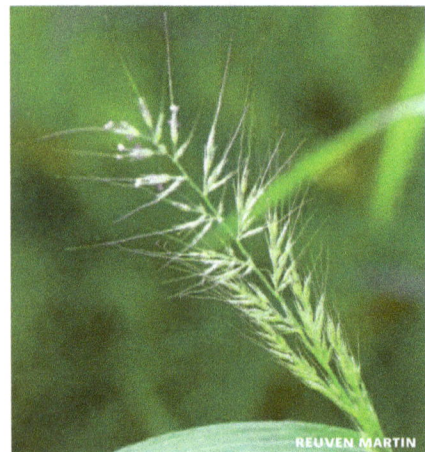

Muhlenbergia frondosa

from ends of stems and leaf axils (where partly enclosed by sheaths), the branches erect to spreading, with spikelets from near base to tip. Spikelets 1-flowered; glumes 2–3 mm long, tipped with a short awn; lemma 3–4 mm long, usually with an awn to 1 cm long, short-hairy at base. Aug–Sept.

HABITAT floodplain forests, streambanks, thickets, shores; also somewhat weedy in disturbed areas such as along railroads.

STATUS MW-FACW | NCNE-FACW | GP-FACW

Muhlenbergia glomerata (Willd.)

Trin. ♦ MARSH-MUHLY

DESCRIPTION Native perennial grass, spreading from rhizomes. Stems upright, 3–9 dm long, sometimes with a few branches from base, dull and finely hairy between nodes. Leaves flat, lax, 5–15 cm long and 2–6 mm wide; sheaths smooth; ligule fringed, to 0.5 mm long. Head a narrow, crowded, cylindric panicle, 2–10 cm long and 5–10 mm wide, the lower clusters of spikelets often separate from one another. Spikelets 1-flowered, often purple-tinged, 5–6 mm long; glumes nearly equal, longer than the floret, tipped with an awn 1–5 mm long; lemma lance-shaped, 2–3 mm long, with long, soft hairs at base. Aug–Sept.

HABITAT swamps, wet meadows, marshes, springs, open bogs, fens, calcareous shores.

STATUS MW-FACW | NCNE-OBL | GP-FACW

Muhlenbergia mexicana (L.) Trin.

WIRESTEM MUHLY

DESCRIPTION Native perennial grass, from scaly rhizomes. Stems

upright, 2–8 dm long, sometimes branched from base; dull and finely hairy between nodes. Leaves flat, lax, 5–20 cm long and 2–5 mm wide; sheaths smooth; ligule entire to fringed, to 1 mm long. Head a narrow, densely flowered panicle, 5–15 cm long and 2–10 mm wide, from ends of stems and leafy branches. Spikelets 1-flowered, green or purple, 2–3 mm long; glumes nearly equal, lance-shaped, 3–4 mm long, about as long as floret, tipped with a short awn about 1 mm long; lemma lance-shaped, 2–3 mm long, unawned or with an awn to 7 mm long. Aug–Sept.

SYNONYM *Muhlenbergia foliosa.*

HABITAT swamps, floodplain forests, thickets, wet meadows, marshes, springs, fens and streambanks.

STATUS MW-FACW | NCNE-FACW | GP-FACW

Muhlenbergia richardsonis (Trin.)

Rydb. ♦ MAT-MUHLY

DESCRIPTION Loosely clumped, native perennial grass, rooting from lower nodes and forming mats. Stems very slender, erect or ± horizontal at base, 2–6 dm long. Leaves upright, usually inrolled, 1–5 cm long and 1–2 mm wide; sheaths smooth; ligule 2–3 mm long. Head a narrow panicle, 2–8 cm long. Spikelets 1-flowered, uncrowded, green or gray-green, 2–3 mm long; glumes nearly equal, ovate, to half as long as floret; lemma lance-shaped, smooth,

Muhlenbergia glomerata

Muhlenbergia richardsonis

2–3 mm long tipped with a short point. Flowering July–Sept.

HABITAT low prairie, wet meadows, marshes and seeps; often where brackish.

STATUS MW-FAC | NCNE-FACW | GP-FAC; Wisc (END), Mich (THR).

Muhlenbergia sylvatica Torr.
FOREST MUHLY

DESCRIPTION Native perennial grass, spreading by rhizomes. Stems erect, or sprawling when old, 4–10 dm long, coarse-hairy between nodes. Leaves flat, lax, upright to spreading, 5–15 cm long and 2–6 mm wide; sheaths smooth; ligule fringed, 1–3 mm long. Head a slender panicle, often nodding, 5–20 cm long and 2–7 mm wide. Spikelets 1-flowered, 2–4 mm long, at ends of stalks about 3 mm long; glumes nearly equal, sharp-tipped, shorter than lemma; lemma 2–4 mm long, short hairy at base, tipped with an awn 5–15 mm long. Aug–Sept.

HABITAT streambanks, shaded wet areas.

STATUS MW-FACW | NCNE-FACW | GP-FACW

Muhlenbergia uniflora (Muhl.) Fern.
♦BOG MUHLY

DESCRIPTION Clumped, native perennial grass. Stems very slender, 2–4 dm long, often ± horizontal and rooting at base. Leaves flat, crowded near base of plant, 5–10 cm long and to 1 mm wide; sheaths ± smooth, compressed; ligule ragged, about 1 mm long. Head a loose, open panicle, 7–20 cm long and 2–4 cm wide, the branches threadlike. Spikelets 1-flowered (rarely 2-

DOUG MCGRADY

Muhlenbergia uniflora

flowered), oval, purple-tinged, 1–2 mm long; glumes about equal, ovate, to half the length of spikelet; lemma 1–2 mm long, unawned.

HABITAT wetland margins, open sandy shores.

STATUS MW-OBL | NCNE-OBL

Panicum PANIC-GRASS

Annual or perennial grasses. Heads narrow to open panicles (ours). Spikelets small, with 1 fertile flower; glumes usually unequal, the first glume membranous, usually very small, second glume green, about as long as spikelet; sterile lemma similar to second glume, enclosing the palea and sometimes a staminate flower, fertile lemma whitish, smooth.

Panicum PANIC-GRASS

1 Sheaths hairy .... *P. flexile*
1 Sheaths ± smooth...................... 2
2 Spikelets smooth, nerves evident .... *P. rigidulum*
2 Spikelets covered with bumps, nerves ± absent. .... *P. verrucosum*

Panicum flexile (Gattinger) Scribn.
WIRY WITCH-GRASS

DESCRIPTION Slender native annual grass. Stems erect, 2–7 dm long, branched from base, hairy at nodes. Leaves erect, smooth or sparsely hairy, 10–30 cm long and 2–6 mm wide. Head a narrow panicle, 10–20 cm long and about a third as wide, the branches threadlike, upright to spreading. Spikelets lance-shaped, 3–4 mm long, with 1 fertile flower; first glume about half as long as second glume and sterile lemma. Flowering Aug–Sept.

HABITAT sandy and gravelly shores, marshes; often where calcium-rich.

STATUS MW-FACW | NCNE-FACW | GP-FACW

Panicum rigidulum Nees
MUNRO GRASS

DESCRIPTION Densely clumped, native perennial grass. Stems 5–15 dm long. Leaves crowded near base, upright, 2–4 dm long and 5–10 mm wide, sometimes longer than the panicle, margins finely rough-hairy; sheaths smooth, flattened; ligule mem-

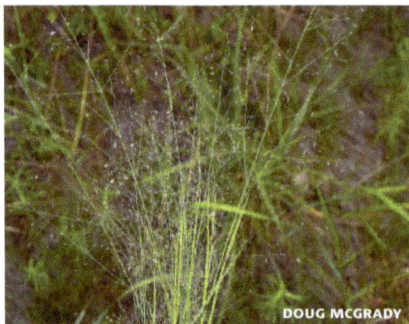

branous, ragged. Head a narrow to open panicle, 1–3 dm long, from ends of stems and leaf axils, the spikelet-bearing branches mostly on upper sides of panicle branches. Spikelets lance-shaped, green to purple, 2–3 mm long, with 1 fertile flower; glumes and sterile lemma pointed at tip, with conspicuous veins; first glume about half as long as spikelet; second glume and sterile lemma about equal, longer than the fruit. Flowering July–Aug.

SYNONYMS *Coleataenia rigidula, Panicum agrostoides.*

HABITAT pond margins, streambanks, ditches and swales.

STATUS MW-FACW | NCNE-FACW | GP-FACW

Panicum verrucosum Muhl.
WARTY PANIC-GRASS

DESCRIPTION Native annual grass; plants bright green. Stems erect or spreading, single or few together, smooth, 3–10 dm long. Leaves flat, lax, 5–20 cm long and 5–10 mm wide, fringed with sparse hairs at base; sheaths ± smooth; ligule a fringe of hairs 2–3 mm long. Head an open, spreading panicle, 5–25 cm long, often with smaller panicles from leaf axils. Spikelets with 1 fertile flower, obovate, about 2 mm long, covered with small bumps; first glume triangular, less than 1 mm long; second glume and sterile lemma nearly equal to fruit. Flowering Aug–Sept.

SYNONYM *Kellochloa verrucosa.*

HABITAT moist sandy shores.

STATUS MW-FACW | NCNE-FACW | GP-FACW; Mich (THR).

Phalaris CANARY-GRASS

Phalaris arundinacea L.
◆REED CANARY-GRASS

DESCRIPTION Tall perennial grass, mostly introduced; spreading by rhizomes and typically forming large, dense colonies. Stems stout, smooth, 5–15 dm long. Leaves flat, smooth, 1–2 dm long and 1–2 cm wide; sheaths smooth; ligule membranous, 3–8 mm long. Head a narrow, densely flowered panicle, 5–25 cm long, often purple-tinged, becoming straw-colored

with age, the branches short and upright to ascending. Spikelets 4–6 mm long, breaking above glumes, with 1 fertile flower and 2 small sterile lemmas below; glumes nearly equal, longer than fertile floret, lance-shaped, tapered to tip or short-awned, 3-veined; fertile lemma ovate, 3 mm long, shiny; palea as long as lemma. Flowering June–July.

HABITAT wet meadows, shallow marshes, ditches, shores and streambanks; common to abundant.

STATUS MW-FACW | NCNE-FACW | GP-FACW

NOTE Reed canary-grass is an aggressive, competitive wetland species, to the detriment of other plants. Most populations are probably non-native strains, originally introduced from Eurasia as a pasture plant and now widely naturalized.

Phragmites COMMON REED

Phragmites australis (Cav.) Trin.
◆COMMON REED

DESCRIPTION Tall, stout perennial reed, from deep, scaly rhizomes, or the rhizomes sometimes exposed and creeping over the soil; often forming large colonies. Stems erect, hollow, 2–4 m long and 5–15 mm wide near base, the internodes often purple. Leaves flat, long, 1–3 cm wide; sheaths open; ligule white, 1 mm long. Head a large, plumelike panicle, purple when young, turning yellow-brown with age, 15–

Phalaris arundinacea

40 cm long, much-branched, the branches angled or curved upward. Spikelets 3-7-flowered, linear, 10-15 mm long, breaking above the glumes; the stem within the spikelet (rachilla) covered with long silky hairs, these longer than the florets and becoming exposed as the lemmas spread after flowering; glumes unequal, the first glume half the length of second glume. Grain (seed) seldom produced. Aug–Sept.

SYNONYM *Phragmites communis.*

HABITAT fresh to brackish marshes, shores, streams, ditches, occasional in tamarack swamps; sometimes in shallow water.

STATUS MW-FACW | NCNE-FACW | GP-FACW

NOTE two subspecies in the Upper Midwest, one native (**subsp.** *americanus,* whose distribution is poorly understood) and one introduced and invasive (**subsp.** *australis*). The native strain is non-invasive, its lower stems are usually exposed, shiny and often reddish; in non-native strains, lower stems are usually covered by the leaf sheath.

1 Plants rarely forming a monoculture; ligules 1-1.7 mm long; lower glumes 3-6.5 mm long; upper glumes 5.5-11 mm long; lemmas 8-13.5 mm long; leaf sheaths deciduous, exposing stems in winter . . . . subsp. *americanus*

1 Plants invasive and often forming a monoculture; ligules 0.4–0.9 mm long; lower glumes 2.5-5 mm long; upper glumes 4.5-7.5 mm long; lemmas 7.5-12 mm long; leaf sheaths not deciduous, stems not exposed in winter . . . . subsp. *australis*

ANDREY ZHARKIKH

Phragmites australis

Poa BLUEGRASS

Perennial, loosely clumped or rhizomatous grasses. Leaves mostly near base, flat to folded, the tip keeled similar to the bow of a boat; sheaths partly closed, ligules membranous. Head an open panicle. Spikelets small, with 2 to several flowers breaking above the glumes; glumes nearly equal, the first glume usually 1-veined, the second glume 3-veined; lemmas often with a tuft of distinctive cobwebby hairs at base; palea nearly as long as lemma.

Poa BLUEGRASS

1 Keel of lemma silky-hairy; lemma nerves without hairs . . . . *P. alsodes*
1 Keel of lemma and some or all nerves hairy **2**
2 Plants with rhizomes . . . . *P. pratensis*
2 Plants without rhizomes **3**
3 Panicle branches single or in groups of 2; sheaths rough-to-touch; ligules less than 2 mm long; rare . . . . *P. paludigena*
3 Panicle branches in groups of 3-5; sheaths smooth; ligules 3-5 mm long; widespread . . . . *P. palustris*

Poa alsodes A. Gray
♦ GROVE BLUEGRASS

DESCRIPTION Loosely clumped, native perennial, rhizomes absent. **Stems** slender, 3–8 dm long. **Leaves** lax, 5–20 cm long and 2–5 mm wide; sheaths smooth, ligule 1–3 mm long.

Head a lax, open panicle, 10–20 cm long, the branches becoming widely spreading, mostly in groups of 4–5, with l to few spikelets near tip of branch; base of panicle sometimes remaining enclosed by sheath. **Spikelets** ovate, 2-3-flowered, 3–5 mm long; glumes nearly equal, 2–4 mm long; lemmas 2–4 mm long, with cobwebby hairs at base. **Flowering** May–July.

HABITAT alder thickets, swamp hummocks, most common in moist deciduous or mixed conifer-deciduous forests.

STATUS MW-FACW | NCNE-FAC

Poa paludigena Fernald & Wieg.
MARSH BLUEGRASS

DESCRIPTION Native perennial grass, without rhizomes. Stems single or in small clumps, slender and weak, 2–6 dm long. Leaves upright, to 10 cm long and 1–2 mm wide; sheaths finely rough-hairy; ligule flat-topped, about 1 mm long. Head a loose, open panicle, 5–12 cm long, the lower branches in groups of 2, with a few spikelets above middle. Spikelets 2–5-flowered, 4–5 mm long, glumes lance-shaped, the first glume to 2 mm long, the second glume 2–3 mm long; lemma 3–4 mm long, with cobwebby hairs at base. Flowering June–July.

HABITAT swamps, alder thickets, sedge meadows, open bogs, cold springs; usually in sphagnum moss and often under black ash (*Fraxinus nigra*).

STATUS MW-OBL | NCNE-OBL; Minn (THR), Wisc (THR), Mich (THR).

Poa palustris L. ◆ FOWL BLUEGRASS

DESCRIPTION Loosely clumped, native perennial grass. Stems smooth, 4–12 dm long, reclining at base and rooting from lower nodes, lower portion often purple-tinged. Leaves flat, upright to spreading, 1–4 mm wide, rough-to-touch; sheaths smooth; ligule 2–5 mm long.

Head a loosely spreading panicle (narrow when emerging from sheath), 1–3 dm long, the branches in mostly widely separated groups along panicle stem (rachis). Spikelets 2–4-flowered, 2–5 mm long and 1–2 mm wide; glumes nearly equal, lance-shaped, 2–3 mm long, often purple; lemma 2–3 mm long, often purple on sides, with cobwebby hairs at base. Flowering June–Sept.

HABITAT wet meadows, marshes, shores, streambanks, ditches and low prairie; also moist woods.

STATUS MW-FACW | NCNE-FACW | GP-FACW

Poa pratensis L.
◆ KENTUCKY BLUEGRASS

DESCRIPTION Introduced perennial grass, spreading by rhizomes and forming a sod. Stems erect, 3–10 dm long. Leaves flat or folded, 1–4 mm wide, margins sometimes somewhat rough-to-touch; sheaths smooth; ligule 0.5–2 mm long. Head an open, pyramid-shaped panicle, 5–15 cm long, the branches spreading to ascending, the lowest branches in groups of 4–5. Spikelets 2–5-flowered, green or purple-tinged, compressed, 3–5 mm long and 2–3 mm wide; glumes unequal, lance-shaped, 2–4 mm long, roughened on keels; lemma 2–4 mm long, with an obvious tuft of cobwebby hairs at base, often purple-tinged on sides. Flowering May–Aug.

HABITAT all types of moist to dry places; not

MATT LAVIN

Poa palustris

DOUGLAS GOLDMAN

Poa pratensis

usually in very wet situations. Kentucky bluegrass was introduced from Europe for lawns and pastures, now naturalized across most of USA.

STATUS MW-FAC | NCNE-FACU | GP-FACU

Puccinellia ALKALI-GRASS

Clumped, smooth perennial grasses, usually in brackish habitats. Leaves mostly from base of plants, flat to inrolled. Head an open panicle, the branches upright to spreading. Spikelets several-flowered, oval to linear, nearly round in section, breaking above the glumes; glumes unequal, the first glume 1-veined, the second glume 3-veined; lemmas rounded on back, often short-hairy at base; palea shorter to about as long as lemma.

Puccinellia ALKALI-GRASS

1 Lower panicle branches horizontal or angled downward when mature; lemma broad, not tapered to the blunt or rounded tip. . . . .*P. distans*

1 Lower panicle branches usually angled upward; lemma narrow, tapered to a rounded tip. . . . . *P. nuttalliana*

Puccinellia distans (Jacq.) Parl.
♦EUROPEAN ALKALI-GRASS

DESCRIPTION Clumped, smooth, introduced perennial grass. Stems erect or reclining at base, 1–5 dm long. Leaves flat to slightly inrolled, 1–3 mm wide; ligule about 1 mm long. Head a loose, pyramid-shaped panicle, 5–15 cm long, the branches in groups, the lower branches angled downward. Spikelets 3–7-flowered, 4–6 mm long; glumes ovate, 1–2 mm long; lemmas about 2 mm long, smooth or short-hairy at base. Flowering May–Aug.

HABITAT adventive in brackish waste areas and ditches along salted highways.

STATUS MW-OBL | NCNE-FACW | GP-FACW

Puccinellia nuttalliana (Schultes) A. Hitchc. NUTTALL'S ALKALI-GRASS

DESCRIPTION Clumped, perennial grass. Stems slender, erect, 2–8 dm long. Leaves flat or often inrolled, 1–3 mm wide; ligule 1–3 mm long. Head an open panicle, 5–25 cm long, the branches ascending to spreading, rough-to-touch, to 10 cm long, the spikelets mostly above middle of branch. Spikelets 3–9-flowered, slender, 4–7 mm long, glumes lance-shaped, 1–3 mm long; lemmas oblong, 2–3 mm long, with tiny hairs at base. Flowering June–July.

SYNONYM *Puccinellia airoides*.

HABITAT moist flats, sometimes in shallow water, often where salty or disturbed.

STATUS MW-OBL | NCNE-OBL | GP-OBL; adventive in Wisc; main range of species w USA.

Scolochloa SPRANGLETOP

Scolochloa festucacea (Willd.) Link
SPRANGLETOP, WHITETOP

DESCRIPTION Tall native perennial grass, spreading by thick rhizomes and forming colonies. Stems erect, hollow, 1–2 m long and 3–5 mm wide near base, usually with a few suckers and roots from lower nodes. Leaves flat or slightly inrolled, 3–10 mm wide, tapered to a sharp tip, upper surface rough-to-touch; sheaths smooth; ligule

Puccinellia distans

white, ragged at tip, 4–7 mm long. Head a loose, open panicle, 15–20 cm long, the branches ascending, the lowest branches much longer than upper. Spikelets 3–4-flowered, purple or green, becoming straw-colored, 7–10 mm long, breaking above glumes; glumes unequal, lance-shaped, the first glume 3-veined, 4–7 mm long, the second glume 5-veined, 6–9 mm long; lemmas lance-shaped, about 6 mm long; palea as long as lemma. Flowering June–July.
SYNONYM *Fluminea festucacea*.
HABITAT shallow water, marshes; a species of the northern Great Plains.
STATUS MW-OBL | NCNE-OBL | GP-OBL

Spartina CORDGRASS

Coarse perennial grasses, spreading by long scaly rhizomes. Stems stout and erect. Leaves flat to inrolled, tough, rough-to-touch; sheaths smooth; ligule a fringe of hairs. Head of several to many 1-sided spikes in racemes at ends of stem, the spikes upright to appressed. Spikelets 1-flowered, flattened, overlapping in 2 rows on 1 side of the rachis, breaking below the glumes; glumes unequal, 1–2-veined, with rough hairs on the keel; lemma with pronounced midvein and 2 faint lateral veins; palea about as long as lemma.

Spartina CORDGRASS
1 Plants 1–2 m tall; leaf blades flat (at least near base), more than 5 mm wide . . . . *S. pectinata*
1 Plants to 1 m tall; leaf blades inrolled or flat, 2–5 mm wide 2
2 Leaf blades usually flat; spikes erect; w Minn (and adventive in LP of Mich) . . . . *S. gracilis*
2 Leaf blades usually inrolled; spikes ascending or spreading; se Mich. *S. patens*

Spartina gracilis Trin.
ALKALI CORDGRASS
DESCRIPTION Native perennial grass, from rhizomes. Stems 4–8 dm long. Leaves usually inrolled, 10–20 cm long and 2–4 mm wide. Head a spikelike raceme of 4–8, 1-sided spikes, the spikes 2–5 cm long, appressed to the raceme stem (rachis). Spikelets 1-flowered, 6–9 mm long; glumes and lemma fringed with hairs on keel, the first glume half as long as second; lemma nearly as long as second glume. Flowering July–Sept.
SYNONYM *Sporobolus hookerianus*.
HABITAT wet meadows, shores, flats and seeps; often where brackish.
STATUS MW-FACW | NCNE-FACW | GP-FACW

Spartina patens (Aiton) Muhl.
SALTMEADOW CORDGRASS
DESCRIPTION Native perennial grass, from long rhizomes. Stems slender, tough, 3–9 dm long. Leaves inrolled or flat near base, 5–30 cm long and 1–4 mm wide. Head a raceme of 2–7 spikes, the spikes 1.5–5 cm long, upright but not appressed, somewhat separate from one another along the raceme stem (rachis). Spikelets 1-flowered, 8–12 mm long, fringed with hairs on keels; first glume 2–6 mm long, half as long as floret, second glume lance-shaped, 8–12 mm long; lemma 5–8 mm long.
SYNONYM *Sporobolus pumilus*.
HABITAT Adventive in salt marshes in se Mich, disjunct from main range along Atlantic coast..
STATUS MW-FACW | NCNE-FACW | GP-FACW

Spartina pectinata Link
♦ PRAIRIE CORDGRASS
DESCRIPTION Native perennial grass, with scaly rhizomes. Stems tough, 1–2 m long. Leaves flat to inrolled, 3–10 mm wide, margins very rough. Head a spikelike raceme of mostly 10–30, 1-sided spikes, the spikes upright to sometimes appressed, 3–10 cm long. Spikelets 1-flowered, 8–11 mm long, fringed with hairs on keels; first glume nearly as long as floret, tapered to tip or with an awn 1–5 mm long, second glume longer than floret, tipped with an awn 2–8 mm long; lemma 7–9 mm long, shorter than second glume. Flowering July–Sept.
SYNONYM *Sporobolus michauxianus*.
HABITAT shallow marshes, wet meadows, sandy shores, ditches, low prairie.
STATUS MW-FACW | NCNE-FACW | GP-FACW

Sphenopholis WEDGE-GRASS

Sphenopholis obtusata (Michx.) Scribn.
WEDGE-GRASS

DESCRIPTION Native clumped perennial (sometimes annual) grass; plants smooth to rough-hairy. **Stems** slender, 2–10 dm long. **Leaves** upright to spreading, flat, rough-to-touch, 2–7 mm wide; ligule membranous, ragged at tip, 1–4 mm long. **Head** a dense, shiny, spikelike panicle, 5–20 cm long, the spikes often (in part) separate from one another. **Spikelets** 2-flowered, 3–4 mm long, unawned, breaking below the glumes; glumes 2–3 mm long, the first glume linear, 1-veined, the second glume broader, 3–5-veined; lemma 2–3 mm long, 1-veined; palea linear, about as long as lemma. **Flowering** June–Aug.

SYNONYM *Sphenopholis intermedia.*

HABITAT Low prairie, wet meadows, gravelly shores, streambanks, wetland margins; also in moist woods.

STATUS MW-FAC | NCNE-FAC | GP-FAC

Torreyochloa
FALSE MANNA GRASS

Torreyochloa pallida (Torr.) Church
♦ PALE FALSE MANNA GRASS

DESCRIPTION Native perennial grass. **Stems** slender, weak, usually reclining at base, 3–10 dm long. **Leaves** flat, soft, 3–8 mm wide; sheaths open; ligule 3–9 mm long. **Head** a pale green, open panicle, 5–15 cm long, the branches upright, becoming spreading. **Spikelets** 4–7-flowered, oval in outline, 5–7 mm long; glumes rounded at tip, 1–3 mm long; lemmas 2–3 mm long, 5-

veined, finely hairy, the tip rounded and ragged. **Flowering** June–Aug.

SYNONYMS *Glyceria pallida, Puccinellia fernaldii, Puccinellia pallida.*

HABITAT marshes, pond margins, thickets, forest depressions; often in shallow water.

STATUS MW-OBL | NCNE-OBL | GP-OBL

Zizania WILDRICE

Large annual grasses (ours) of marshes and shallow water, with tall stems, wide flat blades, and fleshy yellow roots. Sheaths open, not inflated; ligules membranous or scarious. Spikelets 1-flowered, articulated at the base, readily deciduous, nearly terete, unisexual, the staminate on the lower, the pistillate on the upper branches of the large panicle. Glumes absent. Lemma of the staminate spikelet thin, herbaceous, linear, acuminate or short-awned, 5-nerved. Pistillate spikelets inserted in a cup-shaped excavation at the summit of the pedicels; lemma firm at maturity, prominently 3-ribbed, awned.

Zizania WILDRICE

1 Pistillate inflorescence branches usually divaricate at maturity; pistillate lemma thin and membranous and at least sparsely hispid-scabrous between the strong nerves; widest leaves 1.5–4.5 cm wide . . . .
. . . . *Zizania aquatica*

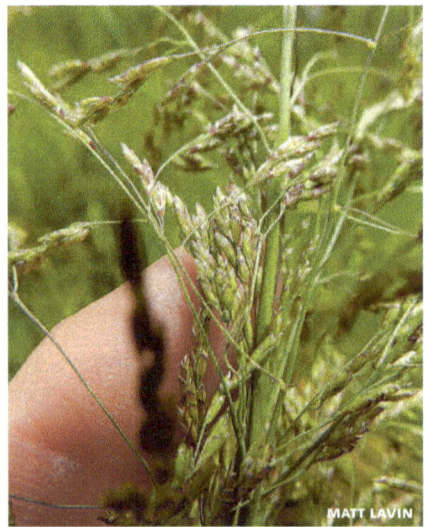

Spartina pectinata

Torreyochloa pallida

1 Pistillate inflorescence branches usually appressed at maturity, or with 1 to few, somewhat spreading branches; pistillate lemma firm and tough, scabrous-hispid only on the nerves and at most at the base and apex; widest leaves 0.5–1.7 cm wide .. . .*Zizania palustris*

Zizania aquatica L.
ANNUAL WILDRICE

DESCRIPTION Large, native annual emergent grass, with fleshy yellow roots. **Stems** single or few together, 1–3 m long. **Leaves** flat, 1–4 cm wide, smooth or finely hairy, usually floating on water surface early in season, becoming upright; sheaths short-hairy at top, smooth below; ligule membranous, entire or with a jagged margin, 10–15 mm long. **Head** a panicle, 3–6 dm long, the branches 10–20 cm long; staminate and pistillate flowers separate on same plant, the staminate flowers on lower panicle branches, pistillate flowers on upper branches, the staminate portion becoming spreading, branches of pistillate portion remaining upright. **Spikelets** 1-flowered, round in section, breaking as a unit from the stalk; glumes absent; staminate spikelets straw-colored to purple, 6–12 mm long, hanging downward from branches, lemma linear, tapered to tip or tipped with an awn to 3 mm long, early deciduous; pistillate spikelets linear, purple or light green, lemma awl-shaped, 1–2 cm long, tapered to a slender awn 3–6 cm long. Grain cylindric, dark brown to black, 1–2 cm long. **Flowering** July–Sept.
HABITAT Shallow water (up to 1 m deep) or mud of streams, rivers, lakes, ponds; where water is slightly flowing and not stagnant; soils vary from muck to silt, sand, or gravel, with best establishment of plants on a layer of soft silt or muck several cm thick.
STATUS MW-OBL | NCNE-OBL | GP-OBL; Mich (THR).

Zizania palustris L.
♦ NORTHERN WILD RICE

DESCRIPTION Large, native annual emergent grass, with fleshy yellow roots. **Stems** to 3 m tall, usually at least partly immersed in water; sheaths glabrous or with scattered hairs; ligules 3–15 mm long. **Leaves**
0.5–1.7 cm wide. **Head** a panicle 25–60 cm long; staminate and pistillate flowers separate on same plant, the staminate flowers on lower panicle branches, pistillate flowers on upper branches; staminate branches ascending or divergent; pistillate branches mostly appressed or ascending, a few sometimes divergent. **Spikelets** staminate spikelets 6–17 mm long, lanceolate, acuminate or awned, the awns to 2 mm long. Pistillate spikelets 8–33 mm long, lanceolate or oblong, leathery or indurate, lustrous, glabrous or with lines of short hairs, tips usually hirsute and abruptly narrowed, awned, the awns to 10 cm long; lemmas and paleas remaining clasped at maturity. Grain 6–30 mm long. **Flowering** July–Sept.
HABITAT as in *Zizania aquatica*, but much more common.
STATUS MW-OBL | NCNE-OBL | GP-OBL
NOTE rangewide, *Zizania palustris* grows mostly to the north of *Z. aquatica*, but their ranges overlap in the Midwest region. Two varieties occur in the Upper Midwest:

1 Lower pistillate branches with 9–30 spikelets; pistillate part of the inflorescence 10–40 cm or more wide, the branches ascending to widely divergent; plants 1–3 m tall; blades 10–40 mm wide or more . . . . var. *interior*

1 Lower pistillate branches with 2–8 spikelets; pistillate part of the inflorescence usually less than 10 cm wide, the branches appressed or ascending, or a few branches somewhat divergent; plants to 2 m tall; blades 3–21 mm wide . . . . . . var. *palustris*

JEAN-MARIE VAN DER MAREN

Zizania palustris

Zizania palustris is the source of commercial wild rice (California is the nation's largest producer); in the Upper Midwest region, harvesting is most common in Minnesota, especially by Native Americans where large areas of lakes and shallow marshes may be dominated by this plant. The grain is also an excellent food for waterfowl. Many of our populations are intentional introductions.

Pontederiaceae
Pickerel-Weed Family

MOSTLY PERENNIAL, aquatic or emergent herbs. Leaves alternate, stalkless and straplike, or with a petiole and broad blade. Flowers perfect (with both staminate and pistillate parts), regular or irregular, single from leaf axils or in spikes or panicles, subtended by leaflike bracts (spathes), light yellow, white or blue-purple, perianth of 6 petal-like lobes, usually joined near base to form a tube; stamens 3-6, the filaments attached to throat of perianth tube; ovary superior, 3-chambered, style 1. Fruit a many-seeded capsule inside the spathe, or a 1-seeded, achene-like utricle.

Pontederiaceae PICKEREL-WEED FAMILY

1   Flowers yellow, regular, solitary; leaves less than 1 cm broad, borne along leafy stem, usually submersed. *Heteranthera*

1   Flowers blue-violet, borne in a dense inflorescence; leaves (except for one large bract in *Pontederia*) arising near base of plant, mostly emersed and over 1 cm broad.................................**2**

2   Plants rooted in soil; petioles not inflated and bulbous; widespread native species. ....*Pontederia*

2   Plants free-floating, petioles modified into bulbous, inflated 'floats'; invasive in se Mich ....*Eichhornia*

Eichhornia
WATER-HYACINTH

Eichhornia crassipes (Mart.) Solms
♦WATER-HYACINTH

DESCRIPTION Floating aquatic herb, invasive but not typically over-wintering in our climate. Leaves broad, the petiole usually spongy-inflated; leaf blade suborbicular to broadly elliptic, leathery, to 1.5 dm long and wide; submersed leaves (when present) long and narrow. Flowers in a loose terminal spike; showy, light blue to bluish purple; stamens 6, the 3 upper all included, the 3 lower more exserted. Fruit a many-seeded capsule.

STATUS MW-OBL | NCNE-OBL

NOTE also spelled *Eichhornia*.

Heteranthera
MUD-PLANTAIN

Heteranthera MUD-PLANTAIN

1   Flowers blue-purple or white; leaves with a petiole and blade, emersed or floating....
.............................*H. limosa*

1   Flowers light yellow; leaves linear and straplike, not differentiated into petiole and blade, usually underwater ..*H. dubia*

Heteranthera dubia (Jacq.) MacMill.
♦WATER STAR-GRASS

DESCRIPTION Native aquatic perennial herb, with lax stems and leaves, or plants sometimes exposed and forming small, leafy rosettes. Stems slender, forked, often rooting at lower nodes, to 1 m long. Leaves alternate, linear, flat, translucent, rounded at tip or tapered to a point, 2-12 cm long and 2-6 mm wide, the midrib and veins inconspicuous; petioles absent. Flowers 1, opening on water surface, light yellow, enclosed in a spathe from upper leaf axils, the spathe membranous, 2-5 cm long, surrounding much of the slender perianth tube; perianth tube often curved, 2-8 cm

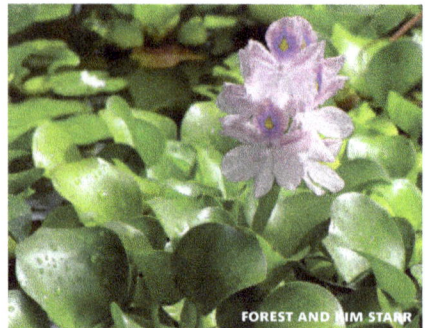

Eichhornia crassipes

long, the 6 perianth segments linear, 4–6 mm long; stamens 3, all alike. Fruit a many-seeded capsule about 1 cm long. Flowering July–Sept.

SYNONYM *Zosterella dubia*.

HABITAT shallow water, muddy shores of ponds, lakes, streams and marshes; popular as an aquarium plant.

STATUS MW-OBL | NCNE-OBL | GP-OBL

Heteranthera limosa (Swartz) Willd.
SMALLER MUD-PLANTAIN

DESCRIPTION Small native annual herb. Stems much-branched from base, 1–3 dm long, short when exposed, longer and sprawling when in water. Leaves with blade and petiole, the blades usually emersed, ovate to oval, 2–6 cm long and 1–3 cm wide, tapered to a rounded tip, base rounded or flat across; petioles 5–15 cm long, with a membranous sheath at base. Flowers 1, enclosed by a spathe; spathe folded, 2–4 cm long, abruptly narrowed at tip, enclosing the tubular portion of the perianth, the flower and spathe at end of a stout stalk arising from stem; perianth segments usually blue-purple, sometimes white, lance-shaped, 5–10 mm long, the perianth tube 1–4 cm long, the lobes ± equal, 5–15 mm long, the 3 upper lobes with a yellow spot at base; stamens 3, the 2 lateral stamens short, yellow, the center stamen longer and blue or yellow. Flowering June–Sept.

HABITAT shallow water or mud of ponds and marshes.

STATUS MW-OBL | NCNE-OBL | GP-OBL; Minn (THR).

Pontederia PICKEREL-WEED

Pontederia cordata L.
♦ PICKEREL-WEED

DESCRIPTION Native perennial emergent herb, spreading from rhizomes and forming colonies. Stems stout, upright, to 12 dm long, with 1 leaf. Leaves lance-shaped to ovate, 5–20 cm long and 2–15 cm wide, heart-shaped at base; petioles 3–7 cm long, sheathing on stem. Flowers blue-purple (rarely white), many in a spike 5–15 cm long, subtended by a bractlike spathe 3–6 cm long; perianth funnel-like, the tube 6 mm long, 2-lipped above, upper lip with 3 ovate lobes, lower lip with 3 slender, spreading lobes, the lobes 7–10 mm long. Fruit a 1-seeded utricle, 5–10 mm long. Flowering June–Sept.

SYNONYM *Pontederia lanceolata*.

HABITAT shallow water (to 1 m deep) of lakes, ponds, rivers and swamps.

STATUS MW-OBL | NCNE-OBL | GP-OBL

Potamogetonaceae
Pondweed Family

THIS TREATMENT includes *Zannichellia*, previously included in a separate family. (For more details on this often difficult group, see **Appendices A and B**, *Potamogeton* **and** *Stuckenia* **Conspectus**, page 440–446.)

Heteranthera dubia

Pontederia cordata

Potamogetonaceae PONDWEED FAMILY

1 Submersed leaves opposite or whorled, floating leaves absent . . . .
. . . . *Zannichellia palustris*

1 Submersed leaves alternate, floating leaves (sometimes present) alternate or opposite **2**

2 Flowers 2, at first enclosed in sheathing leaf base, the peduncle elongating and often spiraled or coiled at its base; fruit long-stalked; stipular sheath lacking free ligule at summit (the stipule wholly adnate to the leaf blade and merely rounded at the summit); leaf blade terete . . . . *Ruppia cirrhosa*
. . . . (Ruppiaceae, see p. 401)

2 Flowers several to many in a peduncled head or spike; perianth of 4 tepals; fruit ± sessile; stipular sheath absent (stipules entirely free from leaf) or with a short ligule-like extension if stipules fused to the leaf blade **3**

3 Stipules adnate to the leaves for 10–30 mm or more (at least on the larger leaves), adnate for ca. 2/3 of the length of the stipule; leaves all submersed, filiform to narrowly linear (up to 2.5 mm wide) . . . . . .
. . . . *Stuckenia*

3 Stipules free from the leaves or adnate for less than half the length of the stipule (adnate for 5 mm or less except in *P. robbinsii*); leaves submersed or floating, filiform to ovate, oblong, or elliptic . . . .
. . . . *Potamogeton*

Potamogeton PONDWEED

Aquatic perennial herbs, with only underwater leaves or with both underwater and floating leaves, from rhizomes or tubers, sometimes reproducing and over-wintering by free-floating winter buds. Stems long, wavy, anchored to bottom by roots and rhizomes. Leaves alternate, or becoming opposite upward in some species, simple, with an open or closed sheath at base. Underwater leaves usually linear and threadlike, sometimes broader, margins often wavy, usually stalkless. Floating leaves, if present, oval or ovate, stalked, with a waxy upper surface. Flowers perfect, regular, green to red, in stalked spikes at ends of stems or from leaf axils, usually raised above water surface, the spikes with few to many small flowers; perianth of 4 sepal-like bracts; stamens 4. Fruit a 4-parted, beaked ach-

ene. The narrow-leaved pondweeds (leads 7–16 in Group 2 key), although important as a group as waterfowl food, are often difficult to positively identify in the field, the distinguishing features being somewhat hard to see.

Potamogeton Groups

1 Plants with underwater leaves only, these all alike . . . . Group 1

1 Plants with 2 kinds of leaves: broad floating leaves and broad or narrow underwater leaves . . . . Group 2

Potamogeton Group 1

Plants with underwater leaves only, these all alike.

1 Leaves broad, lance-shaped to oval or ovate, never linear **2**

1 Leaves linear **7**

2 Leaf margins wavy-crisped, finely toothed . . . . *P. crispus*

2 Leaf margins flat or sometimes wavy, entire (or rarely finely toothed at tip) **3**

3 Base of leaf blade tapered, not clasping stem **4**

3 Base of leaf blade clasping stem **5**

4 Plants green, upper leaves stalked, leaf margins finely toothed near tip . . . .
. . . . *P. illinoensis*

4 Plants red-tinged, upper leaves ± stalkless, leaf margins entire . . . . *P. alpinus*

5 Stems whitish; leaves 10–30 cm long; fruit 4–5 mm long . . . . *P. praelongus*

5 Stems green; leaves 1–12 cm long; fruit 2–4 mm long **6**

6 Leaves ovate, mostly 1–5 cm long, margins flat; stipules small or absent; plants drying olive-green . . . . *P. perfoliatus*

6 Leaves lance-shaped, mostly more than 5 cm long; margins wavy-crisped; stipules conspicuous, persisting as shreds; plants drying light green . . . . *P. richardsonii*

7 Stipules joined with lower part of leaf to form a sheath at least 1 cm long **8**

7 Stipules free from leaf, or rarely joined to leaf base for only 1–2 mm **9**

8 Leaves 4–8 mm wide, auricled at base, margins finely toothed . . . . *P. robbinsii*

8 Leaves threadlike, rarely to 3 mm wide, not auricled, margins entire (these 3 species are now placed in genus *Stuckenia,* which see) . . . . *Stuckenia*, p. 399

9 Plants with slender creeping rhizomes . 10

9 Plants with short rhizomes or rhizomes absent (plants often rooting at lower nodes of stem)............................ 11

10 Flower clusters on stalks at ends of stems, the stalks mostly 5–25 cm long; leaves threadlike, narrower than stems ....
.... *P. confervoides*

10 Flower clusters on stalks from leaf axils, the stalks less than 3 cm long; leaves linear, wider than stems .... *P. foliosus*

11 Leaves 9- to many-veined (with 1–2 main veins and many finer ones) ....
.... *P. zosteriformis*

11 Leaves 1-7-veined 12

12 Leaves without glands at base . *P. foliosus*

12 At least some of leaves with pair of glands at base.............................. 13

13 Leaves with 5-7 nerves.......... *P. friesii*

13 Leaves with 3 (rarely 1 or 5) nerves..... 14

14 Leaves gradually tapered to a bristlelike tip 15

14 Leaves rounded at tip or tapered to a point, not bristle-tipped 16

15 Leaf margins rolled under; widespread species................... *P. strictifolius*

15 Leaf margins flat, not rolled under; uncommon..................... *P. hillii*

16 Leaves 1-4 mm wide, rounded at tip; body of achene 2.5-4 mm long.. *P. obtusifolius*

16 Leaves to 2.5 mm wide, usually tapered to a sharp tip; body of achene to 2 mm long .
.... *P. pusillus*

Potamogeton Group 2

Plants with 2 kinds of leaves: broad floating leaves and broad or narrow underwater leaves.

1 Underwater leaves broad, never narrowly linear 2

1 Underwater leaves linear or threadlike .. 7

2 Floating leaves with 30–55 nerves; underwater leaves with 30–40 nerves ....
.... *P. amplifolius*

2 Floating leaves with fewer than 30 nerves; underwater leaves with less than 30 nerves ....

3 Underwater leaves with more than 7 nerves, all leaves stalked 4

3 Underwater leaves mostly with 7 nerves, at least the lower leaves stalkless 5

4 Base of floating leaves ± heart-shaped....
.... *P. pulcher*

4 Base of floating leaves tapered or rounded, not heart-shaped. .... *P. nodosus*

5 Margins of underwater leaves finely toothed near tip .... *P. illinoensis*

5 Margins of underwater leaves entire 6

6 Plants red-tinged; underwater leaves 5–20 cm long and at least as wide as floating leaves, mostly on main stem ... *P. alpinus*

6 Plants green; underwater leaves 3–8 cm long and narrower than floating leaves, often numerous on short branches from leaf axils .... *P. gramineus*

7 Spikes of 1 kind only; fruits not (or only slightly) compressed; stipules not joined with leaf base........................ 8

7 Spikes of 2 kinds: those in axils of lower underwater leaves on short stalks; those in axils of upper or floating leaves often emersed on long stalks; fruit flattened; stipules of leaves (or at least some of lower leaves) joined with leaf base.......... 12

8 Floating leaves less than 1 cm wide and less than 2 cm long .... *P. vaseyi*

8 Floating leaves more than 1 cm wide and more than 2 cm long 9

9 Underwater leaves flat and tapelike, 2–10 mm wide .... *P. epihydrus*

9 Underwater leaves round in cross-section, often reduced to a petiole, mostly less than 1.5 mm wide 10

10 Blade of floating leaves oval, tapered to base; fruit 3-keeled .... *P. nodosus*

10 Blade of floating leaves ovate to nearly heart-shaped at base; fruit barely keeled .
.... 11

11 Floating leaves mostly 3–10 cm long; spikes 3–6 cm long .... *P. natans*

11 Floating leaves 2–5 cm long; spikes 1–3 cm long .... *P. oakesianus*

12 Underwater leaves hair-like, to only about 0.3 mm wide, acute to long-tapering at tip; tips of floating leaves acute ....
.... *P. bicupulatus*

12 Underwater leaves hair-like but slightly wider (more than about 0.5 mm), leaf tips obtuse to acute; floating leaf tips rounded 13

13 Underwater leaves blunt-tipped; floating leaves with a small notch at tip *P. spirillus*

13 Underwater leaves tapered to a pointed tip; floating leaves not notched at tip ....
.... *P. diversifolius*

Potamogeton alpinus Balbis.
♦ RED PONDWEED

DESCRIPTION Native aquatic perennial herb; plants red-tinged. Stems round in section, unbranched or sometimes branched above, to 1 m long and 1–2 mm wide. Underwater leaves linear lance-shaped, 4–20 cm long and 5–15 mm wide, 7–9-veined, usually rounded at tip, narrowed to a stalkless base. Floating leaves often absent, if present, thin, obovate, 4–6 cm long and 1–2 cm wide, 7- to many-veined, rounded at tip, tapered to a narrow base; stipules not joined to leaf base, membranous, 1–3 cm long and to 1.5 cm wide. Flowers in cylindric spikes, 1–3 cm long, with 5–9 whorls of flowers, on stalks 6–15 cm long and about as thick as stem. Achenes yellow-brown to olive, flattened, 3 mm long, the beak short. Flowering July–Sept.
HABITAT shallow to deep (usually cold) water of lakes and streams.
STATUS MW-OBL | NCNE-OBL | GP-OBL

Potamogeton amplifolius Tuckerman
♦ BIGLEAF PONDWEED

DESCRIPTION Native aquatic perennial herb. Stems round in section, usually unbranched, to 1 m or more long and 2–4 mm wide. Underwater leaves upper underwater leaves ovate, folded and sickle-shaped, 8–20 cm long and 2–7 cm wide, many-veined; lower underwater leaves lance-shaped, to 2 cm wide, often not folded, usually decayed by fruiting time, many-veined; petioles 1–5 cm long. Floating leaves usually present at flowering time, ovate 5–10 cm long and 3–6 cm wide, many-veined, rounded at tip or abruptly tapered to a sharp tip, rounded at base; petioles 5–15 cm long; stipules open and free of the petioles, 5–12 cm long, long-tapered to a sharp tip. Flowers in dense cylindric spikes, 3–6 cm long in fruit, on stalks 6–20 cm long, widening near tip. Achenes green-brown to brown, 4–5 mm long, beak to 1 mm long. Flowering July–Aug.
HABITAT shallow to deep water of lakes and rivers.
STATUS MW-OBL | NCNE-OBL | GP-OBL

Potamogeton bicupulatus Fern.
SNAILSEED PONDWEED

DESCRIPTION Native aquatic perennial herb; plants delicate, spreading by rhizomes. Stems compressed. Leaves both underwater and floating, or the floating leaves sometimes absent; ± spirally arranged. Underwater leaves light green to rarely

Potamogeton alpinus

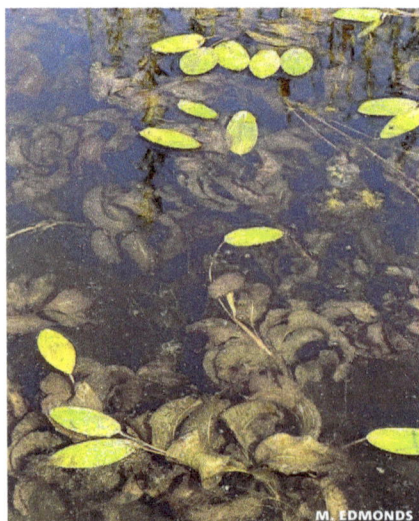

Potamogeton amplifolius

brown, hair-like, linear, 1.5–11 cm long and less than 0.5 mm wide, base slightly tapered, basal lobes absent, not clasping stem, stalkless, margins entire, not crisped; veins 1; stipules joined to blade for less than 1/2 stipule length. **Floating leaves** variable, stalked, oval, to 2 cm long and 1 cm wide, upper surface light green, base tapered or rounded, apex acute to long tapered; veins 3–7. Flowers in unbranched heads. **Achenes** greenish brown, somewhat keeled, beak absent; very small but have a noticeably bumpy surface(visible to the naked eye) due to 3 rows of sculpted ridges around the rim of the tiny, disk-shaped seed, similar to the closely related *P. diversifolius*. **Flowering** early summer–fall.

SYNONYM *Potamogeton diversifolius* var. *trichophyllus*.

HABITAT soft water lakes (water low in dissolved minerals).

range uncommon in **STATUS** MW-OBL | NCNE-OBL; Minn (END), Mich (THR).

NOTE *Potamogeton bicupulatus* is similar to *P. diversifolius*. Both species have extremely fine, hair-like underwater leaves, but in *P. diversifolius,* these leaves very slightly wider and the floating leaves somewhat rounded at tip.

Potamogeton confervoides Reichenb.
ALGA PONDWEED

DESCRIPTION Native aquatic perennial herb, from a long rhizome. **Stems** slender, to 8 dm long, branched, the branches forking. **Leaves** many, all underwater, flat, bright green, 2–5 cm long and about 0.3 mm wide, tapered to a hairlike tip, l-veined; stipules short-lived, 1–5 cm long. Leaves so delicate that they resemble greenish colored hair in the water. **Flowers** in a short spike 5–10 mm long, at end of an erect stalk 5–20 cm long. **Achenes** 2–3 mm long, with a sharp keel. **Flowering** June–Aug.

HABITAT shallow water of lakes, kettle-hole ponds and peatlands.

STATUS MW-OBL | NCNE-OBL; Wisc (THR)

NOTE unique among our pondweeds in its much-branched stems with linear leaves, and the flower spike atop an elongate, leafless stalk.

Potamogeton crispus L.
♦CURLY PONDWEED

DESCRIPTION Introduced aquatic perennial herb. **Stems** compressed, with few branches, to 8 dm long and 1–2 mm wide. **Leaves** all underwater, oblong, 3–9 cm long and 5–10 mm wide, rounded at tip, slightly clasping at base, stalkless, 3–5-veined, margins wavy-crisped, finely toothed; stipules 4–10 mm long, slightly joined at base, early shredding. **Flowers** in dense cylindric spikes, 1–2 cm long, appearing bristly in fruit from long achene beaks; on stalks 2–6 cm long. **Achenes** brown, 2–3 mm long, with a beak 2–3 mm long. **Flowering** April–June.

HABITAT shallow to deep water of lakes (including Great Lakes) and rivers; introduced and sometimes invasive; pollution-tolerant.

STATUS MW-OBL | NCNE-OBL | GP-OBL

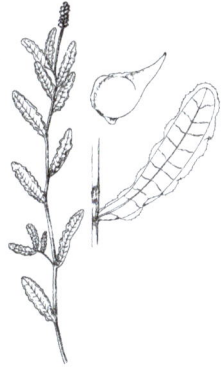

Potamogeton crispus

Potamogeton diversifolius Raf.
COMMON SNAILSEED PONDWEED

DESCRIPTION Native aquatic perennial herb. **Stems** slender, flattened or round in section, branched, to 15 dm long and 1 mm wide. **Underwater leaves** linear, flat, 1–10 cm long and 0.2–1.5 mm wide, 1-veined (sometimes 3-veined), stalkless; stipules 2–18 mm long, joined to leaf blade for less than half their length. **Floating leaves** sometimes absent, oval, leathery, 5–40 mm long and 4–20 mm wide, acute to rounded at tip, rounded at base, 3- to many-veined, the veins sunken on leaf underside; petioles 1–3 cm long; stipules not joined to leaf base, 5–20 mm long. **Flowers** in spikes of 2 types, the underwater spikes 3–6 mm long, on stalks 2–10 mm long; emersed spikes cylindric, 1–3 cm long, on stalks 5–30 mm long. **Achenes** olive to yellow, round and flattened, spiraled on surface, winged, the beak tiny. **Flowering** June–Sept.

SYNONYM *Potamogeton capillaceus.*

HABITAT shallow water of ponds.

STATUS MW-OBL | NCNE-OBL | GP-OBL; Minn (END)

NOTE *Potamogeton bicupulatus* similar to *P. diversifolius;* both species have extremely fine, hair-like underwater leaves, but in *P. diversifolius,* these leaves very slightly wider and the floating leaves somewhat rounded at tip.

Potamogeton epihydrus Raf.
♦RIBBONLEAF-PONDWEED

DESCRIPTION Native aquatic perennial herb. **Stems** slender, compressed, sparingly branched, to 2 m long and 1–2 mm wide. **Underwater leaves** linear, ribbonlike, 10–20 cm long and 3–8 mm wide, with a translucent strip on each side of midvein forming a band 1–3 mm wide, 5–13-veined, stalkless; stipules 1–3 cm long, not joined to leaf. **Floating leaves** usually present and numerous, opposite, oval to obovate, 3–8 cm long and 1–2 cm wide, mostly obtuse to bluntly abruptly short-awned at the tip, 11–25-veined, tapered to flattened petioles; stipules free, 1–3 cm long. **Flowers** in dense, cylindric spikes 2–3 cm long, on stalks 2–6 cm long and about as thick as stem.

Achenes olive to brown, 2–3 mm long; beak tiny. **Flowering** July–Sept.

HABITAT water to 2 m deep in lakes, ponds and rivers.

STATUS MW-OBL | NCNE-OBL | GP-OBL

Potamogeton epihydrus

Potamogeton friesii

Potamogeton foliosus Raf.
LEAFY PONDWEED

DESCRIPTION Native aquatic perennial herb. Stems compressed, much-branched, to 8 dm long and 1 mm wide. Leaves all underwater, linear, 1–8 cm long and 1–2 mm wide, 1–3-veined, stalkless; stipules free, 0.5–2 cm long, glands usually absent at base of stipules. Flowers in rounded to short-cylindric spikes, 2–7 mm long, with 1–2 whorls of flowers, on stalks 5–12 mm long, widened at tip. Achene green-brown, 1.5–3 mm long, winged, the beak to 0.5 mm long. Flowering June–Aug.
HABITAT shallow to deep water of lakes, ponds, rivers and streams.
STATUS MW-OBL | NCNE-OBL | GP-OBL

Potamogeton friesii Rupr.
◆ FRIES' PONDWEED

DESCRIPTION Native aquatic perennial herb. Stems compressed, branched, 1–1.5 m long, to 1 mm wide. Leaves all underwater, linear, 3–7 cm long and 1.5–3 mm wide, tip rounded with a short slender point, tapered to the base, 5–7-veined, stalkless, margins flat or becoming rolled under; stipules free, 5–20 mm long, fibrous, often shredding above, 2 glands present at base of stipule. Flowers in cylindric spikes, 8–16 mm long, with 2–5 whorls of flowers, on stalks 1.5–6 cm long. Achenes olive-green to brown, 2–3 mm long, beak flat, short. Flowering June–Aug.
HABITAT shallow to deep water of lakes, ponds, rivers and streams.
STATUS MW-OBL | NCNE-OBL | GP-OBL

Potamogeton gramineus L.
◆ VARIABLE PONDWEED

DESCRIPTION Native aquatic perennial herb. Stems slender, slightly compressed, much-branched, to 8 dm long and 1 mm wide. Underwater leaves variable, linear to lance-shaped or oblong lance-shaped, 3–9 cm long and 3–12 mm wide, 3–7-veined, tapered to a stalkless base. Floating leaves usually present, oval, 2–6 cm long and 1–3 cm wide, 11–19-veined, rounded at base; petioles 2–10 cm long, shorter to longer than blade; stipules free, persistent, 1–4 cm long. Flowers in dense, cylindric spikes, 1.5–4 cm long, the stalks thicker than stem, 2–10 cm long. Achenes dull green, 2–3 mm long. Flowering June–Aug.
HABITAT shallow to deep water of streams, ponds and lakes (including Great Lakes).
STATUS MW-OBL | NCNE-OBL | GP-OBL

Potamogeton hillii Morong
HILL'S PONDWEED

DESCRIPTION Native aquatic perennial herb, rhizomes ± absent. Stems slender, slightly compressed, much-branched, to 1 m long. Leaves all underwater, linear, 3–7 cm long

Potamogeton gramineus

and 1–2 mm wide, 3-veined, the lateral veins nearer margins than midvein; stipules white or cream-colored, free, 1–2 cm long, becoming fibrous. Flowers in rounded spikes, 4–8 mm long, with 1 (sometimes 2) whorls of flowers, on stalks 5–15 mm long. Achenes flattened, 2–4 mm long, the beak 0.5 mm long.

HABITAT shallow water of ponds and streams, often where calcium-rich.

STATUS NCNE-OBL; Mich (THR).

Potamogeton illinoensis Morong
♦ILLINOIS PONDWEED

DESCRIPTION Native aquatic perennial herb. Stems nearly round in section, usually branched, to 2 m long and 2–5 mm wide. Underwater leaves lance-shaped to obovate, 6–20 cm long, 2–4 cm wide, 9–17-veined, tapered to a broad, flat petiole, 2–4 cm long; stipules free, persistent, 3–8 cm long.

Floating leaves sometimes absent, opposite, lance-shaped to oval, 5–14 cm long and 2–6 cm wide, 13- to many-veined, often short-awned from the rounded tip, rounded to wedge-shaped at base; petioles 3–10 cm long, shorter than blades. Flowers in dense

cylindric spikes, 2–6 cm long, on stalks 4–20 cm long, usually wider than stem. Achenes olive-green, 3–4 mm long, the beak short, blunt. Flowering July–Sept.

HABITAT shallow to deep water of lakes and rivers.

STATUS MW-OBL | NCNE-OBL | GP-OBL

Potamogeton natans L.
♦FLOATING PONDWEED

DESCRIPTION Native aquatic perennial herb. Stems slightly compressed, usually unbranched, 0.5–2 m long and 1–2 mm wide. Underwater leaves reduced to linear, bladeless, expanded petioles (phyllodes), these often absent by flowering time, 10–30 cm long and 1–2 mm wide. Floating leaves ovate to oval, 4–10 cm long and 2–5 cm wide, usually tipped with a short point, rounded to heart-shaped at base, many-veined; petioles usually much longer than blades, the blade often angled at juncture

Potamogeton illinoensis

Potamogeton natans

with petiole; stipules free, 4–10 cm long, persistent or shredding with age. Flowers in dense cylindric spikes, 2–5 cm long, stalks thicker than the stem, 6–14 cm long. Achenes green-brown to brown, 3–5 mm long, with a loose, shiny covering, the beak short. Flowering June–Aug.

HABITAT usually shallow water (to 2 m deep) of ponds, lakes, rivers and peatlands.

STATUS MW-OBL | NCNE-OBL | GP-OBL

Potamogeton nodosus Poiret
♦ LONGLEAF PONDWEED

DESCRIPTION Native aquatic perennial herb. Stems round in section, branched, to 2 m long and 1–2 mm wide. Underwater leaves commonly decayed by fruiting time, lance-shaped to linear, translucent, 10–30 cm long and 1–3 cm wide, 7–15-veined, gradually tapered to a petiole 4–10 cm long. Floating leaves oval, thin, 5–12 cm long and 1–5 cm wide, tapered at both ends, many-veined; petioles somewhat winged, 5–20 cm long and 2–3 mm wide, usually longer than blades; stipules free, those of underwater leaves often absent by flowering time, those of floating leaves persistent, 3–10 cm long. Flowers in dense cylindric spikes, 2–6 cm long, on stalks 3–15 cm long and thicker than stem. Achenes red-brown to brown, 3–4 mm long, the beak short. Flowering July–Aug.

HABITAT shallow water to 2 m deep, mostly in rivers; lakes.

STATUS MW-OBL | NCNE-OBL | GP-OBL

Potamogeton oakesianus J. W. Robbins
OAKES' PONDWEED

DESCRIPTION Native aquatic perennial herb. Stems slender, often much-branched, to 1 m long. Underwater leaves bladeless, petiole-like, 0.5–1 mm wide, often persistent. Floating leaves oval, 3–6 cm long and 1–2 cm wide, rounded at base, 12- to many-veined; petioles 5–15 cm long; stipules free, 2.5–4 cm long. Flowers in cylindric spikes, 1.5–3 cm long, on stalks 3–8 cm long and wider than stem. Achenes 2–4 mm long, with a tight, dull covering, the beak flat.

HABITAT ponds and streams, peatland pools.

STATUS MW-OBL | NCNE-OBL

NOTE similar to **floating pondweed** (*Potamogeton natans*) but plants smaller and fruit ± smooth on the sides (vs. depressed in *Potamogeton natans*).

Potamogeton obtusifolius Mert. & Koch ♦ BLUNTLEAF-PONDWEED

DESCRIPTION Native aquatic perennial herb, rhizomes ± absent. Stems very slender, compressed, much-branched, to 1 m long. Leaves all underwater, linear, stalkless, often red-tinged, 3–10 cm long and 1–4 mm wide, rounded at tip, the midvein broad, base usually with pair of translucent glands; stipules free, white, 1–2

Potamogeton nodosus

Potamogeton obtusifolius

cm long. Flowers in thick cylindric spikes, 8–14 mm long, on slender, upright stalks 1–3 cm long. Achenes 2–3 mm long, the beak rounded, 0.5 mm long

HABITAT lakes, ponds and streams, peatland pools.

STATUS MW-OBL | NCNE-OBL | GP-OBL

Potamogeton perfoliatus L.
CLASPING-LEAVED PONDWEED, REDHEAD-GRASS

DESCRIPTION Native aquatic perennial herb. Stems slender, to 2.5 m long, often much-branched. Leaves all underwater, ovate to nearly round or sometimes lance-shaped, 1–7 cm long and 5–30 mm wide, tip often very finely toothed, base heart-shaped and clasping stem, stalkless; stipules free, soon decaying. Flowers on underwater cylindric spikes, 1–5 cm long, on upright stalks 1–7 cm long and about as wide as stem. Achenes 2–3 mm long, the beak short, curved.

HABITAT lakes and streams; more common in ne USA.

STATUS MW-OBL | NCNE-OBL

Potamogeton praelongus Wulfen
♦ WHITESTEM-PONDWEED

DESCRIPTION Native aquatic perennial herb. Stems white-tinged, compressed, branched,

Potamogeton praelongus

to 2–3 m long and 2–4 mm wide, the shorter internodes often zigzagged. Leaves all underwater, lance-shaped, 10–30 cm long and 1–4 cm wide, with 3–5 main veins, rounded and hoodlike at tip, base ± heart-shaped and clasping stem, stalkless, margins entire and gently wavy; stipules free, white, 1–3 cm long, fibrous at tip. Flowers in dense, cylindric spikes 2–5 cm long; stalks erect, 1–4 dm long, as wide as stems. Achenes green-brown, swollen, 4–5 mm long, the beak rounded, 0.5 mm long.

Flowering June–Aug.

HABITAT shallow to deep water of lakes (including Great Lakes), streams.

STATUS MW-OBL | NCNE-OBL | GP-OBL

Potamogeton pulcher Tuckerman
SPOTTED PONDWEED

DESCRIPTION Native aquatic perennial herb. Stems round in section, unbranched, black-spotted, usually less than 5 dm long. Underwater leaves thin, narrowly lance-shaped, 8–15 cm long and 1–3 cm wide, base tapered to a short petiole, margins wavy; the lowest leaves often thick and spatula-shaped. Floating leaves alternate, clustered at top of stem on short branches, ovate, 4–8 cm long and 2–5 cm wide, many-veined, the base somewhat heart-shaped; petioles black-spotted, 2–8 cm long; stipules free, to 6 cm long. Flowers in dense cylindric spikes, 2–4 cm long, on stalks 5–10 cm long and slightly wider than stem. Achenes 4–5 mm long, the beak broad and blunt.

HABITAT muddy shores and shallow water of lakes.

STATUS MW-OBL | NCNE-OBL | GP-OBL; Wisc (END), Mich (END).

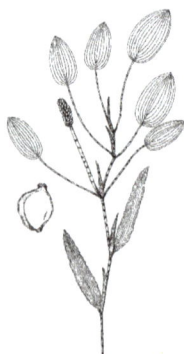

Potamogeton pusillus L.

♦ SLENDER *or* SMALL PONDWEED

DESCRIPTION Native aquatic perennial herb, rhizomes ± absent. Stems very slender, round in section, usually freely branched, 2–10 dm long and about 0.5 mm wide. Leaves all underwater, linear, 1–7 cm long and 0.5–2 mm wide, tapered to a stalkless base, the midvein broad; stipules free, boat-shaped, brown-green, 4–10 mm long and 2x width of leaf base, soon decaying, glands sometimes present at stipule base. Flowers in short-cylindric spikes 2–10 mm long, the flowers in 1–3 whorls, on slender, upright stalks 1–5 cm long. Achenes green to brown, 1–2 mm long, the beak flat. Flowering June–Aug.

HABITAT shallow water (to 2 m deep) of lakes and ponds, occasionally in streams.

STATUS MW-OBL | NCNE-OBL | GP-OBL

NOTE includes plants sometimes separated as *Potamogeton berchtoldii* Fieber.

Potamogeton richardsonii (Ar. Benn.)
Rydb. ♦ CLASPING-LEAVED PONDWEED

DESCRIPTION Native aquatic perennial herb. Stems brown to yellow-green, round in section, sparingly to freely branched, mostly 3–10 dm long and 1–2.5 mm wide, the shorter internodes rarely zigzagged. Leaves all underwater, lance-shaped, 5–12 cm long and 1–2.5 cm wide, with 13 or more prominent veins, base heart-shaped and clasping stem, stalkless, margins entire and gently wavy; stipules free, 1–2 cm long, soon shredding into white fibers. Flowers in dense cylindric spikes 1.5–4 cm long, on stalks 2–20 cm long, the stalks strongly curved when in fruit. Achenes green to brown, 2–4 mm long, the beak short. Flowering July–Aug.

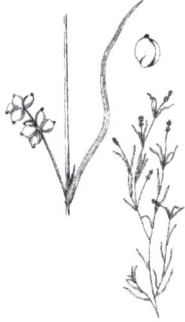

HABITAT shallow to deep water of lakes (including Great Lakes), streams.

STATUS MW-OBL | NCNE-OBL | GP-OBL

NOTE similar to **redhead-grass** (*Potamogeton perfoliatus*) but leaves narrower and often longer, and the stipules persisting as fibers, vs. soon decayed in *P. perfoliatus*.

Potamogeton robbinsii Oakes

♦ FERN-PONDWEED

DESCRIPTION Native aquatic perennial herb, rhizomes not tuberous. Stems few-branched below, much-branched above, to 1 m long. Leaves all underwater, crowded in 2 ranks, linear, 4–10 cm long and 3–7 mm wide, tapered to a pointed tip,

Potamogeton pusillus

STEFAN LEFNAER

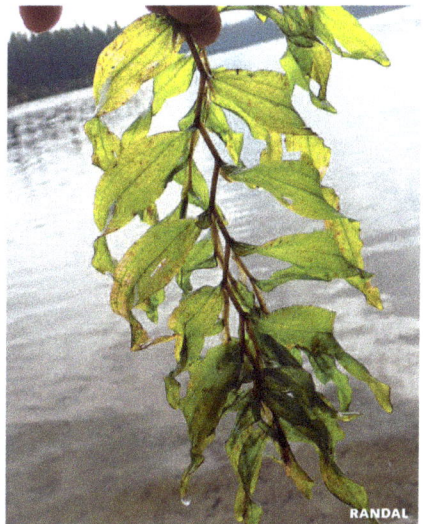

Potamogeton richardsonii

RANDAL

abruptly narrowed at base, with rounded auricles where joined with stipule, midvein pronounced, margins pale; stipules joined to leaf for 5-15 mm, soon decaying into fibers. Flowers on underwater, cylindric spikes 1-2 cm long, with 3-5 separated whorls of flowers, the inflorescence often branched into 5-20 stalks, 2-5 cm long, at ends of stems. Achenes rarely produced, 3-5 mm long, the beak thick, somewhat curved; reproduction most commonly by stem fragments which root from the nodes. Flowering July-Aug.
HABITAT shallow to deep water of lakes, ponds and streams.
STATUS MW-OBL | NCNE-OBL | GP-OBL

Potamogeton spirillus Tuckerman
NORTHERN SNAILSEED PONDWEED

DESCRIPTION Native aquatic perennial herb. Stems compressed, to 1 m long, branched, the branches short and often curved. Underwater leaves 1-8 cm long and 0.5-2 mm wide, rounded at tip, stalkless; stipules joined for most of length. Floating leaves if present, 1-4 cm long and 5-12 mm wide, 5-13-veined, the veins sunken on underside of blade, petioles 2-4 cm long; stipules free. Flowers in 2 types of spikes, the underwater spikes

Potamogeton robbinsii

round, with 1-8 fruits, ± stalkless in the leaf axils; emersed spikes longer, cylindric, to 8-12 mm long, on stalks from leaf axils. Achenes 1-3 mm long, flattened, winged, spiraled on surface, the beak absent.
HABITAT shallow water of lakes and ponds.
STATUS MW-OBL | NCNE-OBL | GP-OBL
NOTE *Potamogeton spirillus* is similar to *P. diversifolius*, but the underwater leaves are typically blunt-tipped, and the floating leaves have a small notch at tip.

Potamogeton strictifolius Ar. Bennett
STRAIGHT-LEAVED PONDWEED

DESCRIPTION Native aquatic perennial herb. Stems slender, slightly compressed, unbranched or branched above, to 1 m long and 0.5 mm wide. Leaves all underwater, linear, upright, 1-6 cm long and 0.5-2 mm wide, 3-5-veined, the veins prominent on underside, tapered to stalkless base, margins often rolled under; stipules free, white, shredding at tip, 5-20 mm long; 2 glands present at base of stipules. Flowers in cylindric spikes 6-15 mm long, with 3-5 whorls of flowers, on stalks 1-5 cm long. Achenes green-brown, 2 mm long, the beak broad, rounded. Flowering June-Aug.
HABITAT shallow to deep water of lakes and rivers.
STATUS MW-OBL | NCNE-OBL | GP-OBL

Potamogeton vaseyi Robbins
VASEY'S PONDWEED

DESCRIPTION Native aquatic perennial herb. Stems threadlike, 2-10 dm long, much-branched, the upper branches short. Underwater leaves transparent, linear, 2-6 cm long and to 1 mm wide, tapered to a sharp tip, 1-veined or rarely with 2 weak lateral nerves, stalkless; stipules free, linear, white, 1-2 cm long, sometimes with 2 glands at base. Floating leaves on flowering plants only, opposite,

obovate, leathery, 8-15 mm long and 4-7 mm wide, 5-9-veined, the veins sunken on underside, petiole about as long as blade. Flowers in cylindric spikes 3-8 mm long, with 1-4 whorls of flowers, on stems 1-3 cm long. Achenes 2-3 mm long, the beak short. HABITAT shallow to deep water of ponds. STATUS MW-OBL | NCNE-OBL | GP-OBL; historically known from Mich.

Potamogeton zosteriformis Fernald
♦ FLATSTEM-PONDWEED

DESCRIPTION Native aquatic perennial herb, rhizomes ± absent. Stems strongly flattened, sometimes winged, freely branched, to 1 m long and 1-3 mm wide. Leaves all underwater, linear, 5-20 cm long and 3-5 mm wide, 15- to many-veined, tapered to a tip, or sometimes with a short, sharp point, slightly narrowed to the stalkless base; stipules free, white, shredding with age, 1-4 cm long. Flowers in cylindric spikes, 1-2.5 cm long, with 7-11 whorls of flowers, on curved stalks 2-6 cm long. Achenes dark green to brown, 4-5 mm long, the beak short and blunt. Flowering July-Aug.

Potamogeton zosteriformis

HABITAT shallow to deep water of lakes (including Great Lakes) and streams.
STATUS MW-OBL | NCNE-OBL | GP-OBL

Stuckenia FALSE PONDWEED
Stuckenia is a small genus of perennial aquatic herbs, now segregated from *Potamogeton*. In *Stuckenia*, the stipules are joined to the blade for 2/3 to nearly the entire length of the stipule; in *Potamogeton*, the stipules in most species are free, or if adnate, joined for well less than half the length of the stipule. Also, submersed leaves of *Potamogeton* are translucent, flat, and without grooves or channels; in *Stuckenia*, submersed leaves are opaque, channeled, and turgid.

Stuckenia **FALSE PONDWEED**

1 Leaves gradually tapered to tip; rhizomes tuber-bearing; stigmas raised on a tiny style .... *S. pectinata*

1 Leaves rounded, blunt-tipped or tipped with a short, sharp point, stigmas inconspicuous, broad and not raised **2**

2 Plants short, to 0.5 m long; sheaths tight around stem; spikes with 2-5 whorls of flowers .... *S. filiformis*

2 Plants large and coarse, 2-5 m long; sheaths enlarged to 2-5 times diameter of stem; spikes with 5-12 whorls of flowers........ .... *S. vaginata*

Stuckenia filiformis (Pers.) Börner
THREADLEAF-PONDWEED

DESCRIPTION Native aquatic perennial herb, from a long, tuber-bearing rhizome. Stems ± round in section, branched from base, mostly unbranched above, 1-5 dm or more long and 1 mm wide. Leaves all underwater, narrowly linear, 5-10 cm long and 0.2-2 mm wide, 1-veined; stipules 1-3 cm long, joined to base of leaf blade, forming a tight sheath around stem. Flowers in un-

derwater spikes, 1–5 cm long, with 2–5 separated whorls of flowers, on slender stalks 2–12 cm long. Achenes olive-green, 2–3 mm long, the beak flat, tiny. Flowering July–Aug.
SYNONYM *Potamogeton filiformis.*
HABITAT mostly shallow water (to 1 m) in lakes (including Great Lakes) and rivers.
STATUS MW-OBL | NCNE-OBL | GP-OBL

Stuckenia pectinata (L.) Böerner
♦SAGO-PONDWEED

DESCRIPTION Native aquatic perennial herb, the rhizomes tipped with a white tuber important in the diet of waterfowl. Stems slender, round in section, 3–10 dm long and 1–2 mm wide much-branched and forking above, fewer branched near base. Leaves all underwater, threadlike to narrowly linear, 3–12 cm long and 0.5–1.5 mm wide, stalkless; stipules joined to base of blade for 1–3 cm, forming a sheath around stem. Flowers on underwater, cylindric spikes 1–5 cm long, with 2–5 whorls of flowers, on lax, threadlike stalks to 15 cm long. Achenes yellow-brown, 3–4 mm long, the beak to 0.5 mm long. Flowering June–Sept.
SYNONYM *Potamogeton pectinatus.*
HABITAT shallow to deep water of lakes, ponds and streams; tolerant of brackish water.
STATUS MW-OBL | NCNE-OBL | GP-OBL

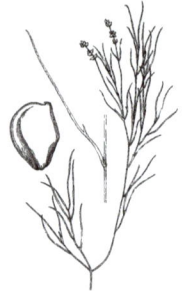

Stuckenia vaginata (Turcz.) Holub
BIGSHEATH-PONDWEED

DESCRIPTION Native aquatic perennial herb, rhizomes tipped by a tuber 3–5 cm long.

Stems round in section, much-branched above, to 1.5 m long and 1–2 mm wide. Leaves all underwater, crowded in 2 ranks, threadlike to narrowly linear, 2–20 cm long and 0.5–2 mm wide, with 1 main vein; stipules joined to base of leaf for 1–5 cm and sheathing stem, the sheaths on main stem inflated 2–4x wider than the stem. Flowers in spikes 3–6 cm long, with 5–12 spaced whorls of flowers, on lax, slender stalks to 10 cm long, the stalks often much shorter than upper leaves. Fruit dark green, 3 mm long, the beak short or nearly absent. Flowering July–Aug.
SYNONYM *Potamogeton vaginatus.*
HABITAT cold-water streams and lakes.
STATUS MW-OBL | NCNE-OBL | GP-OBL

Zannichellia
HORNED PONDWEED

Zannichellia palustris L.
♦HORNED PONDWEED

DESCRIPTION Perennial aquatic herb, with creeping rhizomes and often forming extensive

Stuckenia pectinata

ANDREY ZHARKIKH

Zannichellia palustris

CHRISTIAN FISCHER

underwater mats. Stems slender and delicate, wavy, 0.5–5 dm long, branched from base. Leaves simple, opposite (or upper leaves appearing whorled), threadlike, 2–8 cm long and 0.5 mm wide, stalkless; stipules membranous and soon deciduous. Flowers small, produced underwater, either staminate or pistillate, separate on plant but from same leaf axil, with 1 staminate flower and usually 4 (varying from 1–5) pistillate flowers at each node, surrounded by a membranous, spathelike bract; petals and sepals absent; staminate flower a single anther. Fruit a brown to red-brown, crescent-shaped nutlet, gently wavy on margins, 2–3 mm long, tipped by a beak 1–2 mm long; the fruits mostly 2–6 per node. Flowering June–Aug.

HABITAT submerged in fresh or brackish water of streams, reservoirs, muddy lake and pond bottoms, marshes and ditches. STATUS MW-OBL | NCNE-OBL | GP-OBL

Ruppiaceae
Ditch-Grass Family

Ruppia
DITCH-GRASS, WIDGEON-GRASS

Ruppia spiralis L. ex Dumort.
⬧DITCH-GRASS, WIDGEON-GRASS

DESCRIPTION Native aquatic perennial herb. Stems slender, round in section, white-

tinged, wavy, to 6 dm long, branching at base and with short branches above, the internodes often zigzagged. Leaves simple, alternate or opposite, stalkless, threadlike, mostly 5–25 cm long and 0.5 mm wide, 1-veined, with a sheathing stipule at base. Flowers very small, perfect, in small, 2-flowered spikes from leaf axils, the spikes enclosed by the leaf sheath at flowering time, the flower stalks elongating and usually coiling as fruits mature; sepals and petals absent; stamens 2; pistils typically 4 (varying from 2–8), raised on a slender stalk in fruit and becoming umbel-like. Achene an olive-green to black, ovate drupelet, 2–3 mm long. Flowering July–Aug.

SYNONYM *Ruppia cirhosa.*

HABITAT lakes and ponds, often where brackish.

STATUS MW-OBL | NCNE-OBL | GP-OBL

NOTE ***Ruppia maritima*** reported for Mich UP, distinguished by its non-coiled flower stalks.

Scheuchzeriaceae
Scheuchzeria Family

Scheuchzeria POD-GRASS

Scheuchzeria palustris L.
⬧POD-GRASS, RANNOCH-RUSH

DESCRIPTION Native perennial rushlike herb,

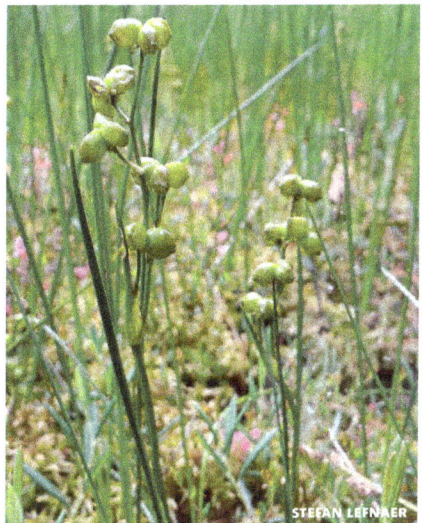

Ruppia spiralis

Scheuchzeria palustris

from creeping rhizomes. Stems 1 to several, 1–4 dm long, remains of old leaves often persistent at base of plant. Leaves alternate, several from base and 1–3 along stem, 1–3 dm long and 1–3 mm wide, the stem leaves smaller; lower part of blade half-round in section, with an expanded sheath at base, upper portion of blade flat, with a small pore at leaf tip. Flowers perfect, regular, green-white, in a several-flowered raceme 3–10 cm long, the flowers on stalks 1–2.5 cm long; tepals 6, in 2 series, ovate, 2–3 mm long; stamens 6. Fruit a group of 3 (rarely to 6) spreading follicles, 5–10 mm long, each with 1–2 seeds; seeds brown-black, 4–5 mm long. Flowering May–June.

HABITAT wet, sphagnum moss peatlands.

STATUS MW-OBL | NCNE-OBL | GP-OBL

Typhaceae
Cat-Tail Family

FAMILY now includes genus *Sparganium* from former family Sparganiaceae (discontinued under APG III).

Typhaceae CAT-TAIL FAMILY

1 Pistillate flowers in one to several spherical heads; perianth of greenish sepals; leaves strongly keeled (3-angled in cross-section) ....*Sparganium*

1 Pistillate flowers in an elongate densely flowered spike; perianth of white hairs; leaves flat-elliptic in cross-section . *Typha*

Sparganium BUR-REED

Perennial sedgelike herbs, floating or emergent in shallow water, from rhizomes and forming colonies. Stems stout, usually erect, unbranched, round in section. Leaves long, broadly linear, sheathing stem at base. Flowers crowded in round heads, the heads with either staminate or pistillate flowers; staminate heads few to many, borne above pistillate heads in a unbranched or sparsely branched inflor-escence; the pistillate heads 1 to several, from leaf axils or borne above axils on upper stem; sepals and petals reduced to chaffy, spatula-shaped scales, these appressed to the achenes in the mature pistillate heads; staminate flowers with

mostly 3–5 stamens; pistillate flowers with a 1–2-chambered pistil, stigmas 1 or 2. Fruit a beaked, nutletlike achene, stalkless or short-stalked.

Sparganium BUR-REED

1 Plants large, about 1 m tall; leaves usually erect; stigmas 2; achenes broadly oblong pyramid-shaped ....*S. eurycarpum*

1 Plants smaller, leaves erect or floating; stigmas 1; achenes slender 2

2 Fruiting heads about 1 cm wide; staminate head 1 (often absent by fruiting time); achene beaks less than 1 mm long........ ....*S. natans*

2 Fruiting heads 1.5 cm or more wide; staminate heads 2 or more; achene beaks 2 mm or more long 3

3 Fruiting heads 1.5–2 cm wide; leaves mostly flat.................................4

3 Fruiting heads larger mostly 2–3 cm wide; leaves often keeled 5

4 Staminate heads several, separate from the pistillate heads; achene not shiny ....*S. fluctuans*

4 Staminate heads usually 1 (sometimes 2) and near upper pistillate head; achene shiny...................*S. glomeratum*

5 Fruiting heads or branches all from leaf axils 6

5 At least some fruiting heads or branches borne above leaf axils 7

6 Inflorescence unbranched or the branches short and sometimes with 1–2 staminate heads; achenes dull, with a beak 3–4 mm long ....*S. americanum*

6 Inflorescence branched, the branches jointed, with 3 or more staminate heads; achenes shiny, with a beak 5–7 mm long .. ..*S. androcladum*

7 Leaves floating; achene beak 1–3 mm long ....*S. angustifolium*

7 Leaves usually stiffly erect and emersed; achene beak 3–5 mm long ... *S. emersum*

Sparganium americanum Nutt.
♦AMERICAN BUR-REED

DESCRIPTION Native perennial herb. Stems stout, erect, mostly unbranched, 3–10 dm long. Leaves linear, flat to somewhat keeled, to 1 m long and 4–12 mm wide; leaflike bracts on upper stem

shorter than leaves, widened at base. Inflorescence usually unbranched, or with a few, straight branches; pistillate heads stalkless, 2–4 on main stem, sometimes with 1–3 on branches, 2 cm wide when mature; scales widest at tip; staminate heads 3–10 on main stem, sometimes with 1–5 on branches. Achenes widest at middle, tapered to both ends, dull brown, 3–5 mm long, the beak straight, 2–4 mm long. Flowering July–Aug.

HABITAT marshes, shallow water, streambanks.

STATUS MW-OBL | NCNE-OBL | GP-OBL

Sparganium androcladum (Engelm.) Morong BRANCHED BUR-REED

DESCRIPTION Native perennial herb. Stems stout, erect, branched, 4–10 dm long. Leaves linear, keeled, triangular in section near base, 4–8 dm long and 5–12 mm wide; bracts leaflike, upright, shorter than leaves, slightly widened at base. Inflorescence often branched, the branches zigzagged; pistillate heads stalkless, 2–4 on main stem, absent or occasionally 1 near base of branches, 3 cm wide when mature; scales spatula-shaped, widest at tip; staminate heads 5–8 on main stem, 3 or more on branches. Achenes oval, shiny light brown, 5–7 mm long, often slightly narrowed at middle, the beak straight, 4–6 mm long. Flowering July–Aug.

SYNONYM *Sparganium lucidum.*

FRITZ REYNOLDS

Sparganium americanum

HABITAT marshes, lakeshores, fens.

STATUS MW-OBL | NCNE-OBL | GP-OBL

Sparganium angustifolium Michx.

♦ NARROW-LEAVED BUR-REED

DESCRIPTION Native perennial herb. Stems long and usually floating. Leaves floating, mostly 2–3 mm wide, often wider at base. Inflorescence unbranched; pistillate heads 1–3, shiny, about 2 cm wide, the lowest stalked, the upper pistillate heads stalkless; scales spatula-shaped, ragged at tip; staminate heads 2–6, close together above pistillate heads. Achenes spindle-shaped, 5–7 mm long, dull brown except at red-brown base, abruptly contracted to a beak 1–3 mm long. Flowering July–Aug.

HABITAT lakes, ponds and shores.

STATUS MW-OBL | NCNE-OBL | GP-OBL

Sparganium emersum Rehmann

DWARF BUR-REED

DESCRIPTION Native perennial herb. Stems usually erect, sometimes lax and trailing in water, 2–6 dm long. Leaves linear, yellow-green, flat to keeled, 3–7 dm long and 3–6 mm wide, usually longer than stems; bracts leaflike, erect, barely widened at base. Inflorescence unbranched, 1–2 dm long; pistillate heads 1–4, stalkless or lowest head often stalked, at least 1 head on stem above leaf axils, 1.5–2.5 cm wide when mature; scales spatula-shaped, widest at tip; staminate heads usually 2–5, 1.5–2 cm wide at flowering time. Achenes widest at middle, tapered to both ends, 4–5 mm long, shiny olive-green, the beak 3–5

MARKO VAINU

Sparganium angustifolium

mm long. Flowering June–Aug.
SYNONYM *Sparganium chlorocarpum.*
HABITAT shallow water or mud of marshes, streams, ditches, open bogs, ponds.
STATUS MW-OBL | NCNE-OBL | GP-OBL

Sparganium eurycarpum Engelm.
◆ COMMON *or* GIANT BUR-REED

DESCRIPTION Native perennial herb. Stems stout, branched, 4–10 dm long. Leaves linear, bright green, keeled, 8–
10 dm long and 5–12 mm wide; bracts leaflike, slightly widened at base. Inflorescence 1–3 dm long, branched from the bract axils; lower branches with 1 pistillate head and several staminate heads, main stem and upper branches with 6–10 staminate heads; pistillate heads 2–6, 1.5–2.5 cm wide in fruit, scales spatula-shaped; staminate heads numerous, 1–2 cm wide. Achenes oblong pyramid-shaped, 6–8 mm long, the top flattened, 4–7 mm wide, brown to golden-brown, the beak 2–4 mm long. Flowering June–Aug.
HABITAT usually in shallow water of marshes, streams, ditches, ponds and lakes, often with cat-tails (*Typha*).
STATUS MW-OBL | NCNE-OBL | GP-OBL

Sparganium fluctuans (Morong) Robinson ◆ FLOATING BUR-REED

DESCRIPTION Native perennial herb. Stems slender, floating, to 15 dm long. Leaves float-

ing, linear, flat, translucent, 3–10 mm wide, underside with netlike veins; bracts leaflike, short, widened at base. Inflorescence usually branched, the main stem with 2–4 staminate heads, the branches with 1 pistillate head near base and 2–3 staminate heads above; pistillate heads 2–4, 1.5–2 cm wide when mature, scales oblong; staminate heads to 1 cm wide. Achenes obovate, 3–4 mm long, sometimes narrowed near middle, brown, the beak curved, 2–3 mm long.
HABITAT shallow water of ponds and lakes.
STATUS MW-OBL | NCNE-OBL | GP-OBL

Sparganium glomeratum Laest.
◆ CLUSTERED BUR-REED

DESCRIPTION Native perennial herb. Stems stout, floating or erect, 2–4 dm long. Leaves linear, ± flat, 3–8 mm wide; bracts leaflike, widened at base. Inflorescence usually unbranched; pistillate heads several, clustered on the stem, stalkless, 1.5–2 cm wide when mature, scales narrowly oblong; staminate heads 1–2 above the pistillate heads and continuous with them on stem. Achenes widest at middle, tapered to both ends, 3–8 mm long, slightly narrowed below the middle, shiny brown, the beak ± straight, 1–2 mm long.
HABITAT shallow water of marshes and bogs.
STATUS MW-OBL | NCNE-OBL | GP-OBL; Wisc (THR).

Sparganium natans L.
◆ SMALL BUR-REED

DESCRIPTION Native perennial herb. Stems usually long and floating, sometimes shorter and upright, 1–3 dm or
more long. Leaves linear, dark green, thin, flat, 2–6 mm wide; bracts leaflike, short,

Sparganium eurycarpum

Sparganium fluctuans

somewhat widened at base. Inflorescence unbranched; pistillate heads 2-3, from bract axils, stalkless or the lowest sometimes short-stalked, 1 cm wide when mature; scales spatula-shaped, widest at tip; staminate heads usually 1 (rarely 2). Achenes broadly oval, 3-4 mm long, dull green-brown, the beak 1-2 mm long.

SYNONYM *Sparganium minimum.*
HABITAT shallow water, pond margins.
STATUS MW-OBL | NCNE-OBL | GP-OBL

Typha CAT-TAIL

Large reedlike perennials, from fleshy rhizomes and forming colonies. Stems erect, unbranched, round in section, sheathed for most of length by overlapping leaf sheaths. Leaves mostly near base of plant, alternate in 2 ranks, erect, linear, spongy. Flowers tiny, either staminate or pistillate, separate on same plant; petals and sepals reduced to bristles. Staminate flowers usually of 3-5 stamens, bristles absent or 1-3 or more. Female flowers intermixed with some sterile flowers; pistil 1, raised on a short stalk (gynophore), with numerous bristles near base, the bristles longer than pistil; small bracts (bractlets) also sometimes present, these intermixed with the bristles, slender but with a widened brown tip. Heads in a single, dense, cylindric spike, with staminate flowers above pistillate, the staminate and

pistillate portions of the spike unalike, contiguous in common cat-tail (*Typha latifolia*) or separated in narrow-leaved cat-tail (*Typha angustifolia*); the mature spike brown and fuzzy in appearance due to the crowded stigmas and gynophore bristles. Fruit a yellow-brown achene, 1-2 mm long, the style persistent, long and slender with an expanded stigma.

NOTE A hybrid between *Typha angustifolia* and *Typha latifolia* is termed *Typha* ×*glauca* Godr. Usually larger than either parent, staminate and pistillate portions of hybrid plants are usually separated by a space to 4 cm long. The staminate portion of the spike is light brown, 0.5-2 dm long and about 1 cm wide at flowering time; the pistillate portion is dark brown, 10-20 cm long and 1-2 cm wide. Since *Typha* ×*glauca* is sterile, reproduction is vegetative from rhizomes. The hybrid occurs in Upper Midwest region in s Wisc, and LP of Mich (especially common in Lake Erie marshes of se Mich), and may be found wherever populations of *Typha angustifolia* and *Typha latifolia* overlap.

Typha CAT-TAIL
1 Staminate and pistillate portions of spike usually separated; leaves to 1 cm wide; stigmas long and slender, pale brown .... *T. angustifolia*

Sparganium glomeratum

Sparganium natans

1 Staminate and pistillate portions of spike usually contiguous, not separated; leaves mostly 1–2 mm wide; stigmas broad and flattened, dark brown . . . . *T. latifolia*

Typha angustifolia L.
◊NARROW-LEAVED CAT-TAIL

DESCRIPTION Perennial emergent herb. **Stems** erect, 1–2 m long. **Leaves** upright, flat, 4–10 mm wide. **Flowers** either staminate or pistillate, on separate portions of the spike, separated by an interval of 2–10 cm; staminate portion 7–20 cm long and 7–15 mm wide, staminate bractlets brown; pistillate portion of spike dark brown, 10–20 cm long and 1–2 cm wide; each flower with 1 bristlelike bractlet, these flat and brown at the widened tip, gynophore hairs brown-tinged at tips; stigmas pale brown, linear, 1 mm long. **Fruit** 5–7 mm long, subtended by many fine hairs, the hairs slightly widened and brown at tip. June. **HABITAT** marshes, lakeshores, streambanks, roadside ditches, pond margins, usually in shallow water; more tolerant of brackish conditions than common cat-tail (*Typha latifolia*); introduced and invasive..
STATUS MW-OBL | NCNE-OBL | GP-OBL

Typha latifolia L.
◊COMMON CAT-TAIL

DESCRIPTION Perennial emergent herb. **Stems** erect, 1–2.5 m long. **Leaves** upright, mostly 1–2 cm wide. **Flowers** either staminate or pistillate, the staminate and pistillate portions of spike normally contiguous, rarely separated by 3–4 mm; staminate portion 5–15 cm long and 1.5–2 cm wide at flowering time, staminate bractlets white; pistillate portion of spike dark brown, 10–15 cm long and 2–3 cm wide when mature, pistillate bractlets absent, gynophore hairs white; stigma lance-shaped, becoming dark brown, less than 1 mm long. **Fruit** 1 cm long, with many white, linear hairs from base. **Flowering** June. **HABITAT** marshes, lakeshores, streambanks, ditches, pond margins, usually in shallow water; less tolerant of brackish conditions than narrow-leaved cat-tail (*Typha angustifolia*).
STATUS MW-OBL | NCNE-OBL | GP-OBL

Xyridaceae
Yellow-Eyed Grass Family

Xyris YELLOW-EYED GRASS
Perennial rushlike herbs. Stems erect, leafless, straight or sometimes ridged. Leaves all from base of plant, upright to

Typha angustifolia

Typha latifolia

spreading, linear, often twisted, usually dark green. Flowers small, perfect, yellow, from base of tightly overlapping bracts or scales, in rounded or cylindric heads at ends of stems; sepals 3, petals 3; stamens 3; style 3-parted. Fruit an oblong, 3-chambered capsule.

Xyris **YELLOW-EYED GRASS**

1 Plants swollen and hard at base .. *X. torta*
1 Plants flattened and soft at base **2**
2 Leaves 5 mm or more wide; upper flower scales with a green spot, 2–3 mm long, near center .... *X. difformis*
2 Leaves to 2 mm wide; flower scales without central green spot .... *X. montana*

Xyris difformis Chapman
♦ COMMON YELLOW-EYED GRASS

DESCRIPTION Native perennial herb. Stems leafless, 1.5–6 dm long, lower stem round in section and twisted, upper stem compressed and straighter, with 2 prominent ridges. Leaves linear, not twisted, l–5 dm long and 5–15 mm wide, widened to a soft base. Flowers yellow, in round to ovate spikes 0.5–1 cm long; scales ovate, entire; lateral sepals shorter than scales, the margins finely fringed from middle to tip; petals obovate, 4 mm long. Seeds 0.5 mm long. Flowering July–Aug.

HABITAT sandy or peaty lakeshores, sphag-num peatlands, floating sedge mats.
STATUS MW-OBL | NCNE-OBL | GP-OBL

Xyris montana H. Ries
♦ NORTHERN YELLOW-EYED GRASS

DESCRIPTION Native, densely clumped perennial herb. Stems leafless, 0.5–3 dm long, round in section, straight or lower part of stem slightly twisted. Leaves narrowly linear, flat or only slightly twisted, 5–20 cm long and 1–2 mm wide, rough, dark green, red-purple at base. Flowers yellow, in ovate spikes less than 1 cm long; scales obovate, finely fringed at tip; lateral sepals about as long as scales, linear, margins entire or finely hairy near tip. Seeds 1 mm long.

HABITAT wet, sandy shores, pools in sphag-num peatlands.
STATUS MW-OBL | NCNE-OBL | GP-OBL

Xyris torta J. E. Smith
♦ TWISTED YELLOW-EYED GRASS

DESCRIPTION Native perennial herb. Stems leafless, 1.5–8 dm long, spirally twisted, ridged. Leaves linear, twisted, 2–5 dm long and 2–5 mm wide; outer leaves shorter, tinged purple-brown and swollen and bulblike at base. Flowers yellow, in cylindric spikes 1–2.5 cm long; scales oblong; lateral sepals linear, about as long as scales,

Xyris difformis

Xyris montana

tips of scales and lateral sepals with tuft of short, red-brown hairs; petals obovate, 4 mm long. Seeds 0.5 mm long. Flowering June–Aug.

HABITAT wet sandy shores.

STATUS MW-OBL | NCNE-OBL | GP-OBL; Minn (END).

Xyris torta (leaves)

Xyris torta

CHRISTIAN FISCHER

Muskgrass or **Stonewort** (*Chara* spp., Characeae) is a macro-algae easily confused with vascular aquatic plants. Plants of *Chara* are gray-green, and have a gritty texture and strong musky odor when rubbed. When dried, *Chara* will turn whitish due to calcium deposits. Although not having true roots, *Chara* will loosely attach itself to the bottom of lakes and ponds via rhizoids (threadlike structures), and will sometimes form extensive colonies in shallow water, especially where calcium-rich. More than 30 species of *Chara* have been identified in the United States. Another stonewort, ***Nitella***, is similar but plants have no skunky odor, and its stems and branches are typically bright green and smooth to the touch.

Key to Groups and Families

THE FOLLOWING KEYS will aid in identifying unknown wetland and aquatic plants of the Upper Midwest region. In cases where the key is specific to a particular genus or species, these are noted in italics. Begin identifying an unknown plant by first using the **Key to Groups and Families** below. After identifying the family, turn to the page number for the family and continue keying the plant using the genus and species keys provided after the family description.

NOTE see page 420 for a simplified key to the region's **woody plants** (trees, shrubs and vines); page 422 for a key to the region's strictly **aquatic plants**.

KEY TO GROUPS AND FAMILIES

1 Plants herbaceous (not woody), reproducing by spores . . . .
. . . . . . **KEY 1. FERNS AND FERN ALLIES**

1 Trees, shrubs or herbs; reproducing by seeds (conifers, angiosperms) 2

2 Trees or shrubs with needlelike or scalelike, usually evergreen leaves; seeds in a dry cone, not inside an ovary. . . . .
. . . . **KEY 2. CONIFERS**

2 Trees, shrubs, or herbs; leaves usually deciduous; seeds inside an ovary which matures into the fruit (angiosperms) 3

3 Plant families with unusual or specialized features: Araceae ('duckweeds'), Asteraceae, Orchidaceae, and Apocynaceae ('milkweeds') will key here. 4

3 'Typical' trees, shrubs, or herbs. 7

4 Plants very small, floating in still water or stranded on wet shores; stem and leaf not differentiated; roots absent or 1–several from leaf underside . . . . **ARACEAE** p. 244

4 Plants larger; stems and leaves present. . 5

5 Flowers clustered into a head on a plate-like receptacle, the head resembling a single flower. . . . . **ASTERACEAE** p. 59

5 Flowers not as above; stamens, style and stigma are highly modified and joined into a special structure at center of flower . . . 6

6 Flowers irregular, the lower petal different than the other 2 petals (or rarely the upper petal different); ovary inferior . . . .

. . . . **ORCHIDACEAE** p. 344

6 Flowers regular, the 5 petals alike; ovary superior; fruit a podlike follicle containing many seeds, the seeds tufted with hairs . .
. . . . **APOCYNACEAE** p. 57

7 Herbaceous plants with undivided, usually narrow leaves, main veins parallel; petals and sepals in 3s or multiples of 3 . . . .
. . . . **KEY 3. MONOCOTS**

7 Plants herbaceous or woody; leaves simple or divided, veins usually in a net-like pattern; petals and sepals usually equal to or in multiples of 2, 4 or 5 . . . .
. . . . **KEY 4. DICOT GROUP KEYS**

KEY 1. FERNS AND FERN ALLIES

1 Leaves less than 1 cm long, sometimes scale-like, usually numerous and closely spaced along the stem (in separate whorls in Equisetaceae) 2

1 Leaves more than 1 cm long (often much longer), single or few in number (except in the aquatic *Isoetes*); if more numerous, then borne singly along a rhizome, or forming a crown or tuft 5

2 Leaves scale-like, blackish or brown, with whitish margins, in separated whorls of 3–30; stems jointed, hollow . . . .
. . . . **EQUISETACEAE** p. 25

2 Leaves green, not scale-like, alternate or opposite; stems solid. 3

3 Plants free floating on water; tiny (duckweed -ized) . . . **SALVINIACEAE** p. 37

3 Plants terrestrial, rooted in soil or occasionally on rocks 4

4 Plants resembling a moss (almost 2-dimensional) or new growth of a Juniper or a miniature clubmoss. Spores (of two sizes) hidden beneath the expanded parts of green leaves (sporophylls) in fertile stems that are often the only erect part of the plant . . . . **SELAGINELLACEAE** p. 38

4 Plants always 3-dimensional, resembling large mosses, miniature spruce/fir trees or the ultimate branches of *Thuja* (northern white cedar). Spores in discrete bean-shaped sporangia in leaf axils or in strobili subtended by usually non-green bract-like sporophylls . . . . . . **LYCOPODIACEAE** p. 31

5 Leaves ± round in cross section composed of 4 hollow tubes (seen in cross section),

quill-like, rather abruptly expanded at base to enclose sporangia .. ISOETACEAE p. 29

5 Leaves with a flat blade, not quill-like with 4 hollow tubes, not expanded at the very base to contain a sporangium 6

6 Photosynthetic blades 4-parted in clover-like fashion, often floating on water . . . . MARSILEACEAE p. 32

6 Photosynthetic blades not divided or clover-like, not floating on water 7

7 Sporangia fused laterally into 2-rowed, long-stalked spike-like structure borne on a stalk separate from the sterile blade; vegetative part of leaf entire . . . . OPHIOGLOSSACEAE . . . . (Ophioglossum) p. 34

7 Sporangia not fused, separate from the sterile blade or not; sterile parts of leaves usually variously divided (if not, then bearing elongated sori) 8

8 Spores in globose sporangia to 1 mm wide; indusia absent; borne singly on special fertile branches that turn yellow and eventually shrivel at or before spore release 9

8 Spores in sporangia (excluding stalks) less than 0.4 mm wide; indusia present or absent; sporangia in round or elongated clusters (sori) on the undersides or along the margins of regular, green blades, or sometimes on separate non-green, hard and persistent fertile fronds 10

9 Leaves usually 1 per plant, often less than 30 cm tall; fertile portions of leaves inserted near base of sterile portion; spores whitish or yellow . . . . OPHIOGLOSSACEAE p. 34

9 Leaves several to many per plant, in a cluster, normally more than 30 cm long; fertile portion of leaves either apical, in center of leaf, or leaves dimorphic with sporangia on separate fertile leaves; spores green . . . . OSMUNDACEAE p. 36

10 Sori borne inside hard, inrolled pinnae segments in separate stiff, long persistent fertile leaves that become hard and brown upon spore release . ONOCLEACEAE p. 33

10 Sori borne on regular green leaves, these sometimes somewhat different in form from the vegetative leaves, but not hard, brown, or persistent 11

11 Sori elongated; indusia elongate and attached on one side to form a flap . . . . ATHYRIACEAE p. 21

11 Sori round; indusia attached at a point, borne under the sori, hood-like, or absent 12

12 Leaf margins ciliate with whitish hairs; plants colonial from long-creeping rhizomes (fronds spaced along the rhizome) . . . . . THELYPTERIDACEAE p. 39

12 Leaf margins without whitish cilia (margins may have bristle-tipped teeth, or the blade surface may have brownish hairs); plants solitary from essentially erect rhizomes or forming clumps or patches from short-creeping rhizomes (fronds separated by less than ca. 5 mm along the rhizome) 13

13 Blade margins (especially at the apex of the pinnules) with teeth sharply acute, acuminate, or narrowed to a bristle-like tip; fronds firm, persistent into the fall or evergreen; indusia peltate or attached laterally at one point, ± persistent . . . . DRYOPTERIDACEAE p. 22

13 Blade margins with blunt teeth; fronds delicate, arising early in the season and often deciduous by late summer; indusia hood-like, attached on one side and arching over the sorus, fragile and not persisting . . . . CYSTOPTERIDACEAE p. 22

KEY 2 - CONIFERS

Division Spermatopsida (Flowering Plants), Subdivision Gymnosperm

1 Leaves scalelike, pressed flat to the stem . . . . CUPRESSACEAE . . . . (Thuja occidentalis) p. 41

1 Leaves needlelike, often borne in clusters . . . . PINACEAE p. 41

KEY 3 - MONOCOTS

Division Spermatopsida (Flowering Plants), Class Monocotyledoneae ("Monocots")

1 Sepals and petals (perianth) absent or reduced to scales or bristles, never petal-like in size or color 2

1 Sepals and petals present 17

2 Flowers in axils of chaffy scales larger than the flower; sepals and petals absent or reduced to bristles or small scales; flowers in regular heads or spikes 3

2 Flowers not in chaffy bracts, or if bracts present, the flower equal or larger in size and not hidden by the bracts 5

3 Plants with a basal cluster of narrow linear leaves and an upright stalk tipped by a single, button-like head . . . . ERIOCAULACEAE (Eriocaulon) p. 325

3 Plants with leaves on stem (these sometimes reduced to scales), or plants with several to many heads or spikes ...**4**

4 Leaves usually 2-ranked around stem; stems usually hollow, round, or flat, and never triangular in cross-section; leaf sheath usually open on side opposite blade ....**POACEAE** p. 359

4 Leaves usually 3-ranked (sometimes reduced to scales); stems usually solid and pithy and triangular in cross-section; leaf sheath usually closed **CYPERACEAE** p. 251

5 Plants aquatic, leaves submerged or floating and becoming limp when withdrawn from water; flowers underwater, floating, or sometimes raised slightly above water surface **6**

5 Plants of land or shallow water, leaves and flowers normally above water surface..**11**

6 Flowers small and inconspicuous, single to several from leaf axils...................**7**

6 Flowers in heads or spikes **10**

7 Leaves alternate, or sometimes uppermost leaves opposite **8**

7 Leaves all opposite (or whorled) **9**

8 Widespread plants of mostly fresh water; flowers with 4 sepals and 4 stamens...... ....**POTAMOGETONACEAE** p. 387

8 Uncommon plants of brackish water ....**RUPPIACEAE** (*Ruppia cirrhosa*) p. 401

9 Leaves 1–4 cm long, tapering from base to a long point, margins with minute, spiny teeth; flowers with 1 ovary ....**HYDROCHARITACEAE** (*Najas*) p. 327

9 Leaves 3–10 cm long, threadlike, margins not spiny; flowers usually with 4 ovaries . ...**POTAMOGETONACEAE** ....(*Zannichellia*) p. 400

10 Flowers in globe-shaped heads; upper heads staminate and deciduous, lower heads pistillate and persistent ....**TYPHACEAE** (*Sparganium*) p. 402

10 Flower heads or spikes all alike; flowers perfect (with both staminate and pistillate parts) ....**POTAMOGETONACEAE** p. 387

11 Flowers tiny, in spikes or heads surrounded by a large white or co-lored bract (spathe); leaves broad...........**ARACEAE** p. 244

11 Flowers not surrounded by a large spathe; leaves ± linear **12**

12 Flowers many, in long spikes **13**

12 Flowers in globe-shaped heads, racemes, or loose, open clusters **15**

13 Spike appearing to be from side of stem .. ..**ACORACEAE** (*Acorus*) p. 239

13 Spike erect at end of stem **14**

14 Top of spike slender, made up of staminate flowers; lower spike broader and of pistillate flowers ....**TYPHACEAE** ....(*Typha*) p. 405

14 Spike uniform; flowers both staminate and pistillate ....**JUNCAGINACEAE** ....(*Triglochin*) p. 340

15 Flowers either staminate or pistillate, with staminate in upper clusters, pistillate in lower clusters ....**TYPHACEAE** ....(*Sparganium*) p. 402

15 Flowers with both staminate and pistillate parts.................................**16**

16 Flowers with 1 ovary; fruit a 3-parted capsule ......**JUNCACEAE** (*Juncus*) p. 331

16 Flowers with 3 or 6 ovaries; fruit separating into segments when mature ....**JUNCAGINACEAE** (*Triglochin*) p. 340

17 Flowers either staminate or pistillate, on same or separate plants **18**

17 Flowers with both staminate and pistillate parts.................................**20**

18 Flowers with 3 green sepals and 3 white or pinkish petals....**ALISMATACEAE** p. 240

18 Sepals and petals ± same color........**19**

19 Aquatic plants, mostly underwater; stamens 3–12 ....**HYDROCHARITACEAE** p. 325

19 Land plants; stamens 3–6 ....*traditional* **LILIACEAE** p. 341

20 Ovary inferior (located below sepals and petals) **21**

20 Ovary or ovaries superior (attached above sepals and petals); plants aquatic and on land **23**

21 Aquatic plants; leaves underwater or floating....**HYDROCHARITACEAE** p. 325

21 Plants emergent in shallow water or on wet soils **22**

22 Stamens 3 ....**IRIDACEAE** p. 329

22 Stamens 6 . *traditional* **LILIACEAE** p. 341

23 Flowers with 1 ovary.................**24**

23 Flowers with 2 or more ovaries **28**

24 Flowers radially symmetric (regular) ..**25**

24 Flowers irregular, not radially symmetric. ....**PONTEDERIACEAE** p. 386

25 Sepals green; petals variously colored..**26**

25 Sepals and petals with same appearance.. ....**27**

26 Stamens 3 (or sometimes 2) ....**XYRIDACEAE** (*Xyris*) p. 406

26 Stamens 6 .. *traditional* **LILIACEAE** p. 341

27 Petals 6, united into a tube; stamens 3; plants submerged or on muddy shores ... .**PONTEDERIACEAE** p. 386

27 Stamens 6; plants of drier sites. . . . .
. . . . *traditional* **LILIACEAE** p. 438
28 Leaves from base of plant and alternate on stem; flowers with 3 pistils . . . .
. . . . **SCHEUCHZERIACEAE**
. . . . (*Scheuchzeria palustris*) p. 401
28 Leaves all from base of plant; pistils 3 to many **29**
29 Flowers with 3 green sepals and 3 white or pinkish petals; flowers in panicles or umbels . . . . **ALISMATACEAE** p. 240
29 Flowers with 6 pink petals, the outer 3 smaller and darker pink than the inner; flowers in an umbel . . . . . **BUTOMACEAE**
. . . . (*Butomus umbellatus*) p. 250

KEY 4 - DICOT GROUP KEYS
Division Spermatopsida (Flowering Plants), Class Dicotyledoneae ("Dicots")

1 Trees, shrubs, or woody vines . **GROUP 1**
1 Plants herbaceous **2**
2 Flowers either staminate or pistillate (imperfect) . . . . **GROUP 2**
2 Flowers with both staminate and pistillate parts (perfect) **3**
3 Petals and sepals absent . . . . **GROUP 3**
3 Petals and sepals present **4**
4 Flowers with either sepals or petals but not both **5**
4 Flowers with both sepals and petals **6**
5 Ovary inferior (below sepals and petals) . .
. . . . **GROUP 4**
5 Ovary superior (above sepals and petals) .
. . . . **GROUP 5**
6 Flowers with 2 or more ovaries . **GROUP 6**
6 Flowers with 1 ovary; styles or stigmas 2 or more **7**
7 Ovary inferior . . . . **GROUP 7**
7 Ovary superior **8**
8 Stamens many and more than petals . . . **9**
8 Stamens few; same or fewer than number of petals **10**
9 Flowers regular . . . . **GROUP 8**
9 Flowers irregular . . . . **GROUP 9**
10 Flower petals separate, not joined . . . .
. . . . **GROUP 10**
10 Petals united **11**
11 Flowers regular and stamens same number as petals or petal lobes . . . . **GROUP 11**
11 Flowers either irregular, or number fewer than corolla lobes . . . . **GROUP 12**

DICOT GROUP 1
Trees, shrubs, and woody vines.

1 Leaves opposite or whorled **2**
1 Leaves alternate **15**
2 Flowers fully developed before leaves expand **3**
2 Flowers develop with or after leaves expand **6**
3 Flowers with both petals and sepals **4**
3 Only petals or sepals present (not both), or absent **5**
4 Trees, petals separate, not joined; flowers either staminate or pistillate; stamens 8; ovary superior . . . . **SAPINDACEAE**
. . . . (*Acer*) p. 228
4 Shrubs; flowers perfect, with both staminate and pistillate parts; stamens 5; ovary inferior . . . . **CAPRIFOLIACEAE**
. . . . (*Lonicera*) p. 102
5 Stamens usually 2; ovary not lobed . . . . . .
. . . . **OLEACEAE** (*Fraxinus*) p. 163
5 Stamens usually 8; ovary distinctly 2-lobed
. . . . **SAPINDACEAE** (*Acer*) p. 228
6 Leaves compound **7**
6 Leaves simple **10**
7 Plants vinelike and trailing . . . .
. . . . **RANUNCULACEAE** (*Clematis*) p. 196
7 Plants not vinelike **8**
8 Petals present . . . . **ADOXACEAE**
. . . . (*Sambucus*) p. 44
8 Petals absent **9**
9 Stamens 2; ovary not lobed . . . **OLEACEAE**
. . . . (*Fraxinus*) p. 163
9 Stamens 8; ovary 2-lobed **SAPINDACEAE**
. . . . (*Acer negundo*) p. 228
10 More stamens than petals or petal lobes **11**
10 Stamens equal or less than number of petals or petal lobes. **13**
11 Petals joined . . . . **ERICACEAE** p. 114
11 Petals separate **12**
12 Stamens 8-10 . . . . **LYTHRACEAE**
. . . . (*Decodon verticillatus*) p. 154
12 Stamens numerous . . . . . **HYPERICACEAE**
. . . . (*Hypericum kalmianum*) p. 138
13 Petals separate . . . . **CORNACEAE**
. . . . (*Cornus*) p. 108
13 Petals joined **14**
14 Flowers in many-flowered, globe-shaped heads; leaves entire; shrub. . **RUBIACEAE**
. . . . (*Cephalanthus occidentalis*) p. 215
14 Flowers not in dense heads; leaves entire, toothed, or lobed . . . .
. . . . **CAPRIFOLIACEAE** p. 101

15 Plants with either staminate or pistillate flowers but not both on same plant (dioecious)........................ **16**

15 Plants with staminate and pistillate flowers on same plant; flowers either perfect (with both staminate and pistillate parts), or imperfect (flowers single sex only)..... **24**

16 Woody climbing vines .... **VITACEAE** .... (*Vitis riparia*) p. 239

16 Trees or shrubs **17**

17 Flowers in catkins; individual flowers small **18**

17 Flowers not in catkins; flowers often large and showy **19**

18 Trees or shrubs; twigs not covered with resinous dots .... **SALICACEAE** p. 218

18 Shrubs; twigs densely covered with resinous dots .... **MYRICACEAE** .... (*Myrica gale*) p. 159

19 Small shrub; leaves cylindric, short, 3–8 mm long; rare in ne Minn and Mich UP, mostly near Lake Superior ..**ERICACEAE** .... (*Empetrum nigrum*) p. 116

19 Mostly larger shrubs; leaves much larger **20**

20 Sepals and petals absent or not different from one another (tepals); tepals 6 and petal-like, yellow; stamens 9 ....**LAURACEAE** (*Lindera benzoin*) p. 147

20 Female flowers with sepals and petals, although sepals sometimes very small . **21**

21 Flowers in a large terminal panicle .... **ANACARDIACEAE** .... (*Toxicodendron*) p. 50

21 Flowers 1 to several from leaf axils..... **22**

22 Style very short... **AQUIFOLIACEAE** p. 58

22 Style relatively long.................... **23**

23 Style long, curved at tip; tree of se Wisc and Mich LP .... **NYSSACEAE** .... (*Nyssa sylvatica*) p. 162

23 Style of pistillate flower divided above its middle; shrubs or small trees .... **RHAMNACEAE** p. 204

24 Flowers small, either staminate or pistillate, never perfect; staminate flowers in catkins or round heads **25**

24 Flowers mostly perfect, sometimes large and showy, not grouped into catkins or globe-shaped heads **29**

25 Staminate flowers in crowded, globe-shaped heads; tree of s Wisc and Mich LP .... **PLATANACEAE** .... (*Platanus occidentalis*) p. 182

25 Staminate flowers in cylindric catkins . **26**

26 Female flowers 1, or several in clusters . **27**

26 Female flowers more numerous; cone-like, or arranged into catkins or heads **28**

27 Tree, leaves pinnately compound; s LP of Mich .... **JUGLANDACEAE** .... (*Carya laciniosa*) p. 140

27 Tree, leaves not compound; margin wavy or lobed; widespread .... **FAGACEAE** .... (*Quercus*) p. 124

28 Female flowers 2 or 3 behind each catkin scale................. **BETULACEAE** p. 88

28 Female flowers 1 behind each catkin scale .... **MYRICACEAE** (*Myrica gale*) p. 159

29 Petals and sepals absent or in a single series, the petals and sepals not distinct from one another..................... **30**

29 Petals and sepals present and clearly distinct from each other.............. **36**

30 Stamens about 10, petals tiny ....**NYSSACEAE** (*Nyssa sylvatica*) p. 162

30 Stamens as many as number of petal and sepal lobes **31**

31 Style 1, sometimes branched **32**

31 Styles 2–3........................... **35**

32 Vines ....**VITACEAE** (*Vitis riparia*) p. 239

32 Shrubs or small trees **33**

33 Flowers in clusters at ends of branches ... . **CORNACEAE** (*Cornus*) p. 108

33 Flowers from leaf axils or on short lateral branches........................... **34**

34 Style not branched, stigma 1 .... **AQUIFOLIACEAE** p. 58

34 Style 2–4-parted, stigmas 2–4.......... .... **RHAMNACEAE** p. 204

35 Shrub, stems stout, very prickly; uncommon plant of Isle Royale and north shore of Lake Superior ....**ARALIACEAE** .... (*Oplopanax horridus*) p. 59

35 Tree, stems and twigs not prickly; widespread in region ..**ULMACEAE** p. 231

36 Ovaries 3 or more; stamens more than 10 .... **ROSACEAE** p. 206

36 Ovary 1 .... **37**

37 Petals 1 .... **FABACEAE** .... (*Amorpha fruticosa*) p. 122

37 Petals more than 1; flowers ± regular ... **38**

38 Petals joined **39**

38 Petals free **41**

39 Stamens more in number than corolla lobes .... **ERICACEAE** p. 114

39 Stamens equal to number of corolla lobes **40**

40 Style well-developed ..**ERICACEAE** p. 114

40 Style short, the stigma ± stalkless .... **AQUIFOLIACEAE** p. 58

41 Ovary inferior...................... **42**

41 Ovary superior **45**

42 Stamens 2x or more than number of petals **43**

42 Stamens equal number of petals **44**

43 Style 1 . . . . **ERICACEAE** (*Vaccinium*) p. 119

43 Styles 2–5 . . . . **ROSACEAE** p. 206

44 Petals 4 . . . . **CORNACEAE** p. 108

44 Petals 5 . . . . **GROSSULARIACEAE**
. . . . (*Ribes*) p. 129

45 Leaves short and cylindric, less than 1 cm long . . . . **ERICACEAE**
. . . . (*Empetrum nigrum*) p. 116

45 Leaves flat and broader **46**

46 Leaves compound . . . . **ANACARDIACEAE**
. . . . (*Toxicodendron*) p. 50

46 Leaves simple **47**

47 Stamens number more than petals . . . .
. . . . **ERICACEAE** p. 114

47 Stamens number as many as petals **48**

48 Flowers in clusters at ends of stems; leaf margins rolled under, leaf underside densely hairy . . . . **ERICACEAE**
. . . . (*Rhododendron*) p. 118

48 Flowers from leaf axils; leaf margins not rolled under, leaves not densely hairy below **49**

49 Style 3-parted; stamens opposite the petals . . . . **RHAMNACEAE** (*Rhamnus*) p. 204

49 Style very short; stamens alternate with the petals . . . . **AQUIFOLIACEAE** p. 58

DICOT GROUP 2

Herbaceous plants; flowers either staminate or pistillate (imperfect).

1 Leaves absent or reduced to small scales . . . . **AMARANTHACEAE** (*Salicornia*) p. 49

1 Leaves not reduced to scales **2**

2 Aquatic plants; leaves dissected into threadlike segments **3**

2 Leaves not dissected into threadlike segments **4**

3 Leaves pinnately dissected . . . .
. . . . **HALORAGACEAE** p. 133

3 Leaves palmately dissected . . . .
. . . . **CERATOPHYLLACEAE**
. . . . (*Ceratophyllum*) p. 108

4 Leaves simple, not compound **5**

4 Leaves compound **17**

5 Leaves all from base of plant **6**

5 Leaves all or mostly along stem **7**

6 Flowers in short or long spikes . . . .
. . . . **PLANTAGINACEAE** p. 174

6 Flowers in small panicles . . . .
. . . . **POLYGONACEAE** (*Rumex*) p. 188

7 Leaves whorled or opposite **8**

7 Leaves alternate **13**

8 Flowers single **9**

8 Flowers in clusters from leaf axils or at ends of stems **10**

9 Leaves whorled . . . . . **PLANTAGINACEAE**
. . . . (*Hippuris vulgaris*) p. 178

9 Leaves opposite . . . . . **PLANTAGINACEAE**
. . . . (*Callitriche*) p. 175

10 Flower clusters from leaf axils **11**

10 Flowers in clusters at ends of stems **12**

11 Leaf margins entire . . . .
. . . . **PLANTAGINACEAE** p. 174

11 Leaf margins toothed . . . .
. . . . **URTICACEAE** p. 232

12 Stamens 3, style 1 . . . .
. . . . **CAPRIFOLIACEAE** p. 101

12 Stamens 10, styles mostly 5 . . . .
. . . . . **CARYOPHYLLACEAE** (*Silene*) p. 104

13 Sepals and petals present **14**

13 Petals absent **16**

14 Plants vining and with tendrils . . . .
. . . . **CURCURBITACEAE** p. 111

14 Plants not vining; tendrils absent **15**

15 Flowers in a panicle at end of stem . . . .
. . . . **MALVACEAE** (*Napaea*) p. 157

15 Flowers single . **ROSACEAE** (*Rubus*) p. 211

16 Sepals dry and chaffy . . . .
. . . . **AMARANTHACEAE** p. 47

16 Sepals often absent, herbaceous . . . .
. . . . **AMARANTHACEAE**
. . . . (*Atriplex, Chenopodium*) p. 48

17 Leaves divided into 3 leaflets . . . .
. . . . **RANUNCULACEAE** (*Clematis*) p. 196

17 Leaves divided into more than 3 leaflets **18**

18 Stem leaves alternate **RANUNCULACEAE**
. . . . (*Thalictrum*) p. 203

18 Stem leaves opposite . . . .
. . . . **CAPRIFOLIACEAE** p. 101

DICOT GROUP 3

Herbaceous plants; flowers with both staminate and pistillate parts (perfect); flowers without petals or sepals.

1 Plants underwater or sometimes stranded on shores . . . . **HALORAGACEAE** p. 133

1 Land plants or of shallow water only **2**

2 Leaves deeply lobed or compound . . . .
. . . . **RANUNCULACEAE** p. 194

2 Leaves simple, margins entire **3**

3 Leaves whorled, linear . . . .
. . . . **PLANTAGINACEAE**
. . . . (*Hippuris vulgaris*) p. 178

3 Leaves alternate, heart-shaped . . . .
. . . . **SAURURACEAE**
. . . . (*Saururus cernuus*) p. 229

DICOT GROUP 4

Herbaceous plants; flowers with both staminate and pistillate parts (perfect); petals and sepals not different (in one series); ovary inferior.

1 Stamens number more than sepals and petals 2
1 Stamens less than or equal to number of sepals and petals 4
2 Stamens many; stipules large ROSACEAE .... (*Sanguisorba canadensis*) p. 214
2 Stamens 4–8; plants in water or on wet soils; stipules ± absent 3
3 Leaves broadly ovate . SAXIFRAGACEAE ..... (*Chrysosplenium americanum*) p. 230
3 Leaves linear or divided into linear segments .... HALORAGACEAE p. 133
4 Leaves from base of plant or alternate... 5
4 Leaves opposite or whorled 10
5 Stamens 4 or less; perianth lobes 3 or 4 . 6
5 Stamens 5; perianth lobes 5 9
6 Leaves with large stipules .... ROSACEAE .... (*Sanguisorba canadensis*) p. 214
6 Stipules absent 7
7 Stamens 3; perianth lobes 3 .... HALORAGACEAE p. 133
7 Stamens 4; perianth lobes 4 8
8 Leaves broadly ovate, margins with rounded teeth .... SAXIFRAGACEAE ..... (*Chrysosplenium americanum*) p. 230
8 Leaves linear or lance-shaped, margins entire .. ONAGRACEAE (*Ludwigia*) p. 167
9 Flowers single in leaf axils; leaf margins entire .... SANTALACEAE .... (*Geocaulon lividum*) p. 227
9 Flowers in heads or umbels; leaves usually dissected............... APIACEAE p. 51
10 Flowers densely clustered into heads at ends of stems .... CORNACEAE .... (*Cornus*) p. 108
10 Flowers not in dense heads............ 11
11 Leaves whorled .... RUBIACEAE .... (*Galium*) p. 215
11 Leaves opposite 12
12 Flowers in clusters at ends of stems; stamens 3 ..... CAPRIFOLIACEAE p. 101
12 Flowers single or few from leaf axils; stamens 4 13
13 Style 1 .. ONAGRACEAE (*Ludwigia*) p. 167
13 Styles 2 .... SAXIFRAGACEAE ..... (*Chrysosplenium americanum*) p. 230

DICOT GROUP 5

Herbaceous plants; flowers with both staminate and pistillate parts (perfect); petals and sepals not different (in 1 series); ovary superior.

1 Each flower with more than 1 ovary; these sometimes joined up to their middle, free above 2
1 Flowers with 1 ovary 3
2 Ovaries of flowers joined up to their middle .... SAXIFRAGACEAE .... (*Penthorum sedoides*) p. 171
2 Ovaries distinct, not joined .... RANUNCULACEAE p. 194
3 Stamens more than 2x number of perianth lobes.................................. 4
3 Stamens 2x or fewer than perianth lobes 5
4 Leaves all at base of plant, modified into water-holding pitchers .... SARRACENIACEAE .... (*Sarracenia purpurea*) p. 229
4 Leaves not pitcher-like, large and heart-shaped, usually floating on water surface . ... NYMPHAEACEAE p. 160
5 Styles 2 or more...................... 6
5 Styles absent or 1.................... 10
6 Leaves small and scalelike; plants of salty habitats .... AMARANTHACEAE .... (*Salicornia*) p. 49
6 Leaves not scalelike.................... 7
7 Leaves opposite or whorled............ 8
7 Leaves alternate 9
8 Leaf margins with shallow, rounded teeth .... SAXIFRAGACEAE ..... (*Chrysosplenium americanum*) p. 230
8 Leaf margins entire .... CARYOPHYLLACEAE p. 104
9 Stipules present at base of each leave and sheathing stem .... POLYGONACEAE .... p. 184
9 Stipules absent AMARANTHACEAE p. 47
10 Stamens more numerous than perianth lobes 11
10 Stamens as many or fewer than number of perianth lobes....................... 12
11 Leaves opposite .... LYTHRACEAE .... (*Ammania*) p. 153
11 Leaves alternate .... BRASSICACEAE (*Cardamine*) p. 94
12 Flowers in clusters at ends of stems .... CARYOPHYLLACEAE p. 104
12 Flowers in clusters from leaf axils...... 13
13 Perianth with 4 lobes LYTHRACEAE p. 153
13 Perianth with 5 lobes .... PRIMULACEAE .... (*Lysimachia maritima*) p. 190

DICOT GROUP 6

Herbaceous plants; flowers with both staminate and pistillate parts (perfect); petals and sepals present; each flower with 2 or more ovaries.

1 One style for each flower 2
1 Styles as many as ovaries, or styles absent
   4
2 Petals ± separate, not joined; stamens many; ovaries 5 or more . . . .
   . . . . MALVACEAE p. 156
2 Petals joined; stamens 2–5; ovaries 4 . . . 3
3 Leaves opposite; stamens 2 or 4 . . . .
   . . . . LAMIACEAE p. 140
3 Leaves alternate; stamens 5. . . . .
   . . . . BORAGINACEAE p. 91
4 Flowers irregular . . . . RANUNCULACEAE
   . . . . (*Aconitum columbianum*) p. 195
4 Flowers regular. 5
5 Sepals 3; petals 3 . CABOMBACEAE p. 98
5 Sepals and petals each more than 3 6
6 Leaves round, attached to petiole at center; flowers single, large, 1–2 dm wide . . . .
   . . . . NELUMBONACEAE
   . . . . (*Nelumbo lutea*) p. 160
6 Leaves not attached to petiole at center . 7
7 Sepals not joined to form a cup. . . . .
   . . . . RANUNCULACEAE p. 194
7 Sepals joined and cuplike 8
8 Pistils as many or more than number of petals . . . . ROSACEAE p. 206
8 Pistils fewer than petals 9
9 Leaves entire or shallowly lobed . . . .
   . . . . SAXIFRAGACEAE p. 230
9 Leaves compound . . . . ROSACEAE
   . . . . (*Agrimonia*) p. 206

DICOT GROUP 7

Herbaceous plants; flowers with both staminate and pistillate parts (perfect); petals and sepals present; ovary inferior, 1 in each flower.

1 Stamens more than number of petals . . . 2
1 Stamens less than or equal to number of petals 6
2 Style 1 3
2 Styles 2 or 3 5
3 Plants of moist soils ONAGRACEAE p. 164
3 Aquatic plants or plants of wet, muddy shores 4
4 Flowers small, not showy, white or green .
   . HALORAGACEAE (*Myriophyllum*) p. 133

4 Flowers large and showy, yellow . . . .
   . . . . ONAGRACEAE (*Ludwigia*) p. 167
5 Styles 2 . . . . SAXIFRAGACEAE p. 230
5 Styles 3 . . . . MONTIACEAE
   . . . . (*Montia chamissoi*) p. 159
6 Petals separate 7
6 Petals joined for most of their length. . . 14
7 Petals 2; stamens 2 . . . . . . ONAGRACEAE
   . . . . (*Circaea alpina*) p. 164
7 Petals 4 or 5; stamens 4 or 5 8
8 Petals 4 9
8 Petals 5 11
9 Plants with underwater leaves, these dissected into narrow segments . . . .
   . . . . HALORAGACEAE
   . . . . (*Myriophyllum*) p. 133
9 Underwater leaves absent, or if present, the leaves entire 10
10 Leaves with normal blade ONAGRACEAE
   . . . . (*Ludwigia*) p. 167
10 Leaves reduced to small scales . . . .
   . HALORAGACEAE (*Myriophyllum*) p. 133
11 Leaves entire 12
11 Leaves dissected or compound 13
12 Flowers in panicles. . . . .
   . . . . SAXIFRAGACEAE p. 230
12 Flowers in heads or umbels . . . .
   . . . . APIACEAE p. 51
13 Flowers in narrow, spikelike racemes . . . .
   . . . . ROSACEAE (*Agrimonia*) p. 206
13 Flowers in umbels. . . . . . . APIACEAE p. 51
14 Stem leaves alternate 15
14 Stem leaves alternate or whorled, or leaves all from base of plant 17
15 Flowers irregular . . . CAMPANULACEAE
   . . . . (*Lobelia*) p. 99
15 Flowers regular. 16
16 Flowers small, 2–3 mm wide . . . .
   PRIMULACEAE (*Samolus parviflorus*) p. 194
16 Flowers larger . . . . . . CAMPANULACEAE
   . . . . (*Campanula aparinoides*) p. 99
17 Leaves in whorls of 3–8 . . . . . RUBIACEAE
   . . . . (*Galium*) p. 215
17 Leaves opposite along stem or all from base of plant 18
18 Stamens 3 . . . . . CAPRIFOLIACEAE p. 101
18 Stamens 4 or 5 . . . . CAPRIFOLIACEAE
   . . . . (*Linnaea borealis*) p. 102

DICOT GROUP 8

Herbaceous plants; flowers with both staminate and pistillate parts (perfect); petals and sepals present; flowers regular, with a single superior ovary; stamens number more than petals or corolla lobes.

1 Stamens more than 2x number of petals. **2**
1 Stamens 2x number of petals or less.....**4**
2 Aquatic plants with large leaves from base of plant........**NYMPHAEACEAE** p. 160
2 Land plants with alternate or opposite leaves along stem......................**3**
3 Leaves alternate .....**MALVACEAE** p. 156
3 Leaves opposite ..**HYPERICACEAE** p. 136
4 Stamens less than 2x number of petals ..**5**
4 Stamens exactly 2x number of petals....**9**
5 Leaves opposite or whorled; styles 2–5 ..**6**
5 Leaves alternate; style 1...............**8**
6 Flowers yellow ...**HYPERICACEAE** p. 136
6 Flowers not yellow....................**7**
7 Stamens grouped into 3 clusters of 3 each .....**HYPERICACEAE** (*Triadenum*) p. 139
7 Stamens not grouped together.......... ....**CARYOPHYLLACEAE** p. 104
8 Sepals 4; petals 4 ..**BRASSICACEAE** p. 92
8 Sepals 5; petals 5 ....**FABACEAE** ....(*Senna hebecarpa*) p. 123
9 Leaves simple, margins entire or shallowly toothed ....**MELASTOMATACEAE** ....(*Rhexia*) p. 157
9 Leaves compound or divided to base...**10**
10 Leaves 3-parted; styles 5 . **OXALIDACEAE** ....(*Oxalis montana*) p. 171
10 Styles 1; pinnately compound . **FABACEAE** ....(*Senna hebecarpa*) p. 123
11 Style 1 ....**12**
11 Styles 2 or more **13**
12 Ovary lobed, each lobe tipped by a style .. ..**SAXIFRAGACEAE** p. 230
12 Ovary not lobed, the styles all from tip of ovary **14**
13 Flowers yellow ....**HYPERICACEAE** ....(*Triadenum*) p. 139
13 Flowers white or pink-tinged ....**CARYOPHYLLACEAE** p. 104

DICOT GROUP 9

Herbaceous plants; flowers with both staminate and pistillate parts (perfect); petals and sepals present; flowers irregular, with a single superior ovary; stamens number more than petals or corolla lobes.

1 Sepals petal-like, or long and spurlike ...**2**
1 Sepals not petal-like, often green **3**
2 One sepal a spur or sac; leaf margins shallowly toothed...................... ....**BALSAMINACEAE** p. 87
2 Sepals not spurlike; leaves entire......... ....**POLYGALACEAE** p. 183
3 Lower 2 petals joined for their length along lower margins, enclosing the stamens.... ....**FABACEAE** p. 122
3 Lower 2 petals not joined, or petals 1 ...**4**
4 Leaves compound ....**FABACEAE** ....(*Senna hebecarpa*) p. 123
4 Leaves simple, entire to lobed **5**
5 Styles 2 ....**SAXIFRAGACEAE** ....(*Micranthes pensylvanica*) p. 231
5 Style 1 ....**LYTHRACEAE** (*Lythrum*) p. 155

DICOT GROUP 10

Herbaceous plants; flowers with both staminate and pistillate parts (perfect), with a single superior ovary; petals and sepals present; petals separate, not joined; stamens fewer or equal to number of petals.

1 Leaves opposite **2**
1 Leaves alternate or all from base of plant **9**
2 Sepals 2 or 3; petals 2 or 3 . **ELATINACEAE** ....(*Elatine*) p. 114
2 Sepals and petals each 4–6 or more **3**
3 Style 1................................**4**
3 Styles 2–5............................**6**
4 Flowers with well-developed hypanthium (cuplike structure around ovary) ....**LYTHRACEAE** p. 153
4 Hypanthium absent...................**5**
5 Stamens opposite the petal lobes........ ....**PRIMULACEAE** p. 190
5 Stamens alternate with petal lobes....... ....**GENTIANACEAE** p. 125
6 Ovary and capsule divided into 4–5 parts . ...**LINACEAE** p. 152
6 Ovary and capsule 1-parted **7**
7 Flowers yellow ....**HYPERICACEAE** ....(*Hypericum*) p. 136
7 Flowers white to pink **8**
8 Petals joined at their base ....**GENTIANACEAE** ....(*Sabatia angularis*) p. 129
8 Petals free to base...................... ....**CARYOPHYLLACEAE** p. 104
9 Styles 2 or more **10**
9 Styles 1 or absent **11**

10 Leaves all from base of plant, covered with sticky, stalked glands . . . . **DROSERACEAE** p. 112

10 Leaves from stem, smooth. . . . . **LINACEAE** p. 152

11 Hypanthium tube-shaped . **LYTHRACEAE** . . . . (*Lythrum*) p. 155

11 Hypanthium absent **12**

12 Flowers irregular, sometimes spurred . . . . **VIOLACEAE** p. 235

12 Flowers regular, spur absent **13**

13 Flowers single at ends of stems . . . . **CELASTRACEAE** (*Parnassia*) p. 107

13 Flowers in a cluster at ends of stems . . . . **PRIMULACEAE** (*Lysimachia*) p. 191

11 Leaves compound or dissected into leaflets **12**

11 Leaves entire, toothed, or with shallow lobes only **13**

12 Leaves divided into 3 leaflets . . . . **MENYANTHACEAE** . . . . (*Menyanthes trifoliata*) p. 158

12 Leaves pinnately compound . . . . **POLEMONIACEAE** (*Polemonium*) p. 183

13 Leaves small and scalelike; corolla 4-lobed . . . . . . **GENTIANACEAE** (*Bartonia*) p. 125

13 Leaves not scalelike; corolla deeply 5-lobed . . . . **PRIMULACEAE** . . . . (*Samolus valerandi*) p. 194

DICOT GROUP 11

Herbaceous plants; flowers with both staminate and pistillate parts (perfect), with a single superior ovary; petals and sepals present; petals joined; stamens equal to number of petal lobes.

1 Leaves all from base of plant; flowers at end of naked stalk **2**

1 Leaves mostly along stem **3**

2 Flowers 4-parted, chaffy, in spikes or heads . . . . **PLANTAGINACEAE** p. 174

2 Flowers 5-parted, not chaffy, in umbel at end of stem . . . . **PRIMULACEAE** . . . . (*Primula mistassinica*) p. 193

3 Ovary deeply parted into 2 or 4 sections . **4**

3 Ovary not parted **5**

4 Leaves opposite . . . . . **LAMIACEAE** p. 140

4 Leaves alternate . . **BORAGINACEAE** p. 91

5 Leaves opposite or whorled. **6**

5 Leaves alternate **11**

6 Flowers in crowded heads or spikes; corolla 4-lobed **7**

6 Flowers mostly not in dense heads (sometimes in short racemes); corolla 4–12-lobed **8**

7 Flowers chaffy; leaves linear . . . . **PLANTAGINACEAE** p. 174

7 Flowers petal-like; leaves not linear . . . . **VERBENACEAE** (*Phyla lanceolata*) p. 234

8 Stamens attached opposite the corolla lobes . . . . **PRIMULACEAE** p. 190

8 Stamens alternate with corolla lobes **9**

9 Corolla lobes 4 or 6–12 . . . . **GENTIANACEAE** p. 125

9 Corolla lobes 5 . . . . **10**

10 Stigmas 1. . . . . **GENTIANACEAE** p. 125

10 Stigmas 3 . . . . . **POLEMONIACEAE** p. 182

DICOT GROUP 12

Herbaceous plants; flowers with both staminate and pistillate parts (perfect), with a single superior ovary; petals and sepals present; petals joined. Flowers irregular, or stamens less than number of petal lobes.

1 Corolla base a spur or sac. **2**

1 Corolla not spurred or saclike. **4**

2 Sepals 2-lobed . . . . . **LENTIBULARIACEAE** . . . . (*Utricularia*) p. 148

2 Sepals joined, deeply 5-lobed **3**

3 Leaves all from base of plant; flowers single atop a naked stalk. . **LENTIBULARIACEAE** . . . . (*Pinguicula*) p. 148

3 Leaves mostly along stem . . . . **PLANTAGINACEAE** p. 174

4 Leaves alternate or all from base of plant 5

4 Leaves opposite or whorled. **6**

5 Stamens 2. . . . . **PLANTAGINACEAE** . . . . (*Veronica*) p. 180

5 Stamens 4 . . . **PLANTAGINACEAE** p. 174

6 Plants usually aromatic when rubbed; stems often 4-sided; ovary deeply 4-parted . . . . **LAMIACEAE** p. 140

6 Plants not strongly scented; stems rarely 4-angled; ovary not deeply 4-parted **7**

7 Stamens 2 . . . . **8**

7 Stamens 4 . . . . **9**

8 Flowers in spikes or racemes at ends of stems, or 1 or 2 together from leaf axils . . . . **PLANTAGINACEAE** p. 174

8 Flowers in spikes from leaf axils . . . . **ACANTHACEAE** . . . . (*Justicia americana*) p. 44

9 Corolla ±regular . . . . **VERBENACEAE** . . . . (*Verbena hastata*) p. 234

9 Corolla lipped or irregular **10**

10 Flower in clusters from leaf axils .... **VERBENACEAE** (*Phyla lanceolata*) p. 234

10 Flowers in clusters at ends of stems, or single from leaf axils . **LAMIACEAE** p. 140

Trees, Shrubs, and Vines

The following simplified key includes many of the trees, shrubs and vines occurring in wetlands of the Upper Midwest. In some cases, only the genus is identified.

TREES

1 Leaves needlelike or scalelike (**Conifers**) 2

1 Leaves broad and flat..... (**Hardwoods**) 5

2 Leaves in bundles of 10 or more and shed in the fall . **TAMARACK** (*Larix laricina*) p. 42

2 Leaves single and persistent............3

3 Leaves overlapping and scalelike .... **NORTHERN WHITE CEDAR** .... (*Thuja occidentalis*) p. 41

3 Leaves needlelike or strap-shaped 4

4 Leaves stiff, 4-sided in cross-section .... **SPRUCE** (*Picea*) p. 42

4 Leaves soft, flat in cross-section .... **BALSAM FIR** (*Abies balsamea*) p. 42

5 Leaves compound (with 3 or more leaflets) 6

5 Leaves simple 8

6 Leaves alternate; southern LP of Mich.... .... **SHELLBARK-HICKORY** .... (*Carya laciniosa*) p. 140

6 Leaves opposite; mostly widespread species............................. 7

7 Leaflets 3-5; fruit a paired samara .... **BOXELDER** (*Acer negundo*) p. 228

7 Leaflets 7-11; samaras single .... **ASH** (*Fraxinus*) p. 163

8 Leaves opposite..... **MAPLE** (*Acer*) p. 228

8 Leaves alternate 9

9 Leaves not toothed or lobed; Mich LP, se Wisc .... **BLACK GUM** .... (*Nyssa sylvatica*) p. 162

9 Leaves toothed or lobed or both 10

10 Leaves toothed but not lobed..........11

10 Leaves lobed 15

11 Leaves asymmetrical at base (one lobe lower than other) ...... **AMERICAN ELM** .... (*Ulmus americana*) p. 231

11 Leaves symmetrical at base (lobes equal) 12

12 Leaves at least 4x as long as wide .... **WILLOW** (*Salix*) p. 220

12 Leaves less than 4x as long as wide 13

13 Leaves as wide as long or wider .... **COTTONWOOD** (*Populus*) p. 218

13 Leaves longer than wide 14

14 Leaf margins doubly toothed (the teeth themselves toothed)....... **RIVER BIRCH** .... (*Betula nigra*) p. 90

14 Leaf margins singly toothed .... **COTTONWOOD** (*Populus*) p. 218

15 Leaves palmately shallowly lobed and veined; fruit a round ball .... **SYCAMORE** .... (*Platanus occidentalis*) p. 182

15 Leaves shallowly to deeply lobed; if lobed more than halfway to middle, then also bristle-tipped; fruit an acorn .... **OAK** .... (*Quercus*) p. 124

SHRUBS AND VINES

1 Leaves evergreen and persistent on plant 2

1 Leaves deciduous, shed in the fall....... 8

2 Leaves narrow, less than 5 mm wide 3

2 Leaves broader, 5 mm or more wide..... 4

3 Plants small and trailing; leaves elliptic, pointed or blunt-tipped; fruit a red cranberry; plants of sphagnum peatlands. **CRANBERRY** (*Vaccinium macrocarpon, V. oxycoccos, V. vitis-idaea*) ......pp. 120, 121

3 Plants forming mats; leaves narrow with inrolled margins; fruit a black berry; uncommon in ne Minnesota and Michigan UP................ **BLACK CROWBERRY** .... (*Empetrum nigrum*) p. 116

4 Leaves opposite or in whorl of 3 .... **BOG-LAUREL, SHEEP-LAUREL** ... (*Kalmia polifolia, K. angustifolia*) p. 117

4 Leaves alternate 5

5 Leaf underside with a dense covering of white or brown hairs 6

5 Leaf underside with only scattered hairs or with small scales 7

6 Leaf underside with brown hairs; flowers cream-colored, in upright clusters .... **LABRADOR-TEA** ..... (*Rhododendron groenlandicum*) p. 118

6 Leaf underside with short hairs; flowers white to pink, urn-shaped and drooping .... **BOG-ROSEMARY** .... (*Andromeda polifolia*) p. 115

7 Upright shrubs; leaf underside with scales; flowers 5-parted.......... **LEATHERLEAF** .... (*Chamaedaphne calyculata*) p. 115

7 Small, trailing shrubs, leaf underside with scattered brown bristly hairs; flowers 4-parted .... **CREEPING SNOWBERRY** .... (*Gaultheria hispidula*) p. 116

8 Leaves opposite or whorled............ 9

8 Leaves alternate on stem............. 16

9 Leaves compound, divided into leaflets **10**

9 Leaves simple **11**

10 Vines . . . . **CLEMATIS** (*Clematis*) p. 196

10 Shrubs . . . . **ELDER** (*Sambucus*) p. 44

11 Most leaves opposite, some sub-opposite **BASKET WILLOW** (*Salix purpurea*) p. 226

11 All leaves opposite **12**

12 Leaf margins distinctly lobed . . . . **SQUASH-BERRY, WITHEROD, HIGH-** . . . **BUSH CRANBERRY** (*Viburnum*) p. 46

12 Leaf margins not lobed **13**

13 Leaves with translucent glandular dots on upper surface **KALM'S ST. JOHN'S-WORT** . . . . (*Hypericum kalmianum*) p. 138

13 Leaves without translucent glandular dots **14**

14 Leaves opposite or in whorls of 3 (or occasionally 4), flowers numerous, in a ball-shaped head atop a long stalk; fruits brown and nutlike . . . . **BUTTONBUSH** . . . . (*Cephalanthus occidentalis*) p. 215

14 Leaves strictly opposite; fruit white or colored and berrylike **15**

15 Leaf lateral veins noticeably curved toward tip; flowers white, 4-parted, stalkless with 4 white bracts, or in stalked clusters at ends of branches . **DOGWOOD** (*Cornus*) p. 108

15 Leaf lateral veins not curved toward tip; flowers light yellow, 5-parted, borne in pairs from leaf axils . . . . . **HONEYSUCKLE** . . . . (*Lonicera*) p. 102

16 Leaves compound, divided into leaflets **17**

16 Leaves simple **22**

17 Leaflets 3 or 5, palmate or pinnate **18**

17 Leaflets 6 or more, pinnate **20**

18 Leaflets 3; margins entire or with a few coarse teeth; flowers many in a branched inflorescence; fruit a whitish drupe; prickles absent . **COMMON POISON-IVY** . . . . (*Toxicodendron radicans*) p. 50

18 Leaflets 3 or 5; margins entire or coarsely toothed; flowers white, pink, or yellow; prickles sometimes present on stems . . **19**

19 Leaf margins coarsely toothed; fruit a berry . . . . **RASPBERRY, BLACKBERRY OR** . . . . **DEW-BERRY** (*Rubus*) p. 211

19 Leaflets narrow with entire margins, leaflets mostly 5, with upper 3 joined at base . . . . **SHRUBBY CINQUEFOIL** . . . . (*Dasiphora fruticosa*) p. 208

20 Stems with a pair of prickles at each node. . . . . **SWAMP-ROSE** (*Rosa palustris*) p. 211

20 Stems without prickles **21**

21 Low, much-branched shrub less than 1 m high; leaflets 1–2 cm long, narrow; flowers yellow and showy, 1–2.5 cm wide; fruit a capsule . . . . **SHRUBBY CINQUEFOIL** . . . . (*Dasiphora fruticosa*) p. 208

21 Taller shrubs; leaves 5 or more cm long; flowers small and green-yellow; fruit white and berrylike . . . . **POISON-SUMAC** . . . . (*Toxicodendron vernix*) p. 51

22 Leaves deeply or shallowly lobed **23**

22 Leaves not lobed **27**

23 Woody vines . . . . **RIVERBANK GRAPE** . . . . (*Vitis riparia*) p. 239

23 Shrubs **24**

24 Stems without thorns or prickles **25**

24 Stems thorny or prickly **26**

25 Stems with bark peeling into papery strips; flowers white, many in terminal clusters; fruit a dry brown pod . . . . **EASTERN NINEBARK** . . . . (*Physocarpus opulifolius*) p. 210

25 Bark not peeling into papery strips; flowers cream white, yellow or green-purple, in small clusters from leaf axils; fruit a red to black berry . . . . . . **CURRANT** (*Ribes*) p. 129

26 Uncommon plant of Isle Royale and other islands near Lake Superior north shore; leaves large, 2–4 dm wide; leave underside and petiole with sharp, stout prickles . . . . **DEVIL'S CLUB** . . . . (*Oplopanax horridus*) p. 59

26 Widespread plants; leaves smaller . . . . **GOOSEBERRY** (*Ribes*) p. 129

27 Leaf margin entire **28**

27 Leaf margin toothed or wavy **33**

28 Leaves with resinous dots on both sides (especially underside) . . . . **BLACK HUCKLEBERRY** . . . . (*Gaylussacia baccata*) p. 116

28 Leaves without resinous dots **29**

29 Leaves and fruit with spicy, lemony odor; Mich LP only . . . . **SPICEBUSH** . . . . (*Lindera benzoin*) p. 147

29 Leaves and fruit not aromatic **30**

30 Large shrubs, often over 1 m tall; buds covered by a single scale; flowers in catkins; staminate and pistillate flowers on separate plants; fruit a capsule . . . . **WILLOW** . . . . (*Salix*) p. 220

30 Bud scales absent or buds covered by 2 or more scales; flowers single or in several to many flowered clusters **31**

31 Smaller shrubs, typically much less than 1 m tall; flowers bell-shaped, waxy white; fruit a blue or blue-black berry with many small seeds . . . . **BLUEBERRY** . . . . (*Vaccinium*) p. 119

31 Taller shrubs, usually 2 m tall or more; flowers small; fruit a red-purple or crimson with several large seeds **32**

32 Leaves tipped with a small, sharp point; flowers and fruit single on very thin, long stalks . . . . MOUNTAIN-HOLLY . . . . (*Ilex mucronata*) p. 58

32 Leaves pointed but without a small, sharp tip; flowers single or several in leaf axils; fruit on short stalks . . . . GLOSSY BUCKTHORN . . . . (*Frangula alnus*) p. 204

33 Buds covered by a single scale; flowers in catkins; fruit a capsule . . . . WILLOW . . . . (*Salix*) p. 220

33 Buds covered by 2 or more scales; flowers various; if catkins present, fruit is hard and nutlike **34**

34 Leaves aromatic when rubbed; leaves with rounded, toothed tip, lower leaf margin entire, dotted on both sides with yellow glands . SWEET GALE (*Myrica gale*) p. 159

34 Leaves not aromatic **35**

35 Young twigs usually with glands; leaf margins coarsely toothed; fruit in conelike, deciduous catkins . . . . SWAMP or . . . . . . BOG BIRCH (*Betula pumila*) p. 90

35 Glands absent; leaves various **36**

36 Leaf margins coarsely double-toothed or wavy . . . . ALDER (*Alnus*) p. 88

36 Leaf margins not double-toothed or wavy **37**

37 Leaf midrib on upper surface of leaf with small dark glands. . . . . CHOKEBERRY . . . . (*Aronia*) p. 207

37 Glands absent from leaf blades or petioles **38**

38 Leaves with very short petioles, usually less than 5 mm long. **39**

38 Leaves with longer petioles, 5-30 mm long **40**

39 Flowers single or several, bell-shaped; fruit a blue to blue-black berry . . . BLUEBERRY . . . . (*Vaccinium*) p. 119

39 Flowers small and numerous in upright, terminal clusters; fruit a persistent capsule . . . . SPIRAEA (*Spiraea*) p. 214

40 Flowers larger, pink-white, 1-4 at ends of short branches, or single and terminal with single flowers from leaf axils; stalks 1-2 cm long . . . . JUNEBERRY . . . . (*Amelanchier bartramiana*) p. 206

40 Flowers small and yellow-green, single or in several-flowered clusters along branches, flower stalks short or absent . **41**

41 Leaf margin with incurved, forward-pointing teeth; stipules persistent, dark-colored; fruit bright red . . . . WINTERBERRY (*Ilex*) p. 58

41 Leaf margin with rounded, forward-pointing teeth; leaves with pronounced raised veins on underside; stipules present, narrow, but falling before fruits mature; fruit purple-black and berrylike . . . . BUCKTHORN (*Rhamnus*) p. 204

Aquatic Plants

The aquatic plant key (adapted from *www.michiganflora.net*) includes free-floating aquatic species, and rooted plants with leaves floating and/or submersed. Grasses, sedges, and rushes with erect stems extending above the water should not be keyed here unless they have leaves floating on water surface.

AQUATIC PLANTS WITH ALL LEAVES FLOATING OR SUBMERSED, OR PLANTS FREE FLOATING

1 Plants without distinct stem and leaves, free-floating at or below surface of water (except where stranded by drop in water level), the segments (internodes) small (up to 1 mm, but in most species much smaller), often remaining attached where budded from parent plant **2**

1 Plants with distinct stem and/or leaves, mostly much larger **3**

2 Plant body once to several times equally 2-lobed or 2-forked . . . . RICCIACEAE . *(a family of liverworts, not treated here)*

2 Plant body not consistently dichotomous. . . . . ARACEAE p. 244

3 Plants normally free-floating, but with leaves borne above the water surface in a rosette, with roots in the water; uncommon introduced species of tropical regions . . **4**

3 Plants with leaves floating, submerged, or absent; mostly native species **5**

4 Leaves densely hairy, sessile in the rosette . . . . ARACEAE (*Pistia stratiotes*) p. 248

4 Leaves glabrous, with inflated petioles . . . . PONTEDERIACEAE . . . . (*Eichhornia crassipes*) p. 386

5 Plants with floating leaves present (blades, or at least their terminal portions, floating on the surface of the water, usually ± smooth and firm in texture, especially compared with submersed leaves, or submersed leaves none) **6**

5   Plants without any floating leaves, entirely submersed (except sometimes for inflorescence and associated bracts) 20

6   Blades of some or all floating leaves on a plant sagittate or deeply lobed at base, or compound, or peltate 7

6   Blades of floating leaves all unlobed (at most subcordate at base), simple; the petiole marginal or absent if leaves ribbon-like 14

7   Floating blades compound (4-foliolate) .. .. MARSILEACEAE (*Marsilea*) p. 32

7   Floating blades simple 8

8   Floating blades (at least some of them) sagittate (the tip and lobes acute) [Note: Plants with sagittate leaves extending above the surface of the water are ***not*** included in this key] .... ALISMATACEAE .... (*Sagittaria*) p. 242

8   Floating (and any other) blades circular to ± elliptic in outline, peltate or rounded at tip with deep sinus at base................. 9

9   Leaves rounded at tip with deep sinus at base 10

9   Leaves peltate....................... 13

10   Plants free-floating, forming interconnected masses with leafless shoots ending in turions; flowers with 3 white petals; introduced and invasive .... HYDROCHARITACEAE .... (*Hydrocharis morsus-ranae*) p. 327

10   Plants with rhizomes rooted in the substrate, lacking leafless turion-bearing shoots; flowers with 4–many white (rarely pink) or yellow petals or petal-like sepals; mostly native species 11

11   Leaves and flowers with long petioles extending to a large fleshy rhizome rooted in soil......... NYMPHAEACEAE p. 160

11   Leaves and flowers with short petioles connecting to thin stems in the water... 12

12   Flowers white; leaves entire .... RANUNCULACEAE (*Caltha natans*) p. 196

12   Flowers yellow; leaves slightly crenate ... . MENYANTHACEAE .... (*Nymphoides peltata*) p. 159

13   Leaves circular, large (1 dm or more in diameter); flowers yellow............... .... NELUMBONACEAE p. 160

13   Leaves elliptic, less than 1 dm in their longest dimension, flowers reddish or white .... CABOMBACEAE p. 98

14   Floating leaves small (less than 1 cm long), crowded in a terminal rosette; submersed leaves distinctly opposite; flowers solitary, axillary .... PLANTAGINACEAE .... (*Callitriche*) p. 175

14   Floating leaves larger, not in a rosette; submersed leaves alternate, basal, or absent; flowers mostly in a terminal inflorescence 15

15   Leaves narrow and ribbon-like, without a distinct petiole (in some species a sheath surrounds the stem) 16

15   Leaves (at least the floating leaves) with ± elliptic blades and distinct petioles 17

16   Leaves ± rounded at tip (even if tapered), the floating portion smooth and shiny, yellow-green to bright green when fresh, occasionally keeled but midvein not or only slightly more prominent than other veins; leaf not differentiated into blade and sheath, the submersed portion similar to the floating but with a finely checkered pattern; flowers and fruit in spherical heads .... TYPHACEAE (*Sparganium*) p. 402

16   Leaves sharply acute at tip, the floating portion not shiny, ± blue-green when fresh, with a midrib; leaf with a sheath around stem and a membranous ligule at junction of sheath and blade; flowers and fruit in spikelets arranged in panicles........... .... POACEAE p. 359

17   Leaves all basal; petals 3, white .... ALISMATACEAE p. 240

17   Leaves cauline, alternate or opposite; petals 4–6, pink, yellow, or dull and inconspicuous (white in the rare *Caltha natans*) 18

18   Flowers single on a pedicel, white........ .... RANUNCULACEAE .... (*Caltha natans*) p. 196

18   Flowers sessile in spikes, pink or dull and inconspicuous........................ 19

19   Venation netted; flowers bright pink, in dense ovoid to cylindrical spikes .... POLYGONACEAE .... (*Persicaria amphibia*) p. 185

19   Venation parallel; flowers dull, in narrow cylindrical spikes .... POTAMOGETONACEAE p. 387

20   Leaves (or leaf-like structures) all basal and simple 21

20   Leaves cauline, simple or compound (basal and dissected in one species)......... 40

21   Leaves flat, several times as broad as thick (widest about the middle, or parallel-sided) 22

21   Leaves filiform or terete or only slightly flattened (especially near the base), elongate and limp to short and quill-like, less than 2x as broad as thick 27

22   Leaf blades not more than 2x as long as broad......... NYMPHAEACEAE p. 160

22 Leaf blades more than 2x as long as broad
.... **23**

23 Leaves stiff and erect or somewhat outcurved, less than 20 cm long **24**

23 Leaves limp, more than 20 cm long, ribbon like **25**

24 Base of leaf somewhat sheathing, with a membranous ligule (as in grasses) at base of spreading blade... **PONTEDERIACEAE** .... (*Pontederia cordata*) p. 387

24 Base of leaf not sheathing and with no ligule. **ALISMATACEAE** (*Sagittaria*) p. 242

25 Midvein not evident, all veins of essentially equal prominence, the tiny cross-veins giving a checkered appearance to the leaf .... **TYPHACEAE** (*Sparganium*) p. 402

25 Midvein (and usually some additional longitudinal veins) evident, the veins not all of equal prominence, not dividing the leaf into tiny rectangular cells **26**

26 Leaves with the central third (or more) of distinctly different pattern (more densely net-veined) than the outer marginal zones; plants dioecious, the staminate flowers eventually freed from a dense inflorescence submersed at base of plant, the pistillate flower single on a long ± spiraled stalk reaching the water surface; plants without milky juice .... **HYDROCHARITACEAE** .... (*Vallisneria*) p. 329

26 Leaves ± uniform in venation, not 3-zoned; plants monoecious, with emergent inflorescence of white-petaled flowers; plants often with milky juice............. ...... **ALISMATACEAE** (*Sagittaria*) p. 242

27 Major erect structures solitary, spaced along a simple or branched rhizome, consisting either of yellowish stems bearing tiny alternate bumps as leaves or of filiform leaves mostly buried in the substrate, and with a few tiny bladder-like organs **28**

27 Major erect structures solitary to densely tufted, consisting of filiform or quill-like leaves or stems, with neither alternate bumps nor bladders **29**

28 Leaves reduced to tiny alternate bumps on stem; bladders absent; flowers sessile, inconspicuous, regular **HALORAGACEAE** .... (*Myriophyllum tenellum*) p. 135

28 Leaves filiform, mostly buried in substrate (only the green tips, incurled when young, protruding); bladders tiny, usually present on the delicate branching rhizomes and buried leaf bases; flowers short-pediceled, showy (yellow or purple), bilaterally symmetrical .... **LENTIBULARIACEAE** .... (*Utricularia*, in part) p. 148

29 Leaves limp when removed from water (though a stiffer straight stem may also be present)............................ **30**

29 Leaves usually firm (stiff when removed from water **33**

30 Leaves (vegetative stems) terete entire length, not expanded at basally or sheathing each other, but each separate and closely surrounded at base by a delicate membranous tubular sheath; rhizome less than 2 mm in diameter; inflorescence a single terminal spikelet **31**

30 Leaves slightly expanded at base for ca. 2–10cm, sheathing the next inner leaf at least dorsally (usually the sheath continued ventrally as an almost invisible membrane), with tiny ligule or pair of auricles at the summit; rhizome various; inflorescence a lateral spikelet or terminal cyme **32**

31 Rhizome reddish, at least on older portions; leaves (vegetative culms) mostly over 2 cm long, limp; fertile culm triangular in cross-section on emersed portion, much larger in diameter than the hair-like vegetative culms; spikelet no thicker than culm .... **CYPERACEAE** .... (*Eleocharis robbinsii*) p. 309

31 Rhizome whitish; leaves often shorter, usually stiffer; fertile culms terete, no larger than the vegetative culms; spikelet distinctly thicker than culm**CYPERACEAE** .... (*Eleocharis acicularis*) p. 305

32 Leaf somewhat flattened or grooved ventrally for at least a few cm above the sheath (± crescent-shaped in cross-section), with 1–5 longitudinal nerves evident; rhizome less than 2 mm in diameter; inflorescence a solitary lateral spikelet on a stiff wiry stem just above or near water surface; flowers without petals and sepals; fruit an achene. **CYPERACEAE** ..... (*Schoenoplectus subterminalis*) p. 320

32 Leaf terete above sheath, with no evident longitudinal veins, but numerous definite septa extending entirely across the blade; rhizome ca. 2–5 mm thick; inflorescence an open cyme of many several-flowered heads on a stout stem (several mm in diameter, over 50 cm tall); flowers with 6 tepals; fruit a capsule .... **JUNCACEAE** .... (*Juncus militaris*) p. 338

33 Leaves filiform throughout, not broader basally or sheathing each other, solitary or in small tufts along a filiform whitish rhizome; inflorescence a single terminal spikelet.................. **CYPERACEAE** .... (*Eleocharis acicularis*) p. 305

33 Leaves linear or tapered from base to tip, or if otherwise uniformly filiform then expanded at base or sheathing each other; inflorescence various 34

34 Leaf in cross-section of 2 hollow tubes, linear (± parallel-sided), broadly rounded at tip; flowers bilaterally symmetrical, in a few-flowered raceme CAMPANULACEAE . . . . (*Lobelia dortmanna*) p. 100

34 Leaf not of 2 hollow tubes, tapered and ± acute (or filiform); flowers regular and in a raceme or solitary, or in a dense head or spike, or plant producing spores at base 35

35 Roots with prominent cross-septate appearance (checkered with fine transverse lines); inflorescence a small whitish or gray head (flowering in shallow or rarely deep water and on wet shores) . . . . ERIOCAULACEAE . . . . (*Eriocaulon aquaticum*) p. 325

35 Roots not distinctly septate or cross-lined; inflorescence not as above 36

36 Leaves abruptly expanded at base to enclose sporangia, often dark green, composed of 4 hollow tubes (in cross-section), surrounding a hard corm-like stem; plant submersed, non-flowering . . . . ISOETACEAE (*Isoetes*) p. 29

36 Leaves gradually and slightly expanded or grooved on one side, but not composed of 4 tubes nor enclosing sporangia; no corm-like stem present; plants not flowering when submersed (except *Subularia*) but only on wet shores 37

37 Leaves somewhat flattened at least basally, widest at base, gradually tapered to sharp tip; plants with buried rhizome or rhizomes absent 38

37 Leaves ± terete, of ± uniform width at least to the middle (or even slightly thicker there before tapering to tip); plants with rhizomes or stolons at, near, or above surface of substrate 39

38 Plants connected by slender rhizomes (to 1 mm wide); leaves often 4 cm or more long, somewhat flattened laterally below, with 2–3 conspicuous hollow tubes visible in cross-section; inflorescence a spreading cyme of solitary to paired 3-merous flowers . . . JUNCACEAE (*Juncus pelocarpus*) p. 338

38 Plants without rhizomes; leaves less than 4 cm long, somewhat flattened dorsiventrally (especially near base), with numerous small hollow areas of irregular size; inflorescence (often submersed) a few-flowered raceme of 4-merous flowers . . . . BRASSICACEAE . . . . (*Subularia aquatica*) p. 97

39 Plants with green arching stolons above substrate; leaves filiform, ± uniform in diameter, to 1 mm thick, truncate at tip . . . . RANUNCULACEAE . . . . (*Ranunculus flammula*) p. 200

39 Plants with delicate, horizontal, white to green stolons at or near surface of substrate (in addition to stouter short rhizome); leaves 1–3 mm thick at middle, then tapering to tip . . PLANTAGINACEAE . . . . (*Littorella*) p. 179

40 Leaves compound, dissected, forked, or deeply lobed 41

40 Leaves simple, unlobed, usually entire (toothed in a few species) 50

41 Leaves apparently in a basal rosette, few . . . . APIACEAE (*Sium suave*) p. 56

41 Leaves definitely cauline: opposite, whorled, or alternate 42

42 Leaves all or mostly opposite or whorled 43

42 Leaves alternate 47

43 Leaves (or whorled branches) rolled inward at tip when young, bearing tiny stalked bladders; flowers emersed, bilaterally symmetrical, purple or yellow . . . . LENTIBULARIACEAE (*Utricularia*) p. 148

43 Leaves not inrolled at tip, without bladders; flowers various and not as above 44

44 Petiole evident (to 15 mm long on well developed leaves), the blade fan-shaped, much dissected; flowers emergent, white. . . . . CABOMBACEAE (*Cabomba*) p. 98

44 Petiole absent or nearly so, the blade pectinate (with straight central axis following midrib, once-pinnatifid or comb-like on both sides) or much dissected or forking once or twice; flowers inconspicuous or yellow 45

45 Leaves once or twice dichotomously forked, the segments sparsely toothed along one edge; flowers inconspicuous, axillary, submersed . . . . CERATOPHYLLACEAE p. 108

45 Leaves not dichotomously forked, the segments entire; flowers emersed, rarely submersed 46

46 Leaves pectinate (comb-like); flowers inconspicuous, usually emersed in a terminal spike . . . . HALORAGACEAE . . . . (*Myriophyllum*) p. 133

46 Leaves with no definite central axis, dissected; flowers emersed in a showy yellow head (usually with at least one pair of serrate opposite leaves below the head) . . . . ASTERACEAE (*Bidens beckii*) p. 63

47 Leaves with a definite central axis (following midvein); flowers various ... **48**

47 Leaves with no definite central axis (except sometimes after initially forking at the stem); flowers emersed, with conspicuous corolla **49**

48 Leaves pectinate (the lateral segments not again branched); flowers inconspicuous, axillary; fruit a nutlet .. **HALORAGACEAE** .... (*Proserpinaca*) p. 135

48 Leaves with lateral segments again divided; flowers with white corollas, in emersed raceme; fruit a silique .... **BRASSICACEAE** .... (*Rorippa aquatica*) p. 96

49 Petiole present (but sometimes very short), ± adnate to a stipular sheath; plants without bladders; flowers regular, white or yellow, with numerous separate carpels forming achenes .... **RANUNCULACEAE** .... (*Ranunculus*) p. 198

49 Petioles and stipular sheaths absent; plants with small stalked bladders on leaves or on separate branches; flowers bilaterally symmetrical, yellow or purplish, with a single pistil producing a capsule .... **LENTIBULARIACEAE** .... (*Utricularia*) p. 148

50 Leaves much reduced and scale-like, not over 7 mm long, never distinctly opposite or whorled **51**

50 Leaves much longer or distinctly opposite or whorled (or both conditions) **52**

51 Leaves tiny, yellowish, merely widely spaced bumps or scales on stem .... **HALORAGACEAE** .... (*Myriophyllum tenellum*) p. 135

51 Leaves to 7 mm long, green or brownish, loosely overlapping (not treated in this work) .... **AQUATIC MOSSES** .... **and LIVERWORTS**

52 Leaves alternate, with ligule-like stipules (in *Ruppia*, these wholly adnate to leaves) **53**

52 Leaves opposite or whorled, without stipules **55**

53 Leaf blades filiform, terete or at least half as thick as broad, and the stipule adnate to leaf base for 10–30 mm or more, forming a sheath around the stem .... **POTAMOGETONACEAE** p. 387

53 Leaf blades distinctly flattened and several times wer than thick (even if narrow), or stipule little if at all adnate to blade (or both conditions) **54**

54 Blades flattened, ribbon-like (up to ca. 5 mm wide), with no definite midrib; flowers solitary, rare, cleistogamous in axils of submersed leaves or emersed plants with 6 bright yellow tepals . **PONTEDERIACEAE** .... (*Heteranthera*) p. 386

54 Blades flattened with a definite midrib, or filiform; flowers in spherical or cylindrical spikes, neither cleistogamous nor with showy yellow tepals .... **POTAMOGETONACEAE** p. 387

55 Leaves nearly filiform, less than 0.5 mm wide, gradually tapered from base to apex but not abruptly expanded basally, smooth; plants perennial by slender rhizomes; flowers axillary, 1 staminate flower (a single stamen) and 2–several carpels at a node; fruit slightly curved and minutely toothed on convex side . **POTAMOGETONACEAE** .... (*Zannichellia*) p. 400

55 Leaves wider; or if filiform, then abruptly expanded basally and with apiculate or toothed margins; plants annual; fruit solitary and ellipsoid **56**

56 Leaves distinctly whorled **57**

56 Leaves opposite (but some species with bushy axillary tufts of leaves which appear whorled appearance) **60**

57 Whorled branches cylindrical, elongate, usually stiff with calcium deposits; plants with distinctive musky odor (a family of algae, see p. 409) .... **CHARACEAE**

57 Whorled structures (true leaves) flattened, short (not over 20 mm long) or elongate and very limp; plants without musky odor **58**

58 Leaves 6–12 (usually 9) in a whorl, less than 2.5 mm wide, 12–25 times longer than wide; flowers bisexual, apetalous, sessile in axils of emersed leaves or bracts .... **PLANTAGINACEAE** (*Hippuris*) p. 178

58 Leaves mostly 3–4 in a whorl, 1–5 mm wide, at most 10–13 times as long; flowers bisexual or unisexual, petals present .. **59**

59 Leaves mostly 3 (rarely 6) in a whorl, very thin and delicate; stem round (not angled), smooth; flowers unisexual, with 3 often pink petals, the pistillate long-stalked from entirely submersed stem .... **HYDROCHARITACEAE** (*Elodea*) p. 325

59 Leaves mostly 4 in a whorl, stiff and firm; stem 4-sided, often with minutely scabrous angles; flowers bisexual, with 3–4 white petals .... **RUBIACEAE** (*Galium*) p. 215

60 Largest leaves 1–4 cm long, with distinct petiole and expanded, entire blade **61**

60 Largest leaves smaller, or sessile, or toothed (or all of these) **62**

61 Leaf blades ± orbicular, with orange to black glandular dots especially beneath; flowers 5-merous with showy yellow petals and superior ovary ...... **PRIMULACEAE** .... (*Lysimachia nummularia*) p. 192

61 Leaf blades ± diamond-shaped, without glandular dots; flowers 4-merous, inconspicuous, with inferior ovary .... **ONAGRACEAE** (*Ludwigia palustris*) p. 168

62 Leaves large, 3–13 cm long, 5–20 mm wide **63**

62 Leaves small (shorter or narrower than the above, or usually both) **64**

63 Leaves sessile and clasping, limp, at most obscurely and remotely toothed; flowers in axillary racemes .... **PLANTAGINACEAE** ...... (*Veronica anagallis-aquatica*) p. 180

63 Leaves sessile, clasping, tapered, or petioled, stiff, often regularly crenate or toothed; flowers various .... *Submersed individuals of normally terrestrial or emergent plants, chiefly in family* **LAMIACEAE**

64 Leaves linear and bidentate at apex when submersed, often becoming obovate, weakly 3-nerved, and not necessarily bidentate toward top of stem (or in floating rosettes); fruit solitary in axils, somewhat heart-shaped, of two 2-seeded segments . ... **PLANTAGINACEAE** (*Callitriche*) p. 175

64 Leaves filiform to orbicular or tapered from base to apex, but uniform on a plant, and if linear not bidentate at apex; fruit various **65**

65 Leaves at least 3 times longer than wide, wider at base than at middle; fruit absent or solitary in axils of leaves and ± ellipsoid **66**

65 Leaves less than 3 times as long as wide, often nearly round **67**

66 Leaves (especially lower ones) ± evenly tapered from broad base to minutely but bluntly bidentate apex, 3–10 times longer than wide, strictly entire, not subtending axillary tufts of leaves or fruit; plant often with a few scattered pale glandular dots on surface toward upper portion .... **PLANTAGINACEAE** .... (*Gratiola aurea*) p. 177

66 Leaves filiform to linear-lanceolate, expanded at base, acute at tip, at least 6 times longer than wide, minutely apiculate to conspicuously toothed on margins, usually subtending axillary tufts of leaves and/or flowers or ellipsoid fruit; plant without glands on surface .... **HYDROCHARITACEAE** (*Najas*) p. 327

67 Stems forming moss-like mats but the erect or ascending tips (above rooted nodes) less than 3 cm long; leaves with at most 1 weak nerve; stipules tiny but usually evident with some leaves; flowers axillary, inconspicuous ..... **ELATINACEAE** p. 114

67 Stems elongate (generally 10–30 cm); leaves more evidently veined; stipules absent; flowers terminal, yellow **68**

68 Stems stiffly erect; leaves weakly pinnately veined (with evident midvein), with reddish to blackish shiny dots or flecks (most prominent on emersed leaves) .... **PRIMULACEAE** ...... (*Lysimachia terrestris, L. thyrsiflora*) .... pp. 192, 193

68 Stems ± lax; leaves 3-nerved, without dark dots or flecks (but emersed leaves have translucent dots) .... **HYPERICACEAE** .... (*Hypericum boreale*) p. 137

Glossary

abaxial On the side away from the axis, usually refers to the underside of a leaf (compare with adaxial).

acaulescent Without an upright, leafy stem.

achene A one-seeded, dry, indehiscent fruit with the seed coat not attached to the mature wall of the ovary.

acid Having more hydrogen ions than hydroxyl (OH) ions; a pH less than 7.

acuminate Tapering to a narrow point, more tapering than acute, less than attenuate.

acute Gradually tapered to a tip.

adaxial On the side toward the axis, usually refers to the top side of a leaf (compare with abaxial).

adnate Fused with a structure different from itself, as when stamens are adnate to petals (compare with connate).

adventive Not native to and not fully established in a new habitat.

alkaline Having more hydroxyl ions than hydrogen ions; a pH greater than 7.

alluvial Deposits of rivers and streams.

alternate Borne singly at each node, as in leaves on a stem.

ament Spikelike inflorescence of same-sexed flowers (either male or female); same as **catkin**.

androgynous Spike with both staminate and pistillate flowers, the pistillate located at the base, below the staminate (compare with gynaecandrous).

angiosperm A plant producing flowers and bearing seeds in an ovary.

annual A plant that completes its life cycle in one growing season, then dies.

anther Pollen-bearing part of stamen, usually at the end of a stalk called a filament.

anthesis The period during which a flower is fully open and functional.

anthocyanic Pigmented with anthocyanins, this usually manifested as a tinging or suffusion of pink, red, or purple.

aphyllopodic Having basal sheaths without blades; with new shoots arising laterally from parent shoot (compare with phyllopodic).

apiculate Having an apiculus.

apiculus An abrupt, very small, projected tip.

appressed Lying flat to or parallel to a surface.

aquatic Living in water.

areole In leaves, the spaces between small veins.

aril A specialized appendage on a seed, often brightly colored, derived from the seed coat.

aristate Tipped with a slender bristle.

armed Bearing a sharp projection such as a prickle, spine, or thorn.

aromatic Strongly scented.

ascending Angled upward.

asymmetrical Not symmetrical.

attenuate Tapering gradually to a prolonged tip.

auricle An ear-shaped appendage to a leaf or stipule.

awl-shaped Tapering gradually from a broad base to a sharp point.

awn A bristle-like organ.

axil Angle between a stem and the attached leaf.

barb Sharp, thorn-like projection.

basal From base of plant.

basic A pH greater than 7.

beak A slender, terminal appendage on a 3-dimensional organ.

beard Covering of long or stiff hairs.

berry Fruit with the seeds surrounded by fleshy material.

biennial A plant that completes its life cycle in two growing season, typically flowering and fruiting in the second year, then dying.

bifid Cleft into two more or less equal parts.

blade Expanded, usually flat part of a leaf or petiole.

bloom A whitish powdery or waxy coating that can be rubbed away.

bog A wet, acidic, nutrient-poor peatland characterized by sphagnum and other mosses, shrubs and sedges. Technically, a type of peatland raised above its surroundings by peat accumulation and receiving nutrients only from precipitation.

boreal Far northern latitudes.

brackish Salty.

bract An accessory structure at the base of some flowers, usually appearing leaflike.

bractlet A secondary bract (*Typha*).

branchlets A small branch.

bristle A stiff hair.

bud An undeveloped shoot, inflorescence, or flower, in woody plants often covered by scales and serving as the overwintering stage.

bulb A group of modified leaves serving as a food-storage organ, borne on a short, vertical, underground stem (compare with corm).

bulbil A bulb-like structure borne in the leaf axils or in place of flowers.

bulblet Small bulb borne above ground, as in a leaf axil.

ca. About, approximately (Latin *circa*).

caducous Falling off early, as stipules that leave behind a scar.

callosity A hardened thickening.

callus A firm, thickened portion of an organ; the firm base of the lemma in the Poaceae.

calcareous fen An uncommon wetland type associated with seepage areas, and which receive groundwater enriched with primarily calcium and magnesium bicarbonates.

calcium-rich Refers to wetlands underlain by limestone or receiving water enriched by calcium compounds.

calyx All the sepals of a flower.

campanulate Bell-shaped.

capillary Very fine, hair-like, not-flattened.

capitate Abruptly expanded at the apex, thereby forming a knob-like tip.

capsule A dry, dehiscent fruit splitting into 3 or more parts.

carpel Fertile leaf of an angiosperm, bearing the ovules. A pistil is made up of one or more carpels.

caruncle An appendage at or near the hilum of some seeds.

caryopsis The dry, indehiscent seed of grasses.

catkin Spikelike inflorescence of same-sexed flowers (either male or female); same as **ament**.

caudex Firm, hardened, summit of a root mass that functions as a perennating organ.

cauline Of or pertaining to the above-ground portion of the stem.

cespitose Growing in a compact cluster with closely spaced stems; tufted, clumped.

chaff Thin, dry scales; in the Asteraceae, sometimes found as chaffy bracts on the receptacle.

cilia Hairs found at the margin of an organ.

ciliate Provided with cilia.

circumboreal Refers to a species distribution pattern which circles the earth's boreal regions.

clasping Leaves that partially encircle the stem at the base.

clavate Widened in the distal portion, like a baseball bat.

claw The narrow, basal portion of perianth parts.

cleistogamous Type of flower that remains closed and is self-pollinated.

clumped Having the stems grouped closely together; tufted.

colony-forming A group of plants of the same species, produced either vegetatively or by seed.

column The joined style and filaments in the Orchidaceae.

coma A tuft of fine hairs, especially at the tip of a seed.

composite An inflorescence that is made up of many tiny florets crowded together on a receptacle; members of the Aster Family (Asteraceae).

compound leaf A leaf with two or more leaflets.

concave Curved inward.

conduplicate Folded lengthwise into nearly equal parts.

cone The dry fruit of conifers composed of overlapping scales.

conifer Cone-bearing woody plants.

connate Two like parts that are fused (compare with **adnate**).

connivent Converging and touching but not actually fused, applies to like organs.

convex Curved outward.

convolute Arranged such that one edge is covered and the other is exposed, usually referring to petals in bud.

cordate With a rounded lobe on each side of a central sinus; heart-shaped.

coriaceous With a firm, leathery texture.

corm A short, vertical, enlarged, underground stem that serves as a food storage organ (compare with bulb).

corolla   Collectively, all the petals of a flower.

corymb   An indeterminate inflorescence, somewhat similar to a raceme, that has elongate lower branches that create a more or less flat-topped inflorescence.

costa (plural costae)   A prominent midvein or midrib of a leaflet.

crenate   With rounded teeth.

crenulate   Finely crenate.

crisped   An irregularly crinkled or curled leaf margin.

crown   Persistent base of a plant, especially a grasses.

culm   The stem of a grass or grasslike plant, especially a stem with the inflorescence.

cuneate   Tapering to the base with relatively straight, non-parallel margins; wedge-shaped.

cyme   A type of inflorescence in which the central flowers open first.

deciduous   Not persistent.

decumbent   A stem that is prostrate at the base and curves upward to have an erect or ascending, apical portion.

decurrent   Possessing an adnate line or wing that extends down the axis below the node, usually referring to leaves on a stem.

dehiscent   Splitting open at maturity.

deltate   Triangle-shaped.

dentate   Provided with outward oriented teeth.

depauperate   Poorly developed due to unfavorable conditions.

dicots   One of two main divisions of the Angiosperms (the other being the Monocots); plants having 2 seed leaves (cotyledons), net-venation, and flower parts in 4s or 5s (or multiples of these numbers).

dioecious   Bearing only male or female flowers on a single plant.

dimorphic   Having two forms.

disarticulation   Spikelets breaking either above or below the glumes when mature, the glumes remaining in the head if disarticulation above the glumes, or the glumes falling with the florets if disarticulation is below the glumes.

discoid   In composite flowers (Asteraceae), a head with only disk (tubular) flowers, the ray flowers absent.

disjunct   A population of plants widely separated from its main range.

disk   In the Asteraceae, the central part of the head, composed of tubular flowers.

dissected   Leaves divided into many smaller segments.

disturbed   Natural communities altered by human influences.

divided   Leaves which are lobed nearly to the midrib.

dolomite   A type of limestone consisting of calcium magnesium carbonate.

dorsal   Underside, or back of an organ.

driftless area   Portions of sw Wisconsin, ne Iowa, and se Minnesota that are not covered by glacial drift.

drupe   A fleshy fruit with a single large seed such as a cherry.

echinate   With spines.

eglandular   Without glands.

elliptic   Broadest at the middle, gradually tapering to both ends.

emergent   Growing out of and above the water surface.

emersed leaf   Growing above the water surface or out of water.

endangered   A species in danger of extinction throughout all or most of its range if current trends continue.

endemic   A species restricted to a particular region.

entire   With a smooth margin.

erect   Stiffly upright.

erose   With a ragged edge.

escape   A cultivated plant which establishes itself outside of cultivation.

evergreen   Plant retaining its leaves throughout the year.

excurrent   With the central rib or axis continuing or projecting beyond the organ.

exserted   Extending beyond the mouth of a structure such as stamens extending out from the mouth of the corolla.

falcate   Sickle-shaped

false indusium   A modified tooth or reflexed margin of a fern leaf that covers the sorus.

fen   An open wetland usually dominated by herbaceous plants, and fed by in-flowing, often calcium- and/or magnesium-rich water; soils vary from peat to clays and silts.

fern   Perennial plants with spore-bearing leaves similar to the vegetative leaves and bearing sporangia on their underside, or the spore-bearing leaves much modified.

fibrous A cluster of slender roots, all with the same diameter.

filament The stalk of a stamen which supports the anther.

filiform Thread-like.

flexuous An elongate axis that arches or bends in alternating directions in a zig-zag fashion.

floating mat A feature of some ponds where plant roots form a carpet over some or all of the water surface.

floodplain That part of a river valley that is occasionally covered by flood waters.

floret A small flower in a dense cluster of flowers; in grasses the flower with its attached lemma and palea.

follicle A dry, dehiscent fruit that splits along one side when mature.

floricane the second-year flowering stem of Rubus (compare with primocane).

genus The first part of the scientific name for a plant or animal (plural genera).

glabrate Nearly glabrous or becoming so.

glabrous Lacking hairs.

gland An appendage or depression which produces a sticky or greasy substance.

glandular Bearing glands.

glaucous Having a bluish appearance.

glumes A pair of small bracts at base of each spikelet the lowermost (or first) glume usually smaller the upper (or second) glume usually longer.

grain The fruit of a grass; the swollen seed-like protuberance on the fruit of some Rumex.

gymnosperm Plants in which the seeds are not produced in an ovary, but usually in a cone.

gynaecandrous Having both staminate and pistillate flowers on the same spike, the staminate located at the base, below the pistillate (compare with androgynous).

gynophore The central stalk of some flowers, especially in cat-tails (*Typha*).

halophyte A plant adapted to growing in a salty substrate.

hastate More or less triangular in outline with outward-oriented basal lobes.

haustorium A specialized, root-like connection to a host plant that a parasite uses to extract nourishment.

hardwoods Loosely used to contrast most deciduous trees from conifers.

herb A herbaceous, non-woody plant.

herbaceous Like an herb; also, leaflike in appearance.

hilum The scar at the point of attachment of a seed.

hirsute Pubescent with coarse, somewhat stiff, usually curving hairs, coarser than villous but softer than hispid.

hispid Pubescent with coarse, stiff hairs that may be uncomfortable to the touch, coarser than hirsute but softer than bristly.

hummock A small, raised mound formed by certain species of sphagnum moss.

humus Dark, well-decayed organic matter in soil.

hybrid A cross-breed between two species.

hydric Wet (compare with mesic, xeric).

hypanthium A ring, cup, or tube around the ovary; the sepals, petals and stamens are attached to the rim of the hypanthium.

imbricate Overlapping, as shingles on a roof.

indehiscent Not splitting open at maturity.

indusium In ferns, a membranous covering over the sorus (plural indusia).

inferior The position of the ovary when it is below the point of attachment of the sepals and petals.

inflorescence A cluster of flowers.

insectivorous Refers to the insect trapping and digestion habit of some plants as a nutrition supplement.

interdunal swale Low-lying areas between sand dune ridges.

internode Portion of a stem between two nodes.

introduced A non-native species.

invasive Non-native species causing significant ecological or economic problems.

involucral bract A single member of the involucre; sometimes called phyllary in composite flowers (Asteraceae).

involucre A whorl of bracts, subtending a flower or inflorescence.

irregular flower Not radially symmetric; with similar parts unequal.

joint A node or section of a stem where the branch and leaf meet.

keel A central rib like the keel of a boat.

lance-shaped   Broadest near the base, gradually tapering to a narrower tip.

lateral   Borne on the sides of a stem or branch.

lax Loose or drooping.

leaf axil The point of the angle between a stem and a leaf.

leaflet One of the leaflike segments of a compound leaf.

lemma In grasses, the lower bract enclosing the flower (the upper, smaller bract is the palea).

lens-shaped   Biconvex in shape (like a lentil).

lenticel Blisterlike openings in the epidermis of woody stems, admitting gases to and from the plant, and often appearing as small oval dots on bark.

ligulate Having a ligule; in the Asteraceae, the strap-shaped corolla of a ray floret.

ligule In grasses and grasslike plants, the membranous or hairy ring at top of sheath between the blade and stem.

linear Narrow and flat with parallel sides.

lip   Upper or lower part of a 2-lipped corolla; also the lower petal in most orchid flowers.

lobed With lobes; in leaves divisions usually not over halfway to the midrib.

local Occurring sporadically in an area.

low prairie   Wet and moist herbaceous plant community, typically dominated by grasses.

margin The outer edge of a leaf.

marl A calcium-rich clay.

marsh Wetland dominated by herbaceous plants, with standing water for part or all the growing season, then often drying at the surface.

megaspore Large, female spores.

mesic Moist, neither dry nor wet (compare with hydric, xeric).

microspore Small, male spores.

midrib The prominent vein along the main axis of a leaf.

mixed forest A type of forest composed of both deciduous and conifer trees.

moat   The open water area ringing the outer edge of a peatland or floating mat.

monecious Having male and female reproductive parts in separate flowers on the same plant.

monocots One of two main divisions of the Angiosperms (the other being the Dicots); plants with a single seed leaf (cotyledon); typically having narrow leaves with parallel veins, and flower parts in 3s or multiples of 3.

muck An organic soil where the plant remains are decomposed to the point where the type of plants forming the soil cannot be determined.

mucro A sharp point at termination of an organ or other structure.

naked Without a covering; a stalk or stem without leaves.

native An indigenous species.

naturalized An introduced species that is established and persistent in an ecosystem.

needle A slender leaf, as in the Pinaceae.

nerve A leaf vein.

neutral A pH of 7.

node The spot on a stem or branch where leaves originate.

nutlet A small dry fruit that does not split open along a seam.

oblanceolate   Reverse lance-shaped; broadest at the apex, gradually tapering to the narrower base.

oblique Emerging or joining at an angle other than parallel or perpendicular.

oblong Broadest at the middle, and tapering to both ends, but broader than elliptic.

obovate Broadly rounded at the apex, becoming narrowed below.

ocrea A tube-shaped stipule or pair of stipules around the stem; characteristic of the Smartweed Family (Polygonaceae).

opposite   Leaves or branches which are paired opposite one another on the stem.

organic Soils composed of decaying plant remains.

oval Elliptical.

ovary The lower part of the pistil that produces the seeds.

ovate Broadly rounded at the base, becoming narrowed above; broader than lanceolate.

palea   The uppermost of the two inner bracts subtending a grass flower (the lower bract is the lemma).

palmate Divided in a radial fashion, like the fingers of a hand.

panicle An arrangement of flowers consisting of several racemes.

papilla (plural: papillae) A short, rounded or cylindrical projections.

pappus The modified sepals of a composite flower which persist atop the ovary as bristles, scales or awns.

parallel-veined With several veins running from base of leaf to leaf tip, characteristic of most monocots.

peat An organic soil formed of partially decomposed plant remains.

peatland A wetland whose soil is composed primarily of organic matter (mosses, sedges, etc.); a general term for bogs and fens.

peltate More or less circular, with the stalk attached at a point on the underside.

pepo A fleshy, many-seeded fruit with a tough rind, as a melon.

perennial Living for 3 or more years.

perfect A flower having both male (stamens) and female (pistils) parts.

perianth Collectively, all the sepals and petals of a flower.

perigynium A sac-like structure enclosing the pistil in Carex (plural perigynia).

petal An individual part of the corolla, often white or colored.

petiole The stalk of a leaf.

phyllary An involucral bract subtending the flower head in composite flowers (Asteraceae).

phyllode An expanded petiole.

phyllopodic Having the basal sheaths blade-bearing; with new shoots arising from the center of parent shoot (compare with aphyllopodic).

pinna The primary or first division in a fern frond or leaf (plural pinnae).

pinnate Divided once along an elongated axis into distinct segments.

pinnule The pinnate segment of a pinna.

pistil The seed-producing part of the flower, consisting of an ovary and one or more styles and stigmas.

pith A spongy central part of stems and branches.

pollen The male spores in an anther.

prairie An open plant community dominated by herbaceous species, especially grasses.

primocane The first-year, vegetative stem in Rubus (compare with floricane).

pro sp. When a taxon is transferred from the non-hybrid category to the hybrid category, the author citation remains unchanged, but may be followed by an indication in parentheses of the original category.

prostrate Lying flat on the ground.

raceme A grouping of flowers along an elongated axis where each flower has its own stalk.

rachilla A small stem or axis.

rachis The central axis or stem of a leaf or inflorescence.

radiate heads In composite flowers, heads with both ray and disk flowers (Asteraceae).

ray flower A ligulate or strap-shaped flower in the Asteraceae, where often the outermost series of flowers in the head.

receptacle In the Asteraceae, the enlarged summit of the flower stalk to which the sepals, petals, stamens, and pistils are usually attached.

recurved Curved backward.

regular Flowers with all the similar parts of the same form; radially symmetric.

rhizome An underground, horizontal stem.

rib A pronounced vein or nerve.

rootstock Similar to rhizome but referring to any underground part that spreads the plant.

rosette A crowded, circular clump of leaves.

samara A dry, indehiscent fruit with a well-developed wing.

saprophyte A plant that lives off of dead organic matter.

scale A tiny, leaflike structure; the structure that subtends each flower in a sedge (Cyperaceae).

scape A naked stem (without leaves) bearing the flowers.

section Cross-section.

secund Flowers mostly on 1 side of a stalk or branch.

sedge meadow A community dominated by sedges (Cyperaceae) and occurring on wet, saturated soils.

seep A spot where water oozes from the ground.

sepal A segment of the calyx; usually green in color.

sheath Tube-shaped membrane around a stem, especially for part of the leaf in grasses and sedges.

shrub A woody plant with multiple stems.

silicle Short fruit of the Mustard Family (Brassicaceae), normally less than 2x longer as wide.

silique Dry, dehiscent, 2-chambered fruit of the Mustard Family (Brassicaceae), longer than a silicle.

simple An undivided leaf.

sinus The depression between two lobes.

smooth Without teeth or hairs.

sorus Clusters of spore containers (plural sori).

spadix A fleshy axis in which flowers are embedded.

spathe A large bract subtending or enclosing a cluster of flowers.

spatula-shaped Broadest at tip and tapering to the base.

sphagnum moss A type of moss common in peatlands and sometimes forming a continuous carpet across the surface; sometimes forming layers several meters thick; also loosely called peat moss.

spike A group of unstalked flowers along an unbranched stalk.

spikelet A small spike; the flower cluster (inflorescence) of grasses (Poaceae) and sedges (Cyperaceae).

sporangium The spore-producing structure (plural sporangia).

spore A one-celled reproductive structure that gives rise to the gamete-bearing plant.

sporophyll A modified, spore-bearing leaf.

spreading Widely angled outward.

spring A place where water flows naturally from the ground.

spur A hollow, pointed projection of a flower.

stamen The male or pollen-producing organ of a flower.

staminode An infertile stamen.

stem The main axis of a plant.

stigma The terminal part of a pistil which receives pollen.

stipe A stalk.

stipule A leaflike outgrowth at the base of a leaf stalk.

stolon A horizontal stem lying on the soil surface.

style The stalklike part of the pistil between the ovary and the stigma.

subspecies A subdivision of the species forming a group with shared traits which differ from other members of the species (subsp.).

subtend Attached below and extending upward.

succulent Thick, fleshy and juicy.

superior Referring to the position of the ovary when it is above the point of attachment of sepals, petals, stamens, and pistils.

swale A slight depression.

swamp Wooded wetland dominated by trees or shrubs; soils are typically wet for much of year or sometimes inundated.

talus Fallen rock at the base of a slope or cliff.

taproot A main, downward-pointing root.

tendril A threadlike appendage from a stem or leaf that coils around other objects for support (as in grapes, *Vitis*).

tepal Sepals or petals not differentiated from one another.

terete Circular in cross-section.

terminal Located at the end of a stem or stalk.

thallus A small, flattened plant structure, without distinct stem or leaves.

thicket A dense growth of woody plants.

threatened A species likely to become endangered throughout all or most of its range if current trends continue.

translucent Nearly transparent.

tree A large, single-stemmed woody plant.

tuber An enlarged portion of a root or rhizome.

truncate Abruptly cut-off.

tubercle Base of style persistent as a swelling atop the achene different in color and texture from achene body.

tundra Treeless plain in arctic regions, having permanently frozen subsoil.

turion A specialized type of shoot or bud that overwinters and resumes growth the following year.

umbel A cluster of flowers in which the flower stalks arise from the same level.

umbelet A small, secondary umbel in an umbel, as in the Apiaceae.

upright Erect or nearly so.

urceolate Constricted at a point just before an opening; urn-shaped.

utricle A small, one-seeded fruit with a dry, papery outer covering.

valve A segment of a dehiscent fruit; the wing of the fruit in *Rumex*.

variety Taxon below subspecies and differing from other varieties within the same subspecies (var.).

vein A vascular bundle, as in a leaf.

velum The membranous flap that partially covers the sporangium in *Isoetes*.

venation The pattern of veins on an organ.

ventral Front side.

ventricose Inflated or distended.

verrucose Covered with small, wart-like projections.

verticil One whorled cycle of organs.

verticillate Arranged in whorls.

villous Pubescent with long, soft, bent hairs, the hairs not crimped or tangled.

vine A trailing or climbing plant, dependent on other objects for support.

viscid Sticky, glutinous.

whorl A group of 3 or more parts from one point on a stem.

wing A thin tissue bordering or surrounding an organ.

woody Xylem tissue (the vascular tissue which conducts water and nutrients).

xeric Dry (compare with hydric, mesic).

A - Flowers

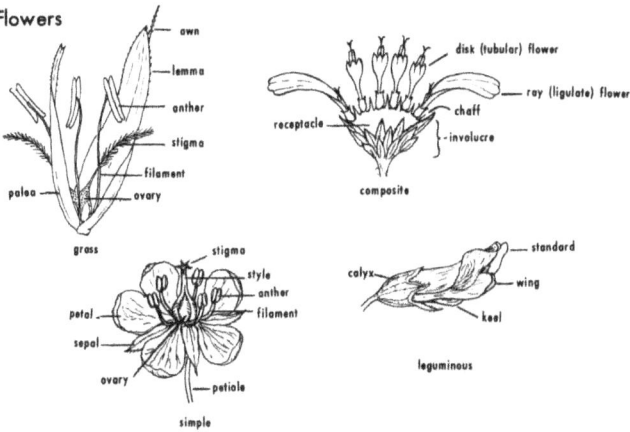

grass — awn, lemma, anther, stigma, filament, palea, ovary

composite — disk (tubular) flower, ray (ligulate) flower, chaff, receptacle, involucre

simple — stigma, style, anther, filament, petal, sepal, ovary, petiole

leguminous — standard, wing, calyx, keel

B - Inflorescences

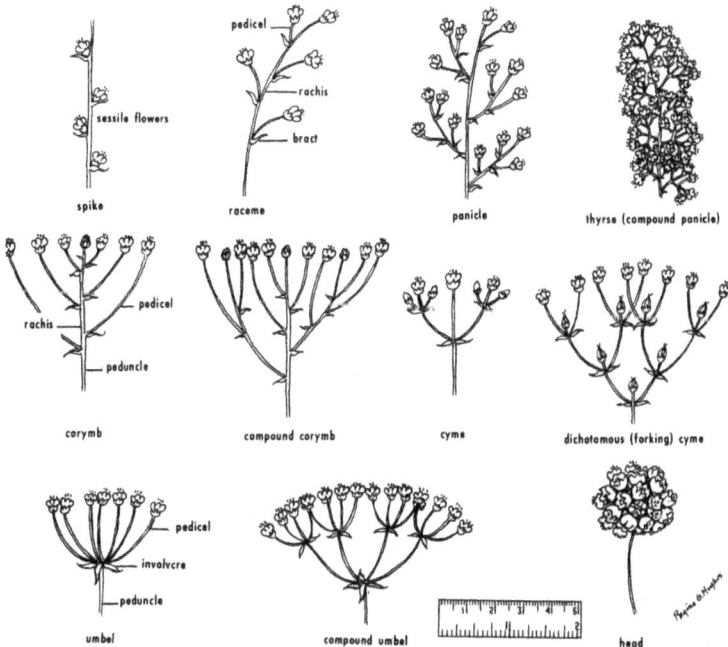

spike — sessile flowers

raceme — pedicel, rachis, bract

panicle

thyrse (compound panicle)

corymb — pedicel, rachis, peduncle

compound corymb

cyme

dichotomous (forking) cyme

umbel — pedicel, involucre, peduncle

compound umbel

head

A - Fruits

capsule

silique

achene

utricle

legume (pod)

drupelets

silicle

nutlets

follicle

samara

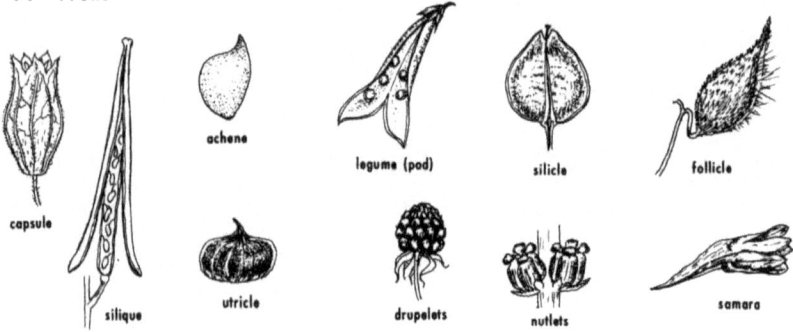

B - Roots and Stems

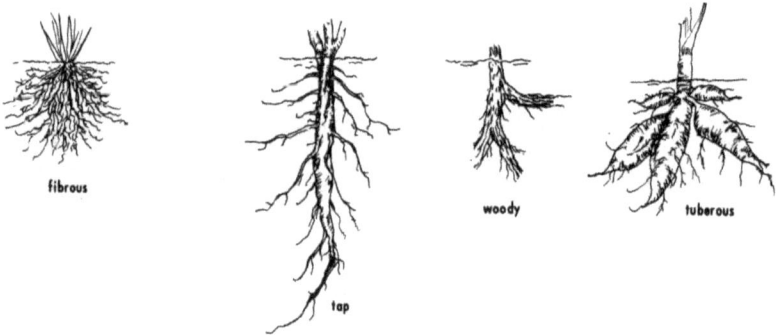

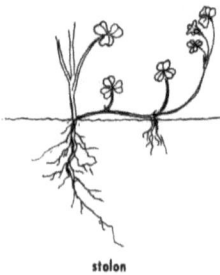

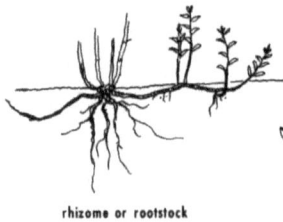

fibrous

tap

woody

tuberous

stolon

rhizome or rootstock

caulescent

acaulescent

A - Simple and Compound Leaves

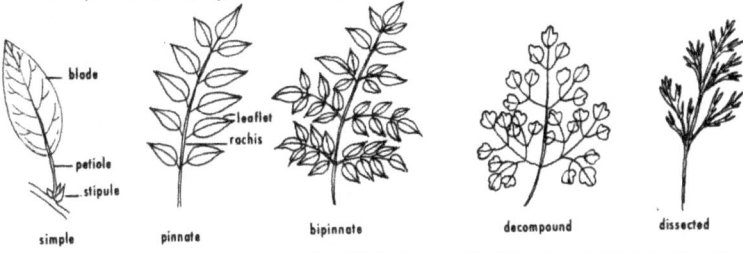

blade
leaflet
rachis
petiole
stipule

simple   pinnate   bipinnate   decompound   dissected

trifoliate   palmate

B - Margins

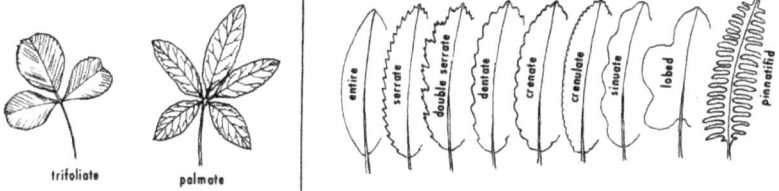

entire   serrate   double serrate   dentate   crenate   crenulate   sinuate   lobed   pinnatifid

C - Shapes

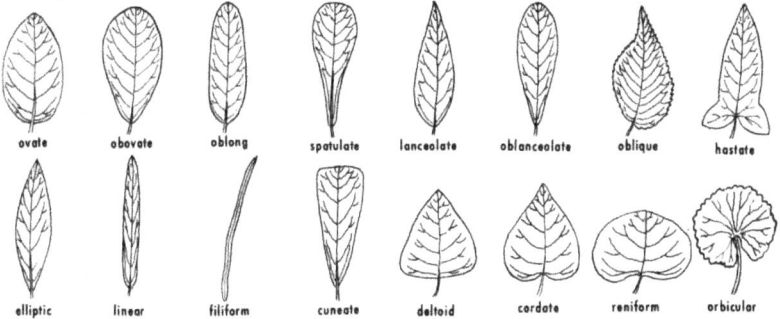

ovate   obovate   oblong   spatulate   lanceolate   oblanceolate   oblique   hastate

elliptic   linear   filiform   cuneate   deltoid   cordate   reniform   orbicular

D - Apices

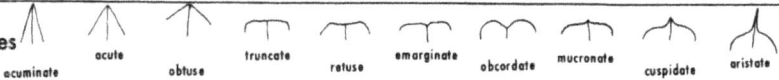

acuminate   acute   obtuse   truncate   retuse   emarginate   obcordate   mucronate   cuspidate   aristate

E - Bases

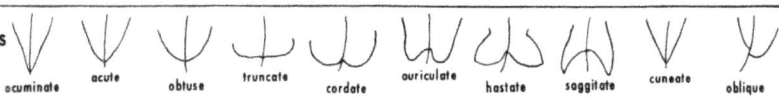

acuminate   acute   obtuse   truncate   cordate   auriculate   hastate   saggitate   cuneate   oblique

F - Attachments

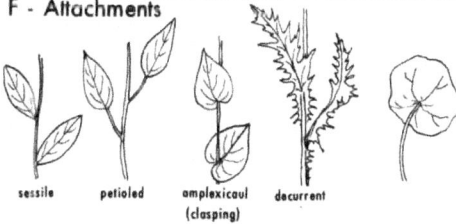

sessile   petioled   amplexicaul (clasping)   decurrent

G - Arrangements

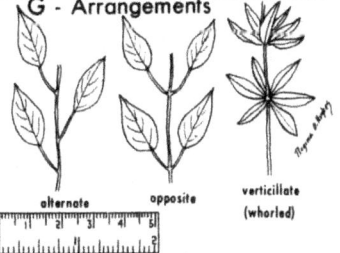

alternate   opposite   verticillate (whorled)

References

Barnes, B., and W. Wagner. 1981. *Michigan Trees*. The University of Michigan Press. Ann Arbor, MI. 383 p.

Billington, C. 1952. *Ferns of Michigan*. Cranbrook Institute of Science Bulletin No. 32. Bloomfield Hills, MI 240 p.

Black, M., and E. Judziewicz. 2009. *Wildflowers of Wisconsin and the Great Lakes Region: A Comprehensive Field Guide*. The University of Wisconsin Press. Madison, WI. 320 p.

Case, F., Jr. 1987. *Orchids of the Western Great Lakes Region*. Cranbrook Institute of Science Bulletin No. 48. Bloomfield Hills, MI. 240 p.

Chadde, S. 2019. *Wetland Plants of the Upper Midwest*. Orchard Innovations Press, Mountain View, AR. 579 p.

Chadde, S. 2016. *Michigan Flora: Upper Peninsula*. Orchard Innovations Press, Mountain View, AR. 590 p.

Chadde, S. 2013. *Minnesota Flora*. Orchard Innovations Press, Mountain View, AR. 788 p.

Chadde, S. 2013. *Wisconsin Flora*. Orchard Innovations Press, Mountain View, AR. 824 p.

Cody, W., and D. Britton. 1989. *Ferns and Fern Allies of Canada*. Publication 1829/E. Research Branch, Agriculture Canada. Ottawa, Canada. 430 p.

Cowardin, L., V. Carter, F. Golet, and E. LaRoe. 1979. *Classification of Wetlands and Deepwater Habitats of the United States*. U.S. Department of the Interior, Fish and Wildlife Service. Washington, DC. 103 p.

Crow, G., and C. Hellquist. 2000. *Aquatic and Wetland Plants of Northeastern North America* (2 vols.). University of Wisconsin Press. Madison, WI.

Crum, H. 1976. *Mosses of the Great Lakes Forest*. University Herbarium, University of Michigan, Ann Arbor, MI. 404 p.

Crum, H. 1988. *A Focus on Peatlands and Peat Mosses*. The University of Michigan Press. Ann Arbor, MI. 306 p.

Crum, H., and L. Anderson. 1981. *Mosses of Eastern North America* (2 vols). Columbia Univ. Press. New York, NY.

Curtis, J. 1971. *The Vegetation of Wisconsin*. The University of Wisconsin Press. Madison, WI. 657 p.

Eastman, J. 1995. *The Book of Swamp and Bog: Trees, Shrubs, and Wildflowers of Eastern Freshwater Wetlands*. Stackpole Books. Mechanicsburg, PA. 237 p.

Eggers, S., and D. Reed. 1997. *Wetland Plants and Plant Communities of Minnesota and Wisconsin* (2nd Ed.). U.S. Army Corps of Engineers, St. Paul District. 264 p.

Fassett, N. 1957. *A Manual of Aquatic Plants*. The University of Wisconsin Press. Madison, WI. 405 p.

Flora of North America Editorial Committee, eds. 1993+. *Flora of North America North of Mexico*. 16+ vols. New York and Oxford. note Additional volumes in the Flora of North America series have been released. See website for current availability: *www.efloras.org*.

Gleason, H., and A. Cronquist. 1991. *Manual of Vascular Plants of Northeastern United States and Adjacent Canada* (2nd Ed.). The New York Botanical Garden. Bronx, NY. 910 p.

Hipp, A. 2008. *Field Guide to Wisconsin Sedges: An Introduction to the Genus Carex (Cyperaceae)*. The Univ. of Wisconsin Press. Madison, WI. 280 p.

Holmgren, N. (editor). 1998. *Illustrated Companion to Gleason and Cronquist's Manual*. New York Botanical Garden. Bronx, NY. 937 p.

Judziewicz E. J., R. W. Freckmann, L. G. Clark, M. R. Black. 2014. *Field Guide to Wisconsin Grasses*. The Univ. of Wisconsin Press. Madison, WI. 356 p.

Kartesz, J.T. 2014. *Floristic Synthesis of North America, Version 1.0*. Biota of North America Program (BONAP). (in press). (*www/bonap.org/*)

Lichvar, R.W., D.L. Banks, W.N. Kirchner, and N.C. Melvin. 2016. *The National Wetland Plant List: 2016 wetland ratings*. Phytoneuron 2016-30: 1-17.

Ownbey, G., and T. Morley. 1991. *Vascular Plants of Minnesota: A Checklist and Atlas*. The University of Minnesota Press. Minneapolis, MN. 306 p.

Reed, P. 1988. *National List of Plant Species that Occur in Wetlands: North Central (Region 3)*. Biological Report 88(26.3). U.S. Department of the Interior, Fish and Wildlife Service. Washington, DC. 99 p.

Smith, W. 2012. *Native Orchids of Minnesota*. The University of Minnesota Press. Minneapolis, MN. 254p.

Smith, W. 2008. *Trees and Shrubs of Minnesota*. The University of Minnesota Press. Minneapolis, MN.

Soper, J., and M. Heimburger. 1982. *Shrubs of Ontario*. The Royal Ontario Museum. Toronto, Ontario. 495 p.

Stensvold, M.C. and D.R. Farrar. 2017. Published online 13 December 2016. *Genetic diversity in the worldwide* Botrychium lunaria *(Ophioglossaceae) complex, with new species and new combinations*. Brittonia 69(2): 148-175..

Swink, F., and G. Wilhelm. 1994. *Plants of the Chicago Region* (4th Ed.). Indiana Academy of Science. Indianapolis, IN. 921 p.

Tryon, R. 1980. *Ferns of Minnesota* (2nd Ed.). The University of Minnesota Press. Minneapolis, MN. 165 p.

Tryon, R., N. Fassett, D. Dunlop, and M. Diemer. 1953. *The Ferns and Fern Allies of Wisconsin*. The University of Wisconsin Press. Madison, WI. 158 p.

U.S. Army Corps of Engineers. 2020. National Wetland Plant List, version 3.5 (*http://wetland-plants.usace.army.mil*) U.S. Army Corps of Engineers, Engineer Research and Development Center, Cold Regions Research and Engineering Laboratory, Hanover, NH.

Voss, E.G. 1972. *Michigan Flora, Part I. Gymnosperms and Monocots*. Cranbrook Institute of Science Bulletin 55 and University of Michigan Herbarium. 488 p.

Voss, E.G. 1985. *Michigan Flora, Part II. Dicots (Saururaceae–Cornaceae)*. Cranbrook Institute of Science Bulletin 59 and University of Michigan Herbarium. 724 p.

Voss, E.G. 1996. *Michigan Flora, Part III. Dicots (Pyrolaceae–Compositae)*. Cranbrook Institute of Science Bulletin 61 and University of Michigan Herbarium. 622 p.

Voss, E.G. and A. A. Reznicek. 2011. *Field Manual of Michigan Flora*. University of Michigan Press. 1008 p.

Wetter, A.W, T.S. Cochrane, M.R. Black, H.H. Iltis, P.E. Berry. 2001. *Checklist of the Vascular Plants of Wisconsin*. Tech. Bulletin No. 192. Dept. Natural Resources, Madison, WI. 258 p.

Wilhelm G and L. Rericha. *Flora of the Chicago Region: A Floristic and Ecological Synthesis*. 2017. Indiana Academy of Science. 1390 p.

Wright, H., B. Coffin, and N. Aaseng (editors). 1992. *The Patterned Peatlands of Minnesota*. University of Minnesota Press. Minneapolis, MN. 327 p.

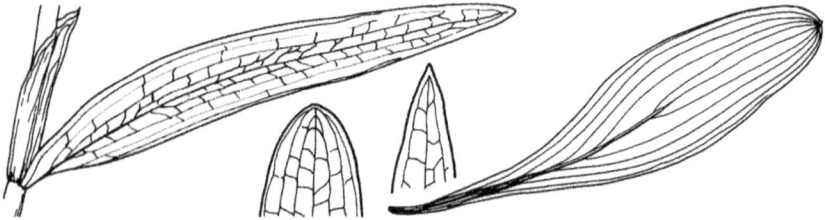

Potamogeton alpinus RED PONDWEED

· IDENTIFICATION **Plants with underwater leaves, floating leaves present or absent.** Fruit smooth, hard, and eggshell-like; submersed leaves sessile, often with a reddish cast; floating leaves (if present) delicate and with blades tapering gradually to the petiole.

• •

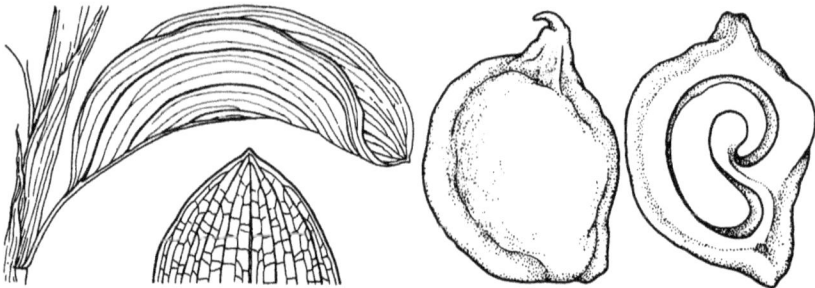

Potamogeton amplifolius BIGLEAF PONDWEED

· IDENTIFICATION **Plants with broad floating leaves and broad underwater leaves.** Submersed leaves numerous, large, arcuate; sometimes confused with *P. pulcher, P. nodosus*, or *P. illinoensis*, but generally *P. amplifolius* has many more veins in the submersed leaves (19-37) and in the floating leaves (29-51) than those species. Also separated from the highly variable P. illinoensis by the submersed leaf tip: sharp in *P. illinoensis*, blunt in *P. amplifolius*.

• •

Potamogeton confervoides ALGA PONDWEED

· IDENTIFICATION **Plants with underwater leaves only;** leaves slender, delicate, grass-green in color, gradually tapering to a slender tip. The primary branches are much branched, giving plants a fan-shaped appearance in the water.

• •

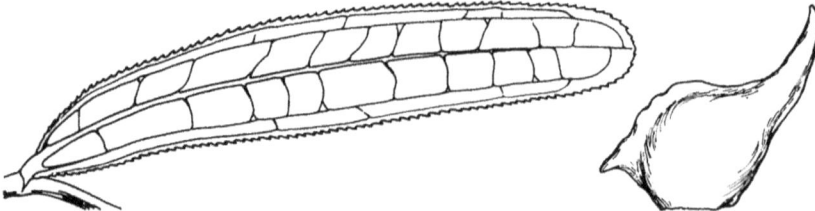

Potamogeton crispus CURLY PONDWEED

· IDENTIFICATION **Plants with underwater leaves only;** our only species with dentate leaf margins.

Potamogeton diversifolius COMMON SNAILSEED PONDWEED
· IDENTIFICATION **Plants with broad floating leaves and linear underwater leaves.** Similar to *P. spirillus;* these two species are recognized by their tiny fruits with walls so thin that the embryo coil is evident.

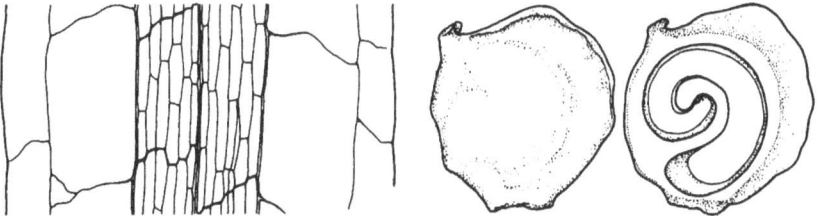

Potamogeton epihydrus RIBBONLEAF-PONDWEED
· IDENTIFICATION **Plants with broad floating leaves and broad underwater leaves.** Submersed leaves ribbonlike, with a broad band of lacunae between the 2 veins closest to the midrib; fruits large with embryo coiled more than 1 complete revolution.

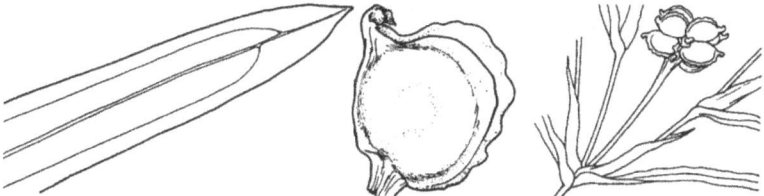

Potamogeton foliosus LEAFY PONDWEED
· IDENTIFICATION **Plants with underwater leaves only;** sterile specimens are difficult to distinguish from *P. pusillus*. Mature fruit distinctive, or only species bearing fruit with a thin, winglike dorsal keel, and the embryo coil with no more than 1 revolution. Leaves are usually less lacunaic and the tip never obtuse, as may occur in the variable *P. pusillus*. Also, stems oval to flattened in cross section, in contrast to the terete stems of *P. pusillus*.

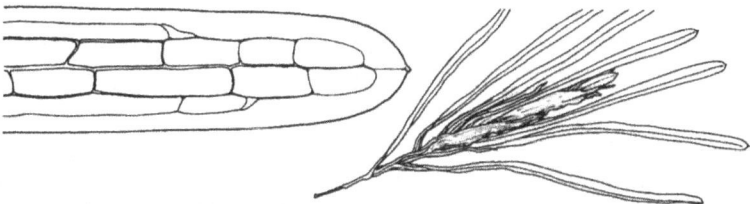

Potamogeton friesii FRIES' PONDWEED
· IDENTIFICATION **Plants with underwater leaves only.** Similar to *P. strictifolius,* and these 2 species distinguished from our other linear-leaved species by presence of firm, fibrous, whitish stipules.

Potamogeton gramineus VARIABLE PONDWEED
· IDENTIFICATION **Plants with broad floating leaves and broad underwater leaves.** Similar to *P. strictifolius,* and these 2 species distinguished from our other linear-leaved species by presence of firm, fibrous, whitish stipules.

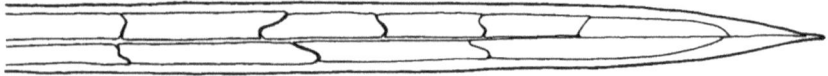

Potamogeton hillii HILL'S PONDWEED
· IDENTIFICATION **Plants with underwater leaves only.** Our only species with leaves more than 1 mm wide that have only 3 veins and a bristle-tipped apex. The mucro at the leaf tip of *P. zosteriformis* may also appear like a short bristle.

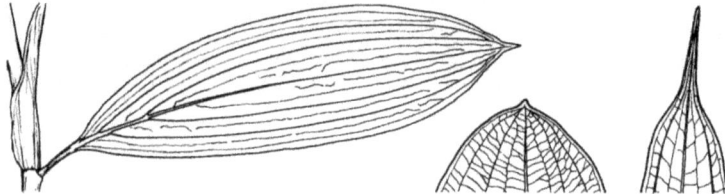

Potamogeton illinoensis ILLINOIS PONDWEED
· IDENTIFICATION **Plants with broad floating leaves and broad underwater leaves;** highly variable. It sometimes approaches *P. gramineus* (with which it may hybridize) but is coarser and less branched. From other species, may be distinguished by the sharp pointed, often mucronate, broad, submersed leaves. Often found with no floating leaves, even when fruiting.

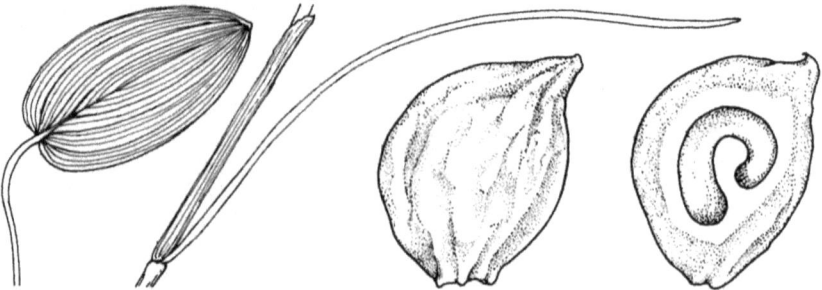

Potamogeton natans FLOATING PONDWEED
· IDENTIFICATION **Plants with broad floating leaves and linear underwater leaves.** Floating leaves cordate, submersed leaves narrowly linear. Submersed leaves decay early and are usually absent by the time fruits appear.

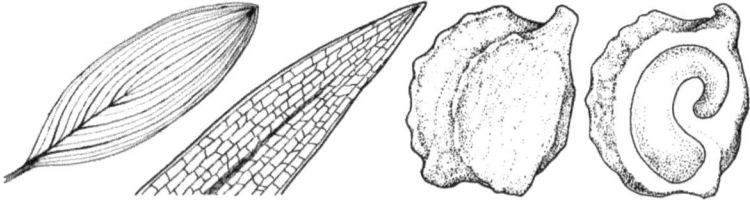

Potamogeton nodosus LONGLEAF PONDWEED

· IDENTIFICATION **Plants with broad floating leaves and narrow underwater leaves;** variable. Floating leaf blades cuneate at base, submersed leaf blades narrowly lanceolate and tapering gradually to each end. Fruits reddish with strongly developed keels.

· ·

Potamogeton oakesianus OAKES' PONDWEED

· IDENTIFICATION **Plants with broad floating leaves and linear underwater leaves.** Similar to *P. natans* but plants smaller.

· ·

Potamogeton obtusifolius BLUNTLEAF-PONDWEED

· IDENTIFICATION **Plants with underwater leaves only.** Distinguished from other linear-leaved species by leaves that are 2-4 mm wide and rounded at tip. Similar to *P. friesii*, from which it differs by having leaves with only 3 prominent veins and with nonfibrous stipules.

· ·

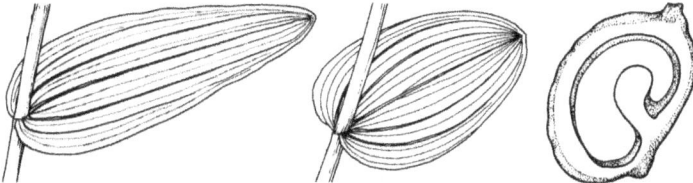

Potamogeton perfoliatus REDHEAD-GRASS

· IDENTIFICATION **Plants with underwater leaves only;** these leaves ovate or sometimes more elongate, similar to leaves of *P. richardsonii*. If fruits absent, the delicate fugaccous stipules are distinctive. Uncommon in Upper Midwest region.

· ·

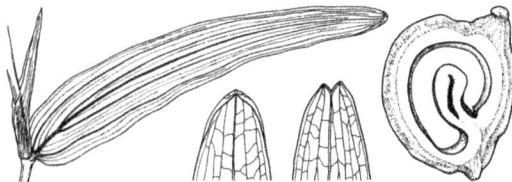

Potamogeton praelongus WHITESTEM-PONDWEED

· IDENTIFICATION **Plants with underwater leaves only;** these leaves large, ovate-oblong, cucullate (boat-shaped) at tip; stems whitish; stipules large, conspicuous. Fruits large, on long peduncles, maturing early summer and sink, so are not seen late in the season. The cucullate leaf apex often splits when flattened.

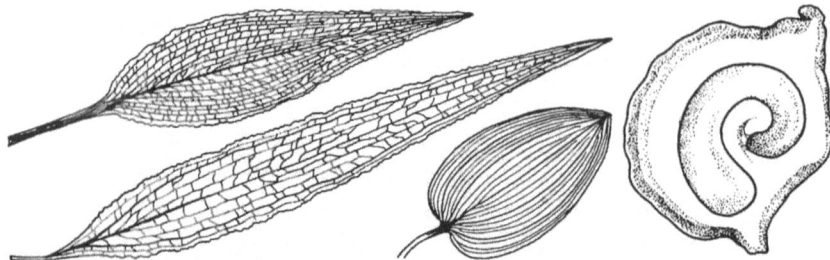

Potamogeton pulcher SPOTTED PONDWEED
· IDENTIFICATION **Plants with broad floating leaves and broad underwater leaves;** our only species with cordate floating leaves and broad submersed leaves.

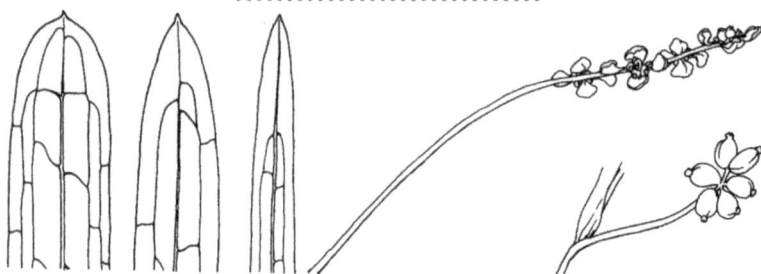

Potamogeton pusillus SLENDER *or* SMALL PONDWEED
· IDENTIFICATION Plants with underwater leaves only; highly variable. Non-fruiting plants similar to *P. foliosus*, which see.

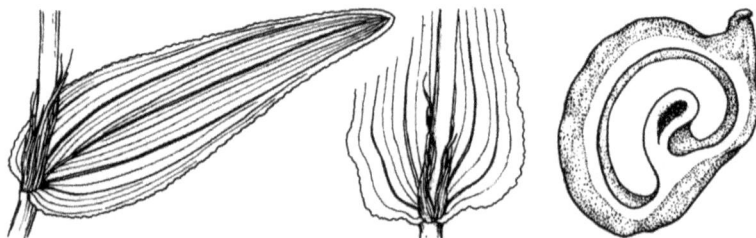

Potamogeton richardsonii CLASPING-LEAVED PONDWEED
· IDENTIFICATION **Plants with underwater leaves only.** Similar to *P. perfoliatus*, distinguished from it by the cavity in the endocarp loop of the fruit, and by the coarse whitish stipules. Sterile hybrids between the two species are reported.

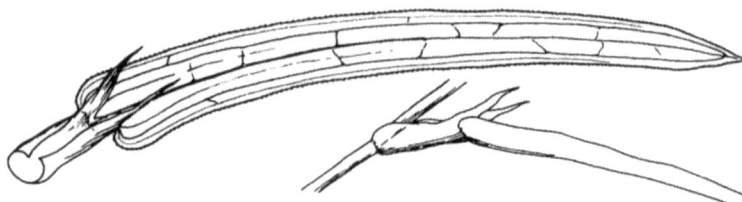

Potamogeton robbinsii FERN-PONDWEED
· IDENTIFICATION **Plants with underwater leaves only;** leaves auriculate at base; leaf tips acute; leaf margins finely serrulate. Also, by the whitish adnate stipular sheaths and acute leaf tips. The leaves are usually stiffly 2-ranked, giving the branches a feathery appearance in the water. Rarely found with fruit.

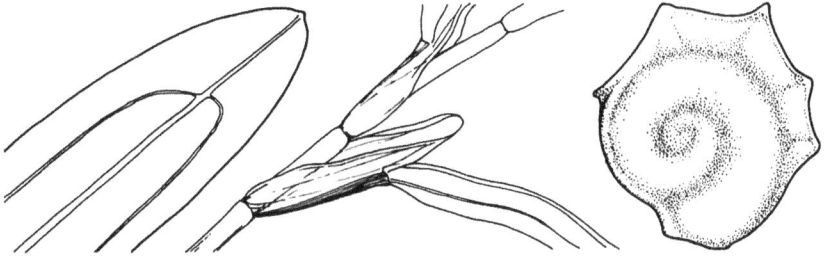

Potamogeton spirillus   NORTHERN SNAILSEED PONDWEED

· IDENTIFICATION **Plants with broad floating leaves and linear underwater leaves.** Similar to *P. diversifolius;* recognized by their tiny fruits with walls so thin that the embryo coil is evident.

Potamogeton strictifolius   STRAIGHT-LEAVED PONDWEED

· IDENTIFICATION **Plants with underwater leaves only;** highly variable. Non-fruiting plants similar to *P. foliosus,* which see.

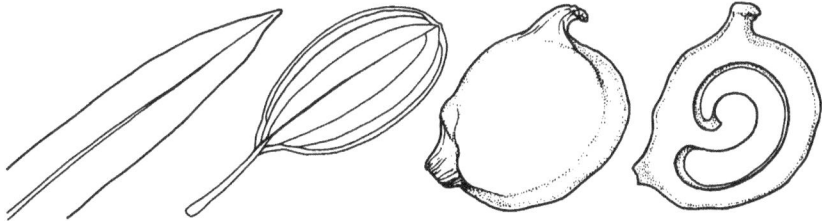

Potamogeton vaseyi   VASEY'S PONDWEED

· IDENTIFICATION **Plants with broad floating leaves and broad underwater leaves.** Our only species with the combination of floating leaves less than 9 mm wide, submersed leaves linear, less than 1 mm wide; and fruits with embryo coil less than 1 complete revolution.

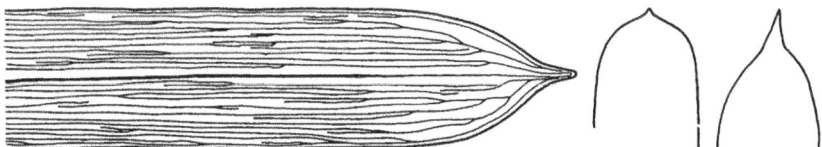

Potamogeton zosteriformis   FLATSTEM-PONDWEED

· IDENTIFICATION **Plants with underwater leaves only.** Stems laterally flattened; leaves linear with 15 or more veins.

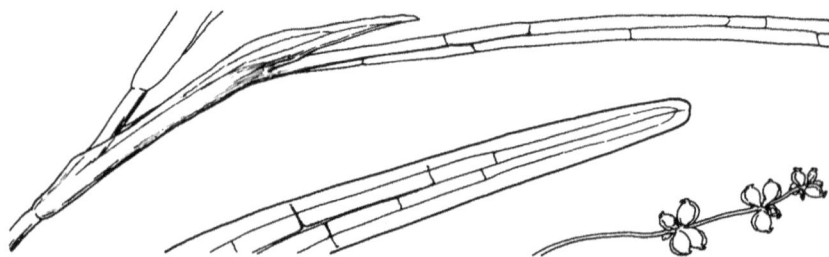

Stuckenia filiformis THREADLEAF-PONDWEED
· IDENTIFICATION Leaf tips obtuse; stipular sheaths adnate, slender.

Stuckenia pectinata SAGO-PONDWEED
· IDENTIFICATION Leaf tips acute to sharp pointed; stipular sheaths adnate, slender.

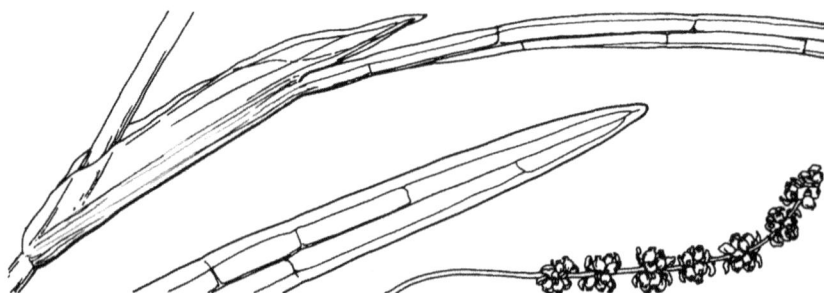

Stuckenia vaginata BIGSHEATH-PONDWEED
· IDENTIFICATION Leaf tips obtuse; stipular sheaths adnate, broad. Upper sterile branches difficult to distinguish from S. *filiformis*.

www.ingramcontent.com/pod-product-compliance
Lightning Source LLC
Chambersburg PA
CBHW052106030426
42335CB00025B/2865